ENCYCLOPEDIC DICTIONARY OF

NAMED PROCESSES IN CHEMICAL TECHNOLOGY

FOURTH EDITION

Joseph Stella, By-Products Plants, 1923/26, Oil on canvas, 61 × 61 cm (24 × 24 in.), Gift of Mr. and Mrs. Noah Goldowsky in memory of Esther Goldowsky, 1955.816, The Art Institute of Chicago. Photography © The Art Institute of Chicago.

ENCYCLOPEDIC DICTIONARY OF

NAMED PROCESSES

IN CHEMICAL

TECHNOLOGY

FOURTH EDITION

Alan E. Comyns

CRC Press
Taylor & Francis Group
Boca Raton   London   New York

CRC Press is an imprint of the
Taylor & Francis Group, an **informa** business

The author wishes to thank Kevin Craig (CRC Press) for designing the cover for the 2nd edition, which was used also, in different colors, in the 3rd and 4th editions. He had been inspired by Joseph Stella's painting which had been proposed, but not accepted, as the frontispiece for the 1st edition, and is now printed in this edition.

Previous editions

1. Dictionary of Named Processes in Chemical Technology, 1993. By Alan E. Comyns. Oxford University Press, Oxford, UK. ISBN 0-19-855385-4

2. Encyclopedic Dictionary of Named Processes in Chemical Technology, 1999. By Alan E. Comyns. 2nd edn., CRC Press, Boca Raton, FL, USA. ISBN 0-8493-1205-1

3. Encyclopedic Dictionary of Named Processes in Chemical Technology, 2007. By Alan E. Comyns. Foreword by Colin A. Russell. 3rd edn., Taylor & Francis, Boca Raton, FL, USA. ISBN 978-0-8493-9163-7

CRC Press
Taylor & Francis Group
6000 Broken Sound Parkway NW, Suite 300
Boca Raton, FL 33487-2742

First issued in paperback 2019

© 2014 by Taylor & Francis Group, LLC
CRC Press is an imprint of Taylor & Francis Group, an Informa business

No claim to original U.S. Government works

ISBN-13: 978-1-4665-6776-4 (hbk)
ISBN-13: 978-0-367-37885-1 (pbk)

Library of Congress Cataloging-in-Publication Data

Comyns, Alan E.
    Encyclopedic dictionary of named processes in chemical technology / Alan E. Comyns. -- Fourth edition.
        pages cm
    "A CRC title, part of the Taylor & Francis imprint, a member of the Taylor & Francis Group, the academic division of T&F Informa plc."
    Includes bibliographical references and index.
    ISBN 978-1-4665-6776-4 (hardcover : acid-free paper)
    1. Chemical processes--Dictionaries. I. Title. II. Title: Named processes in chemical technology.

TP155.7.C664 2014
660'.28--dc23   2013048947

Visit the Taylor & Francis Web site at
http://www.taylorandfrancis.com

and the CRC Press Web site at
http://www.crcpress.com

Dedicated to the generations of industrial chemists and engineers whose ingenuity has given us the materials of civilization.

Contents

Foreword

It is a great honor to write a foreword to the preeminent dictionary of named processes in the chemical industry. Named industrial processes have always been the poor relation of named reactions. On my bookshelf, I have several books devoted to named organic reactions, including ones by J. S. Gowan and T. S. Wheeler (1960), Alexander Surrey (2nd ed., 1961), and R. C. Denny (1969). The *Merck Index* also has an excellent section on named reactions. For industrial processes, "Comyns" stands practically alone, its only rival being the book by Kurt Wagner (1951), which has a much broader brief and is now completely out of date. Yet industrial processes and names are more often mentioned than named organic reactions and far more often by nonchemists. The Haber–Bosch process is far better known than the Cannizzaro reaction or the Michael addition.

At the same time, it is all too easy to confuse two processes or to fail to realize that a process has more than one name. Increasingly, older processes are misunderstood or forgotten altogether. It is not generally known that the Caro of Caro's acid was a completely different chemist to the Caro of the Frank–Caro process, or that the Bayer process has nothing to do with the Bayer chemical company. Even product names can give rise to confusion. Some years ago, a well-known popular writer stated that Buna, an early German synthetic rubber, was an abbreviation of its main components butadiene and natrium (seemingly unaware that Na is the symbol for sodium). Had he consulted this dictionary, he would have soon discovered that sodium was a catalyst rather than a component of Buna. There are no less than four processes associated with Friedrich Bergius, two connected with his well-known oil-from-coal process and two associated with the conversion of wood into sugars. The Merseburg process was actually developed in Oppau on the other side of Germany and later used at Leuna. The name comes from the subsidiary that ran the huge Leuna factory, Ammoniawerke Merseburg. Armed with this book, you will be in no danger of making mistakes or getting confused.

A particular strength of this book is the historical information it contains. The long-vanished Leblanc process is covered in some detail, as are associated processes such as the Deacon and Weldon processes. Comyns is also good at explaining the various nuances of a process or a product name. Take, for example, once again, Buna, which has been the name of several synthetic rubbers and an East German "combinat" (now part of Dow). Here is an author who loves chemical processes and their quirky names—and is deeply knowledgeable about them. The product index at the back is another useful feature. To take synthetic rubber, one can find the processes associated with Ostromislenski, Lebedev, and Reppe, as well as the aldol process. The latter shows the value of the index, for I would not have thought of looking under aldol. The high quality of this book is again demonstrated by the mention in the aldol entry of the "four-step" process, the now obscure name given to this process by IG Farben in the 1920s.

This is an exceedingly useful book for industrial chemists, historians of chemistry and chemical technology, science writers, and journalists. It should be found in any reference library worthy of the name. "Comyns" deserves to be as well known in the field of chemical technology as "Kirk-Othmer." Sadly this is the last edition of "Comyns," so make sure you get it before it becomes unobtainable!

Peter J. T. Morris
Honorary Research Associate
University College London

Acknowledgments

I want to thank my wife Daphne for her patience and support during the 25 years that I have been working on these four editions.

I thank Reg Adams of Artikol for updating many of the titanium dioxide entries.

Introduction

The purpose of this dictionary is to provide concise descriptions of those processes in chemical technology that are known by special names that are not self-explanatory.

The chemical industry is notoriously difficult to define. In addition to its obvious role as a producer of "chemicals" such as sulfuric acid, it may be said to embrace all those industries in which chemical processes are conducted. There is no generally agreed list of such industries, but obvious ones include extractive metallurgy, plastics, paper, ceramics, sewage treatment, and now even electronics. It is this broad spectrum of "chemical technology" that is addressed in this book. It thus includes the gigantic Bessemer process for making steel and the microscopic Manasevit process for applying circuits to silicon chips. The only deliberate omission is food chemistry. The aim has been to include all those processes that are known by special names, of whatever origin. Of course, only a minority of industrial chemical processes are distinguished by the possession of special names, so this book does not include all of even the more important processes. Overviews of the industry are provided by other books, notably the Kirk-Othmer and Ullmann encyclopedias. Many named processes are included in such works, but only a fraction of the names in the present compilation are to be found in them.

The names show a heterogeneous etymology, their origins including inventors, companies, institutions, places, acronyms, abbreviations, and obvious corruptions of the chemical nomenclature. Derivations of the names, where known, are indicated in square brackets.

The names of chemicals used in the entries are the traditional names commonly used in industry today; systematic names have yet to replace many common names and may never do so.

Criteria for the selection of names for inclusion are inevitably subjective, but the intention has been to include all named processes in current commercial use anywhere in the world, and those that have been or are being piloted on a substantial scale. Obsolete processes that have been or might have been important in the past are also included. The coverage is primarily of English names, but some foreign names are included as well.

Process names that combine the name of a company with the name of a chemical (e.g., the Monsanto acetic acid process) have mostly been excluded because they are self-explanatory and can be found in encyclopedias. Some companies (e.g., Lurgi, Texaco) are best known for one process, even though they may have developed many others; in general, only their most famous one is included here under the company name. Their other processes are included if they have special names. Process names that combine the names of two collaborating companies (e.g., Mobil/Badger) have mostly been included. Company names are usually given in the styles in use at the time of their respective inventions, as given in patent applications.

Where two or more processes have the same name, they are distinguished by numbers in parentheses, such as **Parex** (1) and **Parex** (2). Because the numbers are not parts of the names, they are not emboldened.

Generic names are given in lowercase initials.

Names of chemical reactions have mostly been excluded, being adequately defined in standard chemical texts and in the special chemical dictionaries listed in the bibliography. There is remarkably little overlap between reaction names and process names: discoverers of chemical reactions seldom develop them into manufacturing processes. However, some generic process names that combine two or more reactions (e.g., oxychlorination, dehydrocyclodimerization) have been included because they are not generally to be found in any dictionaries. These hybrid names are distinguished by being given lowercase initial letters.

The lengths of the entries have been tailored to reflect importance and topicality. In general, important processes in current use have the longest entries, obsolete processes of minor importance have the smallest, and processes of intermediate importance receive correspondingly intermediate space. There are exceptions: some obsolete processes are given more space because of their technical interest or historical importance; some important current processes are given little space because their essential features are still secret.

References have been chosen that document the origins of the processes and that review the latest developments. Sources that are reasonably accessible have been quoted wherever possible. Journal abbreviations are those used by *Chemical Abstracts*. Patents quoted are generally those of the countries of the original inventions, or their British or US equivalents. Many processes are described in families of patents; only the first or key ones are quoted here. I apologize for the incompleteness of some of the references. Public library services in Britain no longer provide open access to long runs of journals or to older books, so it is no longer practicable to check everything. References to conference proceedings are excluded because they are generally unavailable.

Copyright and trademark laws present special problems for works such as this, protection varying from country to country and from time to time. It would not be feasible to indicate which of the many names in this dictionary are so protected. The inclusion of a protected name, in its registered or commonly used format, has no legal significance and does not indicate that the name has become a generic word. Corporate designations (e.g., Ltd., PLC, Inc., AG, GmbH, SA, and A/S) have been omitted, except in cases in which confusion with personal names might occur. Company names listed are generally those that were current when the developments described were made. Because many companies have changed their names (often using letters instead of words) and their owners, many of these names are not current. Dates of invention given are usually the years in which the relevant patents were filed. Asterisks (*) denote names that are defined elsewhere in this dictionary.

The appendix provides a key to the chemicals and materials whose manufacture is described by one or more of the named processes. Although the dictionary does not purport to be a comprehensive listing of processes for making particular chemicals or materials, reference to the key will identify those processes that have special names, and reference to these entries will often provide general references that will help to identify the unnamed ones as well.

Information on hitherto secret processes developed by government agencies is slowly reaching the public domain, thanks partly to the Freedom of Information Acts in the United States (1966) and the United Kingdom (2000). We now know the significance of COLEX, Runcol, Silex, and several other chemical processes of former military importance.

Many of the processes new to this edition reflect recent trends in the global chemical industry. There is now much less emphasis on new processes for making basic chemicals and more emphasis on pollution prevention and waste disposal. Petrochemical processes, based on petroleum and natural gas, are giving way to biochemical processes, using a variety of sources of biomass. Impending shortages of petroleum have stimulated the investigation of numerous alternatives by many companies and research organizations, including established oil companies.

Alternative carbon sources will require new processes, but in general it is expected that existing refining processes can be adapted to use the new intermediates that will be produced. The alternative carbon sources include algae and lignocellulose. When the first edition of this dictionary was published in 1993, it described six processes whose names started with "bio": in the present edition there are 40 such processes.

Another feature of today's process technology is the development of recycling processes for a wide range of materials and wastes, and improvements in methods for disposing of municipal wastes. Agricultural wastes such as wood chippings and corn stalks are an important source of lignocellulose.

There are a few references to Wikipedia. When Wikipedia was introduced, it was not quoted in academic literature because it was anonymous and might not have been reliable. However, Wikipedia's editorial process has proved robust, and there is now no reason not to refer to it. Many of its articles are definitive statements of the technical state of the art and of the commercial status of the processes described. Of course, as with all web-based literature, there can be no guarantee that it will continue to be available to the general public in the long term (the still unsolved digital permanence problem), so it would be unwise to rely on it completely. For that reason, wherever possible, references to Wikipedia are supported by references to the printed literature.

Author

Dr. Alan Comyns has had an unusually varied career in academic, government, and industrial research laboratories. He graduated with first-class honors in chemistry from the University of London at the age of 19. His PhD work, in the Hughes–Ingold School of Physical–Organic Chemistry at University College London, was followed by postdoctoral studies at the California Institute of Technology and the University of Wisconsin. He has worked at the Atomic Energy Research Establishment, Harwell (United Kingdom); British Titan Products (now Huntsman), United Kingdom; Westinghouse Electric, Pittsburgh, PA; and National Lead (now Kronos), Sayreville, NJ. From 1974 to 1988, he was product research manager, later chief scientist, at Laporte Industries in Widnes (United Kingdom). In the 1980s, he was a part-time visiting lecturer in industrial chemistry at the University of East Anglia. Until recently, he was an independent consultant and author, specializing in market studies for inorganic chemicals and materials. His recent publications include *Fluoride Glasses* (John Wiley & Sons, Chichester, UK, 1989), *Inorganic Peroxides and Peroxy Compounds* (in *Kirk-Othmer's Encyclopedia of Chemical Technology*, 4th ed., Vol. 18), *Laporte and Its Fluoride Businesses* (in *Fluorine Chemistry at the Millennium*, ed. R.E. Banks, Elsevier, 2000), and market research reports on *Titania and the Titanates*; *Fillers, Extenders and White Pigments*; and *Materials for Fuel Cells*. He edits, and largely writes, *Focus on Catalysts,* a monthly newsletter published by Elsevier.

Dr. Comyns was one of the founders of the Applied Solid State Chemistry Group of the Royal Society of Chemistry. He has been chairman of the Industrial Inorganic Chemicals Group and vice president of the Dalton Division of the Royal Society of Chemistry, chairman of the Liverpool Section of the Society of Chemical Industry, chairman of the British Zeolite Association, and chairman of the Chester Archaeological Society. His hobbies include archaeology and collecting antique glass. He edited the guidebook *2000 Years of Building* for The Chester Civic Trust.

A

4A-CAT [**A**ctivity **A**djustment by **A**mmonia **A**dsorption] A method for presulfiding and passivating *hydrocracking catalysts. Developed by EUROCAT in 1989.

Chauvel, A., Delmon, B., and Hölderich, W.F., *Appl. Catal. A: Gen.*, 1994, **115**, 184.

Aachen *See* DR.

A-B [**A**dsorptions-**B**elebsungsverfahren, German, meaning *adsorption-activation process*] A two-stage *Activated Sludge process for treating sewage and industrial wastes. The first stage (A) is highly loaded; the second (B) is low loaded. Such a system can cope with sudden changes in the quantity and quality of effluent feed. Developed in 1983 by B. Bohnke at the Technical University of Aachen and subsequently engineered by Esmil, United Kingdom.

Gray, N.F., *Activated Sludge: Theory and Practice*, Oxford University Press, Oxford, U.K., 1990, 110.
Horan, N.J., *Biological Wastewater Systems*, John Wiley & Sons, Chichester, U.K., 1990, 69.

ABATE A process for removing hydrogen sulfide from sour gases such as landfill gas. The gas, containing oxygen in addition to the hydrogen sulfide, is passed through water containing an iron chelate compound, which oxidizes the hydrogen sulfide to elemental sulfur. Dispersants keep the sulfur in suspension until its concentration reaches 10%. Developed by Dow Chemical, derived from the Dow–Shell *SulFerox process.

Chem. Eng. (NY), 1996, **103**(11), 19.

Abbot-Cox A method of applying vat dyes to cellulosic textiles in package form. The dispersed dye, with a dispersing agent, circulates through the package. The dye becomes substantially transferred to the material by the gradual addition of an electrolyte such as sodium sulfate. When the dye has been transferred to the fabric, it is reduced *in situ*. The color is restored by a mild oxidizing agent such as hydrogen peroxide.

Fischer-Bobsien, C. H., *Internationales Lexicon Textilveredlung + Grenzgebiete*, Rhenus Handelsgesellschaft, Vadus, Liechenstein, 1966, 1123.

ABC Also called Chiyoda ABC. A process for treating heavy hydrocarbons from tar sands by *hydrocracking. Piloted by the Chiyoda Chemical Engineering and Construction Company in the 1980s.

Bowman, C.W., Phillips, R.S., and Turner, L.R., in *Handbook of Synfuels Technology*, Meyers, R.A., Ed., McGraw-Hill, New York, 1984, 5–73.
Marcos, F. and Rosa-Brussin, D., *Catal. Rev. Sci. Eng.*, 1995, **37**(1), 3.

Ab der-Halden A continuous process for distilling coal tar. It is operated under reduced pressure with the heat provided by live, superheated steam. This provides a clean separation of the products, without cracking. Developed in France in the 1920s by C. Ab der-Halden who formed the company PROABD to exploit it. Not to be confused with the Abderhalden reaction in biochemistry.

British Patents 239,841; 253,935.
Hoffert, W.H. and Claxton, G., *Motor Benzole: Its Production and Use*, National Benzole Association, London, U.K., 1957, 38.
Mines, 1957, **12**(53), 223.

Abgas-Turbo-Wascher von Kroll Not a process, but a piece of equipment for scrubbing flue gases with an aqueous suspension of lime. Developed by Walter Kroll GmbH and used in 14 plants in West Germany in 1986.

ABGC [**A**ir **B**lown **G**asification **C**ycle] A coal gasification process developed by the former British Coal. The technology was subsequently sold to Mitsui Babcock Energy Ltd.

Rezaiyan, J. and Cheremisinoff, N.P., *Gasification Technologies*, Taylor & Francis, Boca Raton, FL, 2005, 59.

ACAR *See* Steelmaking.

Accar A direct-reduction ironmaking process using coal and oil as the reductants. Operated at the OSIL plant at Keonjhar, India, from 1983 to 1987. *See* DR.

Accent [**A**queous **c**arbon **c**ompound **e**ffluent **t**reatment] A process for oxidizing organic contaminants in aqueous streams by catalyzed oxidation with sodium hypochlorite. The catalyst is promoted nickel oxide, which retains active oxygen at its surface, as well as adsorbing the organics. Developed by ICI Katalco and first offered in 1998.

Acedox [**Acet**ic **ox**idation] A pulp-bleaching process using peracetic acid as the oxidant. Developed by Eka Nobel in 1994 and first commercialized, in combination with *Lignox, in Sweden in 1995.

ACES 21 [**A**dvanced process for **C**ost and **E**nergy **S**aving] A process for making urea from carbon dioxide and ammonia, developed by Toyo Engineering Corporation. Toyo has been developing and licensing urea processes since 1961. Earlier versions of its urea process were named TR-C1, TR-D, and ACES and were installed in Spain and several Asian countries. The latest version, ACES 21, has been built in China and Indonesia.

Kirk-Othmer's Encyclopedia of Chemical Technology, 4th ed., Suppl., John Wiley & Sons, New York, 1998, 609.
Eur. Chem. News, 1999, **70**(1860), 29.

Acetate A general name for processes for making cellulose acetate fibers. Cellulose is acetylated, dissolved in acetone, and spun into fibers by injecting the solution through orifices into heated chambers. Cellulose mono-acetate is made by acetylating with a mixture of acetic acid, acetic anhydride, and sulfuric acid as the catalyst. Cellulose triacetate is made in a similar fashion, but using perchloric acid as the catalyst and dry spinning from a solution in ethanol–methylene chloride. Cellulose triacetate fibers were first made commercially by Courtaulds in London in 1950.

Peters, R.H., *Textile Chemistry*, Vol. 1, Elsevier, Amsterdam, the Netherlands, 1963, 187.
LaNieve, H.L., in *Handbook of Fiber Chemistry*, 3rd ed., Lewin, M., Ed., CRC Press, Boca Raton, FL, 2006, Chap. 11.

Acetex A process for removing acetylene from ethylene by selective hydrogenation. Developed by IFP in France in 1993.

Chem. Eng. News, 1993, **71**(34), 21.
Chauvel, A., Delmon, B., and Hölderich, W.F., *Appl. Catal. A: Gen.*, 1994, **115**, 186.

Acetica A process for making acetic acid by the heterogeneous carbonylation of methanol in a bubble column reactor. The catalyst is a rhodium carbonyl iodide, anchored by ion-pairing to a polyvinyl pyridine resin. Developed by Chiyoda Corporation and UOP and first described in 1998. Licensed to Guizhou Crystal Organic Chemical Group, China, in 2002; one plant was under construction in 2005.

U.S. Patents 5,334,755; 5,364,963; 5,576,458.
Yoneda, N., Kusano, S., Yasui, M., Pujado, P., and Wilcher, S., *Appl. Catal., A: Gen.*, 2001, **221**(1–2), 260.
Chem. Eng. (N.Y.), 2002, **109**(12), 17.
J. Chem. Eng. Jpn., 2004, **37**(4), 536.

Acetocell An *Organocell process.

Villa, C., Santos, V., and Parajo, J.C., *Ind. Eng. Chem. Res.*, 2003, **42**(2), 349.

Acetosolv A wood pulping and bleaching process that uses hydrogen peroxide and acetic acid. *See* Organosolv.

Eur. Chem. News (Finland Suppl.), 1991, May, 28.
Villa, C., Santos, V., and Parajo, J.C., *Ind. Eng. Chem. Res.*, 2003, **42**(2), 349.
Pye, E.K., in *Biorefineries: Industrial Processes and Products*, Kamm, B., Gruber, P.R., and Kamm, M., Eds., Wiley-VCH, Weinheim, Germany, 2005, **II**, 179.

Acetylene Black A process for making carbon black from acetylene by thermal decomposition at 800°C–1000°C in refractory-lined, water-cooled retorts. Used since the beginning of the twentieth century and still a major industry in 2005.

Claasen, E.J., in *Inorganic Chemicals Handbook*, Vol. 2, McKetta, J.J., Ed., Marcel Dekker, New York, 1993, 510.
Kühner, G. and Voll, M., in *Carbon Black Science and Technology*, Donnet, J.-B., Bansai, R.C., and Wang, M.-J., Eds., Marcel Dekker, New York, 1993, 61.
Buxbaum, G. and Pfaff, G., *Industrial Inorganic Pigments*, 3rd ed., VCH, Weinheim, Germany, 2005, 176.

ACGP [**A**dvanced **C**oal **G**asification **P**rocess] A coal gasification process developed in South Africa by Kynoch Fertilizer and Babcock & Wilcox as a replacement for the *KBW process. Unlike the KBW process, it operates under atmospheric pressure. ACGP had not been commercialized by 2006.

ACH (1) [**A**cetone **C**yan**H**ydrin] A process for making methyl methacrylate via this intermediate. Acetone is reacted with hydrogen cyanide to yield the cyanhydrin. This is then converted to methacrylamide using concentrated sulfuric acid. Methanolysis of this yields methyl methacrylate. Developed by Röhm GmbH Chemische Fabrik, Germany, and ICI, United Kingdom; used in 11 countries in 1990. Other processes have been developed that avoid the coproduction of large quantities of ammonium bisulfate. *See also* Alpha (2).

Porcelli, R.V. and Juran, B., *Hydrocarbon Process. Int. Ed.*, 1986, **65**(3), 39.
Chem. Eng. (N.Y.), 1990, **97**(3), 35.
Eur. Chem. News, 2002, **76**(1992), 20.

ACH (2) [**A**luminum **C**hloro**H**ydrate] This is the common name for some types of basic aluminum chloride, but the name has also been used to designate the process by which such a product is made. Several processes, some of which are proprietary, are used to make the several commercial aluminum chloride products available. In general, it is necessary to introduce an excess of aluminum metal to a chloride solution, so that the atom ratio of aluminum to chlorine is less than three. The aluminum may be introduced as either the metal or the hydrated oxide.

Acheson (1) A process for making silicon carbide from sand and coke, in an electric furnace, at 2200°C–2400°C:

$$SiO + 3C = SiC + 2CO$$

Invented by E.G. Acheson in Monongahela City, PA, in 1892. Acheson was heating clay and carbon by means of an electric arc, in the hope of making diamond. The hard, crystalline product was called carborundum in the mistaken belief that it was a compound of carbon and corundum (alumina). The process and product were patented in 1893 and made on a small scale in Monongahela City using the city's electricity supply. In 1895, the Carborundum Company was formed to exploit the process in Niagara, NY, using hydroelectric power from Niagara Falls. This same process is now operated in many countries. The name Carborundum is a registered trademark owned by the Carborundum Company, NY, and used for several of its refractory products, in addition to silicon carbide.

U.S. Patent 492,767.
Szymanowitz, R., *Edward Goodrich Acheson: Inventor, Scientist, Industrialist*, Vantage Press, New York, 1971.
Mühlhaeuser, O., *J. Am. Chem. Soc.*, 1893, **15**, 411.

Acheson (2) A process for converting carbon articles into graphite, invented by E.G. Acheson in 1895 and commercialized in 1897. This process uses transverse graphitization, unlike the *Castner process, which uses lengthwise graphitization.

U.S. Patents 568,323; 617,979; 645,285.
Szymanowitz, R., *Edward Goodrich Acheson: Inventor, Scientist, Industrialist*, Vantage Press, New York, 1971.

Acid A process for making sodium perborate by reacting sodium borate ("borax") with sodium peroxide and hydrochloric acid:

$$Na_2B_4O_7 + 4Na_2O_2 + 6HCl + 13H_2O = 2Na_2[B_2O_4(OH)_4] + 6H_2O + 6NaCl$$

Operated by the Castner–Kellner Company, Runcorn, England, from 1915 until it was supplanted by the *Duplex (2) process in 1950.

Hardie, D.W.F. and Pratt, J.D., *A History of the Modern British Chemical Industry*, Pergamon Press, Oxford, U.K., 1966, 141.

Acid Bessemer An alternative name for the original *Bessemer steelmaking process in which the furnace is lined with a siliceous refractory. It is suitable only for ores relatively free from phosphorus.

Acid Open Hearth The original version of the *Open Hearth process for steelmaking in which the hearth is made of a siliceous refractory. The process does not remove phosphorus or sulfur, the acid impurities in the iron, so the raw materials must be relatively free from these. Pioneered by C.W. Siemens and F.M.E. and P. Martin at Sireuil, France, in 1864.

British Patent 2,031 (1864).
Barraclough, K.C., *Steelmaking 1850–1900*, The Institute of Metals, London, U.K., 1990, 137.

ACIMET [**ACI**d **MET**hane] A two-stage, anaerobic digestion process for treating municipal wastewaters. In the first stage, organic matter is decomposed to a mixture of acids, aldehydes, and alcohols. In the second, the carbon in this mixture is anaerobically converted to methane. Invented in 1974 by S. Ghosh and D.L. Klass at the Illinois Institute of Gas Technology (IGT), Chicago. First commercialized in 1991 by IGT and DuPage County, IL, at the Woodridge-Greene Valley Wastewater Treatment Plant.

U.S. Patent 4,022,665.
Ghosh, S., Conrad, J.R., and Klass, D.L., *J. Water Pollut. Control Fed.*, 1975, **47**(1), 30.

ACP [**A**dvanced **C**ascade **P**rocess] *See* Hostalen.

ACR [**A**dvanced **C**racking **R**eactor] A *thermal petroleum cracking process, the heat being provided by partial combustion of the feed at 2000°C. Developed by Chiyoda Chemical Engineering & Construction Company, Kureha Chemical Industry Company, and Union Carbide Corporation in the 1970s. A demonstration plant was operated in Seadrift, TX, from 1979 to 1981.

Ishkawa, T. and Keister, R.G., *Hydrocarbon Process. Int. Ed.*, 1978, **57**(12), 109.
Hu, Y.C., in *Chemical Processing Handbook*, Marcel Dekker, New York, 1993, 768.

actiCAT A process for presulfurizing hydrotreating catalysts. Presulfurizing differs from presulfiding in that the products are complex metal oxysulfides, rather than sulfides. A novel organic "matrix" retains the sulfur during the conversion process. Developed by CRI International and offered by that company as a service to the petroleum industry.

Welch, J.G., Poyner, P., and Skelly, R.F., *Oil Gas J.*, 1994, **92**(41), 56.
Blashka, S., Bond, G., and Ward, D., *Oil Gas J.*, 1998, **96**(1), 36.

Actiflo A flocculation–clarification system for treating potable and wastewaters. A variety of flocculating agents, typically ferric chloride, may be used. After flocculation, a fine sand is added and the system mixed, which entraps the flocs. A lamellar separator accelerates the settling process, and the floc is separated from the sand in a hydrocyclone. Developed by Kruger Products, a subsidiary of USFilter, and now offered by Vivendi Water Systems. Operated first in France in 1991; by 2002, nearly 100 units were operating worldwide.

Water Waste Treat., 2002, **45**(4), 26–27.

ACTIFLOW A process for treating raw water. Flocculation of insoluble matter by the addition of a polyelectrolyte takes place within an agitated bed of fine sand. Developed in France by OTV and licensed in the United Kingdom through General Water Processes.

Actiflow is also the name of a Dutch consulting engineering company that has been spun off from the Delft University of Technology.

Actimag A process for reducing metal ions in aqueous solution by metallic iron. The iron is in the form of particles 1 mm in diameter contained in a fluidized bed and kept in violent agitation by means of an alternating magnetic field. The agitation accelerates the reaction and prevents the adhesion of deposits of reduction products. Demonstrated for reducing the cupric ion to metallic copper and the chromate ion to chromic ion. Developed by Extramet, France, in the 1980s and offered in the United Kingdom by Darcy Products.

European Patent 14,109.
Bowden, P., *Water Waste Treat.*, 1989, **32**(7), 21.
Bowden, P., *Processing*, 1990, 27.

Activated MDEA A version of the *MDEA process for scrubbing acid gases from gas streams, in which the aqueous MDEA solution is regenerated by flashing rather than by stripping. Developed by BASF, Germany, in 1971, with the Ralph M. Parsons Company becoming the sole licensor in most of the Western Hemisphere in 1982. The process is now operated in Europe, Canada, and the United States.

Hydrocarbon Process. Int. Ed., 1996, **75**(4), 105.

Activated Sludge A sewage treatment process, developed in the 1920s and soon widely adopted, based on the aeration of wastewater with flocculating biological growth, followed by separation of the treated wastewater. This process removes dissolved and colloidal organic material, suspended solids, some of the mineral nutrients (P- and N-compounds), and some volatile organic compounds. Generally ascribed to H.W. Clark and S.M. de Gage in Massachusetts (1912), followed by E. Arden and M.T. Lockett in Manchester (1914). The first plant was installed in Worcester, England, in 1916.

Arden, E. and Lockett, M.T., *J. Soc. Chem. Ind. (London)*, 1914, **33**(10), 523; (23), 1122.
Ganczarczyk, J.J., *Activated Sludge Process: Theory and Practice*, Marcel Dekker, New York, 1983.

ActivCat Not a process but a range of additives for single-site catalysts for making polyolefins that double the capacity of the catalysts. ActivCats can be used for both polyethylene and polypropylene resin production and standard MAO-/silica-type catalysts production. ActivCats do no alter the properties of MAO/silica catalysts during production. Introduced and trademarked by Albemarle in 2008.

Chem. Eng. World, 2008, **43**(11), 20.

ACTIVOX A hydrometallurgical process for extracting copper from sulfide ores, also suitable for nickel and gold. It involves fine grinding, low temperature, and low pressure. Developed in Australia by Western Minerals Technology, now Norilsk Process Technology. Piloted but not yet licensed.

ADA-ES (1) A process for removing mercury emissions from coal-fired power stations by pretreating the coal. Developed by ADA-ES, Inc. and Arch Coal, Inc. ADA-ES has also developed a method for reducing mercury emissions from power stations using powdered activated carbon. *See* *ACI.

Chem. Eng. (N.Y.), 2010, **117**(8), 13.

ADA-ES (2) A process for removing carbon dioxide from coal-fired power plants and other industrial sources. It uses a regenerable solid adsorbent. A pilot plant was planned for 2013.

ADAM-EVA *See* EVA-ADAM.

Addipol A process for making polypropylene, developed and licensed by Himont, in the United States, and commercialized in 1988. *See also* Spheripol.

Adex A process for removing heavy metals from phosphoric acid by precipitation of their complexes with 2-ethylhexyl dithiophosphate. Developed by Hoechst, Germany.

Becker, P., *Phosphates and Phosphoric Acid*, 2nd ed., Marcel Dekker, New York, 1989, 531.

Adib A process for extracting isobutene from petroleum fractions by reaction with phenol. The reaction takes place in the gas phase, over an acid catalyst, and yields all the mono-, di-, and tri-butyl phenols. Heating this mixture liberates isobutene; the phenol and the catalyst are recovered for reuse. Piloted in Argentina in the 1980s.

Miranda, M., *Hydrocarbon Process Int. Ed.*, 1987, **66**(8), 51.

ADIP [Possibly an acronym of DIPA, diisopropanolamine] A process for removing hydrogen sulfide, mercaptans, carbonyl sulfide, and carbon dioxide from refinery streams by extraction into an aqueous solution of diisopropanolamine or methyl diethanolamine. Developed by Shell and licensed by Shell Global Solutions International BV. More than 400 units were operating in 2000.

Bally, A.P., *Erdoel Kohle Erdgas Petrochemie*, 1961, **14**, 921.
Hydrocarbon Process. Int. Ed., 1975, **54**(4), 79; 1992, **71**(4), 86; 2000, **79**(4), 56.
Kohl, A.L. and Nielsen, R.B., *Gas Purification*, 5th ed., Gulf Publishing Co., Houston, TX, 1997, 53.

Adkins–Peterson The oxidation of methanol to formaldehyde using air and a mixed molybdenum/iron oxide catalyst. Not an engineered process, but the reaction that formed the basis of the *Formox process.

U.S. Patent 1,913,405.
Adkins, H. and Peterson, W.R., *J. Am. Chem. Soc.*, 1931, **53**, 1512.

ADOX *See* CATOX.

ADU [**A**mmonium **Di**Uranate] A process for converting uranium hexafluoride into uranium dioxide for use as a nuclear reactor fuel. The hexafluoride is hydrolyzed in water:

$$UF_6 + 2H_2O = UO_2F_2 + 4HF$$

and the solution treated with ammonia, precipitating ammonium diuranate:

$$2UO_2F_2 + 8HF + 14NH_3 + 3H_2O = (NH_4)_2 U_2O_7 + 12NH_4F$$

that is filtered off and reduced with hydrogen. Developed in the United States in the 1950s.

Büchner, W., Schliebs, R., Winter, G., and Büchel, K.H., *Industrial Inorganic Chemistry*, VCH Publishers, Weinheim, Germany, 1989, 581.

ADVACATE A *flue-gas desulfurization process, similar to *CZD, but using a suspension of fly ash instead of lime. Developed by the University of Texas, the U.S. Environmental Protection Agency, and Acurex Corporation.

Kohl, A.L. and Nielsen, R.B., *Gas Purification*, 5th ed., Gulf Publishing Co., Houston, TX, 1997, 486.

Advanced Sclairtech Also written as AST. A version of *Sclairtech that uses a different *Z–N catalyst. There are two autoclave reactors in series having independent monomer, co-monomer, and catalyst feed systems. Operated by Nova Chemicals at Joffre, Alberta, Canada, in 2002.

Eur. Chem. News, 2002, **76**(2004), 25.

AEROSIL A process for making submicron-sized silica, alumina, or titania powders by the flame hydrolysis of the respective chlorides. The chloride vapor is passed through an oxyhydrogen flame; the reaction is thus a flame hydrolysis, rather than an oxidation, so it is to be distinguished from the *Chloride process for making titanium dioxide pigment. Developed by Degussa in 1941 and operated by that company in Rheinfelden, Germany.

German Patent 870,242.
Ulrich, G.D., *Chem. Eng. News*, 1984, **62**(32), 22.

AFC *See* Compagnie AFC.

AFEX [**A**mmonia **F**iber **Ex**plosion (or **Ex**pansion)] A pretreatment process for lignocellulose prior to enzymatic conversion to ethanol. The wood is exposed to liquid ammonia at 60°C–100°C for a short time and the pressure suddenly released. Invented in 1998 by B.E. Dale and M. Moniruzzaman at Texas A&M University and further developed by Dale at Michigan State University. In 2006, AFEX was regarded as the leading nonenzymatic biomass pretreatment process. In 2008, it was being scaled up by SunEthanol, which changed its name to Qteros in late 2008. In 2012, the Michigan Biotechnology Institute was scaling up the process, with funding from the U.S. DOE.

U.S. Patents 6,106,888; 6,176,176.
Teymouri, F., Laureano-Pérez, L., Alizadeh, H., and Dale, B.E., *Appl. Biochem. Biotechnol.*, 2004, **113/116**, 951.
Henderson, L.A., in *Polymer Biocatalysis and Biomaterials*, Cheng, H.N. and Gross, R.A., Eds., American Chemical Society, Washington, DC, 2005, 23.
Eisberg, N., *Chem. Ind.*, 2006, (17), 25.
New Sci., 29 May 2010, **206**(2762), 22.
Biotech. Fuels, 2010, **3**, 1.
Chem. Ind. (London), 2012, **76**(8), 12.

AGC-21 A process for converting natural gas to liquid fuels in three stages: generation of syngas in a fluidized bed, Fischer–Tropsch synthesis in a slurry bubble column reactor, and *hydrocracking. Piloted in 1997 and proposed for installation in Qatar.

Chang, M., *Appl. Catal. A: Gen.*, 1997, **155**(1), N5.

Agrol A fermentation process for making ethanol from agricultural wastes. It uses thermophilic bacteria at 70°C, which prevents other microorganisms from competing. The wastes are first hydrolyzed by acid. Developed by Agrol Technologies, Guildford, United Kingdom.

Chem. Eng. (N.Y.), 1998, **106**(2), 35.

AhlStage [**Ahl**strom **Stage**] A pulp-bleaching process that economizes on oxidizing agents by first destroying hexenuronic acid derivatives that would otherwise consume them. The derivatives are destroyed by hydrolysis with dilute sulfuric acid. Developed by Ahlstrom Machinery Corporation, Finland, in 1996.

Chem. Eng. (N.Y.), 1996, **103**(12), 17.

AHP [**A**ltair **H**ydrochloride **P**igment process] A process for making titanium dioxide pigment. The ore is leached with hydrochloric acid, and the titanium and iron in the leachate are separated by solvent extraction. The titanium solution is subject to spray hydrolysis and the product calcined. Claimed by Altair to be the "first commercially viable pigment production process in 50 years." Developed by BHP, Australia, from 1993 to 1999 and acquired by Altair Nanotechnologies, Reno, NV, in 1999. In April 2005, Altair signed an agreement with Bateman Engineering to assist in commercializing the process.

U.S. Patents 6,375,923; 6,440,383; 6,548,039.
Eur. Chem. News, 2005, **82**(2143), 28.

AHR [**A**dsorptive **H**eat **R**ecovery] A vapor-phase process for removing water from other vapors by selective adsorption in a bed of zeolite molecular sieve, regenerating the adsorbent by passing a noncondensible gas through it at essentially the same temperature and pressure. The heat of adsorption is stored as temperature rises within the bed and provides the heat required for desorption. Developed by Union Carbide Corporation for energy-efficient drying of petrochemical streams containing substantial amounts of water and for drying ethanol for use in motor fuels. Five units have been licensed by UOP.

Garg, D.R. and Ausikaitis, J.P., *Chem. Eng. Prog.*, 1983, **79**(4), 60.
Garg, D.R. and Yon, C M., *Chem. Eng. Prog.*, 1986, **82**(2), 54.

AH Unibon A process for hydrogenating aromatic hydrocarbons in petroleum fractions to form aliphatic hydrocarbons. Developed by UOP.

AIAG Neuhausen An electrolytic process for making aluminum from an all-fluoride melt. Developed by the Société Suisse de l'Aluminium Industrie at Neuhausen, Germany.

Airco A modification of the *Deacon process for oxidizing hydrogen chloride to chlorine. The copper catalyst is modified with lanthanides and used in a reversing flow reactor without the need for external heat. Developed by the Air Reduction Company from the late 1930s.

U.S. Patents 2,204,172; 2,312,952; 2,271,056; 2,447,834.
Redniss, A., in *Chlorine: Its Technology, Manufacture and Uses*, Sconce, J.S., Ed., Reinhold Publishing, New York, 1962, 252.

Airlift Thermofor Catalytic Cracking Also called Airlift TCC. A continuous catalytic process for converting heavy petroleum fractions to lighter ones. The catalyst granules are moved continuously by a stream of air. Developed by Mobil Oil Corporation, United States, and first operated in 1950. *See also* Thermofor.

Enos, J.L., *Petroleum Progress and Profits*, MIT Press, Cambridge, MA, 1962, Chap. 5.
Unzelman, G.H. and Wolf, C.J., in *Petroleum Processing Handbook*, Bland, W.F. and Davidson, R.L., Eds., McGraw-Hill, New York, 1967, 7, Chap. 3.

AIST A process for converting PET (polyethylene terephthalate) into virgin plastic. The scrap PET is first depolymerized by heating with ethylene glycol and a metal salt catalyst.

Yoshioka, T. and Grause, G., in *Feedstock Recycling and Pyrolysis of Waste Plastics*, Scheirs, J. and Kaminsky, W., Eds., John Wiley & Sons, Chichester, U.K., 2006, Chap. 25.

Ajax An oxygen steelmaking process in which the oxygen is injected into an *Open Hearth furnace through water-cooled lances. Used at the Appleby-Frodingham steelworks, United Kingdom.

Akzo-Fina CFI A process for improving the quality of diesel fuel by *dewaxing, *hydrotreating, and *hydro-cracking. Developed by Akzo Nobel and Fina from 1988.

Absci-Halabi, M., Stanislaus, A., and Qabazard, H., *Hydrocarbon Process. Int. Ed.*, 1997, **76**(2), 49.

Albene [**Al**cohol **benz**ene] A process for making ethylbenzene from aqueous ethanol and benzene. The aqueous ethanol may contain as little as 30% ethanol, such as that obtained by one distillation of liquors from sugar fermentation. The mixed vapors are passed over a catalyst at approximately 350°C. The catalyst (Encilite-2) is a ZSM-5-type zeolite in which some of the aluminum has been replaced by iron. Developed in India jointly by the National Chemical Laboratory and Hindustan Polymers; operated commercially by Hindustan Polymers at Vizag, Andhra Pradesh, since 1989.

Indian Patent 157,390.

Alberger A process for crystallizing sodium chloride from brine. The brine is heated under pressure to 145°C to remove calcium sulfate. Flashing to atmospheric pressure produces fine cubic crystals of sodium chloride, and surface evaporation in circular vessels produces flakes of it. Developed by J.L. and L.R. Alberger in the 1880s. *See also* Recrystallizer.

U.S. Patents 351,082; 400,983; 443,186.
Richards, R.B., in *Sodium Chloride*, D.W. Kaufmann, Ed., Reinhold Publishing, New York, 1960, 270.

Albion *See* NENATECH.

Alcar *See* Alkar.

Alcell [**Al**cohol **cell**ulose] A process for delignifying wood pulp by dissolving it in aqueous ethanol at high temperature and pressure. The by-product furfural is purified and sold. Developed by Repap Technologies, United States, from 1987 to 1997. The company was dissolved in 1997 and the technology acquired by Lignol Innovations Corporation of Vancouver, Canada, which is commercializing it and planning to convert the cellulose to ethanol instead of pulp.

Chem. Eng. (N.Y.), 1991, **98**(1), 41.
Pye, E.K. and Lora, J.H., *Tappi J.*, 1991, **74**, 113.
Stockburger, P., *Tappi J.*, 1993, **76**(6), 71.
Oliet, M., Rodriguez, F., Santos, A., Gillaranz, M.A., Garcia-Ochoa, F., and Tijero, J., *Ind. Eng. Chem. Res.*, 2000, **39**(1), 34.
Aato, C., Pye, E.K., and Gjennestadt, G., *Appl. Biochem. Biotechnol.*, 2005, **121/124**, 871.
Pye, E.K., in *Biorefineries: Industrial Processes and Products*, Kamm, B., Gruber, P.R., and Kamm, M., Eds., Wiley-VCH, Weinheim, Germany, 2005, **II**, 179.

Alceru [**Al**ternative **cell**ulose **Ru**dolstadt] A process for making cellulosic filaments and staple fibers. The cellulose is first dissolved in an aqueous solution of N-methyl morpholine N-oxide (NMMNO) and then spun. Developed by the Thüringische Institut für Textil- und Kunstoff-Forschung e.V. Rudolstadt, Germany, and Zimmer (Frankfurt) from 1987. A pilot plant was built in April 1998 and a commercial plant was planned for installation in Baoding, China in 2005. A superabsorbent version of the fiber has been developed by Stockhausen. The fiber is now made commercially by Sea Cell GmbH, a subsidiary of Zimmer. *See also* Lyocell.

Chem. Week, 1997, **159**(25), 21.
Chem. Eng. (N.Y.), 2003, **110**(9), 23.

ALCET [**A**dvanced **L**ow-**C**apital **E**thylene **T**echnology] A process for separating ethylene from the gases made by cracking naphtha. It replaces the conventional cryogenic stages with a proprietary solvent absorption process. Developed by a consortium of Brown & Root, Advanced Extraction Technologies, and Kinetics Technology International, but not yet commercialized. A demonstration unit was planned for summer 1996.

Chem. Eng. News, 1994, **72**(29), 6.
Hydrocarbon Process. Int. Ed., 1995, **74**(3), 118.
Eur. Chem. News, Chemscope, 1996, **65**, Jun, 8.

Alco An early process for thermally polymerizing refinery gases (mainly C_3 and C_4 hydrocarbons) to yield liquid hydrocarbon mixtures, suitable for blending with gasoline. The process was operated without a catalyst at 480°C–540°C and 50 atm. Developed by the Pure Oil Company, Chicago, and licensed to Alco Products, United States.

Asinger, F., *Mono-olefins: Chemistry and Technology*, translated by Hazzard, B.J., Pergamon Press, Oxford, U.K., 1968, 426.

ALCOA A process proposed for manufacturing aluminum metal by the electrolysis of molten aluminum chloride, made by chlorinating alumina. It requires 30% less power than the *Hall–Héroult process and operates at a lower temperature, but has proved difficult to control. Developed by the Aluminum Company of America, Pittsburgh, in the 1970s and operated in Palestine, TX, from 1976; abandoned in 1985 because of corrosion problems and improvements in the efficiency of the established electrolytic process.

Grjotheim, K., Krohn, C., Malinovsky, M., Matiaskovsky, K., and Thonstad, J., *Aluminum Electrolysis: Fundamentals of the Hall-Héroult Process*, CRC Press, Boca Raton, FL, 1982, 17.
Palmear, I.J., in *The Chemistry of Aluminium, Gallium, Indium, and Thallium*, Downs, A.J., Ed., Blackie, London, U.K., 1993, 87.

Aldip *See* metal surface treatment.

Aldol Also called the four-step process. A process for converting acetylene to synthetic rubber, used on a large scale in Germany during World War II. A four-step synthesis converted the acetylene to butadiene, which was then polymerized by the *Buna process. The four steps were as follows:

1. Hydration of acetylene to acetaldehyde, catalyzed by sulfuric acid and mercuric sulfate
2. Condensation of acetaldehyde to aldol using aqueous alkali (the "aldol condensation")
3. Hydrogenation of aldol to 1,3-butanediol
4. Dehydrogenation of 1,3-butanediol to 1,3-butadiene, catalyzed by sodium phosphate on coke

The process was still in use in East Germany in the 1990s.

Fisher, H.L., in *Synthetic Rubber*, Whitby, G.S., Davis, C.C., and Dunbrook, R.F., Eds., John Wiley & Sons, New York, 1954, 121.
Weissermel, K. and Arpe, H.-J., *Industrial Organic Chemistry*, 4th ed., translated by C.R. Hindley and S. Hawkins, Wiley-VCH, Weinheim, Germany, 2003, 108.

Aldox [**Ald**olization **OXO**] A *hydroformylation process for converting olefins having n carbon atoms to aldehydes having (2n + 2) carbon atoms. The olefins are reacted with carbon monoxide and hydrogen in the presence of an organometallic catalyst. Invented by Esso Research & Engineering Company, United States, in 1954 and operated since 1962 by Humble Oil & Refining Company in Baton Rouge, LA.

British Patents 761,024; 867,799.
Weissermel, K. and Arpe, H.-J., *Industrial Organic Chemistry*, 4th ed., translated by C.R. Hindley and S. Hawkins, Wiley-VCH, Weinheim, Germany, 2003, 140.

Alfaprox A process for extracting protein from green leaves, developed in the 1970s.

Kromus, S., Kamm, B., Kamm, M., Fowler, P., and Narodoslawsky, M., in *Biorefineries: Industrial Processes and Products*, Kamm, B., Gruber, P.R., and Kamm, M., Eds., Wiley-VCH, Weinheim, Germany, 2005, **I**, 257.

Alfene [**Alf**a olef**ene**] Also spelled Alfen. A process for making higher alpha olefins. Ethylene is reacted with triethyl aluminum, yielding high molecular weight aluminum alkyls, and these are treated with additional ethylene, which displaces the higher olefins. Developed by the Continental Oil Company.

Chem. Eng. News, 1962, **40**(16), 68, 70.
Acciarri, J.A., Carter, W.B., and Kennedy, F., *Chem. Eng. Prog.*, 1992, **58**(6), 85.
Weissermel, K. and Arpe, H.-J., *Industrial Organic Chemistry*, 4th ed., translated by C.R. Hindley and S. Hawkins, Wiley-VCH, Weinheim, Germany, 2003, 75.

Alfin An obsolete process for making synthetic rubber by polymerizing butadiene in pentane solution. The catalyst was an insoluble aggregate of sodium chloride, sodium *iso*propoxide, and allyl sodium. The name is actually the name of the catalyst, derived from **al**cohol, used to make the sodium *iso*propoxide and ole**fin**, referring to the propylene used to make the allyl sodium.

Morton, A.A., Magat, E.E., and Letsinger, R.L., *J. Am. Chem. Soc.*, 1947, **69**, 950.
Morton, A.A., *Ind. Eng. Chem.*, 1950, **42**, 1488.

Alfol Also called the Conoco process and the Mühlheim process. The same name is used for the products. A process for making linear primary alcohols, from C_2 to C_{28}, from ethylene. The ethylene is reacted with triethyl aluminum, yielding higher alkyl aluminums. These are oxidized with atmospheric oxygen under mild conditions to aluminum alkoxides, which are then hydrolyzed by water to the corresponding alcohols:

$$2AlR_3 + 3O_2 = 2Al(OR)_3$$

$$2Al(OR_3) + 3H_2O = 6ROH + Al_2O_3$$

Invented by K. Ziegler at the Max Planck Institut für Kohlenforschung, Mühlheim/Ruhr, Germany. Operated in the United States by Conoco since 1962 and in Germany by Condea Chemie since 1964. *See also* Epal.

German Patent 1,014,088.
East German Patent 13,609.
Belgian Patent 595,338.
Ziegler, K., Krupp, F., and Zosel, K., *Angew. Chem.*, 1955, **67**, 425.
Lobo, P.A., Coldiron, D.C., Vernon, L.N., and Ashton, A.T., *Chem. Eng. Prog.*, 1962, **58**(5), 85.
Hydrocarbon Process, 1963, **42**(11), 140.
Weissermel, K. and Arpe, H.-J., *Industrial Organic Chemistry*, 4th ed., translated by C.R. Hindley and S. Hawkins, Wiley VCH, Weinheim, Germany, 2003, 76, 210.

AlgaSORB A process for removing toxic heavy metals from aqueous wastes by the use of algae supported on silica gel.

Veglio, F. and Beolchini, F., *Hydrometallurgy*, 1997, **44**, 301.
Veglio, F., Beolchini, F., and Toro, L., *Ind. Eng. Chem. Res.*, 1998, **37**(3), 1105.

Alkacid A process for removing sulfur compounds from gas streams. All the sulfur compounds are first catalytically hydrogenated to hydrogen sulfide using a cobalt–molybdena catalyst. The hydrogen sulfide is then absorbed in an aqueous solution of an amino acid salt. Heating this solution regenerates the hydrogen sulfide as a concentrate, which is then treated by the *Claus process. Invented by IG Farbenindustrie in 1932; by 1950, 50 plants were operating in Europe, the Middle East, and Japan. *See also* Alkazid.

U.S. Patent 1,990,217.
Lühdemann, R., Noddes, G., and Schwartz, H.G., *Oil Gas J.*, 1959, **57**(32), 100.
Kohl, H.L. and Nielsen, R.B., *Gas Purification*, 5th ed., Gulf Publishing Co., Houston, TX, 1997, 397.

Alkad A process for improving the safety of *alkylation processes using hydrofluoric acid as the catalyst. A proprietary additive curtails the emission of the acid aerosol that forms in the event of a leak. Based on the observation of G. Olah in the early 1990s, liquid polyhydrogen fluoride complexes (of amines such as pyridine) depress the vapor pressure of HF above alkylation mixtures. Developed by UOP and ChevronTexaco and operated at Texaco's refinery at El Dorado, TX, since 1992. A competing process is *ReVAP.

U.S. Patent 5,073,674.
Sheckler, J.C., Hammershaimb, H.U., Ross, L.J., and Comey, K.R., III, *Oil Gas J.*, 1994, **92**(34), 60.
Chem. Eng. (N.Y.), 1995, **102**(12), 68.
Detrick, K.A., Himes, J.F., Meister, J.M., and Nowack, F.-M., in *Handbook of Petroleum Refining Processes*, 3rd ed., Meyers, R.A., Ed., McGraw-Hill, New York, 2003, 1.51.

Alkar [**Alk**ylation of **ar**omatics] Also (incorrectly) spelled Alcar. A catalytic process for making ethylbenzene by reacting ethylene with benzene. The ethylene stream can be of any concentration down to 3%. The catalyst is boron trifluoride on alumina. Introduced by UOP in 1958 but no longer licensed by them. Replaced by the *Ethylbenzene process.

Grote, H.W. and Gerald, C.F., *Chem. Eng. Prog.*, 1960, **56**(1), 60.
Hydrocarbon Process., 1963, **42**(11), 141.
Mowry, J.R., in *Handbook of Petroleum Refining Processes*, Meyers, R.A., Ed., McGraw-Hill, New York, 1986, 1–29.

Alkazid A development of the *Alkacid process. The absorbent is an aqueous solution of the potassium salt of either methylamino propionic acid (Alkazid M) or dimethylamino acetic acid (Alkazid DIK). Developed by Davy Powergas, Germany. Over 80 plants were operating in 1975.

Bähr, H., *Chem. Fabrik*, 1938, **11**(23/24), 283.
Unzelman, G.H. and Wolf, C.J., in *Petroleum Processing Handbook*, Bland, W.F. and Davidson, R.L., Eds., McGraw-Hill, New York, 1967, 133.
Process. Int. Ed. 1975, **54**(4), 85.
Speight, J.G., *Gas Processing*, Butterworth Heinemann, Oxford, U.K., 1993, 256.

alkylation Any process whereby an alkyl group is added to another molecule. However, in process chemistry, the word is most commonly used to designate a reaction in which an olefin is added to a saturated aliphatic hydrocarbon or an aromatic compound. In the petroleum and petrochemical industries, this term refers to the conversion of a mixture of light olefins and *iso*butane into a mixture of alkanes suitable for blending into gasoline in order to increase the octane number. An acid catalyst is used. Originally, the acid chosen was anhydrous hydrofluoric or sulfuric acid. Proprietary solid acids, introduced in the 1990s, were easier to dispose of. The product is called alkylate. Alkylation processes with special names described in this dictionary are Alkar, AlkyClean, AlkylClean, Alkylene, FBA, Detal, Detergent Alkylate, Mobil/Badger, Stratco, and Thoma.

AlkyClean [**Alky**late **Clean**] An *alkylation process using a solid acid catalyst. Developed by CB&I, Albemarle, and Neste Oil in the 1990s.

Rothenberg, G., *Catalysis: Concepts and Green Applications*, Wiley-VCH, Weinheim, Germany, 2008, 168.
Chem. Eng. (N.Y.), 2013, **120**(6), 11.

AlkylClean A process for making alkylate (*see* previous entry) that uses a solid acid catalyst in place of the usual HF or H_2SO_4. Developed by ABB in conjunction with Akzo Nobel and Forum Oil & Gas and first demonstrated in 1995.

Technicky Tydenik, 26 Aug 2003, **51**(34), 7.

Alkylene A process for making alkylate (*see* previous two entries) by reacting olefins with isobutene using a proprietary solid acid catalyst called HAL 100. The major constituents of this alkylate are branched trimethylpentanes. Developed by UOP from 1995.

Meyers, R.A., Ed., *Handbook of Petroleum Refining Processes*, 3rd ed., McGraw-Hill, New York, 2003, 1.33.

Alkymax A process for removing benzene from petroleum fractions. They are mixed with light olefin fractions (containing mainly propylene) and passed over a fixed-bed catalyst, which promotes benzene alkylation. The catalyst is "solid phosphoric acid" (SPA), made by mixing a phosphoric acid with a siliceous solid carrier and calcining. Invented in 1980 by UOP.

U.S. Patent 4,209,383.
Chem. Eng. (N.Y.), 1990, **97**, 23.
Hydrocarbon Process. Int. Ed., 1994, **73**(11), 90.

AlkyStar Not a process but a zeolite-based alkylation catalyst claimed to be safer than the traditional acid alkylation catalysts. Used in the *AlkyClean process. Developed by Albemarle, ABB Lummus Global, and Neste Oil.

Hydrocarbon Process., 2007, **86**(9), 32,35.

Allis-Chalmers. *See* DR.

ALMA [**AL**usuisse **M**aleic **A**nhydride] A process for making maleic anhydride by oxidizing n-butane using a fluid bed reactor and a special organic solvent recovery system. The catalyst contains vanadium and phosphorus on iron oxide. Developed jointly by Alusuisse Italia and ABB Lummus Crest. First licensed to Shin-Daikoku Petrochemical Company, Yokkaichi, Japan, in 1988. The world's largest plant was built for Lonza in Ravenna, Italy, in 1994.

> Budi, F., Neri, A., and Stefani, G., *Hydrocarbon Process. Int. Ed.*, 1982, **61**(1), 159.
> Arnold, S.C., Suciu, G.D., Verde, L., and Neri, A., *Hydrocarbon Process. Int. Ed.*, 1985, **64**(9), 123.
> *Chem. Eng. (N.Y.)*, 1996, **103**(11), 17.
> Weissermel, K. and Arpe, H.-J., *Industrial Organic Chemistry*, 4th ed., translated by C.R. Hindley and S. Hawkins, Wiley-VCH, Weinheim, Germany, 2003, 372.

Aloton Also called Büchner. A process proposed for extracting aluminum from clay. Calcined clay is leached with ammonium hydrogen sulfate solution under pressure, and ammonium alum is crystallized from the liquor. Invented by M. Büchner in Hannover-Kleefeld in 1921; piloted in Germany in the 1920s and in Oregon in 1944. The process was never commercialized, but it provided the basis for the *Nuvalon process, which was.

> British Patent 195,998.
> U.S. Patent 1,493,320.
> O'Connor, D.J., *Alumina Extraction from Non-bauxitic Materials*, Aluminium-Verlag, Düsseldorf, Germany, 1988, 159.

Alpha (1) A process for making aromatic hydrocarbons and LPG from C_3 to C_7 olefins. The catalyst is a metal-modified ZSM-5 zeolite. Developed by Asahi Chemical Industries and Sanyo Petrochemical and used since 1993 at Sanyo's Mizushima refinery. Planned to be used at Taiyo Oil's refinery at Shikoku, Japan, from 2011.

> *Eur. Chem. News, Chemscope*, 7 Apr 1994; 4 Jun 1996.
> Nagamori, Y. and Kawase, M., *Micropor. Mesopor. Mater.*, 1998, **21**(4–6), 439.
> *Jpn. Chem. Week*, 2007, **48**(2402), 9.

Alpha (2) A process for making methyl methacrylate, developed by Ineos Acrylics (now Lucite International) since 1990. Ethylene is carbonylated and methylated to produce methyl propionate, which is reacted with formaldehyde to produce methyl methacrylate. The first stage is homogeneously catalyzed by a palladium phosphine complex. The second stage is operated in the gas phase over a catalyst of cesium oxide on silica. Piloted by Davy Process Technology in 2002. The first commercial plant was opened in Singapore in 2008. The second was to be built in Texas by Mitsubishi Rayon for completion in late 2009. Lucite International received the 2009 Kirkpatrick Chemical Engineering Achievement Award for this development. Lucite is now a subsidiary of Mitsubishi Rayon.

> *Ind. Eng. Chem.*, 2002, **76**(1992), 21.
> *Chem. Eng. (N.Y.)*, 2003, **110**(7), 19.
> *Chem. Eng. News*, 2009, **87**(42), 22.
> *Chem. Eng. (N.Y.)*, 2009, **116**(13), 17.

Alphabutol Also called IFP-SABIC. A process for dimerizing ethylene to 1-butene. It operates under pressure at 80°C using a complex Ziegler–Natta catalyst, a titanium alkoxide. Developed by the Institut Français du Pétrole. First operated in Thailand in 1987. Nineteen plants had been licensed by 2005, of which sixteen were operating.

> Commereuc, D., Chauvin, Y., Gaillard, J., Léonard J., and Andrews, J., *Hydrocarbon Process. Int. Ed.*, 1984, **63**(11), 118.
> Hennico, A., Léonard, H.J., Forestiere, A., and Glaize, Y., *Hydrocarbon Process. Int. Ed.*, 1990, **69**(3), 73.
> Chauvel, A., Delmon, B., and Hölderich, W.F., *Appl. Catal. A: Gen.*, 1994, **115**(2), 201.
> Weissermel, K. and Arpe, H.-J., *Industrial Organic Chemistry*, 4th ed., translated by C.R. Hindley and S. Hawkins, Wiley-VCH, Weinheim, Germany, 2003, 67.

Alpha Plant A process for making BTX (mixed benzene, toluene, and xylenes) from C_4 to C_5 raffinates. Developed by Asahi Kasei Chemicals Corporation and commercially available since 1993. *See also* Omega Plant.

> *Chem. Eng. (N.Y.)*, 2004, **111**(7), 13.

Alpha-Sablin [**Alpha Sab**ic **Lin**de] (also written α-**Sablin**) A process for making linear C_4–C_{10} olefins from ethylene. Catalyzed by zirconium carboxylates with aluminum alkyls, used in a bubble column reactor. Developed between 1994 and 2001 by Sabic and Linde. The first commercial plant was completed in Al-Jubail, Saudi Arabia, in 2009. A second plant is expected to come on stream in Nizhnekamsk, Russia, in 2014.

Chem. Eng. News, 5 Nov 2001, **79**(45), 16.
Eur. Chem. News, *Petrochemicals Fact File*, Mar 2002, 6.
Chem. Eng. News, 7 Dec 2009, **87**, 49, 22.
Steinborn, D., *Fundamentals of Organometallic Catalysis*, 2012, Wiley-VCH, Weinheim, Germany, 158.
Chem. Ind. (London), 2012, **76**(8), 20.

Alpha Select A process for making high-purity linear alpha olefins, C_4–C_{10}, by oligomerization of ethylene. The process, developed by IFP and based on its Alphabutol process, is now offered by Axens. Alpha Select operates in the liquid phase at 100°C–150°C and 70–90 bars. The composition of the homogeneous catalytic system used allows the distribution of alpha olefins to be adjusted according to demand. The catalyst is regenerated and recycled. The products are used as comonomers in the production of linear, low-density polyethylene, as monomers to make polyalphaolefins, and as alcohol precursors for plasticizers. Nineteen units for making 1-butene by this process were operating in 2005.

Chimie Hebdo, 15 Feb 1999, (36), 20.
Eur. Chem. News, 29 Mar 1999, 35.
Eur. Chem. News, May 1999 (*Chemscope*), 27–28.
Hydrocarbon Process. Int. Ed., 2003, **82**(3), 73.

AlphOx A process for oxidizing benzene to phenol using nitrous oxide as the oxidant and a zeolite catalyst. Developed by the Boreskov Institute of Catalysis and Solutia and proposed to be commercialized in Pensacola, FL, in 2003; however, this plan was subsequently abandoned. In 2004, the process was relaunched by GTC Technology, Inc. and Solutia, following extensive testing by Solutia in Pensacola.

Eur. Chem. News, 2001, **75**(1969), 53.
Chem. Eng. (N.Y.), 2004, **111**(9), 17.

Alplate *See* metal surface treatment.

Alrak *See* metal surface treatment.

Alstan A process for electroplating aluminum by pretreating the surface with a stannate. Developed by M&T Chemicals.

Di Bari, G.A., *Plat. Surf. Finish.*, 1977, **64**(5), 68.

ALTAM A process for making caprolactam from butadiene and carbon monoxide. Developed by DSM in the late 1990s and subsequently improved by Shell Chemicals, which contributed catalyst know-how. In the first two steps of the process, butadiene undergoes two hydroformylations with carbon monoxide, followed by reductive amination with ammonia and then cyclization to caprolactam. First commercialization was expected in Taiwan. A joint venture with Chiyoda Corporation, to further develop and commercialize the process, was announced in 2002.

Chem. Week, 2000, **162**(32), 17.
Asian Chem. News, 4 Sep 2000, **6**(278), 37.
Eur. Chem. News, 4 Sep 2000, **73**(1923), 46.

Alumet A process for extracting alumina and potassium sulfate from alunite ore (a basic hydrated potassium aluminum sulfate) involving reductive calcination and alkali leaching. Developed and piloted by the Alunite Metallurgical Company, UT, in the mid-1970s but not commercialized. *See also* Kalunite.

O'Connor, D.J., *Alumina Extraction from Non-bauxitic Materials*, Aluminium-Verlag, Düsseldorf, Germany, 1988, 198.

Alumilite *See* metal surface treatment.

ALUREC A process for recovering aluminum from residues obtained from the remelting of aluminum scrap. The material is melted in a rotating furnace heated with natural gas and oxygen. Previous processes involved melting with salt. Developed jointly by AGA, Hoogovens Aluminium, and MAN GHH and offered in 1994.

Alzak A method for electropolishing aluminum using fluoroboric acid. Developed by the Aluminum Company of America. *See also* metal surface treatment.

Amalgam A process for making sodium dithionite by reacting sodium amalgam with sulfur dioxide:

$$2(Hg)Na + 2SO_2 = Na_2S_2O_4 + 2(Hg)$$

Bostian, L.C., in *Speciality Inorganic Chemicals*, Thompson, R., Ed., Royal Society of Chemistry, London, U.K., 1981, 63.

Aman A process for thermally decomposing metal chloride or sulfate solutions in a spray roaster. Used for recovering hydrochloric acid from iron pickle liquors. Developed by J.J. Aman in Israel in 1954.

British Patent 793,700.

AMAR A solvent extraction process for recovering copper. Used in approximately 50 installations worldwide in 1993.

AMASOX *See* Walther.

AMASULF A two-stage process for removing hydrogen sulfide and ammonia from coke-oven gas. In the first stage, hydrogen sulfide is removed by scrubbing with aqueous ammonia; the resulting ammonium sulfide solution is heated in another vessel to expel the hydrogen sulfide:

$$H_2S + 2NH_4OH = (NH_4)_2S + 2H_2O$$

$$(NH_4)_2S + H_2O = H_2S + 2NH_4OH$$

In the second stage, ammonia is removed by scrubbing with water. Developed and licensed by Krupp Koppers, Germany.

AMASULF^{R}PURE A variation on the *AMASULF process in which only the hydrogen sulfide is recovered, not the ammonia.

AMAT *See* Z3A.

aMDEA [activated Methyl DiEthAnolamine] A process for removing CO_2, H_2S, and trace sulfur compounds from natural gas and *syngas. Developed by BASF from the 1960s; more than 200 plants were operating in 2000.

Hydrocarbon Process. Int. Ed., 2000, **79**(4), 56.

American Also known as the Wetherill process and the Direct process. A process for making zinc oxide, in the form of a white pigment, from a zinc oxide ore. The ore is usually franklinite, which is predominately $ZnFe_2O_4$. The ore is mixed with coal and heated in a furnace to approximately 1000°C, forming zinc vapor in a reducing atmosphere. The vapors pass to a second chamber in which they are oxidized with air, forming zinc oxide and carbon dioxide. *See also* French.

Buxbaum, G. and Pfaff, G., *Industrial Inorganic Pigments*, 3rd ed., VCH, Weinheim, Germany, 2005, 90.

Ames (1) A process for making uranium metal by reducing uranium tetrafluoride with calcium or magnesium.

Ames (2) A wet oxidation process for desulfurizing coal, in which the oxidant is oxygen and the sulfur dioxide is absorbed by aqueous sodium carbonate. Developed in the 1970s by the Ames Laboratory of Iowa State University, with funding from the U.S. Department of Energy. *See also* PETC.

IEA Coal Research, *The Problems of Sulphur*, Butterworths, London, U.K., 1989, 20.

AMEX [**AM**ine **EX**traction] A process for the solvent extraction of uranium from sulfuric acid solutions using an amine extractant:

$$UO_2(SO_4)_2{}^{2-} + 2H^+ + 2B \rightarrow (BH^+)_2 + [UO_2 (SO_4)_2{}^{2-}]$$

The amine (B) is a proprietary mixture of C_8 and C_{10} primary alkylamines dissolved in kerosene. The uranium is stripped from the organic solution with an alkaline stripping solution and precipitated as ammonium diuranate. *See also* Dapex.

Chem. Eng. News, 1956, **34**(21), 2590.
Eccles, H. and Naylor, A., *Chem. Ind. (London)*, 1987, (6), 174.
Danesi, P.R., in *Developments in Solvent Extraction*, Alegret, S., Ed., Ellis Horwood, Chichester, U.K., 1988, 204.

Amine Guard A process for extracting acid gases from refinery streams by scrubbing with an alkanolamine. Many such processes have been developed; this one was developed by the Union Carbide Corporation and uses monoethanolamine. It has been used to purify hydrogen produced by *steam reforming. In 1990, over 375 units were operating.

Butwell, K.F., Hawkes, E.N., and Mago, B.F., *Chem. Eng. Prog.*, 1973, **69**(2), 57.
Butwell, K.F., Kubek, D.J., and Sigmund, P.W., *Chem. Eng. Prog.*, 1979, **75**(2), 75.
Kubek, D.K. and Butwell, K.F., in *Acid and Sour Gas Treating Processes*, Newman, S.A., Ed. Gulf Publishing, Houston, TX, 1985, 235.
Hydrocarbon Process. Int. Ed., 1992, **71**(4), 86; 1996, **75**(4), 105.

AMINEX A process for removing hydrogen sulfide and carbon dioxide from gas and LPG streams by circulating an aqueous amine solution through bundles of hollow fibers immersed in them. Developed in 1991 by the Merichem Company, Houston, TX. Four hundred fifty units were operating in 2000.

Hydrocarbon Process. Int. Ed., 1996, **75**(4), 126; 2000, **79**(4), 58.

AMISOL A process for removing sulfur compounds and carbon dioxide from refinery streams by absorption in methanol containing mono- or diethanolamine and a proprietary additive. Developed by Lurgi, Germany, in the 1960s and first commercialized in the early 1970s. Six plants were operating in 1993.

Bratzler, K. and Doerges, A., *Hydrocarbon Process. Int. Ed.*, 1974, **53**(4), 78.
Kohl, H.L. and Nielsen, R.B., *Gas Purification*, 5th ed., Gulf Publishing Co., Houston, TX, 1997, 1231.

Ammonex An ion-exchange process for continuously purifying the water circuits of electric power generators. Ammonia is used to regenerate the cation exchange resins. Developed by Cochrane Environmental Systems in the 1960s and widely used.

Crits, D.J., in *Ion Exchange Technology*, Naden, D. and Streat, M., Eds., Ellis Horwood, Chichester, U.K., 1984, 119.

ammonia soda Also called the Solvay process. A process for making sodium carbonate. The basic process was invented and partially developed in the first half of the nineteenth century by several workers, but the key invention was made by E. Solvay in Belgium in 1861. The first plant was built at Couillet, Belgium, in 1864; thereafter, the process became accepted worldwide, displacing the *Leblanc process. The raw materials are limestone and salt; calcium chloride is a waste product. The overall reaction is

$$CaCO_3 + 2NaCl = CaCl_2 + Na_2CO_3$$

When carbon dioxide is passed into a nearly saturated solution of sodium chloride containing some ammonia, ammonium bicarbonate is formed. The heart of the process is the exploitation of the equilibrium between this bicarbonate and sodium and ammonium chlorides:

$$NH_4HCO_3 + NaCl \rightleftharpoons NaHCO_3 + NH_4Cl$$

In this system, the least soluble component is sodium bicarbonate, which crystallizes out. On calcination, it yields sodium carbonate, and the carbon dioxide is recycled. The ammonia is recovered by adding calcium hydroxide, producing calcium chloride waste and liberating the ammonia for reuse:

$$2NH_4Cl + Ca(OH)_2 = 2NH_3 + CaCl_2 + 2H_2O$$

British Patent 3,131 (1863).
Wood, R.D.E., in *Industrial Inorganic Chemicals: Production and Use*, Thompson, R., Ed., Royal Society of Chemistry, Cambridge, U.K., 1995, 128.
Campbell, W.A., in *Chemistry, Society and Environment*, Russell, C.A., Ed., Royal Society of Chemistry, London, U.K., 2000, 103.

ammoxidation The catalytic oxidation of a mixture of an aliphatic hydrocarbon and ammonia to give an alkyl cyanide:

$$2RCH_3 + 2NH_3 + 3O_2 = 2RCN + 6H_2O$$

This was a development of the *Andrussov process, by which methane yields hydrogen cyanide. In one important version, propylene and ammonia yield acrylonitrile:

$$2CH_2 = CH - CH_3 + 2NH_3 + 3O_2 \rightarrow 2CH_2 = CH - CN + 6H_2O$$

Invented and developed independently in the late 1950s by D.G. Stewart in the Distillers Company and by R. Grasselli in Standard Oil of Ohio. The former used a tin–antimony oxide catalyst; the latter used bismuth phosphomolybdate on silica. Today, a proprietary catalyst containing depleted uranium is used. *See also* Erdölchemie, OSW, Sohio.

Another variation is the catalytic oxidation of toluene with ammonia to produce benzonitrile. Such a process has been developed and is offered for license by Nippon Shokubai Kagaku Kogyo Company. Their plant is in Himeji, Japan.

U.S. Patent 2,904,580.
Wiseman, P., *Chem. Br.*, 1987, **23**, 1198.
Hydrocarbon Process. Int. Ed., 1987, **66**(11), 66.
Sokolovskii, V.D., Davydov, A.A., and Ovsitser, O.Y., *Catal. Rev. Sci. Eng.*, 1995, **37**(3), 425.

ammoximation The conversion of an aldehyde or ketone to its oxime by treatment with ammonia and hydrogen peroxide:

$$R - CHO + NH_3 + H_2O_2 = R - CH = NOH + 2H_2O$$

Cyclohexanone is thus converted to cyclohexanone oxime, an intermediate in the manufacture of Nylon 6. The catalyst is titanium silicalite-2. Commercialized by Enichem, which built a plant in Porto Marghera, Italy, in 1994.

Reddy, J.S., Sivasanker, S., and Ratnasamy. P., *J. Mol. Catal.*, 1991, **69**, 383.
Chem. Br., 1995, **31**(2), 94.

Amoco Amoco Chemicals Company, a subsidiary of Amoco Corporation, formerly Standard Oil Company (IN), is best known in the chemicals industry for its modification of the *Mid-Century process for making pure terephthalic acid. *p*-Xylene in acetic acid solution is oxidized with air at high temperature and pressure. Small amounts of manganese, cobalt, and bromide are used as catalysts. The modification allows the use of terephthalic acid, rather than dimethyl terephthalate, for making fiber. The process can also be used for oxidizing other methylbenzenes and methylnaphthalenes to aromatic carboxylic acids. *See also* Maruzen.

Spitz, P.H., *Petrochemicals: The Rise of an Industry*, John Wiley & Sons, New York, 1988, 327.
Weissermel, K. and Arpe, H.-J., *Industrial Organic Chemistry*, 4th ed., translated by C.R. Hindley and S. Hawkins, Wiley-VCH, Weinheim, Germany, 2003, 398.

AMTREAT An *Activated Sludge process for treating waste liquors rich in ammonia.

Developed and offered by ACWA Services, Skipton, United Kingdom.

AMV A modified process for making ammonia, invented by ICI and announced in 1982. It uses a new catalyst and operates at a pressure close to that at which the synthesis gas has been generated, thereby saving energy. Construction licenses have been granted to Chiyoda Corporation, Kvaerner, and Mannesmann. In 1990, the process was operated in the CIL plant in Ontario, Canada, and then in Henan Province, China.

European Patent 49,967.
Livingston, J.G. and Pinto, A., *Chem. Eng. Prog.*, 1983, **79**(5), 62.
Chem. Eng. (Rugby, Engl.), 1990, 21.
Hydrocarbon Process. Int. Ed., 1991, **70**(3), 134.

ANAMET [**ANA**erobic **MET**hane] An anaerobic biological process for treating industrial effluents containing relatively high concentrations of organic matter. The microorganisms are removed in a lamella separator, in which they slide down inclined plates. Developed by Purac, Sweden, which had installed more than 50 plants by 1992, mostly in the food industry.

anammox [**An**aerobic **amm**onium **ox**idation] A natural microbiological process whereby ammonia is converted to nitrogen gas:

$$NH_4 + NO_2 = N_2 + 2H_2O$$

Several bacteria are involved, and nitrite ion and hydrazine are intermediates. Used in the *ANAMMOX process.

ANAMMOX [**An**aerobic **Amm**onium **Ox**idation] A process for removing nitrogen from wastewaters. Most of the nitrogen is released as nitrogen gas. Developed and commercialized at the Delft University of Technology. The biochemical mechanism of the process was elucidated by M.S.M. Jetting for which he shared the Spinoza Prize in 2012. The first plant operated in 2001 and eleven plants were operating by 2012. *See also* *Anammox.

U.S. Patent 6,485,646.
Schulz, C., *Pollution Solutions*, Sep 2009, **3**(3), 12–13.
Vazquez-Padin, J.R., Fernandez, I., Mosquera-Corral, A., Campos, J.L., and Mendez, R., in *Environmental Technologies to Treat Nitrogen Pollution*, Cervantes, F.J., Ed., IWA Publishing, London, U.K., 2009, 67.

Anatread A hydrometallurgical process for extracting copper from a sulfide ore with ferric chloride solution.

Ancit *See* carbonization.

Andco–Torrax A process for making a fuel gas by the partial oxidation of organic wastes in a vertical shaft furnace. The residue is removed as a liquid slag from the base of the furnace.

Sixt, H., *Chem. Ing. Tech.*, 1981, **53**(11), 844.
Kirk-Othmer's Encyclopedia of Chemical Technology, 3rd ed., Vol. 13, John Wiley & Sons, New York, 1981, 195.

Andrussov A process for making hydrogen cyanide by reacting ammonia, methane, and air at approximately 1000°C over a platinum–rhodium catalyst:

$$2NH_3 + 3O_2 + 2CH_4 = 2HCN + 6H_2O$$

The product gases are freed from ammonia by scrubbing with sulfuric acid, and the hydrogen cyanide is then absorbed in water or diethanolamine. Invented in 1930 by L. Andrussov at IG Farbenindustrie, Germany.

U.S. Patent 1,934,838.
Andrussov, L., *Ber. Dtsch. Chem. Ges.*, 1927, **60**, 2005.
Andrussov, L., *Angew. Chem.*, 1935, **48**, 593.
Andrussov, L., *Chem. Ing. Tech.*, 1955, **27**, 469.
Dowell, A.M., III, Tucker, D.H., Merritt, R.F., and Teich, C.I., in *Encyclopedia of Chemical Processing and Design*, McKetta, J.J. and Cunningham, W.A., Eds., Marcel Dekker, New York, 1988, **27**, 7.

Anglo-Jersey A paraffin isomerization process, catalyzed by aluminum trichloride supported on bauxite. Developed by the Anglo-Iranian Oil Company and Standard Oil Development Company.

Perry, S.F., *Trans. Am. Inst. Chem. Eng.*, 1946, **42**, 639 (Chem. Abstr., **40**, 6792).
Asinger, F., *Paraffins: Chemistry and Technology*, translated by B.J. Hazzard, Pergamon Press, Oxford, U.K., 1968, 708.

Angus Smith *See* metal surface treatment.

Aniline *See* Laux.

ANOF [**AN**odische **O**xidation unter **F**unkentladung—German, meaning anodic oxidation under spark discharging] A process for applying oxide layers to metals in order to provide catalytic surfaces.

German Patent DE 1985 2150.
Patcas, F., Krysmann, W., Hönicke, D., and Buciuman, B.-C., *Catal. Today*, 2001, **69**, 379.
Patcas, F. and Krysmann, W., *Appl. Catal. A: Gen.*, 2007, **316**(2), 240.

Anortal [**Anort**hosite **alum**inum] A process for extracting alumina from anorthosite ore (a calcium aluminosilicate) by leaching with hydrochloric acid, precipitating aluminum trichloride hexahydrate, and calcining this. Developed and piloted by Anortal in Norway in the late 1970s but not commercialized.

Gjelsvik, N., *Light Met. Met. Ind.*, 1980, 133.
O'Connor, D.J., *Alumina Extraction from Non-bauxitic Materials*, Aluminium-Verlag, Düsseldorf, Germany, 1988, 127.

Anox An integrated water-treatment process for removing organic contaminants. The energy is obtained by burning the biogas generated in the process, which contains approximately 70% methane. Developed by W.D. Evers; a demonstration plant was built in France in 1979.

Evers, W.D., *Chimia*, 1979, **33**(6), 217.

ANTHANE/ANODEK [**AN**aerobic Me**THANE**/**AN**aerobic **O. DE K**onickx] A process for generating methane by the anaerobic fermentation of industrial organic wastes. Invented by the Institute of Gas Technology, Chicago; engineered by the Studiebureau O. de Konickx, Belgium, and commercialized since 1977.

Anthracene *See* carbonization.

AO [**A**ut**O**xidation, **A**ir **O**xidation, or **A**nthraquinone **O**xidation] A process for making hydrogen peroxide from hydrogen and oxygen (air) by cyclic oxidation–reduction of an alkyl anthraquinone solution (the working solution). Invented by H.-J. Riedl and G. Pfleiderer in Germany in the mid-1930s; piloted by IG Farbenindustrie in Ludwigshafen during World War II and commercialized in the United Kingdom and the United States during the 1950s. Now virtually the sole manufacturing process:

$$\text{Anthraquinone} + H_2 \xrightarrow[\text{catalyst}]{\text{heterogeneous}} \text{Anthraquinol}$$

$$\text{Anthraquinol } O_2 \rightarrow \text{Anthraquinone} + H_2O_2$$

The anthraquinone derivative is usually 2-ethyl- or 2-pentyl-anthraquinone. The solvent is usually a mixture of two solvents, one for the quinone and one for the quinol. The hydrogenation catalyst is usually nickel or palladium on a support. The hydrogen peroxide is produced at a concentration of 20%–40% and is concentrated by distillation.

U.S. Patents 2,158,525; 2,215,883.
German Patent 671,318.
Bertsch-Frank, B., Dorfer, A., Goor, G., and Süss, H.U., in *Industrial Inorganic Chemicals: Production and Use*, Thompson, R., Ed., Royal Society of Chemistry, Cambridge, U.K., 1995, 176.
Goor, G., in *Catalytic Oxidations with Hydrogen Peroxide as Oxidant*, Strukul, G., Ed., Kluwer Academic Publishers, Dordrecht, the Netherlands, 1993, 13.
Samanta, C., *Appl. Catal. A: Gen.*, 2008, **350**, 133.

A/O A modification of the *Activated Sludge process, designed to maximize the removal of phosphate ion. Developed by Air Products & Chemicals.

Bowker, R.P.G. and Stensel, H.D., *Phosphorus Removal from Wastewater*, Noyes Data, Park Ridge, NJ, 1990, 21.

AOD [**A**rgon **O**xygen **D**ecarburization] A steelmaking process in which a mixture of oxygen and argon is injected into molten iron to reduce the carbon content. Developed by the Union Carbide Corporation in the mid-1970s. By 1989, 90% of the stainless steel made in the United States was made using this process.

Isalski, W.H., *Separation of Gases*, Clarendon Press, Oxford, U.K., 1989, 9.

AO Plus [**A**cid **O**ptimisation **Plus**] A process for making acetic acid by carbonylating methanol. Based on the *Monsanto Acetic Acid process, but an improved catalyst (rhodium with lithium iodide) permits operation at lower levels of water. Developed by Celanese in the 1980s and operated by that company in Clear Lake, TX. Residual iodide in the product is removed by the *Silverguard process.

Plotkin, J., *Eur. Chem. News*, 2003, **78**(2047), 22.

APAC A coal gasification combined cycle process that produces fuel gases, acetylene, and electricity. Limestone is added, which produces calcium carbide, in turn used to generate acetylene by reaction with water. Operated at the Acme power plant, Sheridan, WY.

APOL [**A**lkaline **P**ressure **O**xidation **L**eaching] A process for extracting gold from refractory ores, developed by Davy McKee, Stockton, United Kingdom.

Appleby-Frodingham A process for removing hydrogen sulfide and organic sulfur compounds from coke-oven gas by absorption on iron oxide particles in a fluidized bed at 350°C. The absorbent is regenerated with air at a higher temperature, and the resulting sulfur dioxide is used to make sulfuric acid. Invented by L. Reeve and developed by the South Western Gas Board at Exeter, United Kingdom, in the 1950s; operated at the Appleby-Frodingham steelworks.

British Patent 719,056.
Reeve, L., *J. Inst. Fuel*, 1958, **31**, 319.
Claxton, G., *Benzoles: Production and Uses*, National Benzole & Allied Products Association, London, U.K., 1961, 210.
Kohl, A.L. and Riesenfeld, F.C., *Gas Purification*, 4th ed., Gulf Publishing Co., Houston, TX, 1985, 479.

APR [**A**queous **P**hase **R**eforming] A general name for aqueous processes that convert starches, sugars, and lignocellulosic biomass into hydrogen and hydrocarbons. Developed by R.D. Cortright and J.A. Dumesic at the University of Wisconsin from 2001. Their process is called *BioForming and their company, which is still developing it, is called Virent Energy Systems; it operated a demonstration plant from 2006. The catalyzed process operates at a relatively low temperature (180°C–260°C). Other workers are developing related processes that convert glycerol to fuels.

U.S. Patents 6,699,457; 6,953,873; 6,964,757; 8,017,818.
Cortright, R.D., Davda, R.R., and Dumesic, J.A., *Nature*, 2002, **418**, 64.
Chem. Eng. (N.Y.), 2004, **111**(13), 12.
Huber, G.W., Juben, N.C., Barrett, C.J., and Dumesic, J.A., *Science*, 2005, **308**(5727), 1446.
Huber, G.W. and Dumesic, J.A., *Catal. Today*, 2006, **111**, 119.

APU [**A**dvanced **P**ygas **U**pgrading] A catalytic process for upgrading low-value pyrolysis gasoline, converting it to benzene, toluene, and xylenes, plus LPG. Developed by SK Corporation, South Korea, and licensed through Axens.

Aquacat An *SCWO process for recovering precious metals from aqueous wastes. Developed by Johnson Matthey in 2005.

Aquaclaus A modification of the *Claus process in which hydrogen sulfide is removed from water by reaction with sulfur dioxide. Developed by Stauffer Chemical Company and operated by the Heflin Oil Company in Queen City, TX.

Hayford, J.S., *Hydrocarbon Process. Int. Ed.*, 1973, **52**(10), 95.
Sulphur, 1974, (111), 48.
Chem. Eng. (N.Y.), 1984, **91**(13), 150.

Aquaconversion A process for converting heavy crude petroleum oils and residues into lighter products, which are more easily converted into more valuable products in oil refineries. Intended for use at the well head rather than the oil refinery. Three steps are involved: thermal dissociation of aromatics, dissociation of water giving hydrogen atoms, and addition of these hydrogen atoms to the aromatic fragments to prevent their association. Developed by Foster Wheeler USA Corporation, Intevep, and UOP from 1998. First commercialized in Curacao, Peru, in 1996.

Hydrocarbon Process. Int. Ed., 1997, **76**(12), 36.
Marzin, R., Pereira, P., McGrath, M.J., Feintuch, H.M., Thompson, G., and Houde, E., *Oil Gas J.*, 1998, **96**(44), 79.

Aqua Critox A plug-flow *SCWO system for destroying organic wastes in water. Developed by Chematur Engineering in Sweden and first installed in Johnson Matthey's Brimsdown, United Kingdom, plant in 2002. The water is first heated to 400°C under 250 bars. Oxygen is introduced, and the ensuing oxidations raise the temperature to 600°C. A residence time of only 30–90 s is necessary. No added catalyst is needed, possibly because the process was developed for recovering platinum metals from spent catalysts.

Chem. Eng. (N.Y.), 2001, **108**(8), 21.
Eur. Chem. News, 10 Dec 2001, **75**(1985), 28.
Environ. Bus. Mag., 2002, (75), 39.

AQUAFINING A process for extracting water and other nonsulfur-containing contaminants from petroleum fractions by the use of a proprietary bundle of hollow fibers called a FIBER-FILM contactor. Developed by Merichem Company, Houston, TX, and used in 11 installations in 1991.

Hydrocarbon Process. Int. Ed., 1992, **71**(4), 120; 1996, **75**(4), 126.

Aquarrafin A wastewater treatment process using activated carbon in fixed beds. Developed by Lurgi.

AQUISULF A process for removing hydrogen sulfide from molten sulfur. Developed by Elf in the 1990s and now offered by Lurgi.

Arbiter Previously known as the Sherritt-Gordon ammonia process. A process for leaching copper from sulfide concentrates using ammoniacal ammonium sulfate solution at 85°C and relying on air oxidation. Copper is produced from the leachate by solvent extraction and electrowinning. Sulfur is recovered as ammonium sulfate. Operated on a large scale by the Anaconda Copper Company in Montana from 1974 to 1979. *See* Sherritt-Gordon.

Arbiter, N., *New Advances in Hydrometallurgy*, Institute of Gas Technology, Chicago, IL, 1974.
Kuhn, M.C., Arbiter, N., and Kling, H., *Can. Inst. Min. Met. Bull.*, 1974, **67**, 62.

Arc *See* Birkeland–Eyde.

Arco A process for making isobutene by dehydrating *t*-butanol. The reaction takes place in the gas phase at 260°C–273°C, 14 bar, in the presence of an alumina-based catalyst.

Weissermel, K. and Arpe, H.-J., *Industrial Organic Chemistry*, 4th ed., translated by C.R. Hindley and S. Hawkins, Wiley-VCH, Weinheim, Germany, 2003, 70.

ARDS [**A**tmospheric **R**esidue **D**e**S**ulfurization] A process for upgrading petroleum residues by catalytic hydrogenation.

Hydrocarbon Process. Int. Ed., 1997, **76**(2), 50.

Arex A process for removing aromatic hydrocarbons from petroleum streams by extraction with 1-methyl piperidone (N-methyl caprolactam) at 60°C. Developed by Leuna Werke, Germany.

Chem. Tech. (Leipzig), 1977, **29**, 573.

Arge [**Ar**beits**ge**meinschaft] A version of the *Fischer–Tropsch process, using a fixed catalyst bed. It converts *synthesis gas to a mixture of gasoline, diesel fuel, and waxes. The catalyst is made by adding sodium carbonate solution to a solution of mixed iron and copper nitrates, binding the resulting precipitate with potassium silicate, and reducing it with hydrogen. Used in the *SASOL plant in South Africa since 1955. Developed by Ruhr Chemie–Lurgi.

Mako, P.F. and Samuel, W.A., in *Handbook of Synfuels Technology*, Meyers, R.A., Ed., McGraw-Hill, New York, 1984, 11.

ARI LO-CAT II A new name for *LO-CAT. One hundred nineteen plants were operating in 2000.

Hydrocarbon Process. Int. Ed., 2000, **79**(4), 59.

ARINO [**A**romatic **RIN**g **O**pening] A process for converting "heavy pyrolysis gasoline" into a mixture of light alkanes suitable for *steam cracking. ARINO is essentially a two-step process. In the first step, aromatics are hydrogenated to naphthenes. In the second, the naphthene rings are opened and the products converted to smaller alkanes. The first step is catalyzed by a commercial hydrogenation catalyst containing nickel or a platinum metal. The second step is catalyzed by H-ZSM-5 zeolite containing palladium. Developed by Linde, VEBA Oil, and Süd-Chemie from 2004 in order to reduce the content of aromatic hydrocarbons in motor gasoline.

Ringelahn, C., Burgfels, G., Neumayr, J.G., Seuffert, W., Klose, J., and Kurth, V., *Catal. Today*, 2004, **97**(4), 277.
Slagtern, A., Dahl, I.M., Jens, K.J., and Myrstad, T., *Appl. Catal. A Gen.*, 2010, **375**(2), 213.

Aris [**Ar**omatics **is**omerization] A process for the hydrocatalytic isomerization of C_8 fractions. Developed by Leuna Werke and Petrolchemische Kombinate Schwedt in 1976. The catalyst is platinum deposited in a mixture of alumina and natural mordenite.

Weissermel, K. and Arpe, H.-J., *Industrial Organic Chemistry*, 4th ed., translated by C.R. Hindley and S. Hawkins, Wiley-VCH, Weinheim, Germany, 2003, 333.

Arkenol A complex, integrated process for making ethanol from cellulosic wastes. It uses concentrated sulfuric acid hydrolysis and microbial fermentation. Citric acid and gypsum are by-products. Developed since 1987 by W.A. Farone of Applied Power Concepts in Anaheim, CA. There are nine granted U.S. patents. A small plant was built in Japan in 2002 in association with JGC. Commercial development is now controlled by BlueFire Ethanol that plans to build a large plant in S. California.

> U.S. Patents 5,562,277; 6,054,611.
> *Oil Gas J.*, 2008, **106**(21), 56.

Armco A direct-reduction ironmaking process that used natural gas as the reductant. Operated in Houston, TX, from 1972, dismantled in 1982. *See* DR.

Armour (1) A continuous soapmaking process developed by the Armour Company in 1964.

> Potts, R.H. and McBride, G.W., *Chem. Eng. (N.Y.)*, 1950, **57**(2), 124.

Armour (2) A process for separating fatty acids from acetone by fractional crystallization.

> Potts, R.H. and McBride, G.W., *Chem. Eng. (N.Y.)*, 1950, **57**(2), 124.

Armstrong A process for making titanium metal powder, developed by Don Armstrong at International Titanium Powder (ITP) in Woodbridge, IL. Acquired by Cristal USA in 2008 and commercialized in Ottawa, IL, since 2011. Essentially a continuous version of the Hunter process, in which titanium tetrachloride is reacted with metallic sodium.

> *Adv. Mater. Process.*, Nov 2003, 25.

ARODIS A process for converting light aromatic hydrocarbons to diesel fuel that involves hydrodealkylation and hydrogenation. Developed by the University of New South Wales and BHP Research.

> Jiang, C.J., Trimm, D.L., Cookson, D., Percival, D., and White, N., in *Science and Technology in Catalysis*, Izumi, Y., Aral, H., and Iwamoto, M., Eds., Elsevier, Amsterdam, the Netherlands, 1994, 149.
> *Catal. Today*, 2011, **178**(1), 5.

Arofining A process for removing aromatic hydrocarbons from petroleum fractions by catalytic hydrogenation to naphthenes. Developed by Labofina, France, and licensed by Howe-Baker Engineers.

> *Hydrocarbon Process. Int. Ed.*, 1970, **49**(9), 205.

Aroforming A process for making aromatic hydrocarbons from aliphatic hydrocarbons. Based on the *Aromizing process. Developed by Salutec, Australia, and IFP, France. A demonstration unit with capacity of 500 bbl/day was designed in 1994.

> Mank, L., Shaddick, R., and Minkkinen, A., *Hydrocarbon Technol. Int.*, 1992, 69.
> *Eur. Chem. News, Chemscope*, 1994, 7.
> *Eur. Chem. News*, 1994, **62**(1648), 18.
> Hagen, A. and Roessner, F., *Catal. Revs. Sci. Eng.*, 2000, **42**(4), 403.

Aromax (1) A catalytic process for converting light paraffins to benzene and toluene using a zeolite catalyst. Developed by Chevron Research & Technology Company. Installations were planned for Mississippi, Thailand, and Saudi Arabia.

Aromax (2) Also known as Toray Aromax. A chromatographic process for separating *p*-xylene from its isomers. Similar to the *Parex (1) process, it operates in the liquid phase at 200°C, 15 bar. Developed in 1971 by Toray Industries, Japan.

U.S. Patent 3,761,533.
Otani, S., *Chem. Eng. (N.Y.)*, 1973, **80**(21), 106.
Weissermel, K. and Arpe. H.-J., *Industrial Organic Chemistry*, 4th ed., translated by C.R. Hindley and S. Hawkins, Wiley-VCH, Weinheim, Germany, 2003, 324.

Aromex A process for removing aromatic hydrocarbons from petroleum reformate by extraction with diglycolamine (also called 2-(2-aminoethoxy) ethanol and DGA). Developed by Howe-Baker Engineers. *See also* Econamine.

Jones, W.T. and Payne, V., *Hydrocarbon Process. Int. Ed.*, 1973, **52**(3), 91.
Bailes, P.J., in *Handbook of Solvent Extraction*, Lo, C.C., Baird, M.H.I., and Hanson, C., Eds., John Wiley & Sons, Chichester, U.K., 1983, Chap. 18.2.4.

Aromizing A petroleum reforming process for converting aliphatic to aromatic hydrocarbons. Developed by the Institute Français du Pétrole.

Bonnifay, P., Cha, B., Barbier, J.-C., Vidal, A., Jugin, B., and Huin, R., *Oil Gas J.*, 1976, **74**(3), 48.

Arosat [**Aro**matics **sat**uration] A *hydroprocessing process developed by C-E Lummus.

Arosolvan A solvent extraction process for removing aromatic hydrocarbons from petroleum mixtures using N-methyl pyrrolidone (NMP) containing 12%–14% water at 20°C–40°C. Developed by Lurgi and first used commercially in Japan in 1961.

Eisenlohr, K.-H., *Erdoel Kohle*, 1963, **16**, 530.
Eisenlohr, K.-H. and Grosshaus, W., *Erdoel Kohle*, 1965, **18**, 614.
Oil Gas J., 1966, **64**(29), 83.
Müller, E., *Chem. Ind. (London)*, 1973, 518.

Arosorb A process for extracting aromatic hydrocarbons from refinery streams using a solid adsorbent, either silica gel or activated alumina. Developed by the Sun Oil Company in 1951. The California Research Corporation developed a similar process.

Harper, J.I., Olsen, J.L., and Shuman, FR., Jr., *Chem. Eng. Prog.*, 1951, **48**(6), 276.
Davis, W.H., Harper, J.I., and Weatherly, E.R., *Pet. Refin.*, 1952, **31**(5), 109.
Unzelman, G.H. and Wolf, C.J., in *Petroleum Processing Handbook*, Bland, W.F. and Davidson, R.L., Eds., McGraw-Hill, New York, 1967, 106.

ARP [**A**dvanced **R**eactor **P**rocess] A process for making aluminum metal. A molten slag bath containing alumina and carbon produces aluminum carbide that is reacted with more alumina. Under development by Alcoa and Elkem Technology, but not commercialized by 2011. *See also* Thermical.

Some advanced nuclear reactors share this abbreviation.

Johansen, K., Aune, J.A., Bruno, M., and Schei, A., *Light Metals*, 2003, 401.
Chem. Eng. (N.Y.), 2010, **117**(8), 9.

ARS [**A**dvanced **R**ecovery **S**ystem] An integrated set of engineering modifications for upgrading catalytic crackers for making ethylene, developed jointly by Mobil Chemical Company and Stone & Webster Engineering Corporation. The first plant was planned for an AMOCO plant in South Korea for completion in 1994. Stone & Webster was acquired by the Shaw Group in 2000 and currently offers the ARS.

Eur. Chem. News, 1990, **54**(1434), 22.

Arseno A process for extracting gold from arsenic-containing ores, developed by Arseno Processing. Similar to the *Cashman process.

Yannopoulos, J.C., *The Extractive Metallurgy of Gold*, Van Nostrand Reinhold, New York, 1991, 103.

ART (1) [**A**sphalt **R**esiduum **T**reating] A process for converting heavy petroleum fractions into more easily processed liquid fractions. Developed by Engelhard Corporation and offered by the M.W. Kellogg Company. Three units were operating in 1996.

> *Hydrocarbon Process. Int. Ed.*, 1996, **75**(11), 121.

ART (2) [**A**dvanced **R**eformate upgrading **T**echnology] A process for converting reformate (a petroleum stream) directly to a mixture of benzene, toluene, and xylene. The catalyst is a zeolite promoted by a platinum metal. Developed by Zeolyst International and SK Corporation; operated in Korea by SK Corporation since 2001.

> *Eur. Chem. News*, 2002, **76**(2000), 24.

Arthur D. Little *See* DR.

Arvia An electrochemical method for oxidizing organic pollutants in water using a novel electrically conducting type of activated carbon called Nyex. The Nyex adsorbs the pollutants and they are then electrochemically oxidized. Developed at the University of Manchester and commercialized by Arvia Technology. Trialled by remediation company Geo2 Remediation in Yorkshire in 2011.

> *Chem. Ind. (London)*, 2009, (2), 23.
> *The ENDS Report*, 29 Jun 2011.

ASAM [**A**lkali **S**ulfite **A**nthraquinone **M**ethanol] A process for delignifying wood pulp. Wood chips are digested in sodium hydroxide or sodium carbonate solution, and sodium sulfite is added to remove the lignin. Methanolic anthraquinone is used as a catalyst. Invented by the Department of Chemical Wood Technology at the University of Hamburg. Further developed by the University with Kraftanlage Heidelberg and demonstrated at a plant of Feldmühle in Düsseldorf, Germany, in 1990.

> *Chem. Eng. (N.Y.)*, 1991, **98**(1), 37.
> Patt, R., Kordsachia, O., and Schubert, H.-L., in *Environmentally Friendly Technologies for the Pulp and Paper Industries*, Young, R.A. and Akhar, M., Eds., John Wiley & Sons, New York, 1998, 101.

ASARCO [**A**merican **S**melting **A**nd **R**efining **C**ompany] This large metallurgical company has given its name to a *flue-gas desulfurization process in which the sulfur dioxide is absorbed in dimethylaniline and subsequently desorbed at a higher temperature. Operated in California, Tennessee, and Norway.

> Fleming, E.P. and Fitt, T.C., *Ind. Eng. Chem.*, 1950, **42**(11), 2253.
> Kohl, H.L. and Nielsen, R.B., *Gas Purification*, 5th ed., Gulf Publishing Co., Houston, TX, 1997, 593.

ASCOT [**AS**phalt **CO**king **T**echnology] A process combining deasphalting and decoking, offered by Foster Wheeler, United States.

> U.S. Patent 4,686,027.

ASEA-SKF *See* Steelmaking.

Ashcroft–Elmore A process for extracting tin from its ores. The ore is mixed with coke and calcium chloride and heated in a rotary kiln to 800°C. Stannous chloride, formed by the reaction

$$SnO_2 + C + CaCl_2 = CaO + CO + SnCl_2$$

volatilizes and is condensed in water. The aqueous condensate is neutralized and electrolyzed. Invented by E.A. Ashcroft and S. Elmore and operated in Thailand from 1941 to 1949.

> British Patents 302,851; 602,245; 602,246; 602,247.
> Wright, P.A., *Extractive Metallurgy of Tin*, 2nd ed., Elsevier, Amsterdam, the Netherlands, 1982, 175.

Ashland An industrial process for making aliphatic amines.

Ind. Res., 1988, **47**, 72.

ASR-2 A process for desulfurizing diesel fuel by oxidation with hydrogen peroxide and formic acid. The sulfur compounds are oxidized to sulfones that are removed by adsorption on alumina. Developed by Unipure and Texaco and demonstrated at Valero Energy's refinery at Krotz Spring, LA, from 2002.

Stanislaus, A., Marafi, A., and Rana, M.S., *Catal. Today*, 2010, **153**(1–2), 47.

ASR Sulfoxide [**Al**berta **S**ulfur **R**esearch] A process for removing residual sulfur dioxide and hydrogen sulfide from the tail gases from the *Claus process by wet scrubbing with a solution containing an organic sulfoxide. Elemental sulfur is produced. It had not been piloted in 1983.

ASSR [**A**nti**S**olvent **S**wing **R**egeneration] A process for removing hydrogen sulfide from natural gas by extraction with an anhydrous *t*-alkanolamine. The process uses less energy than processes that recover the hydrogen sulfide by heat. Developed by the Pacific Northwest National Laboratory.

Chem. Eng. (N.Y.), Apr 2011.

AST [**A**dvanced **S**clair **T**echnology] The latest version of an ethylene polymerization process developed by DuPont and acquired by NOVA Chemicals. *See also* *Sclair, *Sclairtech.

Aston A process for making wrought iron, invented by James Aston in the United States in 1925.

Wikipedia, The Free Encyclopedia. Wrought Iron. Revised 10 Jun 2013. Retrieved 20 Jul 2013.

ASVAHL [**AS**sociation pour la **VA**lorisation des **H**uiles **L**ourdes] Not a single process, but an association formed in 1980 by the three French companies—Total, Elf, and IFP—to develop processes for treating heavy oils. The processes they developed were *HYVAHL, *SOLVAHL, and *TERVAHL. These processes are now licensed by Axens.

Chauvel, A., Delmon, B., and Hölderich, W.F., *Appl. Catal. A: Gen.*, 1994, **115**, 186.

ATA [**A**dvanced **T**rans**A**lkylation] A process for converting aromatic hydrocarbons, whose molecules contain more than nine carbon atoms, to a mixture of benzene, toluene, and xylenes. The catalyst is a zeolite promoted by a platinum metal. Developed by Zeolyst International and SK Corporation; operated in Korea by SK Corporation since 1999.

Atgas [**A**pplied **T**echnology Corporation **gas**ification] A coal gasification process in which powdered coal and limestone, mixed with steam and oxygen, are injected into a bath of molten iron at 1400°C. The product gas is a mixture of hydrogen and carbon monoxide, and the sulfur is converted to a calcium sulfide slag. Piloted by the Applied Technology Corporation in the 1970s but not fully developed.

Hebden, D. and Stroud, H.J.F., in *Chemistry of Coal Utilization*, 2nd Suppl. Vol., Elliott, M.A., Ed., John Wiley & Sons, New York, 1981, 1739.

ATHOS A *WAO treatment for oxidizing sewage sludge, which uses pure oxygen, a relatively low temperature, and a dissolved copper catalyst. Demonstrated in Toulouse, France, in 1998.

Luck, F., *Catal. Today*, 1999, **53**(1), 88.
Water Products, 2002, **7**(8), 16.
Pintar, A., *Catal. Today*, 2003, **77**(4), 451.

ATOL [**ATO**chem po**L**ymerization] A gas-phase process for making polyethylene. Developed by Atochem and first commercialized in 1991. It uses a *Ziegler–Natta catalyst containing titanium and magnesium halides. First commercialized at Gonfreville, France, in 1991.

Chauvel, A., Delmon, B., and Hölderich, W.F., *Appl. Catal. A: Gen.*, 1994, **115**, 180.

ATR [**A**uto**T**hermal **R**eforming] A process for making CO-enriched *syngas. ATR combines partial oxidation with adiabatic *steam reforming. Developed in the late 1950s for ammonia and methanol synthesis. Further developed in the 1990s by several companies for making liquid fuels from natural gas. Natural gas is partially oxidized with steam and air or oxygen, and the gaseous products are then equilibrated over a catalyst bed to make a *syngas having an H_2/CO ratio of 2:1.

Christensen, T.S. and Primdahl, I.I., *Hydrocarbon Process. Int. Ed.*, 1994, **73**(3), 39.
Oil Gas J., 1997, **95**(25), 18; 2002, **100**(16), 64.
Bakkerud, P.K., *Catal. Today*, 2005, **106**(1–4), 30.

Atrament *See* metal surface treatment.

ATS [**A**mmonium **T**hio**S**ulfate] A process for removing residual sulfur dioxide from *Claus tail gas by absorption in aqueous ammonia to produce ammonium sulfite and bisulfite. Addition of hydrogen sulfide from the Claus unit produces saleable ammonium thiosulfate. Developed by the Pritchard Corporation and first operated in 1979 by the Colorado Interstate Gas Company at Table Rock, WY. ATS is also offered by Haldor Topsoe.

Zey, A., White, S., and Johnson, D., *Chem. Eng. Prog.*, 1980, **76**(10), 76.
Kohl, H.L. and Nielsen, R.B., *Gas Purification*, 5th ed., Gulf Publishing Co., Houston, TX, 1997, 578.

ATTANE Not a process but a range of ultra low-density grades of polyethylenes developed by Dow Chemical Company.

Attisholz A process for recovering valuable waste products from the *Sulfite process for making wood pulp. This includes methanol, cymenes, and furfural. Developed in Switzerland.

AUC [**A**mmonium **U**ranyl **C**arbonate] A process for converting uranium hexafluoride into uranium dioxide for use as a nuclear reactor fuel. The hexafluoride vapor, together with carbon dioxide and ammonia, is passed into aqueous ammonium carbonate at 70°C, precipitating ammonium uranyl carbonate:

$$UF_6 + 5H_2O + 10NH_3 = (NH_4)_4[UO_2(CO_3)_3] + 6NH_4F$$

The precipitate is filtered off, washed, and calcined in hydrogen in a fluidized bed. Developed by Nukem at Hanau, Germany.

Büchner, W., Schliebs, R., Winter, G., and Büchel, K.H., *Industrial Inorganic Chemistry*, VCH Publishers, Weinheim, Germany, 1989, 581.

Auger A process for chlorinating benzene to chlorobenzene, catalyzed by metallic iron. Invented by V.E. Auger in 1916 and operated in France and Italy in the early twentieth century.

French Patent 482,372.
Ellis, C., *The Chemistry of Petroleum Derivatives*, The Chemical Catalog Co., New York, 1934, 765.

AuPLUS A hydrometallurgical process for extracting gold. The addition of calcium peroxide to the cyanide leaching liquor increases the rate and amount of gold extracted.

AURABON A catalytic process for *hydrotreating and converting the asphaltenes in residual oils and heavy crude oils into lighter products. Developed by UOP.

Bowman, C.W., Phillips, R.S., and Turner, L.R., in *Handbook of Synfuels Technology*, Meyers, R.A., Ed., McGraw-Hill, New York, 1984, 73.
Hydrocarbon Process. Int. Ed., 1997, **76**(2), 45.

AUSCOKE *See* carbonization.

AUSMELT *See* Sirosmelt.

Autofining A fixed-bed catalytic process for removing sulfur compounds from petroleum distillates. This process uses a conventional cobalt–molybdenum hydrodesulfurization catalyst but does not require additional hydrogen. Developed by the Anglo-Iranian Oil Company in 1948.

British Patent 670,619.
U.S. Patent 2,574,449.
McKinley, J.B., in *Catalysis*, Emmett, P.H., Ed., Reinhold, New York, 1957, 405.
Hydrocarbon Process., 1964, **43**(9), 186.
Unzelman, G.H. and Wolf, C.J., in *Petroleum Processing Handbook*, Bland, W.F. and Davidson, R.L., Eds., McGraw-Hill, New York, 1967, 42.

AUTO-PUREX G A process for removing carbon dioxide from air by *PSA, using alumina as the sorbent. Developed by Marutani Chemical Plant & Engineering Company, Japan.

Suzuki, M., in *Adsorption and Ion Exchange: Fundamentals and Applications*, LeVan, M.D., Ed., American Institute of Chemical Engineers, New York, 1988, 121.

Autopurification A wet-scrubbing process for removing hydrogen sulfide from coke-oven gas. The scrubbing liquor was an ammoniacal suspension of ferric ammonium ferrocyanide. The process was developed by ICI, Billingham, United Kingdom, in the 1930s and 1940s but was abandoned in 1947.

Smith, F.F. and Pryde, D.R., *Chem. Ind. (London)*, 1934, **12**, 657.
Craggs, H.C. and Arnold, M.H.M., *Chem. Ind. (London)*, 1947, **66**, 571, 590.

AVADA [AdVanced Acetates by Direct Addition] A process for making ethyl acetate from ethylene and acetic acid using a solid silicotungstic acid catalyst made by BP. Developed from 1997 by BP, Hull, now INEOS. Intended for locations where cheap ethanol is unavailable. The Hull plant, which started up in 2001, is the largest plant in the world for making ethyl acetate.

Platinum Metals Rev., 2001, **45**(1), 12.
Chem. Br., 2001, 38(1), 8.
Eur. Chem. News, 10 Dec 2001, **75**(1985), 27.
Rothenberg, G., *Catalysis: Concepts and Green Applications*, Wiley-VCH, Weinheim, Germany, 2008, 166.
ICIS Chem. Bus., 2011, **280**(13), 47.

Avco An electric arc process for making acetylene from coal and hydrogen. The arc in hydrogen is rotated by a magnetic field in order to spread it out, thereby making better contact with the coal passing through. Developed by V.J. Krukonis at the Avco Corporation in the early 1970s with support from the U.S. Office of Coal Research. Piloted at the rate of 55 kg/h, but not yet commercialized.

Gannon, R.E., Krukonis, V.J., and Schoenberg, T., *Ind. Eng. Chem. Prod. Res. Dev.*, 1970, **9**, 343.

AVENEER A process for making methyl methacrylate that uses all the usual raw materials (ammonia, methane, acetone, methanol) but not sulfuric acid. Developed by Evonik Industries (formerly Degussa) and piloted in Worms, Germany. The first commercial plant is to be built in Mobile, AL, for startup in 2015.

Chem. Ind. (London), 2008, (19), 13.
TCE, 2008, (809), 17.
Chem. Eng. (N.Y.), 2010, **117**(12), 16.

AVH [**A**telier de **V**itrification de la **H**ague] A process for immobilizing nuclear waste, which operated at La Hague, France, based on the earlier *AVM process. Used formerly also in the THORP nuclear reprocessing plant at Sellafield, United Kingdom.

Ojovan, M.I. and Lee, W.E., *An Introduction to Nuclear Waste Immobilisation*, Elsevier, Amsterdam, the Netherlands, 2005, 237.

AVM [**A**telier de **V**itrification de **M**arcoule] A continuous process for immobilizing radioactive waste by incorporation in a borosilicate glass. Developed at Marcoule, France, from 1972, based on the earlier *PIVER process. In 1988, two larger vitrification plants were being designed for installation at La Hague, France. *See* AVH.

Lutze, W., *Radioactive Waste Forms for the Future*, Lutze, W. and Ewing, R.C., Eds., North-Holland, Amsterdam, the Netherlands, 1988, 10, 133.
Ojovan, M.I. and Lee, W.E., *An Introduction to Nuclear Waste Immobilisation*, Elsevier, Amsterdam, the Netherlands, 2005, 235.

Axenis A process for treating wastewaters containing soluble organic pollutants. The water is first fed to a membrane bioreactor in which bacteria oxidize the organic material to CO_2. The effluent is passed through an ultrafiltration unit to remove suspended solids and then through a reverse osmosis unit to remove dissolved inorganic material. Developed by Linde Gases. Commercialization was planned for 2010.

Chem. Eng. (N.Y.), 2009, **116**(9), 11.

Axorb A process for removing carbon dioxide and hydrogen sulfide from gases by scrubbing with an aqueous solution of potassium carbonate containing proprietary additives.

Ayers An early process for making carbon black from oil. The air for combustion is injected tangentially into the furnace, producing swirl, and the atomized oil is injected into this. Invented by J.W. Ayers and developed by Phillips Petroleum Company.

U.S. Patents 2,292,355; 2,420,999.
Shearon, Jr., W.H., Reinke, R.A., and Ruble, T.A., in *Modern Chemical Processes*, Vol. 3, Reinhold Publishing, Washington, DC, 1954, 45.

B

Babcock & Wilcox The Babcock & Wilcox Company developed a number of processes but is perhaps best known for its coal gasification process. This uses a single-stage, two-zone gasifier. In the lower zone, coal and recycled char contact oxygen (or air) and steam at 1650°C–1855°C, and molten slag is removed from the base of the zone. The upper zone is cooled to 900°C. A commercial scale oxygen-blown plant was operated in West Virginia in the 1950s, and a pilot-scale air-blown plant was operated in Ohio in the 1960s. *See also* Bi-Gas.

> Hebden, D. and Stroud, H.F.G., in *Chemistry of Coal Utilization*, 2nd Suppl. Vol., Elliott, M.A., Ed., John Wiley & Sons, New York, 1981, 1724.

Babcock W-D *See* Woodall–Duckham.

BACFOX [**BAC**terial **F**ilm **OX**idation] A process for regenerating hydrometallurgical leach liquors by bacterial oxidation of their iron from Fe^{2+} to Fe^{3+}. Developed by Matthew Hall Ortech Company and used in South Africa.

> *Eng. Min. J.*, 1978, **179**(12), 90.
> Jackson, E., *Hydrometallurgical Extraction and Reclamation*, Ellis Horwood, Chichester, U.K., 1986, 68.

Bachmann A process for making the explosive RDX. Hexamethylene tetramine is nitrated in acetic acid solution, using a mixture of ammonium nitrate and acetic anhydride. Invented by W.E. Bachmann at the University of Michigan during World War II. *See also* K.A., Woolwich.

> Bachmann, W.E. and Sheehan, J.C., *J. Am. Chem. Soc.*, 1949, **71**, 1842.
> *Kirk-Othmer's Encyclopedia of Chemical Technology*, 4th ed., Vol. 10, John Wiley & Sons, New York, 1991–1998, 33.

Backus [Backhaus] A process for purifying carbon dioxide obtained by fermentation, using activated carbon. The carbon beds are reactivated with steam. Invented in 1924 by A.A. Backhaus at the U.S. Industrial Alcohol Company. *See also* Reich.

> U.S. Patents 1,493,183; 1,510,373.

BACOX A bacterial oxidation technology for separating metals and arsenic from various sulfide ores. Invented at Kings College, London, for removing sulfur from coal. It was developed by BacTech Mining Corp. (now REBgold Corp.) who moved it to Australia where it was developed for removing arsenic from gold ores. Now operated in China for gold mining, in Mexico for silver, and in Canada for goldmine cleanup. Natural bacteria are used and the arsenic is converted to ferric arsenate. South African company Mintek has been a partner in some of these projects.

> Corfield, R., *Chem. Ind. (London)*, 2006, (23), 25.

Bacus A scrubbing process for removing olefins and carbon monoxide from gas streams. The adsorbent is a solution of a copper compound that is not deactivated by traces of water. Developed by D. Haase of Herr Haase, Nixon, TX.

> *Chem. Eng. (N.Y.)*, 1995, **102**(3), 19.

Baekeland A process for making organic polymers by reacting phenols with formaldehyde. Based on an observation by A. von Bayer in 1872 and developed into an industrial process by L.H. Baekeland from 1905 to 1909. It was used to make Bakelite, one of the first commercial plastics. The first industrial manufacture began in Germany in 1910. Honored by a National Historic Chemical Landmark of the American Chemical Society at the National Museum of American History in Washington DC.

von Bayer, A., *Ber. Dtsch. Chem. Ges.*, 1872, **5**, 280.
Baekeland, L.H., *Ind. Eng. Chem.*, 1909, **1**, 149.
Kirk-Othmer's Encyclopedia of Chemical Technology, 4th ed., Vol. 18, John Wiley & Sons, New York, 1991–1998, 603.

BAF [**Bi**ological **A**erated **F**ilter] A generic type of sewage treatment process in which the biological medium is supported on a porous matrix. *See* BIOBEAD, BIOCARBONE, BIOFOR, BIOPUR, BIOSTYR, Colox, CTX, FAST, SAFe, STEREAU.

Stephenson, T., Mann, A., and Upton, J., *Chem. Ind. (London)*, 1993, (14), 533.

Balbach A variation of the *Moebius process for electrolytically removing gold from silver, in which the anodes are placed horizontally in wooden trays lined with canvas to retain the slimes. The silver deposits as crystals on graphite cathodes at the base of the cell. Invented by E. Balbach. *See also* Parting, Parkes, Thum, Wohlwill.

BALI [**B**orregaard **A**dvanced **LI**gnin] A process for converting biomass into lignin and fermentable sugars. The first step is a chemical pretreatment that makes the lignin water soluble, leaving the fibers. The fibers are then hydrolyzed enzymically to C_5 and C_6 sugars. Developed by Borregaard Industries, Norway. The first plant is expected to be built in 2015.

Chem. Eng. (N.Y.), 2012, **119**(5), 15.

Balke A process for making niobium by reducing niobium pentoxide with carbon in a carbon crucible, *in vacuo*, at 1800°C.

Bamag-Meguin A process for making nitric acid by oxidizing ammonia. It differs from similar processes in isolating liquid dinitrogen tetroxide as an intermediate. Operated in Silesia in the early twentieth century.

Chilton, T.H., *Strong Water: Nitric Acid, Its Sources, Methods of Manufacture, and Uses*, MIT Press, Cambridge, MA, 1968, 111.

Banox *See* metal surface treatment.

Bardenpho A modification of the *Activated Sludge process, designed for the removal of high levels of nitrogen and phosphorus.

Horan, N.J., *Biological Wastewater Treatment Systems*, John Wiley & Sons, Chichester, U.K., 1990, 234.

Bardet Also called Samica. A process for expanding mica in order to make it into paper. Mica is partially dehydrated by heating, and the hot product is quenched in alkaline water. After drying, the product is immersed in dilute sulfuric acid, which generates gas between the layers, forcing them apart. In this expanded condition, the product can easily be made into a paper.

Barffing *See* metal surface treatment.

Bari-Sol A petroleum *dewaxing process using solvent extraction by a mixture of dichloroethane and benzene.

Unzelman, G.H. and Wolf, C.J., in *Petroleum Processing Handbook*, Bland, W.F. and Davidson, R.L., Eds., McGraw-Hill, New York, 1967, 96.

Barium A process for making hydrogen peroxide by reacting barium peroxide with sulfuric or phosphoric acid:

$$BaO_2 + H_2SO_4 = BaSO_4 + H_2O_2$$

The barium peroxide was made using the *Brin process. The barium was recovered by reducing the sulfate with carbon and then converting this barium sulfide to the oxide via the carbonate:

$$BaSO_4 + 2C = BaS + 2CO_2$$

Phosphoric acid was sometimes used instead of sulfuric acid. The process was first operated in Berlin in 1873 by the Schering Company. In the United Kingdom, it was first operated in 1888 by B. Laporte and Company. It was progressively replaced by the electrolytic process developed between 1908 and 1932. Laporte Chemicals abandoned the barium process in 1950.

Wood, W.S., *Hydrogen Peroxide*, Royal Institute of Chemistry, London, U.K., 1954, 2.
Schumb, W.C., Satterfield, C.N., and Wentworth, R.L., *Hydrogen Peroxide*, Reinhold Publishing, New York, 1955, 14, 115.
Pascal, P., Ed., *Nouveau Traité de Chimie Minérale*, Vol. 13, Masson et Cie, Paris, France, 1960, 528.

Barton A process for making black lead monoxide by atomizing molten lead in air.

BASF/CAN *See* CAN.

BASF/Scholven A pretreatment process for benzole, the product formed by hydrogenating hard coal. The benzole is hydrogenated at 300°C–400°C using a molybdenum or cobalt–molybdenum catalyst. The product is a mixture of aromatic hydrocarbons, suitable for separation by a variety of physical processes. The process was invented by BASF in 1925 and adopted by Scholven-Chemie in 1950. Eleven plants in France and Germany subsequently adopted the process.

Jäckh, W., *Erdoel Kohle*, 1958, **11**, 625.
Reitz, O., *Erdoel Kohle*, 1959, **12**, 339.
Muder, R., *Chemistry of Coal Utilization*, Suppl. Vol., Lowry, H.H., Ed., John Wiley & Sons, New York, 1963, 647.

Bashkirov A process for making aliphatic alcohols by oxidizing paraffins. The reaction is conducted in the presence of boric acid, which scavenges the hydroperoxide intermediates. Borate esters of secondary alcohols are formed as intermediates and then hydrolyzed. Developed in the USSR in the 1950s and now operated in the CIS and in Japan.

Bashkirov, A.N., Kamzolkin, V.V., Sokova, K.M., and Andreyeva, T.P., in *The Oxidation of Hydrocarbons in the Liquid Phase*, Emanuel, N.M., Ed., translation edited by Hopton, J.D., Pergamon Press, Oxford, U.K., 1965, 183.
Ullmann's Encyclopedia of Industrial Chemistry, Online Edition, Dec 2006, Section 2.4.4.

Basic Bessemer *See* Thomas.

Basic Open Hearth A version of the *Open Hearth process for steelmaking, in which the hearth is made from calcined dolomite (calcium and magnesium oxides). The sulfur and phosphorus impurities in the raw materials are converted to basic slag, which is separated from the molten steel. First operated in 1882 at Alexandrovsky, near St. Petersburg, Russia, and at Le Creusot, France. This was the major steelmaking process in the world in the first half of the twentieth century. *See* Thomas.

Barraclough, K.C., *Steelmaking 1850–1900*, The Institute of Metals, London, U.K., 1990, 247.

BASIL [Biphasic Acid Scavenging utilizing Ionic Liquids] A range of catalytic processes for synthetic organic chemistry. These processes use ionic liquids as solvents, from which it is easy to recover the catalysts. Developed by BASF from 2002 and first commercialized in 2004 for the synthesis of alkoxyphenyl phosphines. Awarded the Entec Medal for safety and environmental innovation in 2005. *See also* Difasol.

Eur. Chem. News, 27 Oct 2004, **81**(2117), 20.
Chem. Eng. News, 1 Aug 2005, **83**(31), 34.
Maase, M. and Massonne, K., in *Ionic Liquids IIIB*, Rogers, R.D. and Seddon, K.R., Eds., American Chemical Society, Washington, DC, 2005, 126.
Chem. Eng. News, 24 Apr 2006, **84**(17), 15.
Sheldon, R.A., Arends, I., and Hanefeld, U., *Green Chemistry and Catalysis*, Wiley-VCH, Weinheim, Germany, 2007, 322.

Basil A trademark used by BASF for their range of processes using ionic liquids, which includes *BASIL.

Basset *See* DR.

Batenus A series of processes, including solvent extraction and ion exchange, for recovering metals from scrap batteries. Developed by Pira, Germany, in 1993.

> *Chem. Eng. (N.Y.)*, 1993, **100**(11), 21.

Battersea A pioneering *flue-gas desulfurization process, operated at Battersea Power Station, London, from 1931 until the station was closed. The flue gases were washed with water from the River Thames, whose natural alkalinity was augmented by chalk slurry. One of the problems of this process was cooling of the stack gases, which caused the plume to descend on the neighborhood.

> Rees, R.L., *J. Inst. Fuel*, 1953, **25**, 350.
> *J. Air Pollut. Control Assoc.*, 1977, **27**, 948.
> Kohl, A.L. and Riesenfeld, F.C., *Gas Purification*, 4th ed., Gulf Publishing Co., Houston, TX, 1985, 302.

Bayer A process for making pure alumina hydrate from bauxite, used principally as a raw material for the manufacture of aluminum metal. The ore is digested with hot sodium hydroxide solution, yielding a solution of sodium aluminate. Insoluble impurities are separated off in the form of red mud. The solution is then nucleated with alumina hydrate from a previous batch, causing alumina trihydrate to precipitate:

$$Al_2O_3 + 3H_2O + 2NaOH = 2NaAlO_2 + 4H_2O$$

$$2NaAlO_2 + 4H_2O = Al_2O_3 + 3H_2O + 2NaOH$$

The hydrate is dehydrated by calcination. The presence of carbohydrate degradation products derived from ancient plants seriously interferes with the operation of this process. Invented by K.J. Bayer in Russia in 1887 and now universally used, with minor variations depending on the nature of the ore. The German company Bayer AG was not involved in this invention.

> German Patents 43,977 (1887); 65,604 (1892).
> Misra, C., *Industrial Alumina Chemicals*, American Chemical Society, Washington, DC, 1986, 31.
> Hudson, L.K., *Production of Aluminium and Alumina*, Burkin, A.R., Ed., John Wiley & Sons, Chichester, U.K., 1987, 13.
> McMichael, B., *Ind. Miner. (London)*, 1989, (267), 19.
> Gupta, C.K. and Mukherjee, T.K., *Hydrometallurgy in Extraction Processes*, Vol. 1, CRC Press, Boca Raton, FL, 1990, 129.
> Ellis, A.V., Wilson, M.A., and Kannangara, K., *Ind. Eng. Chem. Res.*, 2002, **41**(12), 2842.
> Ellis, A.V., Wilson, M.A., and Forster, P., *Ind. Eng. Chem. Res.*, 2002, **41**(25), 6493.

Bayer–Bertrams A process for concentrating and purifying waste sulfuric acid by distillation.

> Büchner, W., Schliebs, R., Winter, G., and Büchel, K.H., *Industrial Inorganic Chemistry*, VCH Publishers, Weinheim, Germany, 1989, 117.

Bayer–Hoechst A gas-phase process for making vinyl acetate from ethylene and acetic acid, using a supported palladium catalyst. Developed jointly by Bayer and Hoechst. In 1991, nearly 2 million tons per year of vinyl acetate was made by this process.

> Weissermel, K. and Arpe, H.-J., *Industrial Organic Chemistry*, 4th ed., translated by C.R. Hindley and S. Hawkins, Wiley-VCH, Weinheim, Germany, 2003, 233.

Bayer Ketazine A process for making hydrazine by the reaction of sodium hypochlorite with ammonia in the presence of acetone. Acetone azine is an intermediate. Never commercialized. *See also* Raschig (1).

> Büchner, W., Schliebs, R., Winter, G., and Büchel, K.H., *Industrial Inorganic Chemistry*, VCH Publishers, Weinheim, Germany, 1989, 48.

BayFAME [**Bay**er **F**atty **A**cid **M**ethyl **E**ster] A continuous process for converting fatty acids to their methyl esters for use as diesel fuels. The catalyst is AMBERLYST BD20. Developed by Dow Water and Process Solutions and Bayer Technology Services and announced in 2009.

> *Chem. Eng. News*, 2009, **87**(50), 20.
> *ICIS Chem. Bus.*, 14 Dec 2009.

BAYQIK [**Bay**er **Q**uasi **I**sothermal **K**atalysis] An efficient process for capturing sulfur dioxide from nonferrous metallurgical plants with the production of sulfuric acid. Developed by Bayer in Germany from 2009.

BB-Flex A process for making butadiene from butane. Developed by Asahi Kasei Chemicals.

> *Chem. Eng. (N.Y.)*, Apr 2011, 16.
> *ICIS Chem. Bus.*, 31 Oct 2011.

BBP [**B**abcock **B**orsig **P**ower] *See* Noell (2).

BCD [**B**ase-**C**atalyzed **D**ecomposition] A process for decomposing hazardous organochlorine compounds by treating them in oil, at 300°C–350°C, with sucrose and a high-boiling solvent. Developed by the U.S. Environmental Protection Agency in 1991.

> Kawahara, F.K. and Michalakos, P.M., *Ind. Eng. Chem. Res.*, 1997, **36**(5), 1580.

BCR [**B**io-**C**onversion **R**eactor] A biological process for converting organic waste into organic fertilizer and other agricultural products. Developed by Agronix and offered from 2002.

> *Chem. Week*, 2002, **164**(39), 25.

Beacon A process for recovering carbon from coal gasification. Developed in 1991 by TRW.

> *Chem. Eng. (N.Y.)*, 1984, **91**(13), 157.

Beavon Also called BSR [**Beav**on **S**ulfur **R**emoval]. A process for removing residual sulfur compounds from the effluent gases from the *Claus process. Usually used in conjunction with other processes. Catalytic hydrogenation over a cobalt–molybdena catalyst converts carbonyl sulfide, carbon disulfide, and other organic sulfur compounds to hydrogen sulfide, which is then removed by the *Stretford process. A variation (**BSR/MDEA**), intended for small plants, uses preliminary scrubbing with methyl diethanolamine. Developed by the Ralph M. Parsons Company and Union Oil Company of California in 1971. More than 100 plants were operating in 2000. *See also* SCOT.

> Beavon, D.K., *Chem. Eng. (N.Y.)*, 1971, **78**(28), 71.
> *Hydrocarbon Process. Int. Ed.*, 1996, **75**(4), 106.
> Kohl, H.L. and Nielsen, R.B., *Gas Purification*, 5th ed., Gulf Publishing Co., Houston, TX, 1997, 717.
> *Hydrocarbon Process. Int. Ed.*, 2000, **79**(4), 59.

Béchamp A process for reducing organic nitro-compounds to amines, using iron, ferrous salts, and acetic acid. Invented by A.J. Béchamp in 1854 and still used for making certain aromatic amines.

> Béchamp, A.J., *Ann. Chim. Phys.*, Ser. 3, 1854, **42**, 186.
> Béchamp, A.J., *Annalen*, 1854, **92**, 401.
> Buxbaum, G. and Pfaff, G., *Industrial Inorganic Pigments*, 3rd ed., VCH Publishers, Weinheim, Germany, 2005, 106.

Becher A process for making synthetic rutile (a titanium concentrate) from ilmenite ore. The ore, typically containing 58% titanium dioxide, is first roasted with coal and 0.2%–0.5% of elemental sulfur at 1200°C in a specially designed rotary kiln. This reduces the iron in the ilmenite to the metallic state. After cooling, partially burned coal and ash are removed from the product by screening and magnetic separation. The reduced ilmenite is then agitated in water containing ammonium chloride as a catalyst, and air is blown through. This converts the metallic iron to a flocculent precipitate of iron oxides, which are then removed by wet classification. The synthetic rutile product contains approximately 93% titanium dioxide.

The process was invented and developed in Australia in the 1960s, initially by R.G. Becher at the Mineral Processing Laboratories of the West Australian Government Chemical Centre and later by Western Titanium. In 1998, it was operating in three companies at four sites in Western Australia. Most of the beneficiation is used for the manufacture of titanium pigments; some is used in titanium metal production and in welding rod coatings. Two companies were operating Becher synthetic rutile plants at three sites in Western Australia: Iluka (at Narngulu and North Capel) and Tiwest (at Chandala).

> Australian Patent 247,110.

Bechtel CZD *See* CZD.

Bechtel Seawater A *flue-gas desulfurization process using a suspension of dolomitic lime in seawater. The calcium sulfate produced is returned to the sea. A conceptual design, not known to have been piloted.

> Kohl, H.L. and Nielsen, R.B., *Gas Purification*, 5th ed., Gulf Publishing Co., Houston, TX, 1997, 690.

Behr An obsolete process for separating the drying and nondrying constituents of bodied oils by selective precipitation. The oils are treated with low molecular weight alcohols or ketones, having a medium solvent power, which will not dissolve compounds of molecular weight greater than 900. The polymerized oil settles out, while the nonpolymeric nondrying constituents remain dissolved.

Beilby A process for making potassium cyanide by passing ammonia gas over a molten mixture of potassium carbonate, potassium cyanide, and carbon:

$$K_2CO_3 + C + 2NH_3 = 2KCN + 3H_2O$$

The fused product is poured into molds. The process was invented by G.T. Beilby in Scotland and first used in 1891; in 1900, it was replaced by the *Castner process.

> British Patent 4,820 (1891).
> *Kirk-Othmer's Encyclopedia of Chemical Technology*, 4th ed., Vol. 7, John Wiley & Sons, New York, 1991–1998, 765.

Beja A process for extracting gallium from sodium aluminate solution, as used in the *Bayer process, by means of successive carbonations. Developed by Pechiney in 1946.

> U.S. Patent 2,674,008.
> Palmear, I.J., in *The Chemistry of Aluminium, Gallium, Indium, and Thallium*, Downs, A.J., Ed., Blackie, London, U.K., 1993, 88.
> Wilder, J., Loreth, M.J., Katrack, F.E., and Agarwal, J.C., in *Inorganic Chemicals Handbook*, Vol. 2, McKetta, J.J., Ed., Marcel Dekker, New York, 1993, 942.

Belgian A process for making zinc from zinc oxide by reducing it with carbon in a retort. Invented in 1810.

Belmadur A process for strengthening wood by chemically cross-linking the cellulose fibers. This process enables relatively softwoods to be used in place of tropical hardwoods. The cross-linking agent is a complex polyol, similar to agents used for making textiles crease-resistant, hence the popular name for the product "non-iron wood." Invented in 2002 by H. Militz and A. Krause at the University of Göttingen, Germany, and commercialized by BASF in 2005.

> U.S. Patent 7,595,116.

Bemberg An alternative name for the *cuprammonium process for making artificial silk. Named after J.P. Bemberg, who commercialized the process in the early 1900s in several countries.

Moncrieff, R.W., *Man-made Fibres*, 6th ed., Newnes-Butterworths, London, U.K., 1975, 224.
Kotek, R., in *Handbook of Fiber Chemistry*, 3rd ed., Lewin, M., Ed., CRC Press, Boca Raton, FL, 2006, p. 708.

Bender A continuous process for oxidizing mercaptans in petroleum fractions to disulfides, using a lead sulfide catalyst in a fixed bed. Developed and licensed by Petrolite Corporation; 98 units were operating in 1990. Licensed by Baker Process in 2000.

Waterman, L.C. and Wiley, R.A., *Pet. Refin.*, 1955, **34**(9), 182.
Unzelman, G.H. and Wolf, C.J., in *Petroleum Processing Handbook*, Bland, W.F. and Davidson, R.L., Eds., McGraw-Hill, New York, 1967, 127.
Hydrocarbon Process. Int. Ed., 1996, **75**(4), 108; 2000, **79**(4), 60.

Benfield [**Ben**son and **Field**] A process for removing carbon dioxide, hydrogen sulfide, and other acid gases from industrial gas streams by scrubbing with hot aqueous potassium carbonate containing activators:

$$K_2CO_3 + H_2S \rightleftharpoons KHS + KHCO_3$$

Invented by H.E. Benson in 1952 and then developed with J.H. Field at the U.S. Bureau of Mines. First licensed by the Benfield Corporation of Pittsburgh, PA, subsequently acquired by the Union Carbide Corporation, and now licensed by UOP. The current UOP version includes new solution activators and incorporates zeolites or membrane processes for complete separation of acid gases and minimal loss of product gases. More than 675 plants were operating in 2000. Variations include the Benfield HiPure process and the Benfield LoHeat process. *See also* Carsol, CATACARB, Giammarco-Vetrocoke, HiPure.

Chem. Eng. Prog., 1954, **50**(7), 356.
Benson, H.E., Field, J.H., and Haynes, W.P., *Chem. Eng. Prog.*, 1956, **52**(10), 433.
Hydrocarbon Process. Int. Ed., 1996, **75**(4), 108.
Kohl, H.L. and Nielsen, R.B., *Gas Purification*, 5th ed., Gulf Publishing Co., Houston, TX, 1997, 334.
Hydrocarbon Process. Int. Ed., 2000, **79**(4), 60.

Benfree A process for removing benzene from gasoline by reactive distillation. The benzene is hydrogenated in a distillation column. Developed by the Institut Français du Pétrol and announced in 1998. Three units had been licensed in 2006.

Eur. Chem. News, 19 Jul 1999, **71**(1870), 41.
Pétrole Informations, Jul 1999, (1741), 16.
Chimie Hebdo, 25 Oct 1999, (64), 18.

Bengough–Stuart *See* metal surface treatment.

Benilite [**Ben**eficiation of **il**me**nite**] Also called the Wah Chang process. A process for increasing the titanium content of ilmenite by extracting some of the iron with hot hydrochloric acid. The ore is pretreated by reduction in a rotary kiln. The extraction takes place in a rotating, spherical, iron pressure vessel. The hydrochloric acid is recovered for reuse by the *Woodall–Duckham process. The process originated with the Wah Chang Corporation in the 1960s, the patent by J.H. Chen being filed in 1969. It was further developed by the Benilite Corporation of America, Corpus Christi, TX, now owned by TOR Minerals International (formerly known as Hitox Corporation of America). The product is used as a feedstock for the *Chloride process and as a pigment. The process is now operated by TOR Minerals at Ipoh (Malaysia), by Dhrangadhra Chemical Works (DCW, at Sahupuram, India), and by Kerala Minerals and Metals (KMML, at Sankaramangalam, India). The world's largest Benilite-type plant, operated by Kerr-McGee at Mobile, AL, closed in 2003.

British Patent 1,262,401.
German Patent 2,004,878.
U.S. Patent 3,825,419.
Robinson, S.M., *Polym. Paint Colour J.*, 1986, **176**, 754.

BenSat [**Ben**zene **Sat**uration] A process for removing benzene from C_5 to C_6 petroleum fractions by selective hydrogenation to cyclohexane. Developed by UOP, based on its *HB Unibon process, and first offered for license in 1991. Four units were operating in 2003.

> *Hydrocarbon Process. Int. Ed.*, 1996, **75**(11), 94.
> Sullivan, D.K., *Handbook of Petroleum Refining Processes*, 3rd ed., Meyers, R.A., Ed., McGraw-Hill, New York, 2003, 9.3.

Bensmann A process for recovering lubricating oils by treatment with a strong acid, followed by fuller's earth. Invented in 1926 by N. Bensmann in Germany.

> German Patent 472,184.
> Bensmann, N., *Oel Kohle*, 1933, **1**, 159.

Benson A process for converting methane to ethylene. The methane is reacted with chlorine at a high temperature, yielding hydrogen chloride and ethylene. The hydrogen chloride must be reconverted to chlorine or used in another process. Developed by Hydrocarbon Research, CA, but not commercialized.

> *Eur. Chem. News*, 1982, **39**(1049), 31.
> Wittcoff, H.A., Reuben, B.G., and Plotkin, J.S., *Industrial Organic Chemicals*, 2nd ed., Wiley-Interscience, Hoboken, NJ, 2004, 389.

Benzat A process for improving the quality of gasoline that combines benzene removal, by saturation, with isomerization. Developed by UOP.

> Laredo, G.C., Marroquin, J.O., Castillo, J., Perez-Romo, P., and Navarerette-Bolanos, J., *Appl. Catal., A: Gen.*, 2009, **363**(1–2), 19.

Benzoraffin A hydrofining process for treating naphtha fractions derived from coal. It is a fixed-bed, gas-phase process using a cobalt–molybdenum oxide catalyst. Developed jointly by BASF, Veba-Chemie, and Lurgi, around 1960.

> *Ullmann's Encyclopedia of Industrial Chemistry*, Vol. A12, 5th ed., VCH Publishers, Weinheim, Germany, 1989, 284.

Benzorbon A process for separating and recovering benzene from coke-oven gas and town gas by adsorption on activated carbon. Developed in 1930 by Lurgi.

BenzOUT A process for removing benzene from gasoline. Developed by ExxonMobil (EMRE). It is based on ExxonMobil's ethylbenzene and cumene technologies.

> Laredo, G.C., Marroquin, J.O., Castillo, J., Perez-Romo, P., and Navarerette-Bolanos, J., *Appl. Catal., A: Gen.*, 2009, **363**(1–2), 19.
> *Hydrocarbon Process.*, 2010, **89**(11), 23, 25.

Bergbau-Forschung Also called BF. A process for separating nitrogen from air by selective adsorption over activated carbon in a *PSA system. Developed by Bergbau-Forschung (now Bergwerksverband), Germany. Licensed by Nitrox, United Kingdom, which uses this process in its laboratory gas supply equipment.

> British Patent 2,152,834.
> European Patent 132,756.
> U.S. Patent 4,572,723.
> Knoblauch, K., *Chem. Eng. (N.Y.)*, 1978, **85**(25), 87.

Bergbau-Forschung/Uhde A *flue-gas desulfurization process that uses a movable bed of hot coke. Operated in a power station in Arzberg, Germany, since 1987.

> Wieckowska, J., *Catal. Today*, 1995, **24**(4), 453.

Bergius (1) A coal liquefaction process (also called hydroliquefaction) invented in Germany in 1913 by F. Bergius and subsequently developed by IG Farbenindustrie. The inventor, together with C. Bosch, was awarded the Nobel Prize for chemistry for this invention in 1931. A pilot plant was operated at Rheinau near

Mannheim between 1921 and 1927. The first commercial plant was built at Leuna in 1927. Twelve plants of this type provided much of the aviation fuel used by Germany in World War II. After the war, the process was further developed by the U.S. Bureau of Mines. The process is essentially one of hydrogenation at high pressures and temperatures, catalyzed by an iron oxide catalyst. In Germany, the catalyst was the red mud waste from the *Bayer aluminum process. *See also* Bergius–Pier.

British Patent 18,232 (1914).
German Patents 301,231; 304,348.
Bergius, F., *J. Gasbeleucht*, 1912, **54**, 748.
Storch, H.H., in *Chemistry of Coal Utilization*, 2nd Suppl. Vol., Lowry, H.H., Ed., John Wiley & Sons, New York, 1945, 1750.
James, L.K., Ed., *Nobel Laureates in Chemistry, 1901–1992*, American Chemical Society and Chemical Heritage Foundation, Washington, DC, 1993, 192.

Bergius (2) Also known as Bergius–Willstäter–Zechmeister. A process for extracting sugar from wood by hydrolyzing the cellulose with concentrated hydrochloric acid. Lignin remains undissolved. Developed in the 1920s. The process was hampered by the high cost of HCl recovery. In 2008, an improved version of the process was announced by HCL CleanTech, Israel.

Willstäter, R. and Zechmeister, L., *Ber. Dtsch. Chem. Ges.*, 1913, **40**, 2401.
Bergius, F., *Ind. Eng. Chem.*, 1937, **29**, 247.
Chem. Ind. (London), 2008, (16), 7.

Bergius–Pier An improved version of the *Bergius (1) process, in which the activity of the catalyst was increased by treatment with hydrofluoric acid. Invented by H. Pier and others in the 1930s and used in Germany during World War II.

U.S. Patents 2,154,527; 2,194,186.
Pier, M., *Angew. Chem.*, 1938, **51**, 603.
Pier, M., *Z. Elektrochem. Angew. Phys. Chem.*, 1953, **57**, 456.

Bergius–Rheinau A *saccharification process in which wood is hydrolyzed with concentrated hydrochloric acid to produce sugars for subsequent fermentation to ethanol.

Kent, J.A., Ed., *Riegel's Handbook of Industrial Chemistry*, 9th ed., Van Nostrand Reinhold, New York, 1992, 255.

Bernardini A process for separating fatty acids by fractional crystallization.

Coppa-Zuccari, G., *Oleagineux*, 1971, **26**, 405 (*Chem. Abstr.*, **75**, 128522).

Bertrand A microbiological process for oxidizing aldoses to ketoses. Used in the manufacture of ascorbic acid.

Pigmann, W.W. and Goepp, R.M., *Chemistry of the Carbohydrates*, Academic Press, New York, 1948, 90.

Bertrand Thiel A variation of the *Basic Open Hearth steelmaking process, suitable for ores rich in silicon and phosphorus. Two hearths are used; varying the quantities of lime added to each concentrates most of the silicon and phosphorus in the slag from the first. Developed by E. Bertrand and O. Thiel at Kladno, Bohemia, in 1894; subsequently adopted at Hoesch, Germany (hence the alternative name for the process).

Bertrand, E., *J. Iron Steel Inst. (London)*, Pt. 1, 1897, 115.
Barraclough, K.C., *Steelmaking 1850–1900*, The Institute of Metals, London, U.K., 1990, 281.

Berzelius A tin-smelting process operated by Berzelius Metalhütten at Duisburg-Wanheim, Germany. A mixture of oxide ore, coal, and sodium carbonate is fed continuously into a rotating, tubular kiln with a constriction at the discharge end. The molten metal product collects in the sump at the end, and the slag, which forms an upper layer, flows off.

Mantell, C.L., *Tin: Its Mining, Production, Technology, and Application*, Reinhold, New York, 1949; published in facsimile by Hafner Publishing, New York, 1970, 141.

Bessemer Also called the pneumatic process. A steelmaking process invented by H. Bessemer in Sheffield, England, in 1855 while he was experimenting with the manufacture of gun barrels. He noticed that a draft of air decarbonized iron bars that were lying on the rim of his furnace. Commercialized by associates of Bessemer in Sheffield in 1858 and soon widely adopted. Molten pig iron is contained in a Bessemer converter, which is a pear-shaped vessel with a double bottom, lined with silica, and capable of being tilted on a horizontal axis. Compressed air is blown through the base of the converter, oxidizing most of the dissolved carbon. A similar process was developed in the United States by W. Kelly at around the same time. Kelly made his invention in 1851 but delayed applying for his patent until 1857; although his priority was recognized by the U.S. Patent Office, his name has not become associated with the process.

Bessemer, H., British Patents 2,768 (1855); 630 (1856).
Kelly, W., U.S. Patent 17,628 (1857).
Dennis, W.H., *A Hundred Years of Metallurgy*, Gerald Duckworth, London, U.K., 1963, 98.
Allen, J.A., *Studies in Innovation in the Steel and Chemical Industries*, Kelley, A.M., New York, 1968.
Barraclough, K.C., *Steelmaking 1850–1900*, The Institute of Metals, London, U.K., 1990, 39, 127.
Chem. Eng., 2010, (833), 68–69.

Bethanizing *See* metal surface treatment.

Bethell Also known as the full-cell process. A method for impregnating timber with a creosote preservative. The wood is first degassed under partial vacuum and then impregnated under a pressure of up to 10 atm. *See also* Rüping.

Betterton (1) A process for removing bismuth from lead. A calcium–magnesium alloy is added to the molten lead. The bismuth concentrates in an upper layer and is skimmed off.

Betterton (2) A process for removing zinc from lead by reaction with gaseous chlorine at 400°C. The resulting zinc chloride floats on the molten metal and is skimmed off. Invented in 1928 by J.O. Betterton at the American Smelting and Refining Company.

Dennis, W.H., *A Hundred Years of Metallurgy*, Gerald Duckworth, London, U.K., 1963, 192.

Betterton–Kroll A metallurgical process for removing bismuth from lead. Metallic calcium and magnesium are added to the molten lead, causing precipitation of the sparingly soluble $CaMgBi_2$.

Betts An electrolytic process for refining lead and recovering silver and gold from it. The electrolyte is a solution of lead fluosilicate and hydrofluosilicic acid. The other metals collect as a slime on the anode and are retained there. Developed by A.G. Betts in 1901; first operated at Trail, British Columbia, in 1903; and now widely used in locations having cheap electric power.

Powell, A.R., *The Refining of Non-ferrous Metals*, The Institute of Mining and Metallurgy, London, U.K., 1950.
Dennis, W.H., *A Hundred Years of Metallurgy*, Gerald Duckworth, London, U.K., 1963, 196.

Bextol A catalytic *hydrodealkylation process using an oxide catalyst.

Roebuck, A.K. and Evering, B.L., *Ind. Eng. Chem.*, 1958, **50**, 1135.
Ballard, H.D., Jr., in *Advances in Petroleum Chemistry and Refining*, Vol. 10, McKetta, J.J., Ed., Interscience Publishing, New York, 1965, 219.

BF *See* Bergbau-Forschung, carbonization.

BF/Uhde [Bergbau-Forschung] A *flue-gas treatment that removes sulfur dioxide and nitrogen oxides from power station effluent gases. Sulfur dioxide is first adsorbed on activated coke (made by the German company Bergbau-Forschung). Nitrogen oxides are then converted to elemental nitrogen by the *SCR process. The system was first demonstrated at the Arzberg power station, Germany, in 1988. Engineered by Uhde.

BFB [Bubbling Fluidized Bed] A coal combustion process for developing steam for power generation, developed by Foster Wheeler. It is particularly suitable for burning fuels of low heating value and high moisture and ash content.

BFL *See* carbonization.

BFR [**B**leach **F**iltrate **R**ecycle] A process for almost completely recycling the waste liquors from pulp bleaching using chlorine dioxide. The key to the process is the separation of the sodium sulfate from potassium chloride by crystallization. Developed by Champion International Corporation, Sterling Pulp Chemicals, and Wheelabrator Technologies. It was proposed for installation in North Carolina in 1997.

> *Chem. Eng. (N.Y.)*, 1996, **103**(9), 27.

BFRAC *See* BPX.

Bianchi An early entrained-flow coal gasification process.

Biazzi A continuous, two-phase process for nitrating polyols to form nitrate esters, used as explosives. The nitrating agent is a mixture of anhydrous nitric and sulfuric acids. Used for making nitroglycerine, triethylene glycol nitrate, butanetriol trinitrate, and trimethylolethane trinitrate. Invented by M. Biazzi in Switzerland in 1935. More than 110 plants were sold, worldwide; many are still operating. Today, Biazzi SA is known also for its hydrogenation process and many other aromatic processes. *See also* Meissner, Gyttorp.

> Chilton, T.H., *Strong Water: Nitric Acid, Its Sources, Methods of Manufacture, and Uses*, MIT Press, Cambridge, MA, 1968, 145.

BicarboneR A biological sewage treatment process, using bacteria contained in a fixed bed. Developed by Compagnie Général des Eaux, France, and used in the United States and Japan.

BICHLOR Not a process, but the trade name for a design of bipolar cell for electrolyzing brine. Offered by INEOS Chlor, United Kingdom.

Bifilm A sewage treatment process utilizing two aerated biological filters filled with granular plastic media. Developed by Biwater, United Kingdom, and piloted in 1994.

Bi-Gas [**Bi**tuminous **Gas**] A coal gasification process using a two-stage, entrained-flow slagging gasifier. Developed by Bituminous Coal Research. A 120 tons-per-day pilot plant was built in 1976 at Homer City, PA, under sponsorship from the U.S. Energy Research and Development Administration and the American Gas Association.

> Hegarty, W.P. and Moody, B.E., *Chem. Eng. Prog.*, 1973, **69**(3), 37.
> *Hydrocarbon Process. Int. Ed.*, 1975, **54**(4), 119.
> Hebden, D. and Stroud, H.J.F., in *Chemistry of Coal Utilization*, 2nd Suppl. Vol., Elliott, M.A., Ed., John Wiley & Sons, New York, 1981, 1722.

Binax A process for removing carbon dioxide from methane from landfill by washing with water; absorption takes place under pressure and desorption is done by an air blast at atmospheric pressure. Piloted at a wastewater treatment plant at Modesto, CA, in 1978.

> Henrich, R.A., *Energy Biomass Wastes*, 1983, 879,916 (*Chem. Abstr.*, **99**, 125564).
> Kohl, H.L. and Nielsen, R.B., *Gas Purification*, 5th ed., Gulf Publishing Co., Houston, TX, 1997, 434.

Bio3 A process for oxidizing organic wastes in industrial wastewater by a combination of ozone, hydrogen peroxide, and UV radiation. Intended to be integrated with the *VITOX process by using the Bio3 process's excess of oxygen in the VITOX process.

> *Eur. Chem. News*, 1994, **61**(1607), 27.

BIOBEAD A *BAF process offered by Brightwater.

> Stephenson, T., Mann, A., and Upton, J., *Chem. Ind. (London)*, 1993, (14), 533.

Biobed An anaerobic digestion system for treating industrial organic wastes, developed from the *Biothane system. It uses an expanded granular sludge bed and a three-phase separation system; these features enable it to handle 10 times the hydraulic load of a similar Biothane system. Developed by Gist-Brocades NV (Delft) and licensed in the United Kingdom through Babcock Water Engineering. Twenty systems were operating worldwide in 1996.

Water Waste Treat., 1996, **39**(5), 27.

Biobor HSR [**Bio**logical **Bor**sig **H**ub**S**trahl**R**eaktor] A high-intensity biological process for treating concentrated effluents from farms and food processing. Compressed air forces the waste rapidly up a tower containing moving, perforated discs. Developed by Borsig, Berlin, in the early 1990s. *See also* Deep Shaft.

BIOCARBONE A *BAF process offered by OTV.

Stephenson, T., Mann, A., and Upton, J., *Chem. Ind. (London)*, 1993, (14), 533.

BIOCLAIM A process for removing toxic metals from aqueous systems by use of bacteria supported on poly(ethyleneimine) glutaraldehyde.

Brierly, C.L., *Geomicrobiology*, 1990, **8**, 201.
Vegio, F., Beolchini, F., and Toro, L., *Ind. Eng. Chem. Res.*, 1998, **37**(3), 1107.

Bio-Claus A biological desulfurization process for removing sulfur dioxide from waste gases. There are four stages: scrubbing with a solvent to produce a sulfite solution, biological reduction of the sulfite to sulfide, biological oxidation of the sulfide to elemental sulfur, and separation of the sulfur. Developed by Stork Engineering and first announced in 1997.

Chem. Eng. (Rugby, Engl.), 1997, (638), 24.

BIOCOCOMB [**BIO**mass for **CO-COMB**ustion] A process for adding biomass to coal combustion in a fluidized bed for producing electric power. A cooperative development project has been funded by the EU since 1998 in several European countries.

Biocol A process for removing dyes from aqueous effluents using a microorganism immobilized on activated carbon. The microorganism *Shewanella putrefaciens* was first isolated and characterized at the University of Leeds. The process was developed in Queens University, Belfast, in conjunction with British Textile Technology Group. Piloted from April 2001 at a dyeworks in County Antrim.

Environ. Bus. Mag., 2001, (71), 40.

BioCOP [**Bio**logical **COP**per] A bacterial process for leaching copper from its ores. BioCOP uses an agitated tank for oxidation and leaching of copper sulfides. Developed by BHP Billiton and Codelco, which operated a demonstration plant in Chuquicamata, Chile, from 2003. Several plants have been operated commercially in Chile since then.

U.S. Patent 6,245,125.
Rawlings, D.E., *Pure Appl. Chem.*, 2004, **76**(4), 847.

BioDeNox A process for removing oxides of nitrogen from flue gases. The gases are scrubbed with a proprietary solution containing a chelating agent, which converts the NO_x to a soluble nitrosyl compound. The spent solution is treated with a "hydrocarbon reducing agent" and a culture of bacteria, which convert the nitrosyl compound to elemental nitrogen. The solution is then regenerated. Developed by UOP and Paques Bio Systems in 2002 and demonstrated in the Navajo Refining Company refinery in Artemisia, NM.

Chem. Eng. (N.Y.), 2002, **109**(13), 19.

Biofine A process for making furfural, levulinic acid, and formic acid from lignocellulose. The lignocellulose is hydrolyzed by dilute sulfuric acid in two stages. Developed in the 1980s by S.W. Fitzpatrick and piloted in South Glens Falls, NY, in the 1990s. A commercial unit in Caserta, Italy, fed by bagasse and paper mill sludge, was commissioned in 2006. The process was originally intended to be fed by wastepaper, but it is now envisaged as being of general use for treating municipal wastes and for producing a range of organic chemical intermediates. Won the Presidential Green Challenge Award in 1999.

> U.S. Patents 4,897,497; 5,608,105.
> Hayes, D.J., Fitzpatrick, S., Hayes, M.H.B., and Ross, J.R.H., in *Biorefineries: Industrial Processes and Products*, Kamm, B., Gruber, P.R., and Kamm, M., Eds., Wiley-VCH, Weinheim, Germany, 2005, **I**, Chap. 7.
> *Chem. Eng. News*, 2006, **84**(34), 47.
> Mulder, W. et al., in *Advanced Oil Crop Biorefineries*, Kazmi, A., Ed., Royal Society of Chemistry, Cambridge, U.K., 2012, 147.

BIOFIX A process for removing toxic heavy metals from aqueous wastes by the use of peat moss and a polysulfone.

> Veglio, F. and Beolchini, F., *Hydrometallurgy*, 1997, **44**, 301.
> Vegio, F., Beolchini, F., and Toro, L., *Ind. Eng. Chem. Res.*, 1998, **37**(3), 1105.

BIOFOR A *BAF process offered by Degremont.

> Stephenson, T., Mann, A., and Upton, J., *Chem. Ind. (London)* 1993, (14), 533.

BioForming A proprietary version of *APR developed at the University of Wisconsin since the early 2000s and now under development by Virent Energy Systems. Virent is cooperating with Royal Dutch Shell, Honda, and Cargill. The process won an Innovation Award from ICIS in 2008. Shell and Virent started a demonstration plant in Madison, WI, in 2010, which was operational in June 2011. In June 2011, Virent announced that it expected to have a commercial plant for making *p*-xylene by this process under construction in 2014.

> *Oils Fats Int.*, Jun 2008, **24**(5), 3.
> *ICIS Chem. Bus.*, 2008, **274**(5), 17; 2008, **274**(14), 23.
> *Chem. Week*, 2010, **172**(7), 15.
> *ICIS Chem. Bus.*, 2011, **279** (23), 22.
> Perego, C. and Ricci, M., *Catal. Sci. Technol.*, 2012, published online June 29, 2012.

Biogas A process for generating fuel gas from sewage. Developed by EMS-Inventa and installed in Switzerland and Italy between 1979 and 1983. *See* BIOGAS.

BIOGAS A process for generating methane from wastes and biomass. Developed by the Institute of Gas Technology, Chicago, since 1970, and demonstrated at Walt Disney World, Orlando, FL, in 1990. *See also* Biogas.

BIOHOCH An aerobic wastewater treatment process. Optimum reactor design and a proprietary air injection system achieve 50% of the energy consumption of similar systems. Fifty systems had been installed worldwide by 1994. Licensed by Brown-Minneapolis Tank and Hoechst-Uhde Corporation.

> *Hydrocarbon Process. Int. Ed.*, 1993, **72**(8), 92.

BIOKOP A process for treating liquid effluents containing wastes from organic chemical manufacture. It combines aerobic fermentation, in special reactors known as BIOHOCH reactors, with treatment by powdered activated carbon. Developed originally for treating the effluent from the Griesheim works of Hoechst and engineered by Uhde. *See also* PACT.

> Eisenacher, K. and Neumann, U., *Chim. Ing. Technik*, 1983, **55**(5), 390.

Bioliq A process for converting biomass to an intermediate liquid product that can be converted to liquid fuels. Developed by Lurgi and piloted at the Karlsruhe Institute of Technology in 2009.

> *Chem. Eng. (N.Y.)*, 2009, **116**(2), 12.

Biolysis A sewage treatment process that produces less sludge than the usual *Activated Sludge process. There are two versions: biolysis O, which uses ozone, and biolysis E, which uses bacterial enzymes. Developed in 2001 by Ondeo Degremont. Biolysis O was first demonstrated at Aydoilles, France, in 2001. Biolysis E was based on a process developed by the Japanese company Shinko Pantec.

Chimie Hebdo, 19 Nov 2001, (150), 17.
Water & Wastewater Int., 2002, **17**(2), 23.

BioMeteq A biological process for removing trace metals from oil refinery effluents. The effluent is passed upward through a continuous sand filter. A biofilm of bacteria on the sand particles reduces and precipitates the metals. An air blast scours the precipitates from the sand, which is washed and returned to the process. Developed by Paques BV, the Netherlands, and piloted in Germany in 2005 for removing uranium and selenium.

Chem. Eng. (N.Y.), 2005, **112**(6), 15.

BioNIC [**Bio**logical **NIC**kel] A biological process for leaching nickel from its ores. It comprises bacterial leaching, solvent extraction, and electrowinning. Developed by BHP Billiton at Yabulu from 1999, but not commercialized by 2004 because no ore body of a suitable concentration and size had been identified.

Rawlings, D.E., *Pure Appl. Chem.*, 2004, **76**(4), 847.

BIOPRO A biological process for leaching gold from refractory ores. Developed by Newmont Mining.

BIOPUR A *BAF process offered by Sulzer and John Brown.

Stephenson, T., Mann, A., and Upton, J., *Chem. Ind. (London)*, 1993, (14), 533.

Bio-SR A process for removing hydrogen sulfide from gas streams. Developed by NKK Industries, Japan. The process uses a solution of unchelated iron, regenerated microbiologically.

Oil Gas J., 1994, **92**(21), 58.
Quinlan, M.P., Echterhof, L.W., Leppin, D., and Meyer, H.S., *Oil Gas J.*, 1997, **95**(29), 54.

Biostil A continuous fermentation process for making ethanol. Developed by Chematur Engineering AB, Sweden.

BIOSTYR [**BIO**logical poly**STYR**ene] A biological system for treating effluents containing dissolved organic matter. The microbiological organisms are trapped within rigid, lighter-than-water porous polystyrene granules. The effluent flows upward through a bed of these granules, and air is injected at the base of the bed. Developed in France by OTV and licensed in the United Kingdom through General Water Processes.

Stephenson, T., Mann, A., and Upton, J., *Chem. Ind. (London)*, 1993, (14), 533.

BioSURE A process for removing sulfate from acid mine water drainage by treatment with primary sewage sludge. Developed at Rhodes University, South Africa, and launched in 2005.

Bio-Synfining A process for making liquid fuels from natural oils and fats. It combines thermal depolymerization with catalytic hydrogenation. In 2012, it was being commercialized in Geismar, LA, in conjunction with Dynamic Fuels.

Biothane An anaerobic digestion system for treating industrial organic wastes. The reactor contains an upflow sludge blanket and is operated at approximately 35°C, with the heat provided by burning some of the product gas, which contains 70% methane. Usually, nutrients such as urea and iron must be added. Developed in the early 1970s in the Netherlands by Centrale Suiker Maatschappij; in 1984, Gist-Brocades (Delft) acquired the rights and subsequently licensed the process in the United Kingdom to Esmil. In 1990, more than 70 units had been built worldwide for a variety of industries. *See* Biobed.

Water Waste Treat., 1991, **34**(5), 24.

BioTherm A process for converting dry biomass to a liquid fuel. Developed from 2003 by Dynamotive Energy Systems, Canada.

Rezaiyan, J. and Cheremisinoff, N.P., *Gasification Technologies*, Taylor & Francis, Boca Raton, FL, 2005, 156.

BIOVOX A biological process for removing organic vapors from air streams. Developed by IBS Viridian Ltd., Kent, United Kingdom. First used at the Dr. Martens boot factory in Wellingborough, United Kingdom, in 2000.

Industrial Environ. Manage., 1999, **10**(4), 17.
Eur. Chem. News, 4 Sep 2000, **73**(1923), 29.

BIOX (1) [**BI**ological **OX**idation] A general term for effluent treatment processes employing biological oxidation, such as the *Activated Sludge process.

Capps, R.W., Matelli, G.N., and Bradford, M.L., *Hydrocarbon Process. Int. Ed.*, 1993, **72**(12), 81.

BIOX (2) A process for making diesel fuel from waste animal and vegetable oils by acid-catalyzed, single-phase methanolysis using an excess of methanol and an inert reclaimable cosolvent (e.g., tetrahydrofuran) in a one-pot reaction system. The product is a mixture of the methyl esters of a range of fatty acids. Invented by D. Boocock in Toronto and piloted from 2001 by BIOX Corp. A commercial plant was built in Hamilton, ON, in 2007.

U.S. Patent 6,642,399.
Eur. Chem. News, 2002, **76**(2011), 41.
Demirbas, A., *Biodiesel: A Realistic Alternative for Diesel Engines*, 2008, Springer-Verlag, London, U.K.
Sarin, A., *Biodiesel: Production and Properties*, 2012, Royal Society of Chemistry, Cambridge, U.K., 16.

BIOXIDE A process for removing hydrogen sulfide and other sulfides from wastewaters by introducing nitrate ions, which are metabolized by sulfide-oxidizing microorganisms. Developed in the 1980s by Davis Process Products, now a subsidiary of U.S. Filter.

BioZINC A biological process for leaching zinc from zinc sulfide ores, developed by BHP Billiton.

Birkeland–Eyde Also known as the **arc** process. A process for making nitric acid by passing air through an electric arc, forming nitric oxide, oxidizing this with air, and absorbing the resulting oxides of nitrogen in water:

$$N_2 + O_2 = 2NO$$

$$2NO + O_2 = 2NO_2$$

$$2NO_2 + H_2O = HNO_2 + HNO_3$$

$$3HNO_2 = HNO_3 + 2NO + H_2O$$

The arc is spread into a disc by an electromagnet. Operated in Norway, using hydroelectric power, from 1905 to 1934, when it was made obsolete by the ammonia oxidation process. In the United States, this process was first used in 1917.

Sherwood Taylor, F., *A History of Industrial Chemistry*, Heinemann, London, U.K., 1957, 428.
Chilton, T.H., *Strong Water: Nitric Acid, Its Sources, Methods of Manufacture, and Uses*, MIT Press, Cambridge, MA, 1968, 112.
Haber, L.F., *The Chemical Industry 1990–1930*, Clarendon Press, Oxford, U.K., 1971, 86.
Davies, P., Donald, R.T., and Harbord, N.H., in *Catalyst Handbook*, 2nd ed., Twigg, M.V., Ed., Wolfe Publishing, London, U.K., 1989, 470.
Leigh, G.J., *The World's Greatest Fix: A History of Nitrogen and Agriculture*, Oxford University Press, Oxford, U.K., 2004, 134–139.

BiRON A biological process for removing iron from public water supplies. Developed in the United Kingdom by Biwater Europe Ltd. and piloted in 1994 at a water treatment plant in Ipswich.

Alani, S., *Water Waste Treat.*, 1994, **37**(6), 50.

Bischof An obsolete process for making white lead (basic lead carbonate), invented by G. Bischof around 1900. Lead monoxide was reduced by the carbon monoxide in water gas to form black lead suboxide, oxidized in damp air to lead hydroxide, slurried in dilute acetic acid, and carbonated with carbon dioxide. Piloted in Willesden, London, and commercialized by L. Mond in the Brimsdown White Lead Company. A sample of Bischof's white lead was used by William Holman Hunt in his painting *The Light of the World*, now in St. Paul's Cathedral, London.

Cohen, J.M., *The Life of Ludwig Mond*, Methuen, London, U.K., 1956, 211.

Bischoff A *flue-gas desulfurization process. A slurried mixture of lime and limestone is sprayed into the gas in a spray tower. The calcium sulfite in the product is oxidized by air to calcium sulfate. Piloted in Europe in the 1980s, but by 1997, there were no commercial installations. Lurgi Bischoff is a part of the Lurgi group. The process was offered by Lentjes, Germany, a subsidiary of Lurgi.

IEA Coal Research, *The Problems of Sulphur*, Butterworths, London, U.K., 1989, 100.
Kohl, H.L. and Nielsen, R.B., *Gas Purification*, 5th ed., Gulf Publishing Co., Houston, TX, 1997, 601.

Black Ash One of the two processes comprising the *Leblanc process for making sodium carbonate; the other is the *Salt Cake process. The heart of the process was a rotating kiln made of cast iron, known as a revolver. Invented by G. Elliot and W. Russell in St. Helens, England, in 1853. The name has also been used for a process to extract barium from barium sulfate.

British Patent 887 (1853).

Black Oil Conversion *See* RCD Unionfining.

Blanc A process for extracting alumina from leucite ore (a potassium aluminosilicate), in which the ore is leached with hydrochloric acid, aluminum trichloride is crystallized, and this is calcined to alumina. Invented by G.A. Blanc in 1921 and used in Italy from 1925 to 1943.

U.S. Patent 1,656,769.
O'Connor, D.J., *Alumina Extraction from Non-bauxitic Materials*, -Verlag, Düsseldorf, 1988, 85.

Blaugas An early thermal cracking process for making liquid petroleum gas from petroleum. Developed by the German company Blau in Augsburg from 1905. Not to be confused with blue gas (*see* Water Gas).

BLISS [ButyLene ISomerization System] A process for isomerizing *n*-butenes to *iso*butene. Piloted by Texas Olefins Company in Houston, TX, 1990 to 1992.

Chem. Mark. Rep., 1992, **241**(18), 7.

Bloomery The earliest process for making iron from iron ore, operated from around 1500 BC until the blast furnace was invented around AD 1500. The ore is heated with charcoal in a furnace blown by bellows; the product, known as bloom, is a composite of iron particles and slag. When the bloom is hammered, the slag is expelled to the surface and a lump of relatively pure iron remains. *See also* Catalan.

Barraclough, K.C., *Steelmaking Before Bessemer, Vol. 1, Blister Steel*, The Metals Society, London, U.K., 1984, 15.

Blowdec A process for converting waste plastics into diesel fuel. The plastic is injected into a bed of hot whirling sand. Developed in Slovakia from 1997.

U.S. Patent 6,165,349.
Scheirs, J., in *Feedstock Recycling and Pyrolysis of Waste Plastics*, Scheirs, J. and Kaminsky, W., Eds., John Wiley & Sons, Chichester, U.K., 2006, 429.

Blueprint A reprographic process, based on the photochemical reduction of ferric salts. Paper is impregnated with an aqueous solution of potassium ferricyanide, ammonium ferric citrate, and a gum binder. Exposure to light generates Prussian blue and then Turnbull's blue. Developed in the nineteenth century and widely used for copying line drawings until it was superseded by versions of the *Diazo process and subsequently by xerographic processes. *See* Reprography.

BLUE Tower A process for gasifying biomass by staged reforming. Heat is transferred by alumina balls at ca. 1000°C. The products are high-quality syngas and biochar. Tested in a large plant in Izumo, Japan, in 2012. Offered by Concord Blue Energy, Düsseldorf.

> *Chem. Eng. (N.Y.)*, 2012, **119**(12), 12.

Blumenfeld An early version of the *Sulfate process for making titanium dioxide pigment, in which the nucleation of the precipitation of titania hydrate is accomplished by dilution under controlled conditions. Invented by J. Blumenfeld, a Russian working in London in the 1920s.

> U.S. Patents 1,504,669; 1,504,671; 1,504,672.
> Barksdale, J., *Titanium: Its Occurrence, Chemistry, and Technology*, 2nd ed , Ronald Press, New York, 1966, 278.

BMA [**B**lausaure **M**ethan **A**mmoniak] *See* Degussa.

BMS A process for removing mercury from the effluent from the *Castner–Kellner process. Chlorine is used to oxidize metallic mercury to the mercuric ion, which is then adsorbed on activated carbon impregnated with proprietary sulfur compounds. Developed by Billingsfors Bruk, Sweden.

> Rosenzweig, M.D., *Chem. Eng. (N.Y.)*, 1975, **82**(2), 60.

BOC [**B**lack **O**il **C**onversion] *See* RCD Unionfining.

BOC Unibon A process for upgrading petroleum residues by catalytic hydrogenation. *See also* RCD Unibon.

> Absci-Halabi, M., Stanislaus, A., and Qabazard, H., *Hydrocarbon Process. Int. Ed.*, 1997, **76**(2), 50.

Bofors A process for nitrating organic compounds by mixing the nitrating acid and the substrate rapidly in a centrifugal pump.

Boivin–Loiseau A process for purifying cane sugar. Calcium hydroxide is added to the syrup, and carbon dioxide is passed through it. The precipitated calcium carbonate removes some of the coloring impurities.

> Watson, J.A., *A Hundred Years of Sugar Refining*, Tate & Lyle Refineries, Liverpool, U.K., 1973, 80.

Boliden (1) An obsolete process for reducing sulfur dioxide to elemental sulfur, using hot coke. Operated in Sweden by Boliden from 1933 to 1943. *See also* Trail, RESOX.

> Katz, M. and Cole, R.J., *Ind. Eng. Chem.*, 1950, **42**, 2258.

Boliden (2) A lead extraction process in which a sulfide ore, mixed with coke, is smelted in an electric furnace, air jets forming vortices between the electrodes. Discontinued in 1988 in favor of the *Kaldo process, which uses a rotating furnace.

Boliden/Norzink A method for removing mercury vapor from zinc smelter off-gases by scrubbing with a solution of mercuric chloride:

$$Hg + HgCl_2 = Hg_2Cl_2$$

The precipitated mercurous chloride separates as a sludge. In the original process, some of this mercurous chloride was chlorinated to mercuric chloride for reuse. In a later version of the process, all the mercuric chloride is electrolytically converted to elemental mercury and chlorine. As of 1994, the electrolytic version had been installed in three plants. Developed by Boliden, Sweden, and Norzink, Norway, and now offered for license by Boliden Contech.

Bolkem [**Bol**iden] A process for removing mercury from the sulfuric acid from metallurgical smelters. Addition of aqueous sodium thiosulfate causes mercuric sulfide to be precipitated. Operated at Helsingborg, Sweden, since 1974.

Sundström, O., *Sulphur*, 1975, (116), 37.

Bonderizing *See* metal surface treatment.

Borchers–Schmidt *See* metal surface treatment.

Borstar A catalytic process for polymerizing ethylene or propylene, subdivided into Borstar PE and Borstar PP. The use of two reactors—a loop reactor and a gas-phase reactor—allows better control of molecular weight distribution. The loop reactor operates under supercritical conditions to avoid bubble formation. Either *Ziegler–Natta or metallocene catalysts can be used. The latest version, Borstar PE 2G, uses a single, multizone gas-phase reactor to make polymers that have bimodal molecular weight distributions. Developed by Borealis A/S. The first commercial unit, for polyethylene, was installed in Porvoo, Finland, in 1995. The first polypropylene plant was operated by Borealis in Schwechat, Austria, in 2000. In 2005, Borstar's total capacity for PE and PP was 1.3 million tons.

Eur. Chem. News, 1995, **64**(1688), 37; 1996, **65**(1709), 45.
Chem. Eng. (N.Y.), 1995, **102**(11), 17.
Oil Gas J., 1998, **96**(24), 34.
Eur. Chem. News, 2000, **73**(1929), 26; 2005, **83**(2157), 20.
Soares, J.B.P. and McKenna, T.F.L., *Polyolefin Reaction Engineering*, Wiley-VCH, Weinheim, Germany, 2012, 127.

BOSAC [**BO**fors **S**ulfuric **A**cid **C**oncentrator] A process for recovering sulfuric acid from the production of nitro-compounds. Spent acid is concentrated by distillation, using a heat exchanger with externally heated silica tubes. Developed by Bofors Nobel Chemikur, Sweden.

Douren, L., *Making the Most of Sulfuric Acid*, More, A.I., Ed., British Sulphur, London, U.K., 1982, 317.

Bosch A regenerative, two-stage process for reducing carbon dioxide to oxygen:

$$CO_2 + 2H_2 = 2H_2O + C$$

$$2H_2O = 2H_2 + O_2$$

The first step is conducted over an iron catalyst at 700°C and the second by electrolysis. Proposed for use during prolonged space travel.

Sacco, A., Jr. and Reid, R.C., *Carbon*, 1979, **17**, 459.

Bosch–Meiser A process for making urea from ammonia and carbon dioxide under high temperature and pressure. Invented by C. Bosch and W. Meiser in 1920.

U.S. Patent 1,429,483.

Boss A development of the *Washoe process for extracting silver from sulfide ores, invented in 1861. The chloride solution from the Washoe process was passed continuously through a series of amalgamation pans and settlers.

Bouchet–Imphy *See* DR.

Bower–Barff *See* metal surface treatment.

BP-Hercules *See* Hercules-BP.

BP Polymer Cracking Process *See* SPORT.

BPR [ByProduct Recycle] A process for recycling the chlorine-containing by-products from the manufacture of vinyl chloride, 1,2-dichloroethane, and other chlorinated hydrocarbons. Combustion with oxygen converts 90% of the chlorine to anhydrous hydrogen chloride and 10% to aqueous hydrochloric acid. Developed by BASF and licensed by European Vinyl Corporation.

> *Eur. Chem. News*, 1990, 26.

BPX [Broin Project X] A process for converting starch to ethanol without cooking. Developed by the Broin Companies in Iowa from 2000, now POET Research. Commercialized in 2004 in Emmetsburg, IA.

> A related process, Broin Fractionation (BFRAC), announced in 2005, first separates corn into three fractions—fiber, germ, and endosperm. The endosperm is fermented to produce ethanol and the other fractions are converted to animal feed. U.S. Patent Application 20110111085.

Bradshaw A soapmaking process in which glyceryl esters (fats) are first converted to their corresponding methyl esters by transesterification and then hydrolyzed. Invented by G.B. Bradshaw.

> U.S. Patent 2,360,844.

Brassert A modified blast-furnace process for making iron. Less limestone is added than in the usual process, so the product contains more sulfur. The molten iron product is mixed with sodium carbonate, with which the ferrous sulfide reacts, yielding sodium sulfide and iron oxide. These float on the molten iron and are skimmed off.

Braun A variation on the classic ammonia synthesis process, in which the synthesis gas is purified cryogenically. Widely used since the mid-1960s.

> Isalski, W.H., *Separation of Gases*, Clarendon Press, Oxford, U.K., 1989, 155.

Brennstoff-Technik *See* carbonization.

Bretsznajder A process for extracting aluminum from clays and a variety of aluminous wastes by continuous digestion with concentrated sulfuric acid in an autoclave. Developed in Poland in the 1980s but not yet commercialized.

> O'Connor, D.J., *Alumina Extraction from Non-bauxitic Materials*, Aluminum-Verlag, Dusseldorf, Germany, 1988, 160.

Bridgman Also called Bridgman–Stockbarger. A process for growing large, single crystals. The material is contained in a cylindrical ampule that is slowly lowered through a temperature gradient.

> Stockbarger, D.C., *J. Opt. Soc. Am.*, 1927, **14**, 448.
> Vere, A.W., *Crystal Growth: Principles and Progress*, Plenum Press, New York, 1987, 67.

Brin An obsolete process for making oxygen. Barium monoxide was roasted in air to produce barium peroxide, which was roasted at a higher temperature to produce oxygen:

$$2BaO + O_2 = 2BaO_2$$

$$2BaO_2 = O_2 + 2BaO$$

The process was developed in France by the brothers L. and A. Brin, based on a chemical reaction discovered by J.B.J.D. Boussingault in 1851. Boussingault used temperature cycling between 537°C and 926°C, which caused a progressive loss in the efficiency of the barium oxide. The Brin brothers used pressure cycling at a lower temperature, which overcame this problem. The process was first operated commercially by Brin Frères et Cie in France in 1879 but was abandoned in favor of air liquefaction at the beginning of the twentieth century. In England, Brin's Oxygen Company was incorporated to operate the process in 1886, changing its name to the British Oxygen Company in 1906.

Smith, W., *J. Soc. Chem. Ind. (London)*, 1885, **4**, 568.
Chem Br., 1997, **33**(5), 58.
Aitkin, W., *Chem. Eng. (Rugby, Engl.)*, 1998, (659), 16.

Britannia A process for removing silver from lead, operated by Britannia Refined Metals in England, using ore from the Mount Isa mine in Australia. After initial concentration by the *Parkes process and removal of the zinc by vacuum distillation, the mixture, which contains silver (70%), lead, and some copper, is treated in a bottom-blown oxygen cupel; lead and copper are removed by the injection of oxygen through a shielded lance.

Barrett, K.R. and Knight, R.P., *Silver: Exploration, Mining and Treatment*, The Institute of Mining and Metallurgy, London, U.K., 1988.

Brodie A naphthalene crystallization process.

BSR *See* Beavon.

BSR/MDEA *See* Beavon.

BTG [**B**iomass-**T**o-**G**asoline] A process for converting lignocellulose to gasoline. Developed by Exelus, NJ, and tested in an Asian refinery in 2012.

BTL [**B**iomass **T**o **L**iquids] A general name for processes that convert biomass to liquid fuels. Examples are *BTG, *Carbo-V, *Coskata, *MKS, and *RTP. A version of the *PRENFLO process, originally developed for gasifying coal, can be used as the first stage.

BTO [**B**iomass-**T**o-**O**lefins] A process for converting lignocellulose to olefins developed by Exelus, NJ. Tested with pine sawdust, paulownia, and switchgrass. The biomass is first deconstructed to oligosaccharides that are then hydrolyzed to glucose. This is selectively deoxygenated to short-chain alcohols that are then dehydrated to their respective olefins.

Bubiag An early two-stage coal gasification process.

Bucher A process for making sodium cyanide by the reaction between sodium carbonate and coke:

$$Na_2CO_3 + 4C + N_2 = 2NaCN + 3CO$$

Invented by J.E. Bucher in 1924.
U.S. Patent 1,091,425.
Sittig, M., *Sodium: Its Manufacture, Properties and Uses*, Reinhold Publishing, New York, 1956, 223.
Chilton, T.H., *Strong Water: Nitric Acid, Its Sources, Methods of Manufacture, and Uses*, MIT Press, Cambridge, MA, 1968, 61.

Büchner *See* Aloton.

Budenheim *See* CFB.

Bueb A process for removing cyanogen from coal gas by scrubbing with a concentrated aqueous solution of ferrous sulfate. A complex sequence of reactions occurs, involving the ammonia and hydrogen sulfide that are always present in coal gas; the final product is a blue mud containing ammonium ferrocyanide. This mud is boiled with lime, thereby expelling the ammonia, and several further steps result in the formation of saleable sodium or potassium ferrocyanide. Invented by J. Bueb in 1898 and operated in the United States in the early 1900s.

German Patent 112, 459 (1898).
Muller, M.E., *J. Gas Lighting*, 1910, **112**, 851.
Hill, W.H., in *Chemistry of Coal Utilization*, Vol. 2, Lowry, H.H., Ed., John Wiley & Sons, New York, 1945, 1104.

Bufflex [From **Buff**elsfontein (South Africa), **ex**traction] A process for extracting uranium from its ores, using a solution of an amine (Alamine 336). Bufflex was developed in South Africa and was later replaced by *Purlex. *See also* Eluex.

Eccles, H. and Naylor, A., *Chem. Ind. (London)*, 1987, (6), 174.

Bullard–Dunn *See* metal surface treatment.

Buna [**Bu**tadiene **na**trium] The name has been used for the product, the process, and the company VEB Chemische Werke Buna. A process for making a range of synthetic rubbers from butadiene, developed by IG Farbenindustrie in Leverkusen, Germany, in the late 1920s. Sodium was used initially as the polymerization catalyst, hence the name.

Buna S This was a copolymer of butadiene with styrene; **Buna N** a copolymer with acrylonitrile. The product was first introduced to the public at the Berlin Motor Show in 1936. The trade name **Buna CB** was used for a polybutadiene rubber made by Bunawerke Hüls using a *Ziegler-Natta-type process. Today, **Buna EP** is the name for a range of ethylene–propylene rubbers made by Lanxess (formerly Bayer, which acquired Bunawerke Hüls in 1994).

German Patent 570,980.
Morris, P.J.T., *The American Synthetic Rubber Research Program*, University of Pennsylvania Press, Philadelphia, PA, 1989, 7.

Burgess A coal liquefaction process piloted by the U.S. Office of Synthetic Fuels in Louisiana, MO, in 1949.

Kastens, M., Hirst, L.L., and Chaffe, C.C., *Ind. Eng. Chem.*, 1949, **41**, 870.

Burkheiser Also known as the sulfite–bisulfite process. A complicated process for removing hydrogen sulfide and ammonia from coal gas by absorption in an aqueous solution containing ammonia, iron oxide, and elemental sulfur. The end products are sulfur and ammonia. Invented by K. Burkheiser in 1907 and developed in Germany in the early 1900s.

German Patents 212,209; 215,907; 217,315; 223,713.
Kohl, H.L. and Nielsen, R.B., *Gas Purification*, 5th ed., Gulf Publishing Co., Houston, TX, 1997, 738.

Burton The first commercial process for thermally cracking heavy petroleum fractions to obtain gasoline. Invented in 1912 by W.M. Burton at Standard Oil (Indiana) and operated commercially from 1913 through the 1920s. *See also* Dubbs.

U.S. Patent 1,049,667.
Enos, J.L., *Petroleum Progress and Profits*, MIT Press, Cambridge, MA, 1962, Chap. 1.
Achilladelis, B., *Chem. Ind. (London)*, 1975, 19 Apr, 337.
Flank, W.H., Abraham, M.A., and Matthews, M.A. Eds., *Innovations in Industrial and Engineering Chemistry*, 2009, American Chemical Society, Washington, DC, 104.

BUTACRACKING A process for converting *iso*butanes to *iso*butene, which can then be converted to gasoline-blending components such as methyl *t*-butyl ether. Developed by Kinetics Technology International.

Monfils, J.L., Barendregt, S., Kapur, S.K., and Woerde, H.M., *Hydrocarbon Process. Int. Ed.*, 1992, **71**(2), 47.

Butamax A fermentation process for making isobutanol, for use as a diesel fuel, from sugar. The process depends on a novel yeast that has been genetically engineered to maximize the production of butanols. Developed by Butamax Advanced Biofuels, a joint venture of BP and DuPont. A demonstration plant was being built in Hull, United Kingdom, in 2012.

U.S. Patents 8,178,328; 8,188,250; 8,273,558.

Butamer [**Buta**ne isome**r**ization] A fixed-bed process for converting *n*-butane into *iso*butane; conducted in the presence of hydrogen over a dual-functional catalyst containing a noble metal. Trace amounts of an organic chloride are also needed. Developed by UOP and licensed worldwide since 1959. More than 70 units had been commissioned by 2003.

Hydrocarbon Process. Int. Ed., 1990, **69**(4), 73.
Cusher, N.A., in *Handbook of Petroleum Refining Processes*, 3rd ed., Meyers, R.A., Ed., McGraw-Hill, New York, 2003, 9.7.

BUTENEX A process for separating several C_4 components from C_4 hydrocarbon streams by extractive distillation, using N-formylmorpholine or a mixture of morpholine derivatives. Piloted by Krupp Koppers in 1987. Now licensed by Uhde. Two plants had been built between 1998 and 2005.

Chem. Week, 1987, **141**(9), 24.

Butesom *See* C_4 Butesom.

Butex A process for separating the radioactive components of spent nuclear fuel by solvent extraction from nitric acid solution, using diethylene glycol dibutyl ether (also called Butex, or dibutyl carbitol) as the solvent. Developed by the Ministry of Supply (later the U.K. Atomic Energy Authority) in the late 1940s. Operated at Windscale from 1952 until 1964, when the process was superseded by the *Purex process.

Martin, F.S. and Miles, G.L., *Chemical Processing of Nuclear Fuels*, Butterworth Scientific Publications, London, U.K., 1958, 102.
Howells, G.R., Hughes, T.G., Mackey, D.R., and Saddington, K., in *Proceedings of the Second U.N. International Conference on Peaceful Uses of Atomic Energy*, United Nations, Geneva, Switzerland, 1958, **17**, 3.
British Nuclear Fuels PLC, *Nuclear Fuel Processing Technology*, 1985.

Butomerate A catalytic process for isomerizing *n*-butane to isobutane. Developed by the Pure Oil Company.

Hydrocarbon Process., 1964, **43**(9), 173.
Unzelman, G.H. and Wolf, C.J., in *Petroleum Processing Handbook*, Bland, W.F. and Davidson, R.L., Eds., McGraw-Hill, New York, 1967, 3–51.

BUTOX A biological process for oxidizing bioethanol to acetic acid. Developed by Wacker Chemie. In 2001, they were ready to build a plant but had not done so by 2008. BUTOX is also the name of a proprietary veterinary medicine.

ICIS Chem. Bus., 14 Apr 2008, **273**(15), 17.

BWHP *See* Woodall–Duckham.

BYAS [**BY**pass **A**mmonia **S**ynthesis] An economical process for expanding existing ammonia synthesis plants by introducing the additional natural gas at an intermediate stage in the process. The additional nitrogen in the air, which must also be introduced, is removed by *PSA. Developed and offered by Humphreys and Glasgow, United Kingdom.

C

CAA [Cuprous Ammonium Acetate] A general process for separating alkenes, dialkenes, and alkynes from each other by extraction of their cuprous complexes from aqueous cuprous ammonium acetate into an organic solvent. Exxon used it for separating C_4 fractions containing low concentrations of butadiene. The liquid–liquid extraction processes for butadiene have all been replaced by extractive distillation processes.

U.S. Patents 2,369,559; 2,429,134 (Jasco): 2,788,378 (Polymer Corp.); 2,847,487; 2,985,697; 3,192,282 (Esso).
Morrell, C.E., Palz, W.J., Packie, J.W., Asbury, W.C., and Brown, C.L., *Trans. Am. Inst. Chem. Eng.*, 1946, **42**, 473.
Weissermel, K. and Arpe, H.-J., *Industrial Organic Chemistry*, 4th ed., translated by C.R. Hindley and S. Hawkins, Wiley-VCH, Weinheim, Germany, 2003, 110.

CAB *See* Steelmaking.

CAB-O-SIL Also written Cab-o-Sil. A vapor-phase process for making an AEROSIL silica powder by the flame hydrolysis of silicon tetrachloride. Also a trade name for the product. Developed by Cabot Corp. *See also* *AEROSIL.

CADRE A process for removing and oxidizing volatile organic compounds from gas streams. The compounds are adsorbed on a fixed bed of carbon and then desorbed by a stream of hot air or inert gas. Developed by Vard International, a division of Calgon Carbon Corporation.

Hydrocarbon Process. Int. Ed., 1993, **72**(8), 77.

CAFB [Chemically Active Fluidized-Bed] A coal gasification process intended for producing gas for power generation. Coal particles are injected into a shallow bed of lime particles that trap the sulfur dioxide. The bed particles are regenerated in a second fluidized bed, releasing the sulfur dioxide. Developed in the 1970s by the Esso Petroleum Company, United Kingdom, but not commercialized.

Hebden, D. and Stroud, H.J.F., *Chemistry of Coal Gasification*, 2nd Suppl. Vol., Elliott, M.A., Ed., John Wiley & Sons, New York, 1981, 1668.

Calcilox A process for converting calcium sulfate–sulfite wastes from *flue-gas desulfurization into a disposable, earthy material, by use of a proprietary inorganic additive made from blast furnace slag. Developed by Dravo Corporation of Pittsburgh, PA.

Labovitz, C. and Hoffman, D.C., *Toxic and Hazardous Waste Disposal*, Pojasek, R.B., Ed., Ann Arbor Science Publishers, Ann Arbor, MI, 1979, Vol. 1, Chap. 5.
IEA Coal Research, *The Problems of Sulphur*, Butterworths, London, U.K., 1989, 127.

Calcor A process for making carbon monoxide and hydrogen from natural gas or liquid petroleum gas. It combines *steam reforming with carbon dioxide recovery or recycle. Designed and licensed by Caloric GmbH. Fourteen commercial plants had been built by 2006.

Teuner, S., *Hydrocarbon Process. Int. Ed.*, 1985, **64**(5), 106; 1987, **66**(7), 52.
Hydrocarbon Process. Int. Ed., 1992, **71**(4), 90.

CALIPHOX A catalytic wet oxidation process for destroying organic wastes in water, using atmospheric oxygen as the oxidant. The organic material is preconcentrated by adsorption on activated carbon and then desorbed using hot water. Developed at the National Institute of Chemistry in Slovenia.

Pintar, A., *Catal. Today*, 2003, **77**(4), 451.

Calmet A process for extracting gold from its ores by pressure cyanidation. Developed by the Calmet Corporation in 1983.

Yannopoulos, J.C., *The Extractive Metallurgy of Gold*, Van Nostrand Reinhold, New York, 1991, 164.

Calorizing Also spelled calorising. A proprietary process for protecting the surface of iron or steel by applying a layer of aluminum. Several methods of application may be used: dipping, spraying, or chemical reaction with aluminum chloride. *See also* metal surface treatment.

Calsinter A process for extracting aluminum from fly ash and from *flue-gas desulfurization sludge. The ash is sintered with calcium carbonate and calcium sulfate at 1000°C–1200°C and then leached with sulfuric acid. Developed at Oak Ridge National Laboratory, United States, in 1976, but not known to have been piloted.

Felker, K., Seeley, F., Egan, Z., and Kelmers, D., *CHEMTECH*, 1982, **12**(2), 123.
O'Connor, D.J., *Alumina Extraction from Non-bauxitic Materials*, Aluminium-Verlag, Düsseldorf, Germany, 1988, 102, 262.

CAMERE [**CA**rbon dioxide hydrogenation to form **ME**thanol via a **RE**verse water-gas shift reaction] A process developed in Korea in the 1990s. The preferred catalyst is a mixed oxide of Cu–Zn–Zr–Ga.

Joo, O.-S., Jung, K.-D., Rozovski, Y.A., Lin, G.I., Han, S.-H., and Uhm, S.-J., *Ind. Chem. Res.*, 1999, **38**(5), 1808.

CAN [**C**alcium **A**mmonium **N**itrate] A process for making calcium ammonium nitrate fertilizer. Developed by BASF (hence the alternative process name BASF/CAN) and engineered by Uhde.

CANDID A process for making adiponitrile by reductive dimerization of acrylonitrile. Invented by ICI in 1976 and piloted in the United Kingdom from 1986, but not commercialized.

U.K. Patents 1,546,807; 1,547,431.
Eur. Chem. News, 1990, **54**(1420), 38.

CANMET Hydrocracking A process for demetallizing and converting heavy oils or refinery residues, in the presence of hydrogen and a proprietary additive, into naphtha, middle distillates, and gas oil. Originally developed in the 1970s by the Canada Center for Mineral and Energy Technology (CANMET), a division of the Department of Energy Mines and Resources, Canada; the process is now licensed by Partec Lavalin and Petro-Canada Products. A large demonstration plant, designed and built by Partec Lavalin in Petro-Canada's Montreal oil refinery, has operated successfully since 1986. A variation for treating extracts from tar sands was developed by Petro-Canada Exploration and the Department of Energy Mines and Resources and piloted in Canada in the 1980s. Another variation, for making diesel fuel from vegetable oils, was piloted in Vancouver in 1992.

Canadian Patents 1,094,492; 1,202,588; 1,151,579.
U.S. Patents 4,963,247; 4,969,988.
Menzies, M.A., Silva, A.E., and Denis, J.M., *Chem. Eng. (N.Y.)*, 1981, **88**(4), 46.
Silva, A.E., Rohrig, H.K., and Dufresne, A.R., *Oil Gas J.*, 1984, **82**(13), 81.
Chem. Eng. (N.Y.), 1992, **99**(5), 21.

CANON [**C**ompletely **A**utotrophic **N**itrogen Removal **O**ver **N**itrite] A process for removing nitrogen from sewage sludge in an anaerobic digester. Developed and piloted in the University of Santiago de Compostela from 2001.

Vazquez-Padin, J.R., Fernandez, I., Mosquera-Corral, A., Campos, J.L., and Mendez, R., in *Environmental Technologies to Treat Nitrogen Pollution*, Cervantes, F.J. Ed., IWA Publishing, London, U.K., 2009, 77, 85.

Cansolv A *flue-gas desulfurization process based on the selective absorption of sulfur dioxide in a proprietary, thermally regenerable, organic amine-based solution. Developed by Union Carbide Corporation; the first plant was originally planned for startup at Newburgh, IN, in 1994, although the process was abandoned in 1993. Now licensed by Cansolv Technologies Inc. *See* IIS.

Chem. Mark. Rep., 1991, 11 Nov, 5, 9 Dec, 29.
Kohl, H.L. and Nielsen, R.B., *Gas Purification*, 5th ed., Gulf Publishing Co., Houston, TX, 1997, 595.

CAOS [Continuous Addition Of Starter] A refinement of the *Impact process.

Chem. Week, 2000, **72**(1910), 41.

CAPTOR A modification of the *Activated Sludge sewage treatment system, in which the microorganisms are retained in a reticulated polyether foam. Invented in 1978 at UMIST, Manchester, and developed by Simon-Hartley, United Kingdom.

British Patent 2,006,181.
Cooper, P.F., *Topics in Wastewater Treatment*, Sidgewick, J.M., Ed., Blackwell Scientific, Oxford, U.K., 1985, 48.

CAR [Combined Autothermal Reforming] A *reforming process for making *syngas from light hydrocarbons, in which the heat is provided by partial oxidation in a section of the reactor. Developed by Uhde and commercialized at an oil refinery at Strazske, Slovakia, in 1991.

Chem. Eng. (N.Y.), 1992, **99**(5), 33.
Babik, A. and Kurt, J., *Oil Gas J.*, 1994, **92**(12), 66.

Carbacell [Carbamate cellulose] A process for making rayon filament and staple fiber. Cellulose is reacted with urea in an inert organic solvent at a high temperature to yield cellulose carbamate. This process avoids the environmental problems caused by carbon disulfide in the *Viscose process. Developed by Zimmer in the 1990s and piloted in Germany and Poland. Commercialization was expected by 1999.

Chem. Week, 1997, **159**(25), 21.

Carbochlorination A general name for processes that convert metal oxides to chlorides by heating them with carbon in a chlorine atmosphere. *See* Chloride.

Carbo-Flo An integrated process for treating small volumes of effluent containing agrochemicals or other waste organic materials. Flocculation by proprietary chemicals is used, followed by sand filtration and activated carbon treatment. Developed by ICI in the mid-1980s.

Harris, D.A., Johnson, K.S., and Ogilvy, J.M.E., *Env. Protect. Bull.*, No. 017, I. Chem. E., Rugby, U.K., 1992, 23.

carbonate-looping *See* *CLC.

carbonization A general term for the heat treatment of coal to produce gases, industrial cokes, and domestic smokeless fuels. Many such processes have been developed and many of them have special names. Most of these are outside the scope of this work, but the principal ones are Ancit, Anthracine, Anthracoke, AUSCOKE [AUStralian COKE], BF [Bergbau-Forschung], BFL [Bergbau-Forschung–Lurgi], Brennstoff-Technik, Carbocite, Carbolux, Carmaux, CCC [Consolidation Coal Company], CFRI [Central Fuel Research Institute], Charfuel, Clean Coke, Consol BNR, Coppee [Coppee-Houillères du Bassin du Nord et du Pas de Calais], CRIJ [Coal Research Institute of Japan], DKS [Didier Keihan Sumitomo], Dr. C. Otto, EBV [Eschweiler Bergwerks-Verein], FMC [Food Machinery and Chemical Corp.], Formcoke, GI, HBN [Houillères du Bassin du Nord], HBNPC [Houillères du Bassin du Nord et du Pas de Calais], IGI, INICHAR [Institut National de l'Industrie Charbonnière], INIEX (a development of INICHAR), ISCOR [Iron and Steel Industrial CORporation], Krupp–Lurgi, NCB [National Coal Board], NIPR [National Institute for Pollution and Resources], OZIOLE (same as HBN), Sapozhnikov, Schenk–Wenzel, Stamicarbon, Stevens, Synthracite, Taciuk, Weber, and Wisner.

Denig, F., in *Chemistry of Coal Utilization*, Vol. 1, Lowry, H.H., Ed., John Wiley & Sons, New York, 1945, 774.
Wilson, P.J., Jr. and Clendenin, J.D., in *Chemistry of Coal Utilization*, Suppl. Vol., Lowry, H.H., Ed., John Wiley & Sons, New York, 1945, 395.
Schinzel, W., in *Chemistry of Coal Utilization*, 2nd Suppl. Vol., Elliott, M.A., Ed., John Wiley & Sons, New York, 1981, Chap. 11.

CARBONOX [**CARBO**n-based **NO**$_x$ reduction technology] A clean-coal process developed at the Ohio State University from 2005.

Chem. Eng. News, 2006, **84**(6), 34.

Carbonyl *See* Mond Nickel.

Carbosolvan One of the several processes for absorbing carbon dioxide from gases, using hot potassium carbonate solution. *See also* Benfield, Carsol, CATACARB, Giammarco-Vetrocoke, HiPure.

Linsmayer, S., *Chem. Tech. (Leipzig)*, 1972, **24**(2), 74.

Carbotherm *See* DR.

Carbo-V A process for gasifying biomass developed by Choren Industries in Freiberg. The process was invented in 2006 and sold to Linde Engineering, Dresden, in 2012. A demonstration plant was operated in Freiberg between 2008 and 2009 and produced SunDiesel fuel. The feed can be straw, wood chips, or energy plants. The system comprises a low-temperature gasifier, a high-temperature gasifier, and a system for recycling the coke dust. The product gases are converted to hydrocarbons by the *Fischer–Tropsch process. Choren filed for bankruptcy in 2011 and Linde has not commercialized the process.

U.S. Patents 7,972,392; 7,776,114.
Eur. Chem. News, 2005, **83**(2160), 27.

carbureted water gas *See* Water Gas.

Carburol An early thermal process for cracking petroleum.

Asinger, F., *Mono-olefins: Chemistry and Technology*, translated by B.J. Hazzard, Pergamon Press, Oxford, U.K., 1968, 322.

Carinthian An obsolete lead-smelting process, first operated at Bleiberg, Carinthia, Austria.

Carix An ion-exchange process for purifying water, in which regeneration is accomplished with carbon dioxide. Developed in the 1980s by Kernforschungszentrum, Karlsruhe, Germany.

Ullmann's Encyclopedia of Industrial Chemistry, 5th ed., Vol. A14, VCH Publishers, Weinheim, Germany, 1989, 442.

Carl Still (1) A *Hydrofining process.

Claxton, G., *Benzoles: Production and Uses*, National Benzole & Allied Products Association, London, U.K., 1961, 452.

Carl Still (2) A process for removing hydrogen sulfide from coke-oven gas by scrubbing with aqueous ammonia, itself derived from coke-oven gas. Developed in the 1970s by Firma Carl Still, Germany. Operated at the Armco steel mill at Middletown, OH. *See also* DIAMOX, Still Otto.

British Patent 1,348,937.
Kohl, H.L. and Nielsen, R.B., *Gas Purification*, 5th ed., Gulf Publishing Co., Houston, TX, 1997, 322.

Carnol A hypothetical process for converting carbon dioxide from power station gases into methanol and elemental carbon. In the first step, methane is thermally decomposed into its elements. The carbon is stored, and in the second step, the hydrogen is reacted with more carbon dioxide to produce methanol. Claimed to be cheaper than disposing of the carbon dioxide in the ocean. Developed in the 1990s at the Brookhaven National Laboratory, New York. *See also* Hydrocarb, Hynol.

Steinberg, M., Yuang, D., and Borgwardt, R.H., in *Carbon Dioxide Chemistry: Environmental Issues*, Paul, J. and Pradier, C.H., Eds., Royal Society of Chemistry, Cambridge, U.K., 1994, 189.
Steinberg, M. and Dong, Y., *Int. J. Power Energy Systems*, 2004, **24**(3), 197.

Carom [**Car**bide **arom**atics extraction] A two-stage process for removing aromatic hydrocarbons from petroleum refining streams. In the first stage, the aromatics are removed by liquid–liquid extraction with a proprietary solvent (a mixture of polyalkylene glycols and a glycol ether) at ambient temperature. In the second stage, the aromatics are stripped from the solvent by steam distillation. Developed by Union Carbide Corporation; first commercialized in 1986 and now licensed by UOP. Seven units had been licensed by 2002.

U.S. Patents 4,498,980; 5,022,981.
Stoodt, T.J. and Negiz, A., in *Handbook of Petroleum Refining Processes*, 3rd ed., Meyers, R.A., Ed., McGraw-Hill, New York, 2003, 2.23.

Caron A process for extracting nickel and cobalt from lateritic ores by reductive roasting, followed by leaching with ammoniacal ammonium carbonate solution in the presence of oxygen. Developed by M.H. Caron at The Hague in the 1920s and used in Cuba (where the location of the mine is named Nicaro, after the metal and the inventor) and in Australia.

U.S. Patent 1,487,145.
Gupta, C.K. and Mukherjee, T.K., *Hydrometallurgy in Extraction Processes*, Vol. 1, CRC Press, Boca Raton, FL, 1990, 144.

Carpenter–Evans A catalytic process for removing organic sulfur compounds from synthesis gas by hydrogenation to hydrogen sulfide, which is absorbed by iron oxide. The hydrogenation catalyst is nickel subsulfide, Ni_3S_2. Invented by E.V. Evans and C.C. Carpenter in England around 1913 and operated in three commercial plants.

Carpenter, C.C., *J. Gas Light*, 1913, **122**, 1010; 1913, **123**, 30.
Evans, F.V., *J. Soc. Chem. Ind. (London)*, 1915, **34**, 9.
Kohl, H.L. and Nielsen, R.B., *Gas Purification*, 5th ed., Gulf Publishing Co., Houston, TX, 1997, 1168.

Carrousel An unconventional aerobic treatment system for sewage and industrial effluents, providing efficient oxygenation, mixing, and quiescent flow in an elliptical aeration channel fitted with baffles. Developed in the Netherlands by DHV Raadgevend Ingenieursbureau B.V. and licensed in the United Kingdom by Esmil.

Carsol A process for removing carbon dioxide from gas streams by scrubbing with aqueous potassium carbonate. *See also* Benfield, CATACARB, Giammarco-Vetrocoke, HiPure.

Carter Also known as H.T.S. Carter. A process for making basic lead carbonate pigment (white lead). Lead monoxide, in a slowly revolving drum, is moistened and sprayed with acetic acid. Carbon dioxide is then introduced. Carbonation is subsequently completed in a separate vessel. *See also* Dutch, Thompson–Stewart.

Dunn, E.J., Jr., in *Treatise on Coatings*, **3**(1), Meyers, R.R. and Long, J.S., Eds., Marcel Dekker, New York, 1975, 333.

Carus A process for making potassium permanganate by reacting manganese dioxide with molten potassium hydroxide, in air. Invented by M.B. Carus in 1958 and operated by the Carus Chemical Company at La Salle, IL.

U.S. Patents 2,848,537; 2,940,821; 2,940,822; 2,940,823; 3,172,830.

CAS *See* Steelmaking.

Casale The first synthetic ammonia process, designed by L. Casale, who founded Ammonia Casale of Lugano, Switzerland, in 1921. *See* Claude (1).

> Vancini, C.A., *Synthesis of Ammonia*, translated by L. Pirt, Macmillan Press, Basingstoke, England, 1971, 245.
> *Nitrogen*, 1996, (223), 25.

Cashman A high-pressure process for extracting gold from arsenic-bearing ores, concentrates, and flue dusts.

> Yannopoulos, J.C., *The Extractive Metallurgy of Gold*, Van Nostrand Reinhold, New York, 1991, 103.

Castner (1) A process for making graphite articles, invented by H.Y. Castner in 1893. It uses lengthwise graphitization, unlike the *Acheson (2) process, which uses transverse graphitization.

> U.S. Patent 572,472.

Castner (2) A process for making sodium cyanide. Sodamide is first made by passing ammonia gas over molten sodium. The molten sodamide is then poured over red-hot charcoal, which converts the sodamide first to sodium cyanamide and then to sodium cyanide:

$$2Na + 2NH_3 = 2NaNH_2 + H_2$$

$$2NaNH_2 + C = Na_2CN_2 + 2H_2$$

$$Na_2CN_2 + C = 2NaCN$$

Invented by H.Y. Castner in 1894. Operated first at Frankfurt-am-Main, Germany, in 1899, and thereafter in several other countries, until abandoned in the 1960s in favor of the *neutralization process.

> British Patents 12,218; 12,219 (1894).

Castner (3) A process for making sodium by reducing sodium hydroxide with iron carbide:

$$6NaOH + FeC_2 = 2Na + 2Na_2CO_3 + 3H_2 + Fe$$

Invented by H.Y. Castner and operated by the Aluminium Company at Oldbury, England, in 1888, in order to supply sodium for the manufacture of aluminum. It was abandoned soon after, when the *Hall–Héroult process for aluminum was developed.

> U.S. Patent 342,897.
> Hardie, D.W.F., *A History of the Chemical Industry in Widnes*, Imperial Chemical Industries, Widnes, U.K., 1950, 184.
> Sittig, M., *Sodium: Its Manufacture, Properties and Uses*, Reinhold Publishing, New York, 1956.
> Adam, D., *Chem. Br.* 1999, 36.

Castner (4) A process for making sodium by electrolyzing fused sodium hydroxide. Used in the United Kingdom from the early 1900s until 1952.

> British Patent 13,356 (1890).
> Hardie, D.W.F. and Pratt, J.D., *A History of the Modern British Chemical Industry*, Pergamon Press, Oxford, U.K., 1966, 90.

Castner–Kellner Also called the Chlor-Alkali process. A process for making chlorine and sodium hydroxide by the electrolysis of aqueous sodium chloride in a cell having a mercury cathode. Invented independently in 1892 by H.Y. Castner, an American chemist working in Birmingham, England, and C. Kellner in Austria. First operated in the United States at Saltville, VA, in 1896 and in England by the Castner–Kellner Alkali Company in Runcorn in 1897. Of major importance worldwide in the first half of the twentieth century. Concerns over mercury pollution caused by effluents from this process caused the abandonment of many plants in the 1970s, to be replaced by various diaphragm-based electrolytic processes. *See* diaphragm cell, membrane cell.

British Patents 16,046 (1892); 10,584 (1893).

Fleck, A., *Chem. Ind. (London)*, 1947, **66**, 515.

Sittig, M., *Sodium: Its Manufacture, Properties, and Uses*, Reinhold Publishing, New York, 1956, 21.

Hardie, D.W.F., *Electrolytic Manufacture of Chemicals from Salt*, Oxford University Press, London, U.K., 1959, 19.

MacMullin, R.B., in *Chlorine: Its Manufacture, Properties, and Uses*, Sconce, J.S., Ed., Reinhold Publishing, New York, 1962, 127.

Hocking, M.B., *Modern Chemical Technology and Emission Control*, Springer-Verlag, Berlin, Germany, 1984, 141.

Sella, A., *Chem. World*, 2013, **10**(3), 62.

Cataban A process for removing small amounts of hydrogen sulfide from industrial gas streams by oxidation, in aqueous solution, to elemental sulfur. The oxidant is the ferric ion, in a proprietary chelated form. The solution is regenerated by atmospheric oxygen. The overall reactions are

$$2Fe^{3+} + H_2S = 2Fe^{2+} 2H^+ + S$$

$$2Fe^{2+} + \frac{1}{2}O_2 + 2H^+ = 2Fe^{3+} + H_2O$$

Developed by Rhodia, New York, in the 1970s but abandoned by 1997.

Davis, J.C., *Chem. Eng. (N.Y.)*, 1972, **79**(11), 66.

Kohl, H.L. and Nielsen, R.B., *Gas Purification*, 5th ed., Gulf Publishing Co., Houston, TX, 1997, 804.

CATACARB [**CATA**lyzed removal of **CARB**on dioxide] A process for removing carbon dioxide and hydrogen sulfide from gas streams by absorption in hot potassium carbonate solution containing a proprietary catalyst. Developed and licensed by Eickmeyer and Associates, KS, based on work at the U.S. Bureau of Mines in the 1950s. More than 100 plants were operating in 1997. *See also* Benfield, Carsol, HiPure, Giammarco-Vetrocoke.

U.S. Patents 3,851,041; 3,932,582.

Eickmeyer, A.G., *Chem. Eng. Prog.*, 1962, **58**(4), 89.

Gangriwala, H.A. and Chao, I.-M., in *Acid and Sour Gas Treating Processes*, Newman, S.A., Ed., Gulf Publishing Co., Houston, TX, 1985, 370.

Hydrocarbon Process. Int. Ed., 1992, **71**(4), 91.

Kohl, H.L. and Nielsen, R.B., *Gas Purification*, 5th ed., Gulf Publishing Co., Houston, TX, 1997, 363.

Catacol A catalytic distillation process for petroleum refining, offered by Axens.

Catadiene [**Cata**lytic butadiene] Also spelled Catadien. A version of the *Houdry process for converting mixtures of butane isomers into butadiene by dehydrogenation over an alumina–chromia catalyst. Another version converts propane to propylene. Rapid coking of the catalyst necessitates use of several reactors in parallel, so that reactivation can be carried out continuously. Developed by Houdry and first operated at El Segundo, CA, in 1944. By 1993, 20 plants had been built worldwide. Now licensed by ABB Lummus Global.

Unzelman, G.H. and Wolf, C.J., in *Petroleum Processing Handbook*, Bland, W.F. and Davidson, R.L., Eds., McGraw-Hill, New York, 1967, 149.

Craig, R.G. and Dufallo, J.M., *Chem. Eng. Prog.*, 1979, **75**(2), 62.

Craig, R.G. and Spence, D.C., in *Handbook of Petroleum Refining Processes*, 2nd ed., Meyers, R.A., Ed., McGraw-Hill, New York, 1986, 4–3.

Hydrocarbon Process. Int. Ed., 1991, **70**(3), 142.

Weissermel, K. and Arpe, H.-J., *Industrial Organic Chemistry*, 4th ed., translated by C.R. Hindley and S. Hawkins, Wiley-VCH, Weinheim, Germany, 2003, 112.

CATADOX An oxidation process for destroying carbon compounds in wastewaters. The oxidants are ozone, hydrogen peroxide, and optionally a catalyst. Developed by ESCO International, United Kingdom.

Pollution Solutions, Sep 2009, **3**(3), 16.

Catalan An early ironmaking process in which selected ores were reduced with charcoal and the slag was expelled from the product by hammering while hot. *See also* Bloomery.

Catalloy A gas-phase process for making olefin copolymers, using *Ziegler–Natta catalysts. Catalloy uses a series of three gas-phase reactors, to which monomer is progressively added. The properties of the product can be varied according to the monomer grades used. Developed by Himont and first commercialized in 1990. Now operated by a joint venture of Montell Polyolefins and Japan Polyolefins. *See also* Hivalloy.

Chem. Week, 1992, 13 May, 53.
Plastiques Modernes et Elastomères, 1997, **4**(9), 31.

Catalyst A process for treating industrial off-gases by catalytic oxidation over a titania–vanadia catalyst. Offered by BASF.

Oil Gas J., 1 Oct 1990, (Suppl.), 51.

Catalytica A catalytic process for making methanol from methane via methyl hydrogen sulfate. Both yield and selectivity are good, but the process requires the recycle of large quantities of sulfuric acid and so has proved uneconomic and has never been operated. The catalyst is a platinum complex. If a catalyst could be devised that would make the process a hundred times faster, the process might be economic.

Aasberg-Petersen, K., Hansen, J.-H.B., Christensen, T.S., Dybkjaer, I., Christensen, Madsen, W., and Rostrup-Nielsen, J.R., *Appl. Catal. A: Gen.*, 2001, **221**(1–2), 379.
Wittcoff, H.A., Reuben, B.G., and Plotkin, J.S., *Industrial Organic Chemicals*, 2nd ed., Wiley-Interscience, Hoboken, NJ, 2004, 389.
Conley, B.L., Tenn III, W.J., Young, K.J.H., Ganesh, S.K., Meier, S.K., Ziatdinov, V.R., Mironov, O., Oxgaard, J., Gonzales, J., Goddard III, W.A., and Periana, R.A., *J. Mol. Catal.*, 2006, **251**(1,2), 14.

Catalytic Condensation Also known colloquially as CATCON. A process for oligomerizing olefins or for alkylating aromatic hydrocarbons with olefins. The catalyst is a solid that contains free or combined phosphoric acid. Developed by UOP.

Chem. Eng. (Rugby, Engl.), 1991, (489), 12.

catalytic cracking A process used in petroleum refining to convert high-boiling hydrocarbon fractions to lower-boiling fractions suitable for use as gasoline. First operated on a large scale in 1937 at Sun's Marcus Hook refinery. The first catalysts were natural clays; these were later replaced by synthetic sodium aluminosilicates and then by zeolites, notably zeolite Y. Those processes with special names that are described in this dictionary are Airlift, ARS, Catarole, Demet, Dynacracking, FCC, Houdresid, Houdriflow, Houdriforming, Houdry, MDDW, MSCC, Orthoflow, R2R, SBA-HT, Suspensoid, TCC, Thermofor, THERMOCAT, Ultra-Orthoflow, and Veba-Combi Cracking.

Unzelman, G.H. and Wolf, C.J., in *Petroleum Processing Handbook*, Bland, W.F. and Davidson, R.L., Eds., McGraw-Hill, New York, 1967, Chap. 3.1.
The Petroleum Handbook, 6th ed., Elsevier, Amsterdam, the Netherlands, 1983, 284.

Catalytic Dewaxing Also called CDW. A *hydrocracking process for removing waxes (linear aliphatic hydrocarbons) from petroleum streams by converting them to lower molecular weight hydrocarbons. The catalyst is a synthetic mordenite. Developed by BP. Two units were operating in 1988.

Hardgrove, J.D., in *Encyclopedia of Chemical Processing and Design*, McKetta, J.J. and Cunningham, W.A., Eds., Marcel Dekker, New York, 1967, **15**, 346.
Hydrocarbon Process. Int. Ed., 1988, **67**(9), 69.

catalytic distillation A generic term for processes in which the packing of a distillation column is also a catalyst for the reaction. Developed by CDTECH. In 1996, 54 units were in operation, making several ethers for use as fuel additives. *See also* CD-Cumene, CDETHEROL, CD-HDS, CDHydro, RWD.

U.S. Patent 4,443,559.
Hydrocarbon Process. Int. Ed., 1996, **75**(11), 113.

Catalytic Fast Pyrolysis *See* *CFP.

catalytic hydrogenation In the context of the coal and petroleum industries, this term can mean the conversion either of aromatic compounds to alicyclic compounds or of olefins to saturated aliphatic hydrocarbons. Such processes with special names that are described in this dictionary are BASF/Scholven, COIL, Hydra, Hydrobon, Hydropol, Lignol, Normann, SHP, Unifining, and Unionfining.

catalytic reforming A process for converting linear aliphatic hydrocarbons into a mixture of branched-chain aliphatic hydrocarbons and aromatic hydrocarbons, in order to increase the octane rating of the product. First operated in 1940 at the refinery of the Pan American Oil Refining Corporation in Texas City, using a molybdena–alumina catalyst. Further developed by V. Haensel at Universal Oil Products in the 1940s and still widely used. The catalyst is usually a platinum metal supported on alumina. Hydrogen is a by-product. Those processes with special names that are described in this dictionary are Aromizing, Catforming, Cycloversion, Hydroforming, Iso-Plus Houdriforming, Magnaforming, Octanizing, Orthoforming, Platforming, Powerforming, Rheniforming, SBK, Silamit P3, Sinclair-Baker, SMDS, Sovaforming, SSC, STAR, UGI, and Ultraforming.

Chem. Br., 1981, Nov, 536.
Little, D.M., *Catalytic Reforming*, PennWell Publishing, Tulsa, OK, 1985.
Menon, P.G. and Paal, Z., *Ind. Eng. Chem. Res.*, 1997, **36**(8), 3282.
Wikipedia, The Free Encyclopedia. Catalytic reforming. Revised 17 Jul 2013. Retrieved 20 Jul 2013.

Catapac A technique for loading additional amounts of catalyst in a batch reactor. Developed by Axens.

Catarole Also spelled Catarol. A process for making aromatic hydrocarbons and olefins by cracking petroleum fractions over copper turnings. Invented by C. Weizmann in England in 1940 and developed by Petrochemicals, which used it from 1947 in its refinery at Carrington, United Kingdom, to make ethylene, propylene, and a range of aromatic hydrocarbons.

British Patents 552,115; 552,216; 575,383.
Steiner, H., *J. Inst. Pet.*, 1947, **33**, 410.
Weizmann, C., Bergmann, E., Huggett, W.E., Steiner, H., Sulzbacher, M., Parker, D., Michaelis, K.O., Whincup, S., and Zimkin, F., *Ind. Eng. Chem.*, 1951, **43**, 2312.
Claxton, G., *Benzoles: Production and Uses*, National Benzole & Allied Products, London, U.K., 1961, 90.
King, R., *Chem. Ind. (London)*, 1989, 309.

Catasulf A catalytic process for converting hydrogen sulfide in gas streams to elemental sulfur. The gas, to which a stoichiometric quantity of air or oxygen has been added, is passed over the hot catalyst. Invented in 1983 by BASF. One plant had been built by 1990.

U.S. Patent 4,507,274.
Hydrocarbon Process. Int. Ed., 1992, **71**(4), 91.

CATAZONE [CATAlyzed o**ZONE]** A process for removing traces of organic compounds from groundwater by catalyzed oxidation with ozone. The catalyst is titanium dioxide, and hydrogen peroxide may be added as well. Invented in 1994 by H. Paillard.

European Patent 436,409.
Masten, S.J. and Davies, S.H.R., in *Environmental Oxidants*, Nriagu, J.O. and Simmons, M.S., Eds., John Wiley & Sons, New York, 1994, 534.

CATCON [CATalytic **CON**densation] *See* Catalytic Condensation.

Catforming [Catalytic re**forming]** A *Catalytic Reforming process using a platinum catalyst on a silica–alumina support. Developed by the Atlantic Refining Company and first operated in 1952.

Unzelman, G.H. and Wolf, C.J., in *Petroleum Processing Handbook*, Bland, W.F. and Davidson, R.L., Eds., McGraw-Hill, New York, 1967, 3–27.
Little, D.M., *Catalytic Reforming*, PennWell Publishing, Tulsa, OK, 1985, xv.

Cativa A process for making acetic acid by reacting methanol with carbon monoxide (carbonylation). The catalyst is an iridium complex with carbon monoxide, promoted with ruthenium. Developed jointly by BP Chemicals, Hull, United Kingdom, and the University of Sheffield. First announced in 1996 and installed between 1995 and 1999 in four plants that had been using the former *Monsanto Acetic Acid process. The first plant designed for the process was built by BP Petronas in Malaysia in 2000. A joint venture of BP with Sinopec used the process in a plant expansion in Chongqing, China, in 2005, and another plant was planned for Nanjing, for completion in 2007. Seven plants were operating in 2012.

Chem. Eng. (N.Y.), 1996, **103**(8), 23.
Process Eng., 1996, **77**(7), 21.
Watson, D.J., in *Catalysis of Organic Reactions*, Herkes, F.E., Ed., Marcel Dekker, New York, 1998, 369.
Howard, M.J., Sunley, G.J., Poole, A.D., Watt, R.J., and Sharma, B.K., *Stud. Surf. Sci. Catal.*, 1999, **121**, 61.
Sunley, G.J. and Watson, D.J., *Catal. Today*, 2000, **58**(4), 293.
Jones, J.H., *Platinum Metals Rev.*, July 2000, **44**(3), 94.
Eur. Chem. News, 2002, **77**(2026), 38.
Wikipedia, The Free Encyclopedia. Cativa process. Revised 28 Feb 2013. Retrieved 20 Jul 2013.

CATnap A process for passivating *hydrotreating and *hydrocracking catalysts in petroleum refining. A proprietary mixture of high molecular weight aromatic compounds is added to the catalyst reactors; this forms an inert film on the surfaces of the catalyst and the hardware, preventing oxidation when the catalyst is discharged. Developed in the 1980s by Kashima Engineering Company and Softard Industries.

Appl. Catal. A: Gen., 1993, **102**(2), N19.

CATOFIN [**CAT**alytic **OleFIN**] A version of the *Houdry process for converting mixtures of C_3 to C_5 saturated hydrocarbons into olefins by catalytic dehydrogenation. The catalyst, made by Süd-Chemie, is chromia on alumina in a fixed bed. The reactors operate batchwise, cycling between reaction, purging, and regeneration. Continuous operation is achieved by using several reactors in parallel. Developed by Air Products and Chemicals and licensed through ABB Lummus Global. Fourteen plants were operating worldwide in 2012.

Craig, R.G. and White, E.A., *Hydrocarbon Process. Int. Ed.*, 1980, **59**(12), 111.
Wett, T., *Oil Gas J.*, 1985, **83**(35), 46.
Craig, R.G. and Spence, D.C., in *Handbook of Petroleum Refining Processes*, 2nd ed., Meyers, R.A., Ed., McGraw-Hill, New York, 1986, 4–3.
Hu, Y.C., in *Chemical Processing Handbook*, Marcel Dekker, New York, 1993, 806.
Chem. Week, 1994, **154**(10), 38.
Sanfillipo, D. and Miracca, I., *Catal. Today*, 2006, **111**(1–2), 136.

CATOX [**CAT**alytic **OX**idation] A process for removing organic solvents and other odorous compounds from gaseous effluents, using catalytic oxidation over a proprietary catalyst. The original catalyst was described as a nonprecious metal, mixed-oxide catalyst, known as CK-302; the next generation of catalyst contained both a base metal and a noble metal. Licensed and supplied by Haldor Topsoe and used in the food processing, lacquering, printing, and chemical industries. Related processes are *REGENOX, which uses regenerative heat exchange, and *ADOX. Unrelated to *Watercatox. The name is used also by Nippon Shokubai Company as a trade name for distributed digital control systems for chemical plants.

Kohl, H.L. and Nielsen, R.B., *Gas Purification*, 5th ed., Gulf Publishing Co., Houston, TX, 1997, 1157.

Cat-Ox [**Cat**alytic **Ox**idation] An adaptation of the *Contact process for making sulfuric acid, using the dilute sulfur dioxide in flue gases. A conventional vanadium pentoxide catalyst is used. Developed by Monsanto Enviro-Chemical Systems and operated in Pennsylvania and Illinois in the early 1970s.

Stites, J.G., Jr., Horlacher, W.R., Jr., Bachover, J.L., Jr., and Bartman, J.S., *Chem. Eng. Prog.*, 1969, **65**(10), 74.
Miller, W.E., *Chem. Eng. Prog.*, 1974, **70**(6), 49.
IEA Coal Research, *The Problems of Sulphur*, Butterworths, London, U.K., 1989, 124.
Speight, J.G., *Gas Processing*, Butterworth Heinemann, Oxford, U.K., 1993, 278.
Kohl, H.L. and Nielsen, R.B., *Gas Purification*, 5th ed., Gulf Publishing Co., Houston, TX, 1997, 643.

Catoxid [**Cat**alytic **oxid**ation] A process for destroying organic chlorine compounds, especially from the production of vinyl chloride, by catalytic oxidation in a fluid bed. Developed by the B.F. Goodrich Company in Akron, OH.

German Patents 2,531,981; 2,532,027; 2,532,043; 2,532,052; 2,532,075.
Benson, J.S., *Hydrocarbon Process. Int. Ed.*, 1979, **58**(10), 107.

Catpoly A polymerization process that makes linear olefins for use in making ethers to use as gasoline additives. The catalyst is phosphoric acid on kieselguhr. Developed in 1935 by UOP and later supplanted by IFP's *Polynaphtha Essence process.

Ipatieff, V.N., Corson, B.B., and Egloff, G., *Ind. Eng. Chem.*, 1935, **27**, 1077.
McMahon, J.F., Bednars, C., and Solomon, E., *Adv. Petr. Chem.*, 1963, VII, 285.
Chauvel, A. and G. Lefebvre, G., *Petrochemical Processes*, Vol. 1, Technip, Paris, France, 1992.

Catstill [**Cat**alytic **Distill**ation] A catalytic distillation process for making high-octane gasoline from off-gases from a catalytic cracker. It can also be used for making a mixture of benzene, toluene, and xylenes from the off-gases from a *steam reformer. A demonstration plant was operated at a refinery in the Midwestern United States in 1988 to 1989. Developed and offered for license by Chemical Research and Licensing Company, Houston, TX, now CDTECH.

Chem. Eng. (N.Y.), 1989, **96**(9), 41.

causticization Also called the lime-soda process. A general name for the generation of sodium hydroxide by reacting sodium carbonate with calcium hydroxide.

$$Na_2CO_3 + Ca(OH)_2 = 2NaOH + CaCO_3$$

The process is operated at 80°C–90°C, with a slight excess of the calcium hydroxide. This was the only method used for making sodium hydroxide after the invention of the *Leblanc process and before the introduction of the *Castner–Kellner process around 1890. The process is still used when the demands for chlorine and sodium hydroxide from the Castner–Kellner process are unbalanced and for the regeneration of waste pulping liquors in the manufacture of pulp and paper by the *Kraft process. *See also* Löwig.

Hocking, M.B., *Modern Chemical Technology and Emission Control*, Springer-Verlag, Berlin, Germany, 1985, 129.
Chem. Ind. (London), 1991, 455.

Cazo [Spanish, a copper vessel] An ancient process for extracting silver from sulfide ores. The ore was boiled in a copper pot with salt and water; addition of mercury gave silver amalgam. The copper served as the reducing agent. Described in 1640 by A.A. Barba, who claimed that it had been operated since 1590. Around 1800, this process developed into the Fondon process, in which the raw materials were ground together.

Mellor, J.W., *Comprehensive Treatise on Inorganic and Theoretical Chemistry*, Longmans, Green & Co., London, U.K., 1923, **3**, 303.
Dennis, W.H., *A Hundred Years of Metallurgy*, Gerald Duckworth, London, U.K., 1963, 285.

CBA [**C**old **B**ed **A**dsorption] A variation of the *Claus process, in which the sulfur product is desorbed from the catalyst by a side stream of hot gas from the main process. Developed by AMOCO Canada Petroleum Company and operated in Alberta.

Goddin, G.S., Hunt, E.B., and Palm, J.W., *Hydrocarbon Process. Int. Ed.*, 1974, **53**(10), 122.
Nobles, J.E., Palm, J.W., and Knudtson, D.K., *Hydrocarbon Process. Int. Ed.*, 1977, **56**(7), 143.
Kohl, A.L. and Riesenfeld, F.C., *Gas Purification*, 4th ed., Gulf Publishing Co., Houston, TX, 1985, 449.

C$_4$ Butesom [**But**ene **isom**erization] A process for isomerizing linear butenes to isobutene, catalyzed by a zeolite. The isobutene is intended for use as an intermediate in the production of ethers for use as fuel additives. Developed by UOP in 1992. *See also* C$_5$ Pentesom.

Krupa, S., in *Handbook of Petroleum Refining Processes*, 3rd ed., Meyers, R.A., Ed., McGraw-Hill, New York, 2003, 13.14.

CCC *See* carbonization.

CCG [Catalytic Coal Gasification] A generic name. All such processes require very cheap catalysts. Exxon Engineering Corporation developed such a process in the 1980s, which used a catalyst based on potassium carbonate. Tohoku University, Japan, developed another process using iron salts deposited on coal.

Gallagher, J.E., Jr. and Euker, C.A., Jr., *Energy Res.*, 1980, **4**, 137.
Hirsch, R.L., Gallagher, J.E., Jr., Lessard, R.R., and Wesselhoft, R.D., *Science*, 1982, **251**, 121.
Ohtsuka, Y. and Asami, K., *Catal. Today*, 1997, **39**, 111.

CCL [Catalytic Coal Liquids] A catalytic process developed by Gulf Oil Corporation. The main objective is the production of clean-burning liquid fuels for power plants.

Crynes, B.L., in *Chemistry of Coal Utilization*, Elliott, M.A., Ed., John Wiley & Sons, New York, 1981, 2005.

CCLC [Cascading Closed-Loop Cycle] A heat recovery system using propane as the working fluid. A commercial demonstration was planned for 2006. *See* FFGC.

Chem. Eng. (N.Y.), 2006, **113**(1), 14.
U.S. Patents 6,857,268; 7,096,665.

CCOP [Chlorine-Catalyzed Oxidative Pyrolysis] A process for converting methane into a mixture of ethylene and acetylene. Invented by the Illinois Institute of Technology, Chicago, and under development by Dow Chemical Company in 1991.

Chem. Eng. (N.Y.), 1990, **97**(2), 17.

CCR [Continuous Catalyst Regeneration] A process for regenerating the catalysts in various petroleum reforming processes by continuously withdrawing a portion of the catalyst and regenerating it in a separate (continuous) reactor. Originally introduced by IFP (now Axens) but subsequently adopted and modified by other process developers, including UOP. UOP used vertically stacked regeneration vessels and incorporated CCR into their *Platforming, *Cyclar, and *Oleflex processes. IFP used side-by-side reactors. A further refinement, Sequential Continuous Catalytic Reforming, was first commercialized by UOP in Utah in 2005.

Johnson, J.A., in *Handbook of Petroleum Refining Processes*, 3rd ed., Meyers, R.A., Ed., McGraw-Hill, New York, 2003, 2.5.
Zhou, L., in *Handbook of Petroleum Refining Processes*, 3rd ed., Meyers, R.A., Ed., McGraw-Hill, New York, 2003, 2.35.
Chem. Ind. (London), 2005, (10), 7.

CCU [Catalytic Crude Upgrading] A process developed jointly by UOP, Petrobras, and Albemarle from 2008. UOP provided the technology, equipment, and system design. Albemarle provided an improved *FCC catalyst. Petrobras had already run the process in its pilot plant.

Oil Gas J., 2008, **103**(38), 10.

CD [Catalytic Distillation] A range of petrochemical processes based on *catalytic distillation, developed and offered by Catalytic Distillation Technologies (CDTECH). They include CD*Alky*, CD*Etbe*, CD*Etherol*, CD*Ethers*, CD*IB*, CD*Isotame*, and CD*Mtbe*.

CDAlky An *alkylation process conducted with sulfuric acid in a packed column at a lower temperature than conventional alkylation.

Chem. Eng. (N.Y.), 2013, **120**(6), 11.

CDAlkyl An *alkylation process developed by CB&I.

Chem. Eng. (N.Y.), 2013, **120**(6), 11.

CDAlkyPlus An improved alkylation method that allows operators to alkylate any concentration of isobutylene as much as 100% of the total olefin fraction. The process is said to provide considerable savings in operating and capital costs. It utilizes two commonly used refinery processes that allow inexpensive alkylation of isobutylene as the only olefin source. A simple isobutylene pretreatment process is used upstream of the low-temperature sulfuric acid alkylation process. Developed by CDTECH.

> *Hydrocarbon Process.*, 2007, **86**(5), 23, 27.

CD-Cumene A process for making cumene for subsequent conversion to phenol and acetone. The cumene is made by catalytic alkylation of benzene with propylene in a *catalytic distillation reactor. Developed in 1995 by CDTECH and commercialized in Taiwan.

> *Chem. Eng. (N.Y.)*, 1995, **102**(8), 22.

CDETHEROL A process for making ethers (MTBE, TAME, ETBE) from alcohols. Combines the *Etherol process with *catalytic distillation.

> *Hydrocarbon Process. Int. Ed.*, 1994, **73**(11), 104.

CDHydro [Catalytic Distillation Hydrogenation] A family of petrochemical processes that combine *catalytic hydrogenation with fractional distillation in one unit operation. Most involve the selective hydrogenation of diolefins in C_3 to C_6 hydrocarbon fractions. Developed by CDTECH, a partnership between Chemical Research & Licensing Company and ABB Lummus Crest (now ABB Lummus Global). The first plant was built at Shell's Norco, LA, site in 1994. Ten units were operating in 1997.

> *Hydrocarbon Process. Int. Ed.*, 1996, **75**(11), 129.

CDP A process for destroying dioxins and polychlorinated biphenyls by treatment with a polyethylene glycol and sodium peroxide in a fixed catalyst bed. Developed by Sea Marconi Technologies, Turin, Italy. *See also* KPEG.

> Italian Patents 1,999,283; 2,221,583; 2,444,383.

CDW *See* Catalytic Dewaxing.

CEC [Chisso Engineering Company] A process for removing oxides of nitrogen from flue gases by scrubbing with an aqueous solution containing ferrous ion and ethylenediaminetetraacetic acid (EDTA). An iron nitrosyl compound is formed. Developed by Chisso Engineering Company, Japan, and piloted in France and Japan.

> Kohl, A.L. and Riesenfeld, F.C., *Gas Purification*, 4th ed., Gulf Publishing Co., Houston, TX, 1985, 371.

CED [Conversion Extraction Desulfurization] A process for reducing the sulfur content of diesel fuel. Peroxyacetic acid oxidizes the organic sulfur compounds to sulfones, which are removed by solvent extraction. Developed in 2000 by Petro Star.

> *Chem. Eng. (N.Y.)*, 2000, **107**(4), 17.

Celanese LPO [Liquid Phase Oxidation] A process for making acetic acid by oxidizing *n*-butane in the liquid phase, catalyzed by cobalt acetate. Developed by Hoechst Celanese and operated in the United States and the Netherlands until 1984. *See also* DF.

> Weissermel, K. and Arpe, H.-J., *Industrial Organic Chemistry*, 4th ed., translated by C.R. Hindley and S. Hawkins, Wiley-VCH, Weinheim, Germany, 2003, 174.

Celdecor A process for making paper from straw or bagasse. The fiber is digested in aqueous sodium hydroxide and bleached with chlorine. The essential feature is that the alkali and chlorine are used in the proportions in which they are made by the *Chlor-Alkali process.

> Grant, J., *Cellulose Pulp and Allied Products*, Leonard Hill, London, U.K., 1958, 356.

Celobric An anaerobic wastewater treatment process, suitable for treating high concentrations of organic substances. Developed by Hoechst Celanese Chemicals Group and the Badger Company and used in seven installations in 1993.

> *Hydrocarbon Process. Int. Ed.*, 1993, **72**(8), 92.

CESL [Cominco Engineering Services Ltd.] A family of hydrometallurgical processes for extracting Cu, Ni, Ag, Au, and Mo. They combine pressure oxidizing, solvent extraction, and electrowinning. Chloride ion is used as a catalyst. Developed by CESL, a subsidiary of Teck Cominco Metals, based in BC, Canada. Piloted but not yet licensed.

> U.S. Patents 5,645,708; 5,902,474.

cementation An alchemical term for any reaction that takes place in the solid state. Two examples follow. The name is used also in hydrometallurgy for the electrochemical process by which one metal replaces another in aqueous solution, for example, metallic iron causing precipitation of metallic platinum from a platinum chloride solution (also called "footing").

Cementation (1) The earliest known process for making steel from iron. It originated in the Iron Age, but the first written description was published in Prague in 1574 and the earliest known ironworks using the process was in Nuremberg, Germany, in 1601. The process was operated in Europe between its induction and around 1950, when it was finally replaced by the *Bessemer process. In the United States, it was operated from the mideighteenth century, initially in Connecticut. Wrought iron bars were stacked in alternate layers with lump charcoal in a conical brick kiln and fired for 5–12 days, until the correct carbon content had been achieved. The product was known as "blister steel," because of the appearance of its surface.

> Barraclough, K.C., *Steelmaking Before Bessemer, Vol. 1, Blister Steel*, The Metals Society, London, U.K., 1984, 48.

Cementation (2) An obsolete process for making brass by alloying copper with zinc, introduced as the vapor.

Centaur A process for reducing sulfur dioxide emissions from sulfuric acid plants. An activated carbon with both absorptive and catalytic properties is used. The technology uses fixed beds of Centaur carbon to oxidize sulfur dioxide to sulfuric acid in the pores of the carbon. The sulfuric acid is recovered as dilute sulfuric acid, which is used as make-up water in the sulfuric acid production process. Developed by Calgon Carbon Corporation in the 1990s. Calgon Carbon and Monsanto Enviro-Chem operated a Centaur pilot plant at an existing sulfuric acid plant in 1996.

> *Eur. Chem. News*, 1996, **66**(1723), 19.
> *Sulphur*, 1996, Jul–Aug(245), 13.

Central-Prayon A wet process for making phosphoric acid, similar to the *Prayon process but using extra sulfuric acid in the crystallization of the gypsum in order to minimize losses of phosphoric acid. Developed jointly by the Central Glass Company (Japan) and Société de Prayon (Belgium).

> Gard, D.R., in *Encyclopedia of Chemical Processing and Design*, McKetta, J.J. and Cunningham, W.A., Eds., Marcel Dekker, New York, 1990, **35**, 456.

Centripure *See* De Laval Centripure.

CEP [**C**atalytic **E**xtraction **P**rocessing] Also called Quantum Catalytic Extraction. A process for destroying low-level radioactive wastes by reaction with a molten metal at high temperature. Invented in 1989 by C. Nagel at the Massachusetts Institute of Technology and developed in the early 1990s by Molten Metals Technology, Waltham, MA. The company filed for bankruptcy in 1997, and the technology was acquired by Quantum Catalytic.

Pierce, A. and Chanenchuck, C., *Poll. Prevent.*, 1992, **2**(4), 69.
Chem. Eng. News, 1993, **71**(39), 9.
Chem. Eng. (Rugby, Engl.), 1996, (605), 15.
Chem. Week, 1997, **159**(47), 5.

CER [**C**hlorination with **E**nergy **R**ecovery] A process for making 1,2-dichloroethane by reacting ethylene with chlorine in the presence of a catalyst based on the tetrachloroferrate complex. Developed by Hoechst, Germany, in 1989. *See also* HTC.

CERNOX [**CER**amic **NO**$_x$] A process for destroying NO$_x$ by reaction with ammonia, catalyzed by a zeolite. Developed by Steuler Industriewerke in the 1950s for treating vapors from nitric acid baths used for pickling stainless steel. The zeolite was developed for this process by Mobil Corporation and is still proprietary. The process was introduced to the market in 1982, and by 1988, 100 units had been installed in Europe alone.

Grove, M. and Sturm, W., *Ceram. Eng. Sci. Proc.*, 1989, **10**(3–4), 325.

Cerny A process for crystallizing calcium nitrate tetrahydrate, used as a fertilizer, by cooling the saturated solution with drops of cold petroleum.

Bamforth, A.W., *Industrial Crystallization*, Leonard Hill, London, U.K., 1965.
Mullin, J.W., *Crystallization*, 2nd ed., Butterworths, London, U.K., 1972, 311.

CerOx [**Cer**ium **Ox**idation] An electrochemical process for oxidizing organic materials such as pesticides in wastewater. Cerium ion is an intermediate. This is electrochemically oxidized to Ce^{4+}, which is the actual oxidant.

Eur. Chem. News., 2000, **72**(1908), 30.

Cerphos [**Centre** d'**Ét**udes et de **R**echerche de **Phos**phates Minéraux] A process for making gypsum, suitable for use as plaster, from the waste from the *wet process for making phosphoric acid.

Cetus A four-stage fermentation process for making propylene oxide from glucose. The product is obtained as a dilute aqueous solution. Developed by Cetus Corporation in the 1970s but not commercialized.

U.S. Patents 4,246,347; 4,284,723.
Weissermel, K. and Arpe, H.-J., *Industrial Organic Chemistry*, 4th ed., translated by C.R. Hindley and S. Hawkins, Wiley-VCH, Weinheim, Germany, 2003, 276.

CFB (1) [**C**hemische **F**abrik **B**udenheim] A process for removing cadmium from phosphoric acid by extracting with a solution of a long-chain amine in a hydrocarbon. Developed and operated by the German company for which it is named.

German Patents 3,218,599; 3,327,394.
Becker, P., *Phosphates and Phosphoric Acid*, Marcel Dekker, New York, 1989, 528.

CFB (2) [**C**irculating **F**luidized **B**ed] A metallurgical roasting process, developed originally for the aluminum industry and now used for many other, nonferrous ores. Offered by Lurgi.

CFB (3) [**C**irculating **F**luidized **B**ed] A coal combustion process for developing steam for power generation, developed by Foster Wheeler. The process is suitable for burning many fuels, including waste coal, petroleum coke, and even shredded car tires.

CFC [Controlled Flow Cavitation] A process for accelerating reactions in the liquid phase (optionally containing solid particles) by subjecting them to ultrasonic vibrations that cause cavitation. Developed by Arisdyne Systems and used in biodiesel production.

U.S. Patent 5,492,654.

CFD [Chemical Fluid Deposition] A process for depositing thin films on solid surfaces by a chemical reaction in a liquid such as supercritical carbon dioxide. Superior to *CVD in being capable of operation at almost ambient temperatures. Demonstrated for depositing platinum metal on silicon wafers, polymer substrates, and porous solids by hydrogen reduction of an organoplatinum compound at 80°C.

Watkins, J.J., Blackburn, J.M., and McCarthy, T.J., *Chem. Mater.*, 1999, **11**(2), 213.

CFI A catalytic process for simultaneously *dewaxing and hydrogenating gas oil. Developed by Nippon Ketjen, Fina, and Akzo Nobel.

CFP [Catalytic Fast Pyrolysis] A process for converting lignocellulosic biomass into a mixture of aromatic hydrocarbons, notably *p*-xylene. It uses a fluidized bed at 500°C–1000°C and a modified ZSM-5 zeolite catalyst. Invented by G.W. Huber and colleagues at the University of Massachusetts–Amherst and licensed through Anellotech. The first plant is scheduled for completion in 2014.

U.S. Patent Application 2013 0023706.
Cheng, Y.T., Wang, Z., Gilbert, C.J., Fan, W., and Huber, G.W., *Angew. Chem.*, 2012, 51, 11097.

CFRI [Central Fuel Research Institute] An Indian process for making coke. *See* carbonization.

Jpn. Chem. Week, 1995, **36**(1823), 8.

CGA [Coal Gold Agglomeration] A process for separating gold particles from aqueous slurries of finely ground ores by putting them in contact with slurries of coal particles in oil. The gold particles adhere to the coal particles. When the coal particles have become saturated with gold, they are separated and the coal is burned off. The process was developed as an alternative to the environmentally unacceptable mercury and *Cyanide processes. Invented by BP Minerals and developed in the early 1990s by Davy McKee (Stockton, United Kingdom).

CGCC [Coal Gasification, Combined Cycle] A general name for processes that both gasify coal and produce electricity. One such process has been operated by Louisiana Gasification Technology, at Plaquemine, LA, since 1987. *See also* APAC.

Chem. Mark. Rep., 1990, 19 Nov, 7.

Chamber An obsolete but formerly very important process for making sulfuric acid. Invented by J. Roebuck in Birmingham, England, in 1746, although the patent was not filed in Scotland until 1771. Progressively improved during the nineteenth century and finally abandoned everywhere in favor of the *Contact process by 1980. Essentially, the Chamber process was the gas-phase oxidation of sulfur dioxide to sulfur trioxide, catalyzed by oxides of nitrogen, conducted in a lead-lined chamber, followed by dissolution of the sulfur trioxide in water.

Clow, A. and Clow, N.L., *The Chemical Revolution*, The Letchworth Press, London, U.K., 1952, 140.
Imperial Chemical Industries, *Sulphuric Acid: Manufacture and Use*, Kynoch Press, Birmingham, U.K., 1955, 16.
Campbell, W.A., in *Recent Developments in the History of Chemistry*, Russell, C.A., Ed., Royal Society of Chemistry, London, U.K., 1985, 243.
Roebuck, P., *Chem. Br.*, 1996, **32**(7), 38.

Champion An alternative name for the *English process for extracting zinc from its ores. The Champion family was active in several industries in Bristol, England, in the eighteenth century. In 1738, William Champion was granted a patent for reducing sulfide ores, but the minimal wording of his claim made the patent ambiguous. His brother John was granted a related patent in 1758, which described in more detail the reduction of blende (zinc sulfide ore) with charcoal. The sulfide must first be converted to the oxide before it can be reduced to the metal. The process was first operated commercially at Warmley, near Bristol, around 1740, using calamine (zinc carbonate ore). It was the only process used for extracting zinc until the end of the eighteenth century, when the *Belgian process was introduced.

British Patents 564 (1738); 726 (1758).

Cocks, E.J. and Walters, B., *A History of the Zinc Smelting Industry in Britain*, George G. Harrap, London, U.K., 1968, 7.

Morgan, S.W.K., *Zinc and Its Alloys and Compounds*, Ellis Horwood, Chichester, England, 1985, 17.

Chance Also called Chance–Claus. A process for recovering sulfur from the calcium sulfide residues from the *Leblanc process. Treatment of a suspension of the residues with carbon dioxide generates hydrogen sulfide, which is converted to sulfur dioxide by the *Claus process. The sulfur dioxide is converted to sulfuric acid. Developed by A.M. and J.F. Chance from 1882 to 1887 and widely used until the Leblanc process was superseded by the *Solvay process.

British Patent 8, 666 (1887).

Hardie, D.W.F., *A History of the Chemical Industry in Widnes*, Imperial Chemical Industries, Widnes, U.K., 1950, 126.

Campbell, W.A., in *Chemistry, Society and Environment*, Russell, C.A., Ed., Royal Society of Chemistry, London, U.K., 2000, 97.

Chandelon Also spelled Chandelen. A process for making mercuric cyanate (mercury fulminate) by dissolving mercury in nitric acid and pouring the solution into aqueous ethanol. Developed by Chandelon in Belgium in 1848.

Urbanski, T., *Chemistry and Technology of Explosives*, translated by M. Jurecki and S. Laverton, Pergamon Press, Oxford, U.K., 1964, 150.

Channel Black One of the processes used to make carbon black. In this process, natural gas was incompletely burned in small flames, which impinged on cooled channel irons that were continuously moved and scraped. Invented by L.J. McNutt in 1892 and commercialized that year in Gallagher, PA. The last U.S. plant was closed in 1976. The *Gas Black process works similarly but uses coal tar oils as feedstocks.

U.S. Patent 481,240.

Ellis, C., *The Chemistry of Petroleum Derivatives*, Chemical Catalog, New York, 1934, 237.

Kühner, G. and Voll, M., in *Carbon Black Science and Technology*, Donnet, J.B., Bansai, R.C., and Wang, M.-J., Eds., Marcel Dekker, New York, 1993, 57.

Claasen, E.J., in *Inorganic Chemicals Handbook*, Vol. 1, McKetta, J.J., Ed., Marcel Dekker, New York, 1993, 498.

Buxbaum, G. and Pfaff, G., *Industrial Inorganic Pigments*, 3rd ed., VCH, Weinheim, Germany, 2005, 173.

Chardonnet A process for making "artificial silk" by nitrating cellulose and injecting the nitrate solution into water, thereupon regenerating the cellulose:

$$\left[C_6H_7O_2(OH)_3\right]_n \rightleftharpoons \left[C_6H_7O_2(ONO_2)\right]_n$$

In the original process, the cellulose nitrate itself was used as the fiber (hence, its satirical description as "mother-in-law silk"). The regenerating agent is ammonium hydrosulfide. The basic process was first demonstrated by J.W. Swan in London in 1885 but commercialized by Count L.M.H.B. de Chardonnet ("father of the rayon industry") in France in 1891 and operated there until 1934. The last working factory, in Brazil, burned down in 1949. The other processes for making rayon fibers by regenerating cellulose (*Viscose, *Cuprammonium) gave superior products. *See also* Rayon.

French Patent 165,349.

J. Soc. Dyers Color., 1914, **30**, 199.

Moncrief, R.W., *Man-Made Fibres*, 6th ed., Butterworth Scientific, London, U.K., 1975, 157.

Chem-Char A process for destroying organic wastes by pyrolysis on devolatilized coal char in a reducing atmosphere, followed by secondary combustion of the product gases. Developed at the University of Missouri–Columbia.

Kinner, L.L., McGowin, A., Manahan, S., and Larsen, D.W., *Environ. Sci. Technol.*, 1992, **27**(3), 482.
Chem. Br., 1993, **29**(6), 460.

ChemCoal A *coal liquefaction process, using a phenolic solvent, aqueous alkali, and carbon monoxide.

Porter, C.R. and Rindt, J.R., in *Processing and Utilization of High-Sulfur Coals*, IV, Dugan, P.R., Quigley, D.R., and Attia, Y.A., Eds., Elsevier, Amsterdam, the Netherlands, 1991, 651.

Chemetals A process for reducing manganese oxide to the metal by heating with methane and air. Developed by the Manganese Chemicals Corporation, Baltimore, MD.

U.S. Patent 3,375,097.

Chemex [**Chem**ical **ex**change] A process for separating uranium isotopes, based on the equilibrium between U^{3+} and U^{4+} in aqueous solution. U-238 concentrates in the U^{3+} state and U-235 in the U^{4+}. Uranium in the two valence states is separated by solvent extraction into tributyl phosphate. Developed and piloted in France but not commercialized.

Chemfix A process for solidifying aqueous wastes, converting them to a solid form suitable for landfill. Silicates and a proprietary setting agent are used. Invented by J.R. Connor in 1970 and offered by Chemfix, Kenner, LA. In 1979, 100 million gallons of waste had been treated in this way in the United States.

U.S. Patents 3,837,872; 3,841,102.
Evans, R.J. and Duvel, W.A., Jr., *Pollut. Eng.*, 1974, **6**(10), 44.
Salas, R.K., in *Toxic and Hazardous Waste Disposal*, Pojasek, R.B., Ed., Ann Arbor Science, Ann Arbor, MI, 1979, Chap. 16.

Chemical Vapor Deposition *See* CVD.

ChemicoB-Basic A *flue-gas desulfurization process using magnesium oxide slurry.

Speight, J.G., *Gas Processing*, Butterworth Heinemann, Oxford, U.K., 1993, 281.

Chemithon A process for making detergents by continuous sulfonation of hydrocarbons with sulfur trioxide or oleum. Offered by Lurgi.

Chemo-Trenn A process for absorbing ammonia, carbon dioxide, hydrogen sulfide, and other undesirable gases from coal gas by absorption in a warm solution of a salt of a weak organic acid such as sodium cresylate. Invented by H. Bähr at IG Farbenindustrie, Germany, in 1938.

German Patents 728,102; 741,222.
Kohl, A.L. and Riesenfeld, F.C., *Gas Purification*, 4th ed., Gulf Publishing Co., Houston, TX, 1985, 164.

Chemsweet A semicontinuous process for removing hydrogen sulfide and other sulfur-containing gases from natural gas. The gas is passed through a suspension of zinc oxide in aqueous zinc acetate. The zinc sulfide produced has formerly been buried on land. Licensed by C-E Natco, a division of Combustion Engineering. In 1990, 150 units were operating.

Manning, W.P., *Oil Gas J.*, 1979, **77**(42), 122.
Hydrocarbon Process. Int. Ed., 1992, **71**(4), 94.
Kohl, H.L. and Nielsen, R.B., *Gas Purification*, 5th ed., Gulf Publishing Co., Houston, TX, 1997, 1310.

Chenot An early steelmaking process. Iron ore was first reduced to sponge iron; this was mixed with charcoal, manganese, and resin and compressed into small blocks. These blocks were then melted as in the *Crucible process. Invented by A. Chenot in France in 1846 and operated on a small scale in Sheffield, United Kingdom, in the 1860s, but abandoned after the introduction of the *Open Hearth process.

> British Patent 11,515 (1846).
> Barraclough, K.C., *Steelmaking Before Bessemer, Vol. 2, Crucible Steel*, The Metals Society, London, U.K., 1984, 87, 299.

Chesney A process for extracting magnesium from seawater. Developed by British Periclase in 1937.

> *Chem. Ind. (London)*, 1992, (5), 154.

Chevron (1) An obsolete xylene isomerization process that used a silica–alumina catalyst.

Chevron (2) A process for separating *p*-xylene from its isomers by continuous crystallization, using liquid carbon dioxide in direct contact with the xylene as the refrigerant. Developed by the Chevron Research Company in 1966.

> U.S. Patent 3,467,724.

Chevron WWT [Waste Water Treatment] An integrated process for treating "sour water" from oil refineries, particularly for removing ammonia, hydrogen sulfide, and carbon dioxide. Only physical processes are used: volatilization and condensation under various conditions. Developed by Chevron Research Company and used in 14 plants worldwide in 1985.

> Martinez, D., in *Chemical Waste Handling and Treatment*, Muller, K.R., Ed., Springer-Verlag, Berlin, Germany, 1986, 180.
> Leonard, J.P., Haritatos, N.J., and Law, D.V., in *Acid and Sour Gas Treating Processes*, Newman S.A., Ed., Gulf Publishing, Houston, TX, 1985, 734.

Chinuka An electrochemical process for making titanium metal directly from a titaniferous raw material such as rutile. The rutile is mixed with carbon and heated to above 1400°C in a vacuum, producing titanium oxycarbide. This is electrorefined using a molten sodium/potassium chloride eutectic as the electrolyte. Developed by D. Fray and S. Jiao in Cambridge, United Kingdom. Licensed to White Mountain Titanium Corp, Chinuka Ltd., and La Serena Technologies in 2010. *See also* FFC.

> PCT Patent Application WO2011/015845.
> U.S. Patent Application 2012-01152756.

Chiyoda *See* Thoroughbred, ABC.

Chlor-Alkali *See* Castner–Kellner.

Chlorex (1) A process for extracting lubricating oil stocks from petroleum fractions using β,β-dichloro diethyl ether. Chlorex is also the trademark for this compound. Developed by Standard Oil Company (Indiana) in the early 1930s and used until the 1960s.

> Page, J.M., Buchler, C.C., and Diggs, S.H., *Ind. Eng. Chem.*, 1933, **25**, 418.
> Unzelman, G.H. and Wolf, C.J., in *Petroleum Processing Handbook*, Bland, W.F. and Davidson, R.L., Eds., McGraw-Hill, New York, 1967, 3–89.

Chlorex (2) A process for removing hydrogen chloride from hydrocarbons by aqueous alkaline extraction, using a bundle of hollow fibers. Developed by the Merichem Company, Houston, TX, and used in three installations in Japan in 1991.

> *Hydrocarbon Process. Int. Ed.*, 1996, **75**(4), 126.

Chloride (1) One of the two process used today for making titanium dioxide pigment (the other being Sulfate (1)). Mineral rutile, or another mineral rich in titanium, is chlorinated with coke to produce titanium tetrachloride:

$$TiO_2 + 2Cl_2 + C \rightarrow TiCl_4 + CO_2 (CO)$$

This titanium tetrachloride is reacted with oxygen at approximately 1000°C to yield titanium dioxide crystals around 0.2 μ in diameter, the optimum size for scattering visible light:

$$TiCl_4 + O_2 = TiO_2 + 2Cl_2$$

The oxidation is conducted in a proprietary burner, with extra heat supplied because, although the reaction is exothermic, it is not self-sustaining. The chlorine is recycled. The process was developed by DuPont in the 1940s, and its first plant started operating in 1958. The Chloride process has progressively replaced the older *Sulfate process because it produces less effluent. In 2013, 45% of the world capacity for titanium dioxide production used the Chloride process. *See also* ICON.

Barksdale, J., *Titanium: Its Occurrence, Chemistry, and Technology*, 2nd ed., Ronald Press, New York, 1966, Chap. 21.

Chem. Eng. (Rugby, Engl.), 1991, (497), 33.

Egerton, T.A. and Tetlow, A., in *Industrial Inorganic Chemicals: Production and Use*, Thompson, R., Ed., Royal Society of Chemistry, Cambridge, U.K., 1995, 363.

Buxbaum, G. and Pfaff, G., *Industrial Inorganic Pigments*, 3rd ed., VCH, Weinheim, Germany, 2005, 64.

Chloride (2) *See also* *Leonard.

Chlorine/Hercosett Also called Hercosett. A process for making wool shrink resistant by chlorination followed by resin treatment. The resin is made by Hercules, hence the name. Developed by the International Wool Secretariat.

German Patent 2,018,626.

Smith, P. and Mills, J.H., *CHEMTECH*, 1973, **3**, 748.

Chloroff A process for removing chlorine from organic chlorides, such as polychlorinated biphenyls, by reaction with hydrogen under pressure over a proprietary catalyst. Developed by Kinetics Technology International. *See also* Hi-Chloroff.

chlorohydrination The addition of water and chlorine to an olefin to give a chloroalcohol. Thus, ethylene produces ethylene chlorohydrin:

$$CH_2 = CH_2 + Cl_2 + H_2O = CH_2OH - CH_2Cl + HCl$$

Chloroalcohols are important intermediates. Propylene chlorohydrin is made similarly and is used for making propylene oxide by hydrolysis with either calcium hydroxide or sodium hydroxide. If calcium hydroxide is used, the by-product calcium chloride is useless and must be dumped. If sodium hydroxide is used, the by-product sodium chloride can be recycled to the *Castner–Kellner process.

Chlorsorb Also called **RVG Chlorsorb** [**R**egenerator **V**ent **G**as]. A process for removing chlorine compounds (mainly HCl) from the waste gases from the *CCR Platforming process. The chlorides are reabsorbed on the catalyst. Developed by UOP and now operated in several oil refineries in the United States and Argentina.

Chromizing *See* metal surface treatment.

Chromox [Chromium **ox**idation] A process for destroying organic pollutants in aqueous wastes by oxidation with hydrogen peroxide, catalyzed by Cr^{6+}. Developed by British Nuclear Fuels in 1995, originally for use in nuclear reprocessing.

Chem. Eng. (N.Y.), 1996, **103**(3), 19.

CIC [Carbon In Column] A general name for hydrometallurgical extraction processes using activated carbon as the adsorbent, contained in a series of cylindrical tanks.

Gupta, C.K. and Mukherjee, T.K., *Hydrometallurgy in Extraction Processes*, Vol. 2, 1990, CRC Press, Boca Raton, FL, 73.

CIL [Carbon In Leach] A process for extracting gold from cyanide leach liquors using activated carbon.

Woodhouse, G., in *Trace Metal Removal from Aqueous Solution*, Thompson, R., Ed., Royal Society of Chemistry, London, U.K., 1986.
Gupta, C.K. and Mukherjee, T.K., *Hydrometallurgy in Extraction Processes*, Vol. 2, 1990, CRC Press, Boca Raton, FL, 73.

CIP (1) **[Carbon In P**ulp] A general name for hydrometallurgical extraction processes that use activated carbon in slurries of ground ores. One such process is used for extracting gold from cyanide leach liquors. First operated on a large scale at the Homestake lead mine in South Dakota in 1974. *See also* RIP.

Woodhouse, G., in *Trace Metal Removal from Aqueous Solution*, Thompson, R., Ed., Royal Society of Chemistry, London, U.K., 1986.
McDougall, G.J. and Fleming, C.A., *Ion Exchange and Sorption Processes in Hydrometallurgy*, Streat, M. and Naden, D., Eds., John Wiley & Sons, London, U.K., 1987, Chap. 2.
Gupta, C.K. and Mukherjee, T.K., *Hydrometallurgy in Extraction Processes*, Vol. 2, 1990, CRC Press, Boca Raton, FL, 73.

CIP (2) *See* DR.

Circofer A *DR process for finely divided ores using reducing gases made from coal in a circulating fluidized bed. Under development by Lurgi in 1995.

Chem. Eng. (N.Y.), 1995, **102**(3), 41.

Circored A *DR process using hydrogen made by reforming natural gas. Two fluidized beds are used. Developed by Lurgi in 1995 and first installed in Trinidad in 1998.

Chem. Eng. (N.Y.), 1995, **102**(3), 37; 1996, **103**(9), 25.

CIRCOX A process for treating municipal waste by use of a biofilm supported on a granular solid. Developed by Paques in the Netherlands.

Citrate A process for *flue-gas desulfurization by absorption of the sulfur dioxide in aqueous sodium citrate, reacting with hydrogen sulfide to produce elemental sulfur and recycling the citrate solution. Subsequent modifications involved removing the sulfur dioxide from the citrate solution by steam stripping or by pumping it off and using it to make sulfuric acid. Invented by the U.S. Bureau of Mines in 1968 for use in metal smelters; subsequently piloted at six locations with varying degrees of success. Much of the development work was done by Pfizer, a U.S. citric acid producer. Development was abandoned in the 1980s because of corrosion and plugging of the absorber with sulfur. *See also* Flakt–Boliden.

Chalmers, F.S., *Hydrocarbon Process. Int. Ed.*, 1974, **53**(4), 75.
IEA Coal Research, *The Problems of Sulphur*, Butterworths, London, U.K., 1989, 114.
Kohl, H.L. and Nielsen, R.B., *Gas Purification*, 5th ed., Gulf Publishing Co., Houston, TX, 1997, 563.

Citrex [**Citr**ic acid **ex**traction] An improved version of the *Citrate process, designed by Peabody Engineered Systems.

Vasan, S., *Chem. Eng. Prog.*, 1975, **71**(5), 61.

Citrosolv A two-stage process for removing deposits from steam boilers, using citric acid. The first stage uses ammoniated citric acid at pH 3.5–4 to remove iron oxide; the second uses a solution containing more ammonia, pH 9.5–10, to remove copper oxide, and an oxidant such as sodium nitrite to passivate the surface.

Clandot A process for extracting silver from its ores by extracting with aqueous ferric chloride, precipitating silver iodide by adding zinc iodide, and then reducing the iodide to the metal by heating with zinc. Used in the late nineteenth century.

Clanex A solvent extraction process for converting solutions of the nitrates of actinides and lanthanides into their corresponding chlorides. The extractant is a solution of an aliphatic amine in diethylbenzene.

Leuze, R.E. and Lloyd, M.H., *Prog. Nucl. Energy*, 1970, Ser. III, **4**, 596.

Clark The first water-softening process to be operated on a large scale, based on the addition of calculated quantities of calcium hydroxide. Invented in 1841 by T. Clark at Marischal College, Aberdeen, Scotland. *See also* Porter–Clark.

British Patent 8,875 (1841).
Imperial Chemical Industries, *Ancestors of an Industry*, Kynoch Press, Birmingham, England, 1950, 75.
Nordell, E., *Water Treatment for Industrial and Other Uses*, Reinhold Publishing, New York, 1961, 489.

Claude (1) Also called Claude–Casale. A high-pressure ammonia synthesis process, developed by G. Claude in the 1920s. The Claude and *Casale processes used much higher temperatures and pressures than the Haber–Bosch process, which succeeded them.

Chem. Eng. Prog., 1952, **48**, 468.
Shearon, W.H., Jr. and Thompson, H.L., in *Modern Chemical Processes*, Vol. 3, Reinhold Publishing, New York, 1954, 16.
Chilton, T.H., *Strong Water: Nitric Acid, Its Sources, Methods of Manufacture, and Uses*, MIT Press, Cambridge, MA, 1968, 88.
Vancini, C.A., *Synthesis of Ammonia*, translated by L. Pirt, Macmillan Press, Basingstoke, U.K., 1971, 247.
Aitkin, W., *Chem. Eng. (Rugby, Engl.)*, 1998, (659), 17.
Leigh, G.J., *The World's Greatest Fix: A History of Nitrogen and Agriculture*, Oxford University Press, Oxford, U.K., 2004, 150–153.

Claude (2) A process for fractionally distilling liquid air, based on the original *Linde process but using two stages. Developed by G. Claude, who founded the French company L'Air Liquide.

Claude, G., *C. R. Acad. Sci.*, 1900, **131**, 447.

Claude–Casale *See* Casale.

Claus A process for removing hydrogen sulfide from gas streams by the catalyzed reaction with sulfur dioxide, producing elementary sulfur. The process has two stages: in the first, one third of the hydrogen sulfide is oxidized with air to produce sulfur dioxide; in the second, this sulfur dioxide stream is blended with the remainder of the hydrogen sulfide stream and passed over an iron oxide catalyst at approximately 300°C. The resulting sulfur vapor is condensed to liquid sulfur:

$$H_2S + \tfrac{3}{2}O_2 = SO_2 + H_2O$$

$$SO_2 + 2H_2S = 3S + 2H_2O$$

The process was invented by C.F. Claus in Germany in 1882 but not introduced into the United States until 1943. It is now a major world source of sulfur for conversion to sulfuric acid. The literature is very extensive.

British Patent 5,958 (1883).

Estep, J.W., McBride, G.T., and West, J.R., *Advances in Petroleum Chemistry and Refining*, Vol. 6, Interscience Publishers, New York, 1962.

Misra, C., *Industrial Alumina Chemicals*, American Chemical Society, Washington, DC, 1986, 139.

Downing, J.C. and Goodboy, K.P., in *Alumina Chemicals*, American Ceramic Society, Westerville, OH, 1990, 277.

Nédez, C. and Ray, J.-L., *Catal. Today*, 1996, **27**(1–2), 49.

Kohl, H.L. and Nielsen, R.B., *Gas Purification*, 5th ed., Gulf Publishing Co., Houston, TX, 1997, 670.

Piéplu, O., Sauer, J.-C., Lavalley, O., Legendre, C., and Nédez, C., *Catal. Rev. Sci. Eng.*, 1998, **40**(4), 409, 66.

ClausMaster A physical absorption system for capturing residual sulfur dioxide from the *Claus process. Developed by Monsanto, now MECS (Monsanto Enviro-Chem Systems).

Clausorb A process for removing residual sulfur from the tail gases from the *Claus process.

Sulphur, Jul-Aug 1999 (263), 53–54, 56.

Clauspol [**Claus** **p**olyethylene glyc**ol**] A variation on the *Claus process for removing hydrogen sulfide from gas streams, in which the tail gases are scrubbed with polyethylene glycol to remove residual sulfur dioxide. Clauspol 1500 is a modification of this. Developed by the Institut Français du Pétrole. Forty-eight units had been licensed by 2000. *See* IFP.

Bartel, Y., Bistri, Y., Deschamps, A., Renault, P., Simadoux, J.C., and Dutrian, R., *Hydrocarbon Process. Int. Ed.*, 1971, **50**(5), 89.

Sulphur, 1974, (111), 48.

Hydrocarbon Process. Int. Ed., 1996, **75**(14), 40.

Kohl, H.L. and Nielsen, R.B., *Gas Purification*, 5th ed., Gulf Publishing Co., Houston, TX, 1997, 843.

Hydrocarbon Process. Int. Ed., 2000, **79**(4), 62.

Clayton A continuous soapmaking process, using centrifugation for separating the soap from the oil. Invented by B. Clayton and operated in the United States from the 1930s.

U.S. Patent 2,219,088.

Shearon, W.H., Jr., Seestrom, H.E., and Hughes, J.P., in *Modern Chemical Processes*, Vol. 2, Reinhold Publishing, New York, 1952, 136.

clay treating A general term for processes that remove gum-forming materials from petroleum streams by adsorption on natural clays such as bentonite.

CLC [**C**hemical **L**ooping **C**ombustion] Also called carbonate-looping. A two-stage process for efficient combustion of hydrocarbon gases. It is more efficient than normal one-stage combustion and facilitates carbon dioxide capture. There are two fluidized beds through which particles of a metal/metal oxide circulate. Possible oxygen carrier materials include oxides of Ni, Cu, Mn, and Fe. First proposed in 1951 but not demonstrated until 2003. Piloted at the Technical University of Darmstadt from 2009 to 2012.

Shen, L., Gao, Y., and Xiao, J., *Biomass Bioenergy*, 2008, **32**, 120.

Fan, L.-S., *Chemical Looping Systems for Fossil Energy Conversions*, John Wiley & Sons, Hoboken, NJ, 2010.

Muradov, N.Z., in *Carbon-Neutral Fuels and Energy Carriers*, Muradov, N.Z. and Veziroglu,T.N., Eds., CRC Press, Boca Raton, FL, 2012, 675.

Crow, J.M., *Chem. World*, 2012, **9**(120), 56.

Chem. Eng. (N.Y.), 2013, **120**(1), 11.

Cleanair A process for pretreating gas streams before the Claus and *Stretford processes for removing sulfur compounds. Developed by the J.F. Pritchard Company.

Davis, J.C., *Chem. Eng. (N.Y.)*, 1972, **79**(11), 66.

Sulphur, 1974, (111), 52.

Oil Gas J., 1978, **76**(35), 160.

Clean Coke (1) A process for making coke and chemicals from high-sulfur coals. Developed by the U.S. Steel Corporation with support from the U.S. Department of Energy. *See also* carbonization.

Schowalter, K.A. and Boodman, N.S., *Chem. Eng. Prog.*, 1974, **70**(6), 76.
Iammartino, N.R., *Chem. Eng. (N.Y.)*, 1975, **82**(18), 57.

Clean Coke (2) A process for making briquettes of metallurgical coke by mixing salvaged coke dust with coal dust and a proprietary binder and pyrolyzing it. The gases produced fuel the process, and the liquid produced is mixed with the binder. Invented in 1998 by C.N. Eatough at Covol Technologies with support from the U.S. Department of Energy.

U.S. Patent 5,752,993.

CLEAR [**C**opper **L**each **E**lectrolysis **A**nd **R**egeneration] A process for leaching copper from sulfide ores by boiling with aqueous cupric chloride:

$$CuFeS_2 + 3CuCl_2 = 4CuCl + FeCl_2 + 2S$$

Elemental sulfur precipitates and is recovered. Copper powder is produced by electrolyzing the resulting solution. Developed by G.E. Atwood and C.H. Curtis of Duval Corporation in Tucson, AZ. *See also* Cymet.

U.S. Patents 3,785,944; 3,879,272; 4,025,400.
Atwood, G.E. and Livingston, R.W., *Erzmetall*, 1980, **33**(5), 251.
Gupta, C.K. and Mukherjee, T.K., *Hydrometallurgy in Extraction Processes*, CRC Press, Boca Raton, FL, 1990, 12, 190.

Climax A process for making sodium sulfate from sulfuric acid and sodium chloride. Sulfuric acid is sprayed onto a hot fluidized bed of sodium chloride. The products are granular sodium sulfate and hydrogen chloride gas. Invented in 1967 by C.K. Curtis; later developed and commercialized by C.W. Cannon at the Climax Chemical Company at Midland, NM, in the 1970s. Midland was a favorable location because of the proximity of mineral salt and sulfur from petroleum and the availability of cheap transport of the product from the site.

French Patent 1,549,938.

CLINSULF [**C**arl von **LIN**de **SULF**ur] A variation of the *Claus process in which the heat from the process is used to heat a second catalytic reactor. The process is designed for gases rich in hydrogen sulfide. First commercialized in 1992 and offered by Linde.

Heisel, M.P. and Marold, F.J., *Oil Gas J.*, 1989, **87**(33), 37.
Hydrocarbon Process. Int. Ed., 1996, **75**(14), 110.

CLINSULF DO [**D**irect **O**xidation] A process for recovering sulfur from gas streams containing low concentrations of hydrogen sulfide, under conditions in which the conventional *Claus process is not applicable. The process can recover sulfur over a wider range than direct oxidation scrubbers. Operated in Austria and South Korea.

Chim. Ind. (Milan), 1996, **78**(6), 731.

CLINSULF^{R}SDP [**S**ub-**D**ew-**P**oint] A modified *CLINSULF process using an internally cooled reactor, which maximizes the conversion rate. Five plants were operating in 2000.

Kunkel, J., *Reports on Science and Technology*, Linde, Munich, Germany, 1997, (59), 46.
Sulphur, 1996, (243), 45; (245), 13.
Chem. Eng. (N.Y.), 1997, **104**(11), 126.
Hydrocarbon Process. Int. Ed., 2000, **79**(4), 62.

Clintox A process for removing sulfur dioxide from the tail gases from the *Claus process by washing with a proprietary organic scrubbing agent. Offered by Linde, Munich. Four units were operating in 2000.

Heisel, M.P. and Marold, F.J., *Gas Sep. Purif.*, 1987, **1**, 107.
Hydrocarbon Process. Int. Ed., 1996, **75**(14), 111; 2000, **79**(4), 62.

Clusius A process for separating isotopes by a combination of thermal diffusion and thermal siphoning. Invented in 1938 by K. Clusius and G. Dickel.

Clusius, K. and Dickel, G., *Naturwiss.*, 1938, **2**, 546.
Clusius, K. and Dickel, G., *Nature (London)*, 1939, **144**, 8.

CNA [**C**oncentrated **N**itric **A**cid] A general name for processes that make nitric acid more concentrated than the 70% made in conventional plants. They include the *CONIA and *SABAR processes. *See also* DSN.

Büchner, W., Schliebs, R., Winter, G., and Büchel, K.H., *Industrial Inorganic Chemistry*, VCH Publishers, Weinheim, Germany, 1989, 62.

CNC A chemical–biological process for oxidizing cyanide ion in wastewater. The wastewater is mixed with a proprietary soluble reagent and passed through a biological reactor, in which cyanide is oxidized to carbon dioxide and nitrogen. Developed and offered by Radian Corporation, Austin, TX.

Chem. Eng. (N.Y.), 1989, **96**(3), 17.

CNG [**C**onsolidated **N**atural **G**as] A process for removing acid gases from natural gas and *syngas, using supercritical carbon dioxide. Under development since 1973 by the Consolidated Natural Gas Research Company, with assistance from the U.S. Department of Energy and Helipump Corporation. Liquid carbon dioxide is first used to extract the sulfur compounds. Crystallization at the triple point separates these sulfur compounds from the carbon dioxide. The process is also claimed to be more effective than other acid gas processes in removing a wide range of trace impurities. In 1986, CNG was still being developed, but by 1997, it had still not been commercialized.

U.S. Patent 4,270,937.
Auyang, L., Liu, Y.C., Petrik, M., and Siwajek, L., in *Acid and Sour Gas Processes*, Newman, S.A., Ed., Gulf Publishing, Houston, TX, 1985, 497.
Massey, L.G. and Brown, W.R., in *Recent Developments in Separation Science*, Li, N.N. and Calo, J.M., Eds., CRC Press, Boca Raton, FL, 1986, 15.
Kohl, H.L. and Nielsen, R.B., *Gas Purification*, 5th ed., Gulf Publishing Co., Houston, TX, 1997, 1342.

CO_2 Acceptor A fluidized-bed coal gasification process in which the heat is provided by the exothermic reaction of carbon dioxide with calcium oxide. Developed by the Conoco Coal Development Company in the 1970s.

Hydrocarbon Process. Int. Ed., 1975, **54**(4), 120.
Hebden, D. and Stroud, H.J.F., in *Chemistry of Coal Utilization*, 2nd Suppl. Vol., Elliott, M.A., Ed., John Wiley & Sons, New York, 1981, 1642.

CO_2 Wash A process for removing impurities from landfill gas, leaving only methane and carbon dioxide. The gas is washed with liquid carbon dioxide, isolated from the landfill gas. Developed by Acrion Technology and first demonstrated in Goshen, NY, in 1998.

Wisbrock, W.H., in *Alcoholic Fuels*, Minteer, S., Ed., Taylor & Francis, Boca Raton, FL, 2006, 54.

coal cleaning A general name for processes for removing sulfur from coal. The sulfur is present mostly as iron pyrites (FeS). Such processes with special names that are described in this dictionary are Ames (2), Gravimelt, JPL Chlorinolysis, Ledgemont, Magnex, PETC, SABA, TRW Gravichem, and TRW Meyers.

Coalcon A coal gasification process using a fluidized bed operated with hydrogen. Developed by Union Carbide Corporation and the Chemical Construction Company, based on the work on liquid-phase hydrogenation completed by Union Carbide in the 1950s. A 20 tons-per-day pilot plant was operated in the 1960s, but a planned larger demonstration plant was abandoned because of cost.

Ferretti, E.J., *Chem. Eng. Prog.*, 1976, **72**(8), 62.
Morgan, W.D., *Chem. Eng. Prog.*, 1976, **72**(8), 64.

coal gasification A general name for processes for converting coal to gaseous products. These processes are to be distinguished from *carbonization processes, in which the production of carbonaceous products is the main aim, although some gas may also be produced. Those with special names that are described in this dictionary are Atgas, Babcock & Wilcox, Bi-Gas, Bubiag, CAFB, CGCC, Clean Coke, Coalcon, CO_2 Acceptor, COED, COGAS, Dynagas, H-COAL, HTW, Hydrane, HYFLEX, HYGAS, IG-Hydrogenation, KBW, Kerpely, KILnGas, Koppers Kontalyt, Koppers–Totzek, KR, K-T, LC, Lurgi, Lurgi–Ruhrgas, Mond gas, Morgas, ORC, Otto–Rummel, Panindco, PEATGAS, Pintsch–Hillebrand, PRENFLO, Riley–Morgan, Ruhrgas, Saarburg–OTTO, Shell Coal Gasification, Shell–Koppers, Winkler, Stoic, Synthane, Texaco, Thyssen–Galoczy, U-COAL, U-GAS, Viad, WD-IGI, Wellman, Wellman–Galusha, Winkler, and Woodhall-Duckham (2). *See also* carbonization.

coal liquefaction A general name for processes for converting coal to liquid products, which are usually further converted to liquid fuels. Those with special names that are described in this dictionary are Bergius, Bergius–Pier, Burgess, ChemCoal, COSTEAM, CSD, CSF, Dynagas, Exxon Donor Solvent, IG-NUE, LSE, Pott–Broche, Ruhre/VEBA, SRC, Still, and Synthoil.

COCHAR *See* COED.

CO-C-Iron *See* DR.

Codir A direct-reduction ironmaking process, which uses coal as the reductant in a rotary kiln. Developed by Krupp, Germany. First operated at the Dunswart plant in South Africa in 1973; two plants were operating in India in 1994. *See also* DR, SL/RN.

> *Ullmann's Encyclopedia of Industrial Chemistry*, 5th ed., Vol. A14, VCH Publishers, Weinheim, Germany, 1989, 563.

COED [Char-Oil-Energy Development] Formerly named COCHAR. A coal gasification process, based on carbonization in successive fluidized beds operated with non-oxygen-containing gases. The fluidizing gases were generated by partial combustion of part of the product char. The process was developed by FMC Corporation, under contract to the U.S. Office of Coal Research. It was replaced by *COGAS.

> Dainton, A.D., in *Coal and Modern Coal Processing*, Pitt, G.J. and Millward, G.R., Eds., Academic Press, London, U.K.,1979, 141.
> Probstein, R.F. and Hicks, R.E., *Synthetic Fuels*, McGraw-Hill, New York, 1982, 280.
> Newman, J.O.H., in *Chemicals from Coal: New Developments*, Payne, K.R., Ed., Blackwell Scientific Publications, Oxford, U.K., 1985, 73.

COGAS [COal GASification] A multistage coal gasification process. In the three initial stages, coal is pyrolyzed in fluidized beds to produce oils and gas. In the subsequent stages, the char is gasified with steam. The heat for the pyrolysis is provided by transferring some of the hot gas produced in the gasifiers. The pyrolysis section is similar to that in the *COED process. Developed by the COGAS Development Company, United Kingdom, in the 1970s. A large pilot plant was operated by the British Coal Utilization Research Association, Leatherhead, in 1974.

> Dainton, A.D., in *Coal and Modern Coal Processing*, Pitt, G.J. and Millward, G.R., Eds., Academic Press, London, U.K., 1979, 141.
> Hebden, D. and Stroud, H.J.F., in *Chemistry of Coal Utilization*, 2nd Suppl. Vol., Elliott, M.A., Ed., John Wiley & Sons, New York, 1981, 1701.
> Probstein, R.F. and Hicks, R.E., *Synthetic Fuels*, McGraw-Hill, New York, 1982, 282.

co-generation A general name for processes for making fuel gas and electric power from coal. Those with special names that are described in this dictionary are APAC, GEGas, ICG-GUD, IGCC, STEAG.

COIL [Concurrent OIL] A process for concurrently hydrogenating coal and heavy oil feedstocks. Developed by Hydrocarbon Research.

coking *See* carbonization.

Cold Acid A process for polymerizing isobutene, mainly into dimers and trimers, for making high-octane gasoline blending components. It is catalyzed by 60%–70% sulfuric acid at 25°C–35°C. Developed by the Shell Companies. *See also* Hot Acid.

The Petroleum Handbook, 3rd ed., Shell Petroleum, London, U.K., 1948, 228.

Cold Hydrogenation A process for selectively hydrogenating petroleum fractions made by *steam reforming, in order to produce gasoline. Developed by Bayer and now in use in 70 refineries and chemical complexes worldwide.

Krönig, W., *Hydrocarbon Process. Int. Ed.*, 1970, **49**(3), 121.
Lauer, H., *Erdoel Kohle Erdgas Petrochem.*, 1983, **36**(6), 249.

Cold lime-soda *See* Porter–Clark.

Coldstream A process for recycling cemented carbides by heating them to 1700°C, cooling in nitrogen gas, and blowing them at high speed against a carbide plate.

COLEX [COLumn EXchange] A process for separating lithium isotopes for making lithium deuteride for use in nuclear weapons. A lithium amalgam is first made, using natural lithium. This is agitated with aqueous lithium hydroxide. The desired lithium-6 concentrates in the amalgam, and the lithium-7 remains in the aqueous phase. The process is operated in a multistage contactor until the desired concentration of lithium-6 is reached. The process was operated by the U.S. Atomic Energy Commission at Oak Ridge, TN, from 1955 to 1963. Spillages of mercury from this plant caused severe environmental problems. A related process that was developed in the 1950s at Harwell and Capenhurst, United Kingdom, is mentioned indirectly in Rotherham's biography of Hans Kronberger.

U.S. Department of Energy, *Historical Data*.
Rotherham, L., *Biographical Memoirs of Fellows of the Royal Society*, 1972, **18**, 421.

Colgate-Emery A process for hydrolyzing natural triglycerides (fat splitting). It is a continuous countercurrent process operated at high pressure. Widely used in the United States.

Barnebey, H.L. and Brown, A.C., *J. Am. Oil Chem. Soc.*, 1948, **25**(3), 95.
Brady, C.L., Metcalfek, D., Slaboszewski, D., and Frank, D., *J. Am. Oil Chem. Soc.*, 1988, **65**, 917.

Collin A process for removing hydrogen sulfide from coal gas by absorption in aqueous ammonia. The hydrogen sulfide was regenerated and used for making sulfuric acid. Invented by F. Collin in 1940 and operated in England and Europe in the 1950s.

German Patent 743,088.
Williams, T.H., *Gas World*, 1955, **142**(509), 7.
Claxton, G., *Benzoles: Production and Uses*, National Benzole & Allied Products Association, London, U.K., 1961, 212.
Kohl, A.L. and Riesenfeld, F.C., *Gas Purification*, 4th ed., Gulf Publishing Co., Houston, TX, 1985, 167.

Colox An aerobic biological treatment system for municipal and industrial wastes. The biomass is fixed as a film on granules in a fixed bed. Developed in Texas by Tetra Technologies.

Chem. Mark. Rep., 1991, 2 Dec, 5, 27.
Stephenson, T., Mann, A., and Upton, J., *Chem. Ind. (London)*, 1993, (14), 533.

Combifining A petroleum refining process that removes asphaltenes, sulfur, and metals from residues, before further treatment. The catalyst is an activated petroleum coke in a fluidized bed, operated under hydrogen pressure at 380°C–420°C.

Weisser, O. and Landa, S., *Sulphide Catalysts: Their Properties and Applications*, Pergamon Press, Oxford, U.K., 1973, 352.

CombinOx A process, used in mining operations, for destroying waste cyanide by oxidation with Caro's acid and sulfur dioxide. Developed in 2002 by CyPlus (a subsidiary of Degussa) and Inco.

Eur. Chem. News, 2003, **78**(2045), 32.

COMBISULF A process for removing sulfur compounds from the gases from combined-cycle power plants with integrated coal gasification. Carbonyl sulfide is removed by catalytic hydrolysis, hydrogen sulfide is removed by selective absorption in aqueous MEDA, and the sulfur is recovered from both processes by a modified *Claus unit. Developed by Krupp Koppers and first operated in Spain in 1996. *See* PRENFLO.

Comex An ion-exchange process for removing heavy metals from wastewater by extraction into water-insoluble acids.

Boyadzhiev, L. and Khadzhiev, D., *Khim. Ind. (Sofia)*, 1981, **9**, 404 (*Chem. Abstr.*, **97**, 60331).

Cominco [**Co**nsolidated **Min**ing and Smelting **Co**mpany] A process for absorbing sulfur dioxide from smelting operations. The sulfur dioxide is absorbed in an aqueous solution of ammonium sulfite; regeneration is by acidification with sulfuric acid. The ammonium sulfate by-product is sold. Operated at the Cominco smelter at Trail, Canada, and at other smelters and sulfuric acid plants in the United States. Licensed by the Olin-Mathieson Corporation. The name has also been applied to a lead extraction process.

Kohl, H.L. and Nielsen, R.B., *Gas Purification*, 5th ed., Gulf Publishing Co., Houston, TX, 1997, 569.

Comofining [**Co**balt **mo**lybdenum **refining**] A *hydrorefining process for making lubricating oils. Developed by Lurgi and Wintershall and operated in Salzbergen, Germany.

Conrad, C. and Hermann, R., *Erdoel Kohle*, 1964, **17**, 897.

Compagnie AFC [**A**lais **F**roges **C**amargue] A variation of the *Hall–Héroult process for making aluminum metal, in which the electrolyte is a mixture of aluminum fluoride, sodium fluoride, and barium chloride. The process is operated at a lower temperature than the Hall–Héroult process, and the product has a purity of 99.99% instead of 99.5%. Developed in 1934 by the Cie des Produits Chimiques et Électrometallurgiques Alais, Froges et Camargue, in Salindres, France, now a part of Alcan.

Dennis, W.H., *A Hundred Years of Metallurgy*, Gerald Duckworth, London, U.K., 1963, 155.

Comprex A process for treating sulfide ores by high-temperature pressure leaching.

Nogueira, E.D., Regife, J.M., Redondo, A.L., and Zaplana, M., *Complex Sulfide Ores* (Conf. Proc.), 1980 (*Chem. Abstr.*, **94**, 69161).

Comprimo A version of the *Claus process offered by Comprimo Engineers & Contractors, the Netherlands. In 1983, plants using this process were being installed in Italy, Kuwait, France, and Japan. *See also* SUPERCLAUS.

Eur. Chem. News, 1983, **41**(1104), 22.

Comurhex A process for making uranium dioxide by reducing ammonium diuranate by heating it in hydrogen. Operated by the French company for which it is named.

Concat A process for removing residual sulfur-containing gases from the off-gases from the *Claus process, by oxidation to sulfur trioxide and hot condensation to sulfuric acid. Developed by Lurgi and first operated at Port Sulfur, LA, in 1974.

CONIA A process for making nitric acid simultaneously at two concentrations. *See also* CNA, DSN.

Hellmer, L., *Chem. Eng. Prog.*, 1972, **68**(4), 67.
Büchner, W., Schliebs, R., Winter, G., and Büchel, K.H., *Industrial Inorganic Chemistry*, VCH Publishers, Weinheim, Germany, 1989, 64.

ConocoPhillips Delayed Coking A process for making coke from the residues from petroleum distillation. It is essentially a two-stage process: thermal cracking followed by coking. First developed by Conoco in Oklahoma in the 1950s and now widely used worldwide. In 1999, ConocoPhillips was operating 16 of its own units (some in joint ventures), and 30 units were being operated by other companies under license.

Hughes, G.C., Wohlgenant, M.I., and Doerksen, B.J., in *Handbook of Petroleum Refining Processes*, 3rd ed., Meyers, R.A., Ed., McGraw-Hill, New York, 2003, 12.3.

CONOSOX A complex *flue-gas desulfurization process using potassium carbonate solution as the wet scrubbing medium. The product potassium bisulfite is converted to potassium thiosulfate and then reduced with carbon monoxide to potassium carbonate for reuse. The sulfur is recovered as hydrogen sulfide, which is converted to elemental sulfur by the *Claus process. Developed by the Conoco Coal Development Company and piloted in 1986.

Conox (1) A process for beneficiating sulfide ores by selective flotation. Developed and offered by Lurgi.

CONOx (2) A process for reducing NOx emissions from the *FCC process. It combines a Shell process that redirects the airflow in an *FCC regenerator with an oxygen injection system developed by Praxair that oxidizes CO and destroys NOx precursors. First installed at Valero's Ardmore refinery in 2013.

U.S. Patents 5,266,024; 7,470,412.
Chem. Eng. (N.Y.), 2009, **116**(5), 11.
Pet. Rev., 2010, **64**(767), 8.

Conrad A process for pyrolyzing waste plastics in order to produce useful fuels. An externally fired, auger rotating kiln is used at about 780°C. Developed in the 1990s by Conrad Industries, Chehalis, WA.

U.S. Patent 4,412,889.
Behzadi, S. and Farid, M., in *Feedstock Recycling and Pyrolysis of Waste Plastics*, Scheirs, J. and Kaminsky, W., Eds., John Wiley & Sons, Chichester, U.K., 2006, 537.

Consol BNR *See* carbonization.

Consortium The Consortium für Elektrochemische Industrie, founded by A. Wacker in Germany in 1903, was the corporate research laboratory of Wacker-Chemie. Many processes have been developed in this laboratory, but the one for which it is best known is the *Wacker process for making acetaldehyde; this has also been called the Consortium process.

Contact [From the German *Kontaktverfahren*, meaning catalytic process] The process now universally used for making sulfuric acid. Sulfur dioxide, made either by burning sulfur or by roasting sulfide ores, is oxidized to sulfur trioxide over a heterogeneous catalyst, typically containing vanadia, and then absorbed in sulfuric acid in several stages. The process has a long, complex history, involving various methods for overcoming catalyst poisoning. Invented in 1831 by P. Phillips, Jr., a vinegar manufacturer in Bristol, England. Initially a platinum catalyst was used, but this was replaced by vanadia in 1895. This process operated for many years in competition with the *Chamber process, but eventually displacing it because, on a large scale, it was cheaper and because the Contact yielded oleum, which is necessary for some sulfonation processes. *See also* Mannheim (2), Schröder–Grillo, Tenteleff.

British Patent 6,096 (1831).
Miles, F.D., *The Manufacture of Sulfuric Acid by the Contact Process*, Gurney & Jackson, London, U.K., 1925.
Trickett, A.A., in *Industrial Inorganic Chemicals: Production and Use*, Thompson, R., Ed., Royal Society of Chemistry, Cambridge, U.K., 1995, 93.

ConvEx A process for converting an HF-catalyzed *alkylation plant to one using sulfuric acid. Developed by Stratco.

> *Chem. Eng. (N.Y.)*, 1994, **101**(5), 23.

Cooledge A process for forming tungsten powder into wire, important in the development of the electric light bulb. Developed in 1908.

Coolside A *flue-gas desulfurization process developed by Babcock & Wilcox in the 1980s. Dry hydrated lime is injected into the gas, followed by water sprays. Piloted by Ohio Edison.

> Kohl, H.L. and Nielsen, R.B., *Gas Purification*, 5th ed., Gulf Publishing Co., Houston, TX, 1997, 623.

Copaux A method for extracting beryllium from beryl. The ore is heated with sodium fluorosilicate at 850°C. Leaching with water dissolves the beryllium fluoride, leaving the silica and most of the aluminum fluoride as an insoluble residue. Addition of sodium hydroxide precipitates beryllium as the hydroxide. The process was invented by H. Copaux and has been in use in France since 1915 and in the United Kingdom since World War II.

> Copaux, H., *C.R. Acad. Sci.*, 1919, **168**, 610.
> Bryant, P.S., in *Extraction and Refining of the Rarer Metals*, The Institute of Mining and Metallurgy, London, U.K., 1957, 310.
> Everest, D.A., *The Chemistry of Beryllium*, Elsevier, Amsterdam, the Netherlands, 1964, 104.

Copaux–Kawecki An improved version of the *Copaux process for extracting beryllium from beryl, which permits recovery of the fluorine. Addition of ferric sulfate to the dilute sodium fluoride solution remaining after the separation of the beryllium hydroxide precipitates sodium tetrafluoroferrate, which is then used in place of sodium fluorosilicate.

COPE [Claus Oxygen-based Process Expansion] A modification of the *Claus process, which improves the recovery of the sulfur. The combustion stage uses oxygen instead of air. Introduced in 1985 and now licensed by Air Products and Chemicals and Goar, Allison & Associates. Eleven units were operating in 1995.

> U.S. Patent 4,552,747.
> *Hydrocarbon Process. Int. Ed.*, 1992, **71**(4), 95.
> Schendel, R.L., *Oil Gas J.*, 1993, **91**(39), 63.
> *Hydrocarbon Process. Int. Ed.*, 1996, **75**(4), 112; 2000; **79**(4), 64.
> Kohl, H.L. and Nielsen, R.B., *Gas Purification*, 5th ed., Gulf Publishing Co., Houston, TX, 1997, 691.

Copeland A process for oxidizing organic wastes in a fluidized bed of inert particles. The wastes may be solid, liquid, or gaseous, and the oxidant is air. Inorganic residues are collected as granular solids and the heat generated is normally utilized. Developed and marketed by Copeland Systems, Oak Brook, IL, and used in a wide variety of industries.

COPISA [CO Pressure-Induced Selective Adsorption] A process for separating carbon monoxide from the effluent gases from steel mills by a two-stage *PSA unit. Developed jointly by Kawasaki Steel Corporation and Osaka Oxygen Industry. In the first stage, carbon dioxide is removed by activated carbon. In the second stage, carbon monoxide is removed by adsorption in sodium mordenite.

> Suzuki, M., in *Adsorption and Ion Exchange: Fundamentals and Applications*, LeVan, M.D., Ed., American Institute of Chemical Engineers, New York, 1988, 123.

COPox [CObalt Partial oxidation] A three-stage *GTL process. The first stage produces *syngas by partial oxidation. The second stage converts this to a mixture of hydrocarbons by the *FT process, using a cobalt catalyst in a fluidized-bed reactor. The third stage converts this mixture to the required products: diesel, naphtha, and LPG. Designed by ConocoPhillips and demonstrated in Ponca City, OK, in 2003.

Chem. Eng. (N.Y.), 2004, **111**(5), 23.
Basini, L., *Catal. Today*, 2006, **117**(4), 384.

Coppee *See* carbonization.

COPSA [**CO** **P**ressure-**S**wing **A**dsorption] A process similar to *COPISA but using activated charcoal impregnated with cuprous chloride as the adsorbent for the carbon monoxide. Developed by Mitsubishi Kakoki Kaisha.

Suzuki, M., in *Adsorption and Ion Exchange: Fundamentals and Applications*, LeVan, M.D., Ed., American Institute of Chemical Engineers, New York, 1988, 123.

COREX A two-stage ironmaking process. Iron ore is reduced in a vertical shaft furnace and then melted in a melter gasifier, which also generates reducing gases for the shaft furnace. Developed by DVAI, Düsseldorf and VAI, Linz and first used in 1989 at the Pretoria works of Iscor, South Africa. Four plants were operating in 2002.

Chem. Eng. (N.Y.), 1991, **98**(5), 32.
Downie, N.A., *Industrial Gases*, Blackie Academic, London, U.K., 1997, 302.

Corinthian A common misspelling of *Carinthian.

CORRS [**CO**$_2$ **O**nsite **R**ecovery and **R**ecycling **S**ystem] A process for recovering carbon dioxide from gas mixtures containing >8% CO$_2$ by diffusion through a selective *Generon membrane. Offered by the Messer Group, based on technology developed by the Dow Chemical Company.

Cort A process for making wrought iron by *puddling. Invented by Henry Cort in England in 1784.

Derry, T.K. and Williams, T.I., *A Short History of Technology*, Oxford University Press, Oxford, U.K., 4th impression, 1979, 477.

Cosden A process for polymerizing isobutene to a polyisobutene having a molecular weight between 300 and 2700.

Weissermel, K. and Arpe, H.-J., *Industrial Organic Chemistry*, 4th ed., translated by C.R. Hindley and S. Hawkins, Wiley-VCH, Weinheim, Germany, 2003, 73.

Coskata A process for making fuels from biomass. The biomass is first converted to syngas that is then converted to ethanol by a biological process. First developed at the Oklahoma State University and the University of Oklahoma, then piloted by Coskata from 2006. An agreement between Coskata and General Motors was announced in 2008. A semicommercial plant was operated from 2009 to 2011 and a commercial plant in Boligee, AL, was announced in 2011.

Chem. Ind. (London), Feb 2008, (03), 11.

Coslettizing *See* metal surface treatment.

COSMOS [**C**racking **O**il by **S**team and **MO**lten **S**alts] A catalytic process for cracking petroleum or heavy oils. The catalyst is a molten mixture of the carbonates of lithium, sodium, and potassium. Developed by Mitsui and piloted in 1977.

Hu, Y.C., in *Chemical Processing Handbook*, Marcel Dekker, New York, 1993, 783.

Cosorb [**CO** ab**sorb**] A process for recovering carbon monoxide by absorption in a solution of cuprous aluminum chloride in toluene. Three stages are involved: absorption, desorption, and washing. Invented by Esso Research and Engineering Company and then developed by Tenneco Chemicals in the early 1970s. Piloted in

1976, after which many large-scale plants were built worldwide. The copper solution has also been proposed for transporting acetylene. A variation, referred to as a PSA system (**P**ressure **S**hifting **A**bsorption, not to be confused with *pressure swing adsorption), was developed in Japan jointly by Mitsubishi Kakoki Kaisha and Nichimen Corporation.

U.S. Patents 3,651,159; 3,767,725.

Haase, D.J. and Walker, D.G., *Chem. Eng. Prog.*, 1974, **70**(5), 74.

Walker, D.G., *CHEMTECH*, 1975, **5**, A62, 308.

Haase, D.J., Duke, P.M., and Cates, J.W., *Hydrocarbon Process. Int. Ed.*, 1982, **61**(3), 103.

Keller, G.E., Marcinowsky, A.E., Verma, S.K., and Williamson, K.D., in *Separation and Purification Technology*, Li, N.N. and Calo, J M., Eds., Marcel Dekker, New York, 1992, 64.

Weissermel, K. and Arpe, H.-J., *Industrial Organic Chemistry*, 4th ed., translated by C.R. Hindley and S. Hawkins, Wiley-VCH, Weinheim, Germany, 2003, 25.

COSTEAM A process for obtaining both gas and electric power from coal. The coal is first liquefied by a process that is catalyzed by modifying iron compounds naturally present in some coals. Developed on a laboratory scale by the Pittsburgh Energy Technology Center in the 1970s.

Dainton, A.D., in *Coal and Modern Coal Processing*, Pitt, G.J. and Millward, G.R., Eds., Academic Press, London, U.K., 1979, 172.

Alpert, S.B. and Wolk, R.H., in *Chemistry of Coal Utilization*, 2nd Suppl. Vol., Elliott, M. A., Ed., John Wiley & Sons, New York, 1981, 1934.

Courtaulds Tencel *See* Lyocell.

Cowles An electrothermal process for making aluminum alloys. A mixture of bauxite, charcoal, and the metal forming the alloy (usually copper) was heated in an electric furnace, and the molten alloy was tapped from the base. The process cannot be used for making aluminum alone, because in the absence of the other metal, the product would be aluminum carbide. Invented by the Cowles brothers and operated in Cleveland, OH, in 1884 and later in Stoke-on-Trent, England. The electrical efficiency was poor and the process was superseded by the Hall–Héroult process.

Dennis, W.H., *A Hundred Years of Metallurgy*, Gerald Duckworth, London, U.K., 1963, 145.

CP [**C**ontinuous **P**olymerization] A continuous process for making high-density polyethylene, based on the *Ziegler process but using a much more active catalyst so that deashing (catalyst removal) is not required. Developed by Mitsui Petrochemical Industries and upgraded into its *CX process, which was first licensed in 1976.

C₅ Pentesom [**Pente**ne **isom**erization] A process for isomerizing linear pentenes to isopentenes, catalyzed by a zeolite. The isopentenes are intended as intermediates in the production of ethers for use as fuel additives. Developed by UOP in 1992 and offered for license in 1993. *See also* C₄ Butesom.

Krupa, S., in *Handbook of Petroleum Refining Processes*, 3rd ed., Meyers, R.A., Ed., McGraw-Hill, New York, 2003, 13.13.

CPG [**C**lean **P**ower **G**eneration] A version of the *IGCC process that incorporates the *Texaco coal gasifier. Developed and offered by Humphreys and Glasgow.

CPJ [Named after the inventor, Pierre Jorgensen] A process for upgrading bitumen and heavy oils into lighter, more easily processed hydrocarbon mixtures. The feed is subject to instantaneous thermal shock by the use of superheated steam. Developed in France and piloted in Canada in 2000 by Carbon Resources Ltd.

Oil Gas J., 2000, **98**(47), 56.

CPOX [**C**atalytic **P**artial **O**xidation] A general name for processes that catalytically convert methane to *syngas. A number of industrial companies and universities have published research on this but there are no commercial applications yet. *See also* ***POX***.

Kado, S., Imagawa, K., Kiryu, A., Yagi, F., Minami, T., Kawai, H., Kawai, H., Kawazushi, K., Tomashigi, K., Nakamura, A., and Suehiro, Y., *Catal. Today*, 2011, **171**(1), 97.

CPP [Catalytic Pyrolysis Process] A general term for processes that convert biomass to useful products by heating in the absence of air.

Bulushev, D.A. and Ross, J.P.H., *Catal. Today*, 2011, **171**, 1.

cracking The pyrolysis of petroleum fractions to produce lower molecular weight products. *See* Catalytic Cracking, Thermal Cracking.

Creighton An electrolytic process for reducing sugars to their corresponding polyols. Glucose is thus reduced to sorbitol, mannose to mannitol, and xylose to xylitol. The electrodes are made of amalgamated lead or zinc; the electrolyte is sodium sulfate. Invented in 1926 by H.J. Creighton.

U.S. Patents 1,612,361; 1,653,004; 1,712,951; 1,712,952; 2,458,895.
Creighton, H.J., *Trans. Electrochem. Soc.*, 1939, **75**, 289.
Pigmann, W.W. and Goepp, R.M., *Chemistry of the Carbohydrates*, Academic Press, New York, 1948, 238.
Creighton, H.J., *J. Electrochem. Soc.*, 1952, **99**, 127C.

Cresex [**Cres**ol **ex**traction] One of the *Sorbex processes. This one extracts *p*- or *m*-cresol from mixed cresols and cresols as a class from higher alkyl phenols. By 1990, one plant had been licensed.

Johnson, J.A., in *Handbook of Petroleum Refining Processes*, 3rd ed., Meyers, R.A., Ed., McGraw-Hill, New York, 2003, 10.29.

CRG [Catalytic Rich Gas] A process for making town gas and rich gas from light petroleum distillate (naphtha). The naphtha is reacted with steam over a nickel–alumina catalyst, yielding a gas mixture rich in methane. Developed by British Gas and used in the United Kingdom in the 1960s but abandoned there after the discovery of North Sea gas. In 1977, 13 plants were operating in the United States.

Gas Making and Natural Gas, British Petroleum, London, U.K., 1972, 126.

CRIJ *See* carbonization.

CRISFER A glassmaking process developed by Rhône-Poulenc ca. 1985.

Critox *See* Aqua Critox.

Cros A process for making diammonium phosphate, offered by Davy Corporation.

Crosfield A continuous soapmaking process developed by J. Crosfield & Sons, Warrington, England, and used in the manufacture of the detergent Persil from 1962.

Musson, A.E., *Enterprise in Soap and Chemicals*, Manchester University Press, Manchester, U.K., 1965, 364.
Kirk-Othmer's Encyclopedia of Chemical Technology, 3rd ed., Vol. 21, John Wiley & Sons, New York, 1983, 174.

Cross A high-pressure, mixed-phase, thermal process for cracking petroleum, introduced in the United States in 1924 by the Cross brothers, further developed by the M.W. Kellogg Company, and widely used in the 1920s and 1930s. Eventually, 130 units were built in the United States and abroad.

Spitz, P.H., *Petrochemicals: The Rise of an Industry*, John Wiley & Sons, New York, 1988, 168.

Cross–Bevan–Beadle *See* Viscose.

Crucible *See* Huntsman.

Cryoplus A cryogenic process for separating hydrocarbon gases. Developed and licensed by Technip. In 1992, more than 10 units were operating.

> *Hydrocarbon Process. Int. Ed.*, 1992, **71**(4), 96.

Crystalactor A general range of processes for recovering useful materials from wastewaters. It uses a fluidized-bed pellet reactor and the desired product crystallizes on the seed particles. Developed by Royal Hasking, Amersfoort, the Netherlands.

> *Chem. Eng. (N.Y.)*, 2013, **120**(2), 17.

CrystaSulf A process that uses a nonaqueous solvent–catalyst system to remove sulfuric acid from high-pressure natural gas. This project, part of the GRI Basic Research Program, has been conducted by Radian Corporation. The process was demonstrated in West Texas in 1999.

> *Sulphur*, 1996, (246), 58, 61.
> *Oil Gas J.*, 1997, **95**(29), 54.
> *Hydrocarbon Process. Int. Ed.*, 2000, **79**(4), 64.
> *Sulphur*, 2001, (273), 46.

CrystPX A process for separating *p*-xylene from xylene mixtures by crystallization. Only crystallizers and centrifuges are used. Developed by GTC Technology and offered by that company and Lyondell Chemical.

CSA [Catalytic Solvent Abatement] A process for removing chlorinated solvents from waste gases by catalytic oxidation. Two catalysts are used in series, and the products are carbon dioxide, water, and hydrogen chloride. Developed in Germany by Hoechst and Degussa and licensed by Tebodin in the Netherlands.

> *Hydrocarbon Process. Int. Ed.*, 1993, **72**(8), 77.

CSD [Critical Solvent Deashing] A process for removing insoluble material from coal before liquefaction, using toluene as a solvent under subcritical conditions. Used in the Kerr-McGee and National Coal Board processes.

CSF [Consol Synthetic Fuels] A two-stage *coal liquefaction process. In the first stage, the coal is extracted with process-derived oil and the ash removed. In the second, the extract is catalytically hydrogenated. Piloted by the Consolidation Coal Company, Cresap, WV, from 1963 to 1972. *See also* H-Coal, SRC, Synthoil.

> Alpert, S.B. and Wolk, R.H., in *Chemistry of Coal Utilization*, 2nd Suppl. Vol., Elliott, M.A., Ed., John Wiley & Sons, New York, 1981, 1954.

CSMP [Cupola Surface Melting Process] A process for vitrifying residues from the incineration of municipal wastes. Offered by ML Entsorgungs und Energieanlagen.

CT-121 [Chiyoda Thoroughbred] *See* Thoroughbred.

CT-BISA [Chiyoda Thoroughbred BISphenol-A] A catalytic process for making bisphenol-A from phenol and acetone. The catalyst is an acidic ion-exchange resin. The product is used for making polycarbonate resins. Developed and offered by Chiyoda Corporation, Japan. The first plant was operated in Tobata, Japan, in 1997.

CTL (1) [Coal To Liquids] A general name for processes that convert coal to liquid hydrocarbons, usually for use as fuels. There has been a long history of the development of such processes, notably the *Bergius process used in Germany for making petrol from coal in World War II. These processes are classified as direct or indirect, the latter involving prior gasification to *syngas.

> A partial listing of 11 U.S. projects is given in the 2008 reference.

See also *Kohleoel, *NEDOL, *SRC.
Chem. Eng. (N.Y.), 2006, **113**(2), 16.
Oil Gas J., 2008, **106**(12) 26.
Wikipedia, The Free Encyclopedia. Coal liquefaction. Revised 17 Apr 2013. Retrieved 20 Jul 2013.

CTL (2) [Canadium Titanium Ltd.] A novel process for making titanium dioxide pigments. The ore is leached with hydrochloric acid and the leachate purified by several solvent extraction steps, yielding a titanium oxychloride solution. The solid oxychloride is finally calcined to give the pigment. Developed by Canadium Titanium Ltd. (now a subsidiary of Argex Mining Inc.) in a large pilot plant in Mississauga, ON, from 2002. Argex Titanium is building a commercial plant in Salaberry-de-Valleyfield, Quebec, for completion in 2015.

Canadian Patent 2,513,309.
U.S. Patent 7,803,336.
Focus on Pigments, Jun 2013, 4.
Focus on Pigments, Jun 2013, 4; Dec 2013, 5.

CTM [Coal To Methanol] A general name for processes that convert coal to methanol.

CTO [Coal To Olefins] A general name for processes that convert coal to ethylene and/or propylene, used for making petrochemicals or fuels. Operated in South Africa, using catalysts developed and supplied by Süd-Chemie.

Chem. Week, 2006, **168**(7), 22.

CTX A *BAF process.

Stephenson, T., Mann, A., and Upton, J., *Chem. Ind. (London)*, 1993, (14), 533.

Cumene *See* Hock.

Cumox [Cumene oxidation] A process for making acetone and phenol by oxidizing cumene, based on the *Hock process. This version was further developed and licensed by UOP. Three plants were operating in 1986. UOP now licenses the Allied-UOP Phenol process, which combines the best features of Cumox and a related process developed by the Allied Chemical Corporation.

Pujado, P.R., Salazar, J.R., and Berger, C.V., *Hydrocarbon Process Int. Ed.*, 1976, **55**(3), 91.
Hydrocarbon Process. Int. Ed., 1981, **60**(11), 198.
Pujado, P.R., in *Handbook of Chemicals Production Processes*, Meyers, R.A., Ed., McGraw-Hill, New York, 1986, 1.16-1.

Cupellation An ancient metallurgical operation, still in use, for removing lead and other base metals from silver by blowing air over the surface of the molten metal. The lead oxidizes to lead monoxide (litharge), which floats on the molten silver and is separated off. The molten litharge dissolves the other base metal oxides present. A cupel is the shallow refractory dish in which the operation is conducted.

Old Testament, *Jer.*, **vi**, 29.
Dennis, W.H., *A Hundred Years of Metallurgy*, Gerald Duckworth, London, U.K., 1963, 193.

Cuprammonium A process for making regenerated cellulose fibers. Cellulose, from cotton or wood, is dissolved in ammoniacal copper sulfate solution (Schweizer's reagent, also called cuprammonium sulfate). Injection of this solution into a bath of dilute sulfuric acid regenerates the cellulose as a fiber. The process was invented in 1891 by M. Fremery and J. Urban at the Glanzstoff-Fabriken, Germany, developed there, and subsequently widely adopted worldwide. *See also* Bemberg, Chardonnet, Viscose.

Moncrieff, R.W., *Man-made Fibres*, 6th ed., Newnes-Butterworths, London, U.K., 1975, 224.

Cuprasol Also called EIC. A process for removing hydrogen sulfide and ammonia from geothermal steam by scrubbing with an aqueous solution of copper sulfate. The resulting copper sulfide slurry is oxidized with air, and the copper sulfate is reused. The sulfur is recovered as ammonium sulfate. Developed by the EIC Corporation, MA. Demonstrated by the Pacific Gas & Electric Company at Geysers, CA, in 1979, but finally abandoned.

U.S. Patent 4,192,854.
Chem. Eng. News, 1979, **57**(49), 29.
Kohl, H.L. and Nielsen, R.B., *Gas Purification*, 5th ed., Gulf Publishing Co., Houston, TX, 1997, 853.

Cuprex [**Copper ex**traction] A process for extracting copper from sulfide ores, combining chloride leaching, electrowinning, and solvent extraction. Piloted by a consortium consisting of ICI, Nerco Minerals Company, and Técnicas Reunidas.

Dalton, R.F., Price, R., Ermana, H., and Hoffman, B., in *Separation Processes in Hydrometallurgy*, Davies, G.A., Ed., Ellis Horwood, Chichester, U.K., 1987, 466.

CVD [**C**hemical **V**apor **D**eposition] A general term for any process for depositing a solid on a solid surface by a chemical reaction from reactants in the gas phase. To be distinguished from *physical vapor deposition (PVD), in which no chemical reaction takes place. (For the 10 international conferences held on CVD between 1967 and 1987, see the reference by Stinton et al. in the succeeding text.)

Stinton, D.P., Besman, T.W., and Lowden, R.A., *Am. Ceram. Soc. Bull.*, 1988, **67**, 350.
Hocking, M.G., Vasantasree, V., and Sidky, P.S., *Metallic and Ceramic Coatings*, Longman, Harlow, U.K., 1989.
Journal of Chemical Vapor Deposition, Technomic Publishing, Lancaster, PA, 1992.

CVI [**C**hemical **V**apor **I**nfiltration] A ceramic manufacturing process in which the pores of a matrix are filed by *CVD.

Chiang, Y.-M., Haggerty, J.S., Messner, R.P., and Demetry, C., *Am. Ceram. Soc. Bull.*, 1989, **68**, 420.

CWAO [**C**atalytic **W**et **A**tmospheric **O**xidation] A general name for catalytic processes for destroying waste organic chemicals in water by atmospheric oxidation. First operated by DuPont in the 1950s. Such processes can also be used for selective organic syntheses. *See also* *Zimpro.

U.S. Patent 2,690,425.
Luck, F., *Catal. Today*, 1999, **53**(1), 84.
Pintar, A., *Catal. Today*, 2003, **77**(4), 451.
Levec, J. and Pintar, A., *Catal. Today*, 2007, **124**(3–4), 172.

CX A continuous process for making high-density polyethylene, based on the *Ziegler process, but using a much more active catalyst so that deashing (catalyst removal) is not required. Developed by Mitsui Petrochemical Industries from its *CP process and now licensed by Mitsui Chemicals. First licensed in 1976. By 2005, 41 units were operating or under construction.

CXR [**C**ompact e**X**change **R**eforming] A process for converting natural gas into *syngas. Developed from 1996 jointly by Methanex, Synetix, and ABB Lummus Global but subsequently abandoned.

Chem. Eng. (N.Y.), Jan 2000, **1107**(1), 17.

CyAM [**Cy**anide **AM**monia] A process for reducing the cyanide concentration in the ammonia liquor from coke manufacture, so that the liquor may be fed to an Activated Sludge effluent treatment plant. Developed by the U.S. Steel Corporation and used by that company in two of its coking plants.

Glassman, D., *Ironmaking Conf. Proc.*, 1976, **35**, 121 (*Chem. Abs.*, **89**, 168444).

Cyanamide *See* Frank–Caro.

Cyan-cat A process for destroying gaseous hydrogen cyanide by catalytic oxidation over platinum metal at ca. 300°C.

Martinez, D., in *Chemical Waste Handling and Treatment*, Muller, K.R., Ed., Berlin, Germany, 1986, 133.

Cyanex 301 One of the solvent extraction processes, used together with *UREX, for separating the components of used nuclear fuel. This process uses a complex phosphinic acid [bis(2,4,4-trimethylpentyl)dithiophosphinic acid] made by Cytec Industries, Canada. Its purpose is to separate americium, curium, and lanthanide fission products from the other components.

cyanidation *See* Cyanide.

cyanide Also called cyanidation. A process for extracting gold from crushed rock by putting it in contact with aqueous sodium or potassium cyanide in the presence of air. The gold is converted to an aurocyanide:

$$4Au + 8KCN + 2H_2O + O_2 = 4KAu(CN)_2 + 4KOH$$

The solution is then reduced with metallic zinc:

$$2KAu(CN)_2 + Zn = K_2Zn(CN)_4 + 2Au$$

First patented by J.W. Simpson in 1884 and developed by J.S. MacArthur and R. Forrest and R.W. Forrest in Glasgow in 1887. The first commercial application was at the Crown Mine, New Zealand, in 1889, followed shortly by mines in South Africa and the United States. If zinc dust is used, the process is known as the *Merrill–Crowe process.

Dennis, W.H., *A Hundred Years of Metallurgy*, Gerald Duckworth, London, U.K., 1963, 269.
Yannopoulos, J.C., *The Extractive Metallurgy of Gold*, Van Nostrand Reinhold, New York, 1991, 67, 143.

Cyclar [**Cycl**ization of light hydrocarbons to **ar**omatics] A catalytic process for converting light hydrocarbons, typically mixtures of C_3 and C_4 acyclic aliphatic hydrocarbons, to aromatic liquid products. Also called dehydrocyclodimerization (DHCD) because the mechanism is believed to be dehydrogenation, followed by dimerization, followed by cyclization. Developed jointly by UOP and British Petroleum. The catalyst, which is zeolite ZSM-5 containing gallium, is continuously regenerated in a separate unit. A large demonstration plant was started up in Grangemouth, Scotland, in 1989. The first industrial plant was built at Sabic's plant at Yanbu, Saudi Arabia, in the late 1990s. *See also* CCR.

U.S. Patents 4,175,057; 4,180,689; 4,727,206.
Franck, H.-G. and Stadelhofer, J.W., *Industrial Aromatic Chemistry*, Springer-Verlag, Berlin, Germany, 1988, 88.
Doolan, P.C. and Pujado, P.R., *Hydrocarbon Process. Int. Ed.*, 1989, **68**(9), 72.
Chem. Eng. (Rugby, Engl.), 1991, (491), 12.
Gosling, C.D., Wilcher, F.P., Sullivan, L., and Mountford, R.A., *Hydrocarbon Process. Int. Ed.*, 1991, **70**(12), 69.
Hagen, A. and Roessner, F., *Catal. Rev., Sci. Eng.*, 2000, **42**(4), 403.
Zhou, L., in *Handbook of Petroleum Refining Processes*, 3rd ed., Meyers, R.A., Ed., McGraw-Hill, New York, 2003, 2.29.

Cyclemax CCR An improved version of UOP's *CCR process, operated in India.

Ganapati, M., Ding, R., and Mooley, P.D., *Oil Gas J.*, 2000, **98**(25), 48.

Cyclopol A process for making cyclohexanone from benzene, the intermediates being cyclohexane and cyclohexanol. Developed and licensed by Polimex-Cekop. In 1997, 20% of world demand for cyclohexanone was made by this process.

Haber, J. and Borowiak, M., *Appl. Catal. A: Gen.*, 1997, **155**(2), 292.

CYCLOPOL-bis An improved version of *Cyclopol, first operated in 2003. It uses two reactors, the second being used to decompose the cyclohexyl peroxide under mild conditions.

Gruszka, M. et al., *Przem. Chem.*, 2005, **84**, 493, reported in *Appl. Catal. A: Gen.*, 2005, **296**(1), N2.

Cycloversion A petroleum treatment process that combined *Catalytic Reforming with *hydrodesulfurization. The catalyst was bauxite. The process differed from the *Houdry process in that the catalyst bed temperature was controlled by injecting an inert gas. Developed by the Phillips Petroleum Company and used in the United States in the 1940s.

Pet. Refin., 1960, **39**(9), 205.
Unzelman, G.H. and Wolf, C.J., in *Petroleum Processing Handbook*, Bland, W.F. and Davidson, R.L., Eds., McGraw-Hill, New York, 1967, 61.
Little, D.M., *Catalytic Reforming*, PennWell Publishing, Tulsa, OK, 1985, xiv.

Cycom A combustion process for treating solid industrial wastes. The solids are fed into the top of a vertical, gas-fired cylinder. The ash melts and flows down the walls for recovery. Developed by IGT, Chicago, and piloted with several industrial wastes in the 1990s.

Chem. Eng. (N.Y.), 1996, **103**(4), 25.

Cymet (1) [**Cy**prus **met**allurgical] A process for extracting copper from sulfide ores. It involves leaching, solvent extraction, and electrowinning. Developed by the Cyprus Metallurgical Processes Corporation, Golden, CO. Superseded by *Cymet (2).

Cymet (2) [**Cy**prus **met**allurgical] A process for extracting copper from sulfide ores. Copper is leached from the ore using aqueous ferric and cupric chloride solution:

$$CuFeS_2 + 3FeCl_3 = CuCl + 4FeCl_2 + 2S$$

The copper is crystallized out as cuprous chloride, which is then reduced with hydrogen in a fluidized-bed reactor. Developed by Cyprus Mines Corporation, successor to the Cyprus Metallurgical Processes Corporation, which developed Cymet (1). *See also* CLEAR.

U.S. Patent 3,972,711.
Kruesi, P.R., Allen, E.S., and Lake, J.L., *CIM Bull.*, 1973, **66**, 81.
Eng. Min. J., 1977, **178**(11), 27.
Mukherjee, C.K. and Mukherjee, T.K., *Hydrometallurgy in Extraction Processes*, Vol. 1, CRC Press, Boca Raton, FL, 1990, 12, 188.

Cymex [**Cym**ene **ex**traction] One of the *Sorbex processes, for extracting *p*-cymene or *m*-cymene from cymene isomers. By 1990, one plant had been licensed.

CYTOX An aerobic sewage treatment process using pure oxygen, similar to *UNOX. Developed by the Cytox Equipment Corporation. The name is used also for an industrial biocide and a German company offering biological testing services.

CZD [**C**ombined **Z**one **D**ispersion] Also called Bechtel CZD. A *flue-gas desulfurization process in which a slurry of a chemically reactive form of lime (pressure-hydrated dolomitic lime) is injected into the duct. The water in the slurry evaporates quickly, and the solid product is collected downstream in an electrostatic precipitator. Developed by Bechtel International in the late 1980s, intended for retrofitting to an existing plant.

PCT Patent Application 80/1377.

Czrochralski A process for growing large single crystals. The bulk of the material is melted in a crucible. A single crystal of the same material is lowered onto the surface of the melt and then slowly pulled upward, producing a cylindrical single crystal known as a boule. Invented by J. Czochralski as a method for determining the velocity of crystallization of molten metals.

Czochralski, Z., *Phys. Chem.*, 1917, **92**, 219.
Vere, A.W., *Crystal Growth: Principles and Progress*, Plenum Press, New York, 1987, 67.

D

3D [**D**iscriminatory **D**estructive **D**istillation] A thermal deasphalting process, which uses the same short contact time concept as the *MSCC process and a circulating solid for heat transfer between reactor and generator. 3D is claimed to be most effective on heavy, contaminated whole crude oils or residues. Developed by Bar-Co and now offered by UOP.

> *Eur. Chem. News*, 1995, **64**(1682), 28.
> *Hydrocarbon Process. Int. Ed.*, 1996, **75**(11), 96.

3-DDM A process for making cumene from propylene and benzene by alkylation followed by transalkylation over a dealuminated mordenite catalyst. Operated since 1992 by Dow Chemicals at Terneuzen, the Netherlands.

> Degnan, T.F., Jr., Smith, C.M., and Venkat, C.R., *Appl. Catal. A: Gen.*, 2001, **221**(1–2), 291.

Daicel A process for making propylene oxide by oxidizing propylene with peroxyacetic acid. The peroxyacetic acid is made by reacting ethyl acetate, acetaldehyde (a metal ion catalyst), and air. Developed by Daicel Chemical Industries, Japan, from 1966 and commercialized from 1969 to 1980. *See also* Propylox.

> U.S. Patents 3,654,094; 3,663,574.
> Yamagishi, K., Kageyama, O., Haruki, H., and Numa, Y., *Hydrocarbon Process. Int. Ed.*, 1976, **55**(11), 102.
> Weissermel, K. and Arpe, H.-J., *Industrial Organic Chemistry*, 4th ed., translated by C.R. Hindley and S. Hawkins, Wiley-VCH, Weinheim, Germany, 2003, 273.

Daniell A process for making gas from rosin, the residue from the distillation of turpentine. Used in the United States in the nineteenth century for making gas for lighting. Invented by J.F. Daniell, better known for his invention of an electrical cell.

> Davies, D.I., *Chem. Br.*, 1990, **26**(10), 946.

Dapex [**Di**alkyl**p**hosphoric acid **ex**traction] A process for the solvent extraction of uranium from sulfuric acid solutions using di-(2-ethylhexyl) phosphoric acid (HDEHP). The HDEHP is dissolved in kerosene containing 4% of tributyl phosphate. The uranium is stripped from the organic phase by aqueous sodium carbonate and precipitated as uranyl peroxide (yellow cake). The process was no longer in use in 1988. *See also* Amex.

> *Chem. Eng. News*, 1956, **34**(21), 2590.
> Blake, C.A., Baes, C.F., Jr., and Brown, K.B., *Ind. Eng. Chem.*, 1958, **50**, 1763.
> Danesi, P.R., in *Developments in Solvent Extraction*, Alegret, S., Ed., Ellis Horwood, Chichester, U.K., 1988, 206.
> Johnston, B.E., *Chem. Ind. (London)*, 1988, (20), 658.

DAP-Mn Also called the manganese dioxide process. A *flue-gas desulfurization process using a fluidized bed of manganese dioxide, which becomes converted to manganous sulfate. The adsorbent is regenerated with ammonia.

> Speight, J.G., *Gas Processing*, Butterworth Heinemann, Oxford, U.K., 1993, 303.

DAV A *DR ironmaking process, using coal as the reductant. Operated since 1985 at the Davsteel plant in Vanderbijlpark, South Africa. *See* DR.

Davey A modification of the *Parkes process for removing silver from lead. A water-cooled tray is floated on the molten lead. Invented by T.R.A. Davies in 1970 and operated by Peñarroya in Brazil, France, Greece, and Spain.

> U.S. Patent 4,356,033.
> Davey, T.R.A. and Bied-Charreton, B., *J. Met.*, 1983, **35**(8), 37.

Davison A process for making phosphate fertilizer.

Day–Kesting *See* Kesting.

DCC [Deep Catalytic Cracking] A general term for processes that convert heavy petroleum feedstocks and residues to light hydrocarbons. One such process, for making C_3–C_5 olefins, was developed by the Research Institute of Petroleum Processing, China, and licensed through Stone & Webster. Eight units were operating in China and Thailand in 2005, and another was planned for Saudi Arabia.

> Letzsch, W.S., in *Handbook of Petroleum Refining Processes*, 3rd ed., Meyers, R.A., Ed., McGraw-Hill, New York, 2003, 3.35.

DCH (1) [Direct Contact Hydrogenation] A process that uses catalytic *hydroprocessing to re-refine spent lubricating oils to produce clean fuel, lubricating oil base stocks, or hydrocarbon feedstock for cracking. Developed by UOP but not commercialized as of 1991.

DCH (2) An early name for *HyChlor.

Deacon Also called Deacon–Hurter. A process for oxidizing hydrogen chloride to chlorine, using atmospheric oxygen and a heterogeneous catalyst:

$$4HCl + O_2 = 2H_2O + 2Cl_2$$

The catalyst is cupric chloride, supported on calcined clay. The resulting chlorine is mixed with nitrogen and steam, but that is not disadvantageous when the product is used to make bleaching powder. Invented by H. Deacon in Widnes, England, in 1868, in order to utilize the hydrochloric acid by-product from the *Leblanc process. W. Hurter improved on Deacon's original invention by passing the gases downward through the catalyst column. Used in competition with the *Weldon process, until both processes were made obsolete by the invention of the electrolytic process for making chlorine from brine. More modern variations, using superior catalysts, have been operated in the Netherlands and India. Sumitomo announced a new process using a new catalyst in 2002. A modern version of the process is described in the recent U.S. Patent. *See also* Shell Deacon.

> British Patent 1403, 1868.
> U.S. Patent 85,370, 1868.
> Hardie, D.W.F., *A History of the Chemical Industry in Widnes*, Imperial Chemical Industries, Widnes, U.K., 1950, 67.
> Redniss, A., in *Chlorine: Its Manufacture, Properties and Uses*, Sconce, J.S., Ed., Reinhold, New York, 1962, 251.
> *Hydrocarbon Process. Int. Ed.*, 2002, **81**(2), 27–28.
> Weissermel, K. and Arpe, H.-J., *Industrial Organic Chemistry*, 4th ed., translated by C.R. Hindley and S. Hawkins, Wiley-VCH, Weinheim, Germany, 2003, 221.
> U.S. Patent 8,158,099.

dealkylation In organic chemistry, this word can be used to describe any reaction in which one or more alkyl groups are removed from a molecule. In process chemistry, the word has a more restricted meaning: the conversion of toluene or xylene to benzene. Because the process occurs in an atmosphere of hydrogen, it is also called hydrodealkylation. Named processes that achieve this are *DETOL, *Hydeal, and *Litol.

Dean A process for extracting manganese from low-grade ores by extraction with aqueous ammonium carbamate. Invented by R.S. Dean and used since 1950 by the Manganese Chemical Company, now part of Diamond Shamrock Chemical Company.

> Weiss, S.A., *Manganese: The Other Uses*, Garden City Press, Letchworth, U.K., 1977, 152, 155.

Debatox A rotary kiln system for recycling consumer battery materials. Developed by Sulzer Chemtech. The system first shreds the batteries and then incinerates them. Carbon, plastics, and paper are burned. Dioxins are destroyed in an afterburner, and mercury is condensed in a scrubber. The residual solids, containing zinc, manganese, and iron, can be recycled by standard smelters.

> *Chem. Eng. (N.Y.)*, 1996, **103**(4), 19.

Decompozon [**Decomp**ose **ozon**e] A process for destroying ozone in gas streams by passage through a fixed bed of a proprietary catalyst containing nickel. Developed by Ultrox International, Santa Ana, CA.

DeDiox A process for destroying polychlorinated dioxins and furans in flue gases by catalytic oxidation with hydrogen peroxide. The catalyst is based on silica, and the process is operated at 80°C–100°C. Developed by Degussa from 1994. The business was offered for sale in 1998.

> *Eur. Chem. News*, 1998, **69**(1801), 33.

Deep Shaft A high-intensity biological treatment process for purifying domestic sewage and biodegradable industrial effluents. The waste is circulated very rapidly through a vertical loop reactor, 30–150 m in height, usually installed underground. Compressed air, introduced at the middle of both legs, drives the circulation. Because of its rapid movement, the sludge operates at a much higher density than is normal in *Activated Sludge plants. Developed by ICI in the 1970s as a spin-off of its *Pruteen process. By 1994, more than 60 plants had been installed worldwide. The process was acquired by Davy International in 1995. *See also* VerTech.

> British Patent 1,473,665.
> Dunlop, E.H., in *New Processes of Waste Water Treatment and Recovery*, Ellis Horwood, Chichester, U.K., 1978, 177.
> Gray, N.F., *Activated Sludge: Theory and Practice*, Oxford University Press, Oxford, U.K., 1990, 113.
> Horan, N.J., *Biological Wastewater Treatment Systems*, John Wiley & Sons, Chichester, U.K., 1990, 69.
> *Water Sewage Int.*, 1991, **12**, 30.
> *Chem. Eng. (Rugby, Engl.)*, 1994, (569), 23.

DeFine [**Diole**Fine saturation] A process for converting diolefins to monoolefins by selective dehydrogenation. Developed by UOP in 1984 for use with its *Pacol process. First commercialized in 1986 and now incorporated in all new Pacol plants. Six units were operating in 1990.

> U.S. Patents 4,523,045; 4,523,048; 4,761,509.
> Vora, B., Pujado, P., Imai, T., and Fritsch, T., *Chem. Ind. (London)*, 1990, (6), 187.
> Pujado, P.R., in *Handbook of Petroleum Refining Processes*, 3rd ed., Meyers, R.A., Ed., McGraw-Hill, New York, 2003, 5.13.

de Florez cracking An early gas-phase thermal process for cracking petroleum. Developed by L. de Florez.

> U.S. Patent 1,437,045.
> Ellis, C., *The Chemistry of Petroleum Derivatives*, Chemical Catalog Co., New York, 1934, 112.

Deglor [**De**toxification and **gl**assification **o**f **r**esidues] A process for vitrifying wastes, such as fly ash, by heating to 1400°C in an electric furnace. Some of the heat is provided by radiant heaters, some by passage of electricity through the melt. Developed and piloted by ABB in Switzerland from the 1980s, commercialized in Japan in 1996.

> *Environ. Bus. Mag.*, 1995, (15), 6.

DegOX [**Deg**ussa **OX**idation] A pulp-bleaching process developed by Degussa. The active species is peroxymonosulfuric acid (Caro's acid). The first full-scale commercial trial was held in 1994.

> *Eur. Chem. News*, 1994, **61**(1619), 38.
> *Chem. Eng. (N.Y.)*, 1997, **104**(1), 21; **104**(4), 10.

DeGuide A sugar extraction and purification process in which barium is recycled. Addition of barium hydroxide to molasses precipitates barium saccharate. A slurry of this is treated with carbon dioxide, forming barium carbonate and releasing the sugar. The barium carbonate is reconverted to the hydroxide by a two-stage process involving monobarium and tribarium silicates:

$$2BaCO_3 + BaSiO_3 = Ba_3SiO_5 + 2CO_2$$

$$Ba_3SiO_5 + 2H_2O = BaSiO_3 + 2Ba(OH)_2$$

Dahlberg, H.W. and Brown, R.J., in *Beet Sugar Technology*, McGinnis, R.A., Ed., The Beet Sugar Development Foundation, Fort Collins, CO, 1971, 573.

Degussa Also called BMA. The process by which this large German company is best known is its version of the *Andrussov process for making hydrogen cyanide. Methane and ammonia are reacted in the absence of air, at approximately 1400°C, over a platinum metal catalyst:

$$NH_3 + CH_4 = HCN + 3H_2$$

Heat is provided by passing the gases through externally heated ceramic tubes. The tubes, made of corundum, are lined with catalytic amounts of platinum metals. Developed by Degussa and operated by that company in Mobile, AL; Wesseling, Germany; and Antwerp, Belgium.

German Patents 882,985; 959,364.
Endter, F., *Chem. Eng. Technol.*, 1958, **30**, 305.
Hydrocarbon Process. Int. Ed., 1967, **46**(11), 189.
Dowell, A.M., III, Tucker, D.H., Merritt, R.F., and Teich, C.I., in *Encyclopedia of Chemical Processing and Design*, 4th ed., McKetta, J.J. and Cunningham, W.A., Eds., Marcel Dekker, New York, 1988, **27**, 11.
Weissermel, K. and Arpe, H.-J., *Industrial Organic Chemistry*, 4th ed., translated by C.R. Hindley and S. Hawkins, Wiley-VCH, Weinheim, Germany, 2003, 47.
Wittcoff, H.A., Reuben, B.G., and Plotkin, J.S., *Industrial Organic Chemicals*, 2nd ed., Wiley-Interscience, Hoboken, NJ, 2004, 339.

dehydrocyclodimerization *See* Cyclar.

dehydrogenation A generic name for catalytic processes that remove hydrogen from hydrocarbons. Alkanes are thus converted to alkenes and dialkenes to monoalkenes. Those with special names that are described in this dictionary are Catadiene, CATOFIN, DeFine, Oleflex, Pacol, Phillips (3), and Styro-Plus.

De la Breteque An electrochemical method for extracting gallium from sodium aluminate solution, developed by Schweizerische Aluminium (Alusuisse) in 1955.

Palmear, I.J., in *The Chemistry of Aluminium, Gallium, Indium, and Thallium*, Downs, A.J., Ed., Blackie, London, U.K., 1993, 88.
Wilder, J., Loreth, M.J., Katrack, F.E., and Agarwal, J.C., in *Inorganic Chemicals Handbook*, Vol. 2, McKetta, J.J., Ed., Marcel Dekker, New York, 1993, 946.

De Laval Centripure A continuous soapmaking process. Fat and aqueous sodium hydroxide are fed countercurrently to a vertical reactor through which much of the product is continuously recirculated, thus emulsifying the reactants.

U.S. Patent 2,727,915.
Palmqvist, F.T.E. and Sullivan, F.E., *J. Am. Oil Chem. Soc.*, 1959, **36**, 173.

Delayed Coking A continuous process for increasing the yield of petroleum distillates from petroleum and also generating saleable coke, by thermal cracking of the distillation residues ("residuum") in a complex system of large coking drums. Developed by the Standard Oil Company (Indiana) in the 1930s and first installed in its refinery at Whiting, IN, in 1930. *See* ConocoPhillips Delayed Coking.

> Diwoky, R.J., *Oil Gas J.*, 2002, **100**(35), 130.

Delay for Decay A process for trapping radioactive gases (e.g., xenon, krypton, iodine) from nuclear power plants until their radioactivities have decayed to acceptable levels. Activated carbon is the usual adsorbent, and the gases are first dried with a zeolite.

> Sherman, J.D. and Yon, C.M., in *Kirk-Othmer's Encyclopedia of Chemical Technology*, 4th ed., Vol. 1, John Wiley & Sons, New York, 1991–1998, 561.

delignification A general name for processes that dissolve lignin from wood pulp. Those with special names that are described in this dictionary are Alcell and ASAM.

Demet (1) A catalytic process for cracking petroleum.

> Unzelman, G.H. and Wolf, C.J., in *Petroleum Processing Handbook*, Bland, W.F. and Davidson, R.L., Eds., McGraw-Hill, New York, 1967, 3–14.

DEMET (2) A process for removing metal contaminants from spent *FCC catalysts by a series of pyrometallurgical and hydrometallurgical procedures. These typically include calcination, chlorination, and sulfiding. The demetallized catalyst can be reused. Developed by Chemcat Corporation, LA, in 1988; first operated commercially in New Jersey in 1989. Several versions are called Demet, Demet II, Demet III, and Demet IV. Now operated by Coastal Catalyst Technology.

> U.S. Patent 4,686,197.
> Elvin, F.J., *Hydrocarbon Process. Int. Ed.*, 1989, **68**(10), 71.
> Pavel, S.K. and Elvin, F.J., in *Extraction and Processing for the Treatment and Minimization of Wastes*, Hager, J.P., Ed., The Minerals, Metals & Materials Society, Warrendale, PA, 1993, 1015.
> *Hydrocarbon Process. Int. Ed.*, 1994, **73**(11), 102.
> Yoo, J.S., *Catal. Today*, 1998, **44**(1–4), 27.

Demex [**Dem**etallization by **ex**traction] A process for removing metal compounds from heavy petroleum fractions, after vacuum distillation, by solvent extraction and supercritical solvent recovery. The solvent is typically a mixture of octanes and pentanes. Developed jointly by UOP and the Instituto Mexicano del Petroleo; seven units were operating in 1988.

> *Hydrocarbon Process. Int. Ed.*, 1988, **67**(9), 66.
> Houde, E.J., in *Handbook of Petroleum Refining Processes*, 2nd ed., Meyers, R.A., Ed., McGraw-Hill, New York, 1996, 10.53.

Den A batch process for making the fertilizer "superphosphate." The **den** is the vat into which the mixture of phosphate rock and sulfuric acid is dumped after mixing. There is also a continuous den process. Not to be confused with **DEN** [Deutsch–Englisch–Norwegische Gruppe], a fertilizer cartel operated from 1929 to 1930 in the three countries from which the name derives. *See also* Davison, Oberphos.

> Demmerle, R.L. and Sackett, W.J., in *Modern Chemical Processes*, Vol. 1, Reinhold Publishing, New York, 1950, 33.

Dennis–Bull A process for making phenol by first sulfonating benzene, the benzenesulfonic acid being extracted into water. Invented in 1917 by H. Bull and L.M. Dennis. *See also* Tyrer.

> U.S. Patents 1,211,923; 1,247,499.
> Kenyon, R.L. and Boehmer, N., in *Modern Chemical Processes*, Vol. 2, Reinhold Publishing, New York, 1952, 35.

De Nora An electrolytic process for making chlorine and sodium hydroxide solution from brine. The cell has a mercury cathode and graphite anodes. It was developed in the 1950s by the Italian company Oronzio De Nora,

Impianti Elettrochimici, Milan, based on work by IG Farbenindustrie in Germany during World War II. In 1958, the Monsanto Chemical Company introduced it into the United States in its plant in Anniston, AL. *See also* mercury cell.

Kenyon, R.L. and Gallone, P., in *Modern Chemical Processes*, Vol. 3, Reinhold Publishing, New York, 1954, 207.
Berkey, F.M., in *Chlorine: Its Manufacture, Properties and Uses*, Sconce, J.S., Ed., Am. Chem. Soc. Monogr. No. 154, Reinhold Publishing, New York, 1962, 203.

Denox A generic name for processes for removing nitrogen oxides from flue gases by catalyzed reaction with ammonia. *See also* Thermal DeNOx.

DeNOx (1) A *Denox process for removing nitrogen oxides from the gaseous effluents from nitric acid plants. The oxides are reduced with ammonia, over a catalyst containing potassium chromate and ferric oxide. Developed by Didier Werke in the 1980s.

Chauvel, A., Delmon, B., and Hölderich, W.F., *Appl. Catal. A: Gen.*, 1994, **115**, 179.

DeNOx (2) A *Denox process for removing nitrogen oxides from the gaseous effluents from nitric acid plants. The oxides are reduced by ammonia, over a catalyst containing iron and chromium. Developed by La Grande-Paroisse in the 1970s and used in eight plants in the 1980s.

Chauvel, A., Delmon, B., and Hölderich, W.F., *Appl. Catal. A: Gen.*, 1994, **115**, 179.

DeNOx (3) A *Denox process for removing nitrogen oxides from the gaseous effluents from chemical plants. The catalyst is vanadia on alumina. Developed in the 1970s by Rhône-Poulenc and used in 25 plants by 1994.

French Patent 2,450,784.
Chauvel, A., Delmon, B., and Hölderich, W.F., *Appl. Catal. A: Gen.*, 1994, **115**, 179.

DENOX A *Denox process engineered by Kinetics Technology International for retrofitting on gas turbine engines for generating electric power. First operated in 1995.

Gas Turbine World, Jul/Aug 1997.

DENOX (Shell DENOX) A low-temperature, add-on *SCR system that operates at between 120°C and 350°C. The honeycomb catalyst contains vanadium and titanium.

van der Grift, C.J.G., Woldhuis, A.F., and Maaskant, O.L., *Catal. Today* 1996, **27**(1–2), 23.

Deoxo A family of catalytic processes using a precious metal supported on alumina. The basic reaction is that of oxygen with hydrogen to produce water. Thus, oxygen is removed from hydrogen, or gases containing hydrogen, and hydrogen is removed from oxygen. Another version is used for destroying ozone in the atmosphere of aircraft cabins. Developed and supplied by Engelhard Industries.

Deoxy A process for removing small concentrations of oxygen from natural gas. The gas is passed over a hot catalyst, which converts the oxygen to carbon dioxide. Two systems were operating in 2000.

Hydrocarbon Process. Int. Ed., 1996, **75**(4), 114; 2000, **79**(4), 66.

DEPA-TOP [**D**i-(2-**E**thylhexyl) **P**hosphoric **A**cid and Tri**O**ctyl**P**hosphine Oxide] A process for recovering uranium from *wet-process phosphoric acid, by solvent extraction with a mixture of the two named reagents. Developed at Oak Ridge National Laboratory, TN, and first commercialized in 1978 by Freeport Minerals Corporation and Wyoming Mineral Corporation.

Chem. Eng. (N.Y.), 1979, **86**(26), 87.

DESAL [DESALination] A process for deionizing brackish waters using two weak-electrolyte ion-exchange resins.

Kunin, R., in *Ion Exchange for Pollution Control*, Vol. 1, Calomon, C. and Gold, H., Eds., CRC Press, Boca Raton, FL, 1979, 111.

DESONOX [DEgussa SO_x NO_x] A process for removing SO_x and NO_x from flue gases. The NO_x is reacted with ammonia by the *SCR process, using a zeolite catalyst to give nitrogen, and the SO_2 is catalytically oxidized to SO_3 and converted to sulfuric acid. Developed by Degussa, in association with Lurgi and Lentjes, and demonstrated at the Hafen heat and power station, Münster, Germany, from 1985 to 1988.

Chem. Eng. (N.Y.), 1993, **100**(12), 97.
Chauvel, A., Delmon, B., and Hölderich, W.F., *Appl. Catal. A: Gen.*, 1994, **115**, 186.
Wieckowska, J., *Catal. Today*, 1995, **24**(4), 452.

Desorex A process for removing impurities from industrial gases by adsorption on activated carbon. Offered by Lurgi.

DeSOx A generic name for *flue-gas desulfurization processes. Some of these processes use a limestone slurry that absorbs the sulfur dioxide, forming calcium sulfite, which is then oxidized to saleable gypsum:

$$CaCO_3 + SO_2 = CaSO_3 + CO_2$$

$$2CaSO_3 + O_2 = 2CaSO_4$$

A more recent version is *S-Magyp that uses magnesium hydroxide.

Destec *See* E-Gas.

Destrugas A process for destroying organic wastes by pyrolysis in an indirectly heated vertical retort. The comminuted waste is fed from the top, and raw gas and coke are withdrawn from the bottom.

Martinez, D., in *Chemical Waste Handling and Treatment*, Muller, K.R., Ed., Springer-Verlag, Berlin, Germany, 1985, 148.

Desulco A process for desulfurizing petroleum coke in order to make it suitable for metallurgical use. The particulate coke is passed through a vertical cylindrical reactor, made of graphite, which has a central graphite electrode. Resistive heating to above 2760°C removes most of the sulfur. The name Desulco is also used for the graphite product. Invented by the Superior Graphite Company in 1977.

U.S. Patent 4,160,813.
Wissler, M., *J. Power Sources*, 2006, **156**(2), 146.

DEsulf A process for removing hydrogen sulfide from coke-oven gas by scrubbing with aqueous ammonia. Developed by Didier Engineering.

Hydrocarbon Process. Int. Ed., 1986, **65**(4), 81.

DESUS A *hydrotreating process developed and offered by VEB Petrolchemisches Kombinat Schwedt. Operated in the Schwedt oil refinery, Germany, since 1988.

De Sy *See* DR.

Detal [Detergent alkylation] A process for making "detergent alkylate" (i.e., alkyl aromatic hydrocarbons such as linear alkyl benzenes) as an intermediate for the manufacture of detergents, by reacting C_{10}–C_{13} olefins with benzene in a fixed bed of an acid catalyst. The catalyst composition is proprietary. Developed by UOP and PETRESA as a replacement for their *Detergent Alkylate process, which uses liquid hydrogen fluoride as the catalyst. Demonstrated in a pilot plant in 1991 and first commercialized in Canada in 1996. Four commercial units were operating in 2005. Offered by UOP.

Vora, B., Pujado, P., Imai, T., and Fritsch, T., *Chem. Ind. (London)*, 1990, 19 Mar, 187.
Hydrocarbon Process. Int. Ed., 1991, **70**(3), 130.
Kocal, J.A., Vora, B.V., and Imai, T., *Appl. Catal. A: Gen.*, 2001, **221**(1–2), 299.
Bozzano, A., in *Handbook of Petroleum Refining Processes*, 3rd ed., Meyers, R.A., Ed., McGraw-Hill, New York, 2003, 1.60.

Detergent Alkylate A process for making "detergent alkylate" (i.e., alkyl aromatic hydrocarbons such as linear alkyl benzenes) as an intermediate for the manufacture of detergents, by reacting C_{10}–C_{13} olefins with benzene using liquid hydrogen fluoride as the catalyst. This technology has been commercialized since 1968, and as of 1990, 28 units were operating. *See* Detal.

U.S. Patents 4,467,128; 4,523,048.
Vora, B., Pujado, P., Imai, T., and Fritsch, T., *Chem. Ind. (London)*, 1990, (6), 187.

DETOL [**DE**alkylation of **TOL**uene] A process for making benzene by dealkylating toluene and other aromatic hydrocarbons. Developed by the Houdry Process and Chemical Company and generally similar to its *Litol process for the same purpose. The catalyst is chromia on alumina. Licensed by ABB Lummus Global. Twelve plants had been licensed in 1987.

Hydrocarbon Process. Int. Ed., 1963, **42**(11), 161; 1987, **66**(11), 66.
Lorz, W., Craig, R.G., and Cross, W.J., *Erdoel Kohle Erdgas Petrochem.*, 1968, **21**, 610.

deTOX A process for vitrifying the ash from municipal solid waste incineration. The wastes are melted in a furnace heated with a submerged electric arc, operated with a deep bed and a cold top. Either sand or fly ash is added to ensure the formation of a vitreous phase. Volatile heavy metals are trapped beneath the cold top. Developed in the late 1980s by Dunston Ceramics and Cookson Group but not yet commercialized.

Deville (1) The first commercial process for making aluminum metal. Molten sodium aluminum chloride was reduced by heating with metallic sodium. Invented by H.F. St-Claire Deville and operated circa 1854–1890. Superseded by the *Hall–Héroult process. *See also* Cowles.

Deville, H.F. St.-C., *C. R. Acad. Sci.*, 1854, **38**, 279; **39**, 321.
Chem. Ind. (London), 1992, (11), 403.

Deville (2) An early process for making sodium by reducing sodium carbonate with carbon at or above 1100°C. Developed in 1886 and used until it was superseded by electrolytic processes. *See* Downs and Castner (4).

Fleck, A., *Chem. Ind. (London)*, 1947, (66), 515.

Deville and Debray A process for extracting the platinum metals from their ores. The ore is heated with galena (lead sulfide ore) and litharge (lead oxide) in a reverberatory furnace. The platinum forms a fusible alloy with the metallic lead, which is also formed. Invented by H.F. St-Claire Deville and H.J. Debray.

Partington, H.R., *A History of Chemistry*, Vol. 4, Macmillan, London, U.K., 1964, 499.

Deville–Pechiney An obsolete process for making alumina from bauxite. Bauxite was roasted with sodium carbonate, yielding sodium aluminate:

$$Al_2O_3 + Na_2CO_3 = 2NaAlO_2 + CO_2$$

The product was digested with warm water, which dissolved the sodium aluminate and left the other materials as an insoluble residue. Passage of carbon dioxide through the clarified liquor precipitated aluminum as hydroxide:

$$2NaAlO_2 + CO_2 + 3H_2O = 2Al(OH)_3 + Na_2CO_3$$

All the silica present in the bauxite was converted to insoluble sodium aluminosilicate, which represented a loss of sodium and aluminum. The aluminum hydroxide was calcined to the oxide, and the sodium carbonate solution was concentrated for reuse. The process was developed by H.F. St-Claire Deville in the 1860s; the carbon dioxide stage had been invented earlier by H.L. Le Chatelier. It was superseded by the *Bayer process.

Dennis, W.H., *A Hundred Years of Metallurgy*, Gerald Duckworth, London, U.K., 1963, 147.

DeVOx A catalytic oxidation process for destroying volatile organic compounds in effluent gases. The catalyst contains a nonnoble metal and can easily be regenerated. Typical operating temperatures for 95% VOC conversion are 175°C–225°C for oxygenates and 350°C for toluene. Developed in 1995 by Shell, Stork Comprimo, and CRI Catalysts. First installed in 1996 at Shell Nederland Chemie's styrene butadiene rubber facility at Pernis.

Eur. Chem. News, 1995, **64**(1686), 26; 1996, **66**(1714), 24.
Environ. Bus. Mag., 1996, (17), 7.

dewaxing A general term to describe processes for removing waxes from petroleum streams. These waxes are an ill-defined group of saturated hydrocarbons in the molecular weight range of 225–1000, mostly with unbranched molecules. Those dewaxing processes with special names that are described in this dictionary are Bari-Sol, Catalytic Dewaxing, Endewax, Isodewaxing, MLDW, and MSDW.

De Wet A process proposed for extracting zirconium from zircon in the form of zirconium sulfate tetrahydrate. The sodium silicate by-product should be free from radioactive elements. The key to this process is the formation of an insoluble basic zirconium sulfate as an intermediate. Invented by W. De Wet in South Africa.

Manhique, A., Kwola, Z., and Focke, W.F., *Ind. Eng. Chem. Res.*, 2003, **42**, 777.

DEZ [**Di**Ethyl **Z**inc] A process for preserving books and documents by treatment with diethyl zinc vapor, which neutralizes any residual acidity from the papermaking process. The articles are placed in a low-pressure chamber and suffused by diethyl zinc vapor. This vapor reacts with the moisture in the paper to yield zinc oxide, which neutralizes the acid. The process was developed by Texas Alkyls (a joint company of Akzo and Hercules) and the US Library of Congress. A pilot plant was set up in 1988, capable of treating batches of 300 books, and plans to build two larger plants were announced in 1989. The process was finally abandoned because it was difficult to control and DEZ is difficult to handle. It was replaced by neutralization using microfine magnesia.

Chem. Br., 1989, **25**(10), 975.
Wedinger, R., *Chem. Br.*, 1992, **28**(10), 898.
Knight, B., *Chem. World*, 2006, **3**(1), 26.

DF [**L**ight **D**istillate **F**raction] A process for oxidizing light naphtha (a mixture of C_4–C_8 hydrocarbons) to acetic and other carboxylic acids. It operates in the liquid phase at 150°C–200°C. Developed by Distillers Company (now BP Chemicals) in England in the 1960s; it was still operated there by that company in 1992.

U.S. Patent 2,825,740.
Pennington, J., in *An Introduction to Industrial Chemistry*, Heaton, C.A., Ed., Leonard Hill, London, U.K., 1984, 376.
Weissermel, K. and Arpe, H.-J., *Industrial Organic Chemistry*, 4th ed., translated by C.R. Hindley and S. Hawkins, Wiley-VCH, Weinheim, Germany, 2003, 176.

DH *See* Dortmund-Hoerder.

DHC (1) [**D**ruck-**H**ydrogenium-**C**racken—German, meaning pressure hydrocracking] A vapor-phase *hydrocracking process for making transport fuels from heavy petroleum fractions. Developed and operated in Germany in the 1950s and 1960s.

> Höring, M., Öttinger, W., and Reitz, O., *Erdoel Kohle Erdgas Petrochemie*, 1963, **16**, 361.
> Öttinger, W. and Reitz, O., *Erdoel Kohle Erdgas Petrochemie*, 1965, **18**, 267.

DHC (2) Not a process but the general name for a range of *hydrocracking catalysts offered by UOP. DHC 39 was first commercialized in 1997.

DHCD *See* Cyclar.

DHD A petroleum reforming process operated in Germany from 1940. The catalyst was molybdena on alumina.

> Nonnenmacher, H., *Brennst.-Chem.*, 1950, **31**, 138 (*Chem. Abstr.*, **44**, 8098).

DHDS [**D**iesel deep **H**ydro**D**e**S**ulfurization] A petroleum refining process developed by the Instituto Mexicano del Petroleo (IMP), with plans for it to be in operation at the Pemex refinery at Cadereyta, Mexico, in 1999.

DHR [**D**ruck-**H**ydrogenium-**R**affination—German, meaning hydrorefining] A petroleum refining process developed by BASF in Germany. *See* DHC.

> Öttinger, W., *Erdoel Kohle*, 1953, **6**, 693.

DIAMEX [**DIAM**ide **EX**traction] An alternative to the *TRUEX process for removing transuranic elements from nuclear waste. The extraction agent is malondiamide, which can more easily be burned off than organic phosphates. Developed by the CEA in France.

DIAMOX A process for removing hydrogen sulfide and hydrogen cyanide from coke-oven gas by absorption in aqueous ammonia. Developed by Mitsubishi Chemical Industries and Mitsubishi Kakoki Kaisha and operated in Japan.

> *Hydrocarbon Process. Int. Ed.*, 1986, **65**(4), 81.
> Kohl, A.L. and Nielsen, R.B., *Gas Purification*, 5th ed., Gulf Publishing Co., Houston, TX, 1997, 324.

Dianor [**Dia**mond Alkali, Oronzio de **Nor**a] A process for cracking naphtha to ethylene, adapted for small-scale operation so that polyvinyl chloride could be made by developing nations. Developed in the 1970s by Chem Systems and the two named companies, but not commercialized.

> Spitz, P.H., *Petrochemicals: The Rise of an Industry*, John Wiley & Sons, New York, 1988, 411.

diaphragm cell A family of electrochemical *Chlor-Alkali processes using cells with semipermeable membranes that minimize diffusion between the electrodes. The overall reaction is

$$2NaCl + 2H_2O = 2NaOH + H_2 + Cl_2$$

Chlorine is liberated at the anode and hydrogen at the cathode, where the sodium hydroxide is formed. Development started in the mid-nineteenth century. In 1962, it was stated that several hundred cell designs had been patented and more than 30 types of diaphragm cells were in use in the United States. The diaphragms of the cells used in the 1960s had membranes of asbestos deposited on steel. In the 1970s, diaphragms made of cation-exchange membranes began to be used, and the cells became called *membrane cells. *See also* Castner–Kellner.

Hardie, D.W.F., *Electrolytic Manufacture of Chemicals from Salt*, Oxford University Press, London, U.K., 1959, 24.
Kircher, M.S., in *Chlorine: Its Manufacture, Properties and Uses*, Sconce, J.S., Ed., Reinhold, New York, 1962, 81.
Purcell, R.W., in *The Modern Inorganic Chemicals Industry*, Thompson, R., Ed., The Chemical Society, London, U.K., 1977, 106.

Diazo A family of reprographic processes (including **diazotype**) based on the coupling of diazonium compounds with dye couplers to form colored compounds. Exposure of the diazonium compounds to near-ultraviolet radiation destroys them, so illuminated areas do not develop color.

DiCOM [**Di**gestion and **COM**posting] A biological process for treating the organic fraction of municipal waste. It sequentially combines aerobic treatment with anaerobic treatment in a single vessel, without intermediate process handling. Methane is also produced for use off site. Invented by T. Rice in 2003 and piloted in Perth, Australia, by Organic Resource Technologies. Plans to build a full-scale plant in Perth were announced in 2005.

Australian Patent (2003).
Chem. Eng. (Rugby, Engl.), 2005/6, (774/5), 8.

Dieselmax A petroleum cracking process that combines mild *hydrocracking with *thermal cracking to maximize the production of middle distillate without using more hydrogen than hydrocracking alone. Developed by UOP.

Diesulforming A *hydrodesulfurization process that used a molybdenum-containing catalyst. Developed by the Husky Oil Company and first operated in Wyoming in 1953.

Oil Gas J., 1956, **54**(46), 165.
Unzelman, G.H. and Wolf, C.J., in *Petroleum Processing Handbook*, Bland, W.F. and Davidson, R.L., Eds., McGraw-Hill, New York, 1967, 3–42.

Difasol An improvement on the *Dimersol process for dimerizing propene or butenes. The process utilizes an ionic liquid based on imidazoliniumaluminate and a nickel-based Dimersol catalyst. Developed by IFP in 1999, but not commercialized by 2005.

Eur. Chem. News, 29 Mar 1999, 35.
Eur. Chem. News, May 1999 (*Chemscope*), 27–28.
Chem. Eng. (N.Y.), Mar 2001, **108**(3), 35.
Dyson, P.J. and Geldbach, T.J., *Metal Catalysed Reactions in Ionic Liquids*, Springer, Dordrecht, Germany, 2005, 169.

DIFEX [**DI**methyl **F**ormamide **EX**traction] A process for separating butadiene from cracked petroleum fractions by extracting with dimethyl formamide.

Smeykal, K. and Luetgert, H., *Chem. Tech. (Berlin)*, 1962, **14**, 202.

Dilchill [**Dil**ute, **chill**] A process for dewaxing petroleum by controlled crystallization, with cooling accomplished by the incremental addition of a cold solvent. Developed by Exxon Research and Engineering Company.

Sequeira, A., Jr., in *Petroleum Processing Handbook*, McKetta, J.J., Ed., Marcel Dekker, New York, 1992, 658.

Dimersol A family of processes for dimerizing single or mixed olefins, catalyzed by mixtures of trialkyl aluminum compounds and nickel salts. Developed by IFP (now Axens) and first commercialized in 1977; by 1997, it was used in 26 plants. *See also* the following entries.

Weissermel, K. and Arpe, H.-J., *Industrial Organic Chemistry*, 4th ed., translated by C.R. Hindley and S. Hawkins, Wiley-VCH, Weinheim, Germany, 2003, 84.

Dimersol E A process for making gasoline from ethylene. The catalyst is a soluble *Ziegler-type catalyst containing nickel. Developed by IFP in the 1980s and operated at an undisclosed location since 1988.

Chauvel, A., Delmon, B., and Hölderich, W.F., *Appl. Catal. A: Gen.*, 1994, **115**, 173.

Dimersol G A process for dimerizing propylene to a mixture of isohexenes, suitable for blending into high-octane gasoline. The process is operated in the liquid phase with a dissolved homogeneous catalyst. Developed by IFP and first operated at Alma, MI, in 1977.

Andrews, J.W., Bonnifay, P.L., Cha, B.J., Barbier, J.C., Douillet, D., and Raimbault, J., *Hydrocarbon Process. Int. Ed.*, 1976, **55**(4), 105.
Andrews, J.W. and Bonnifay, P.L., *Hydrocarbon Process. Int. Ed.*, 1977, **56**(4), 161.
Weismantel, G.E., *Chem. Eng. (N.Y.)*, 1980, **87**(12), 77.

Dimersol-X A process for dimerizing mixed butenes to mixed octenes. Selective hydrogenation, catalyzed by a soluble Ziegler catalyst, is used. The spent catalyst is discarded. The process was developed by IFP and first operated at Kashima, Japan, in 1980. BASF has used the process in Ludwigshafen, Germany, since 1985. By 2005, 35 units had been licensed.

Leonard, J. and Gaillard, J.F., *Hydrocarbon Process. Int. Ed.*, 1981, **60**(3), 99.
Boucher, J.F., Follain, G., Fulop, D., and Gaillard, J., *Oil Gas J.*, 1982, **80**(13), 84.

Dimox [**Di**rected **m**etal **ox**idation] *See* Lanxide.

Dipen [**Di**nitrogen **pen**toxide] A process for making dinitrogen pentoxide by oxidizing dinitrogen tetroxide with ozone. The product, dissolved in nitric acid or an organic solvent, is used as a nitrating agent. Developed in the 1990s by QVF Process Systems and the Defence Research Agency, United Kingdom.

Eur. Chem. News, 1995, **64**(1694), 20.

Direct *See* American.

Direct Reduction *See* DR.

disproportionation In industrial chemistry, this term usually means the catalytic conversion of toluene to a mixture of xylene isomers and benzene, but other reactions also are known by this name. Those with special names that are described in this dictionary are MSTDP, MTDP, Raecke, Triolefin, and Xylenes-plus.

DISTAPEX A process for removing aromatic hydrocarbons from pyrolysis gasoline or coke-oven benzole by extractive distillation with added N-methyl pyrrolidone. The operating temperature is at least 170°C. Developed by Lurgi. First announced in 1961; 25 plants had been built by 2005.

Distex A family of extractive distillation processes used in the petroleum industry from 1940. In one such process, furfural is used as the extracting agent for separating butadiene from other C_4 hydrocarbons.

Buell, C.K. and Boatright, G.R., *Ind. Eng. Chem.*, 1947, **39**, 695.

DISTOP An energy-efficient distillation process for separating hydrocarbons using "dividing-wall technology." Developed by BASF and first commercialized in 1985. A "dividing" (or "divided")-wall column is a fractional distillation column containing a vertical wall, which effectively converts of the column into a double column, thereby saving energy.

Chem. Week Suppl., 12 Sep 2001, s8, s9.

DKS *See* carbonization.

D–LM [**D**wight–**L**loyd **M**cWane] A process for prereducing iron ore. A mixture of the ore, noncoking coal, and limestone is pelletized and carbonized. The reduced pellets are then fed to an electric furnace. Commercialized in Mobile, AL. *See also* DR.

Ban, T.E. and Worthington, B.W., *J. Met.*, 1960, **12**, 937.

DMO [**D**irect **M**ethane **O**xidation] A process for converting methane to methanol or synthetic liquid fuels. Under development by Catalytica in 1997.

>   *Oil Gas J.*, 1997, **95**(25), 16.

Dobanol A process for making linear fatty alcohols from *syngas. Developed by Deutsche Shell Chemie.

DOC [**D**irect **OxyCh**lorination] A one-stage process for making vinyl chloride from ethylene and chlorine. Piloted by Hoechst in Germany in 1989.

Doctor *See* Doktor.

DODD [**D**iesel **O**il **D**eep **D**esulfurization] A process for the deep desulfurization of middle petroleum distillates. Introduced by Exxon in 1989, later developed in collaboration with M.W. Kellogg.

>   Zaczepinski, S., in *Handbook of Petroleum Refining Processes*, 2nd ed., Meyers, R.A., Ed., McGraw-Hill, New York, 1996, 8.63.
>   Absci-Halabi, M., Stanislaus, A., and Qabazard, H., *Hydrocarbon Process. Int. Ed.*, 1997, **76**(2), 45.

Doktor Also spelled Doctor. A process for deodorizing oils by converting trace mercaptans to disulfides by the use of alkaline sodium plumbite and sulfur. The reactions are

$$2RSH + Na_2PbO_2 = Pb(RS)_2 + 2NaOH$$

$$Pb(RS)_2 + S = R_2S_2 + PbS$$

The spent solution is regenerated by blowing air through it:

$$PbS + 2O_2 + 4NaOH = Na_2PbO_2 + Na_2SO_4 + 2H_2O$$

No longer used as a manufacturing process but still used as a qualitative test for mercaptans in petroleum fractions. The presence of mercaptans is indicated by a black precipitate of lead sulfide.

>   Unzelman, G.H. and Wolf, C.J., in *Petroleum Processing Handbook*, Bland, W.F. and Davidson, R.L., Eds., McGraw-Hill, New York, 1967, 3–125.

Donau Chemie A process for making saleable gypsum as a by-product of phosphoric acid manufacture. The product is used to make partition panels for buildings.

>   Becker, P., *Phosphates and Phosphoric Acid*, 2nd ed., Marcel Dekker, New York, 1989, 560.

Dored *See* DR.

Dorr One of the two wet processes for making phosphoric acid by the acidulation of phosphate rock; the other is the *Haifa process. The Dorr process uses sulfuric acid. Phosphate rock is primarily apatite, $Ca_5(PO_4)_3F$. The calcium phosphate portion generates orthophosphoric acid and calcium sulfate:

$$Ca_3(PO_4)_2 + 3H_2SO_4 = 3CaSO_4 + 2H_3PO_4$$

The fluoride portion is either removed as gaseous silicon tetrafluoride, if silica is added, or as the sparingly soluble sodium hexafluorosilicate, which remains with the calcium sulfate. There are several variants of the Dorr process, which differ according to the treatment of the calcium sulfate. Some variants produce the dihydrate, gypsum; others produce the hemihydrate. The variants also differ in the concentration of the phosphoric acid produced, but it is never more than 43%. The basic process was patented by Lawes in England in 1842, but the presently used variant was developed by the Dorr-Oliver Company in the 1930s.

Stevens, H.M., *Phosphorus and Its Compounds*, Interscience, New York, 1961.
Beveridge, G.S.G. and Hill, R.G., *Chem. Process. Eng. (London)*, 1968, **49**(7), 61.
Childs, A.F., in *The Modern Inorganic Chemicals Industry*, Thompson, R., Ed., The Royal Society of Chemistry, London, U.K., 1977, 375.

Dorr-Oliver One of the wet processes for making phosphoric acid. *See* Dorr, Jacobs-Dorr.

Dortmund-Hoerder Also called DH. A steelmaking process in which the molten metal is vacuum degassed.

Double Loop A process for removing hydrogen sulfide from natural gas. In the first loop, an organic solvent absorbs the hydrogen sulfide from the gas. In the second, an aqueous solution of an iron chelate converts this to elemental sulfur. Developed by Radian International in the 1990s.

Quinlan, M.P., Echterhof, L.W., Leppin, D., and Meyer, H.S., *Oil Gas J.*, 1997, **95**(29), 54.

double steeping *See* SINI.

Dowa Also called **Dowa Dual Alkali**. A *flue-gas desulfurization process in which the sulfur dioxide is absorbed in a basic aluminum sulfate solution. The product solution is oxidized with air and reacted with limestone to produce gypsum. Developed by the Dowa Mining Company, Japan, in the early 1980s. In 1986, the process was in use in nine plants in Japan and the United States.

Kohl, H.L. and Nielsen, R.B., *Gas Purification*, 5th ed., Gulf Publishing Co., Houston, TX, 1997, 582.

Dow bromine An electrolytic process for extracting bromine from brines. Ferric bromide was an intermediate. Developed by H.H. Dow, founder of the Dow Chemical Company.

Chem. Eng. News, 1997, **75**(27), 43.

Dowlex A low-pressure, solution process for making linear polyethylene using 1-octene as the comonomer. Developed by Dow Chemical Company and used by that company to make products having the same name, but not offered for license. Planned to be used in a plant in Brazil, based on ethanol from sugar cane, expected to startup in 2011.

Chem. Ind. (London), 2007, (14). 11.
Soares, J.B.P. and McKenna, T.F.L., *Polyolefin Reaction Engineering*, Wiley-VCH, Weinheim, Germany, 2012, 120.

Downs A process for making sodium by electrolyzing a molten eutectic mixture of sodium chloride and calcium chloride at 580°C. Invented in 1922 by J.C. Downs at DuPont and widely used ever since. The process was first used in the United Kingdom in 1937.

U.S. Patent 1,501,756.
Sittig, M., *Sodium: Its Manufacture, Properties and Uses*, Reinhold Publishing, New York, 1956, 21.
Hardie, D.W.F., *Electrolytic Manufacture of Chemicals from Salt*, Oxford University Press, London, U.K., 1959, 28.

Dow-Phenol A process for making phenol by oxidizing molten benzoic acid with atmospheric oxygen. The catalyst is cuprous benzoate.

Buijs, W., *Catal. Today*, 1996, **27**(1–2), 159.
Buijs, W., *J. Mol. Catal. A: Gen.*, 1999, **146**(1–2), 237.

Dow seawater A process for extracting magnesium from seawater. Calcined dolomite (CaOMgO, dololime), or calcined oyster shell, is added to seawater, precipitating magnesium hydroxide. This is flocculated, sedimented, and filtered off. For use as a refractory, it is calcined; for the manufacture of magnesium chloride (for use in the manufacture of magnesium metal), it is dissolved in hydrochloric acid. Developed by the Dow Chemical Company and later operated by the Steetley Company in West Hartlepool, United Kingdom, and Sardinia.

Mantell, C.L., in *Chlorine: Its Manufacture, Properties and Uses*, Sconce, J.S., Ed., American Chemical Society Monogr. No. 154, Reinhold Publishing, New York, 1962, 578.

Doxosulfreen A process for oxidizing hydrogen sulfide to elemental sulfur, based on the *Sulfreen process and catalyzed by copper on alumina. The use of a fairly low temperature (90°C–140°C) minimizes the further oxidation to SO_2. Developed by Elf and Lurgi.

Savin, S., Nougarède, J.B., Willing, W., and Bandel, G., *Int. J. Hydrocarbon Eng.*, 1998, **4**, 54.
Savin, S., Legendre, O., Nougarède, J.B., and Nedez, C., *Sulphur*, 2000, **96**, 523.

DPG hydrotreating A process for converting petroleum pyrolysates into high-octane gasoline. Two stages of selective hydrogenation are used. Developed by Lummus Crest. Twenty-eight systems were operating in 1986.

Hydrocarbon Process. Int. Ed., 1986, **65**(9), 85.

DR [**D**irect **R**eduction] A general name for processes for making iron or steel by reducing iron oxide ores at temperatures below the melting points of any of the ingredients. The product is known as DRI [**D**irect-**R**educed **I**ron]. The reducing agent may be carbon in some form, carbon monoxide, hydrogen, or a mixture of these.

The Commission of the European Communities commissioned a literature study of all the DR processes. This study was published by the Verlag Stahleisen mbH, Düsseldorf, updated several times, and then translated into English and published in 1979 by The Metals Society, London. The report classified the processes into three groups: those of great industrial significance, those of limited industrial significance, and those without industrial application. The names of all the processes, in their three groups, are listed as follows. Most of those in the first two categories can be found in their respective alphabetical locations in this dictionary.

Processes of great industrial significance: Allis-Chalmers, Armco, Esso Fior, HIB, HyL, Kawasaki, Kinglor-Metor, Koho, Krupp sponge iron, MIDREX, NSC, SDR, SL/RN, and SPM.

Processes of limited industrial significance: Carbotherm, CIP (2), Echeverria, Elkem, H-D, Hoganas, ICEM, Krupp–Renn, Novalfer, Rotored, and Wiberg-Soderfors.

Processes without industrial application: Aachen rotary furnace reduction, Arthur D. Little, Basset, Bouchet–Imphy, CO-C-iron, De Sy, D–LM, Dored, Electric fluidized bed, Elkem, Finsider, Flame-Smelting, Freeman, Heat-Fast, H-Iron, Jet Smelting, Kalling (Avesta), Kalling (Domnarvets), Madaras, Nu-Iron, Orcarb, ORF, Purofer, Republic Steel, Stelling, Strategic-Udy, Sturzelberg, Ugine, and Voest.

The following additional DR processes have been operated since 1979 and are described in separate entries: Accar, Circored, Codir, DAV, DRC, FASTMET, H-Iron, OSIL, Plasmared, Purofer, SIIL, SPIREX, Tisco, and USKO.

Total world production of DRI in 2012 was 75 million tons, of which 50% were made by the *MIDREX process. The Midrex Corporation publishes an annual report "World Direct Reduction Statistics" that was used to update the information on DR in this dictionary.

Kirk-Othmer's Encyclopedia of Chemical Technology, 4th ed., Vol. 14, John Wiley & Sons, New York, 1991–1998, 855.

Dravo–Ruthner A development of the *Ruthner process for recovering hydrochloric acid and iron oxide from steel pickling liquors. Developed by Ruthner Industrieplanungs, Vienna, with Dravo Corporation, Pittsburgh. *See also* Ruthner.

DRC [**D**irect **R**eduction by **C**oal] A direct-reduction ironmaking process, using coal as the reductant. First operated in Rockwood, TN, in 1978, and now operating in two plants in South Africa and one in China. *See* DR.

Dr. C. Otto *See* carbonization.

Drizo A variation of the glycol process for removing water vapor from natural gas, in which the water is removed from the glycol by stripping with a hydrocarbon solvent, typically a mixture of pentanes and heavier aliphatic hydrocarbons. The process also removes aromatic hydrocarbons. The last traces of water are removed from the triethylene glycol by stripping with toluene in a separate, closed loop. Invented in 1966 by J.C. Arnold, R.L. Pearce, and H.G. Scholten at the Dow Chemical Company. More than 45 units were operating in 2000. The latest version, operated in Hungary, also uses a desiccant bed.

> U.S. Patent 3,349,544.
> *Hydrocarbon Process. Int. Ed.*, 1975, **54**(4), 81.
> Isalski, W.H., *Separation of Gases*, Clarendon Press, Oxford, U.K., 1989, 218.
> Smith, R.S., *Hydrocarbon Process. Int. Ed.*, 1990, **69**(2), 75.
> *Hydrocarbon Process. Int. Ed.*, 1992, **71**(4), 98; 2000, **79**(4), 67.
> *Chem. Eng. (Rugby, Engl.)*, 1993, (549), 17.
> Skiff, T., Szuts, A., and Toth, V.S.A., *Oil Gas J.*, 2002, **100**(44), 60.

Dry Box *See* Iron Sponge.

DRYPAC A *flue-gas desulfurization process in which an aqueous lime suspension is injected into a spray drier. Developed by Flakt Industri, Sweden. In 1986, the process was in use in 17 plants in Sweden, Denmark, Finland, Germany, and the United States.

DS (1) [**D**ual **S**ludge] A two-stage sewage treatment process, combining aerobic and anaerobic processes, which yields methane for use as fuel. Offered by Linde, Munich, Germany.

DS (2) A general name for a process for making iron or steel, which gives a liquid iron product similar to the pig iron produced in a blast furnace. Examples are *SKF and *MIDREX.

DSN [**D**irect **S**trong **N**itric] A general name for processes for concentrating nitric acid from 50% to 70% to 98% by adding dinitrogen tetroxide.

> Fauser, G., *Chem. Met. Eng.*, 1932, **39**, 430 (*Chem. Abstr.*, **26**, 5388).
> Hellmer, L., *Chem. Eng. Prog.*, 1972, **68**(4), 67.
> Newman, D.J. and Klein, L.A., *Chem. Eng. Prog.*, 1972, **68**(4), 62.
> Ohroi, T., Okubo, M., and Imai, O., *Hydrocarbon Process. Int. Ed.*, 1978, **57**(11), 163.
> Marzo, L.M. and Marzo, J.M., *Chem. Eng. (N.Y.)*, 1980, **87**(22), 54.

DSX [**D**irect **S**olvent **E**xtraction] A general term for hydrometallurgical processes having fewer stages. *See also* SSX.

> *TCE (The Chemical Engineer)*, 2008–2009, (810), 34.

D-TOX An oxidative process for destroying unsaturated chlorinated compounds, such as vinyl chloride, in gas streams. Developed by Ultrox International, Santa Ana, CA.

Dual Alkali A *flue-gas desulfurization process. The sulfur dioxide is absorbed in aqueous sodium hydroxide and partially oxidized; this liquor is then treated with calcium hydroxide to regenerate the scrubbing solution and precipitate calcium sulfate. Developed by Combustion Equipment Associates and Arthur D. Little.

> LaMantia, C.R., Lunt, R.R., and Shah, I.S., *Chem. Eng. Prog.*, 1974, **70**(6), 66.

Dualayer Distillate A process for extracting organic acids from petroleum fractions, using an aqueous solution of sodium cresylate. Developed by the Mobil Oil Company in the 1950s.

> Unzelman, G.H. and Wolf, C.J., in *Petroleum Processing Handbook*, Bland, W.F. and Davidson, R.L., Eds., McGraw-Hill, New York, 1967, 3–113.

Dualayer Gasoline A process for extracting mercaptans and other organic sulfur compounds from petroleum distillates by solvent extraction with aqueous sodium or potassium hydroxide and a proprietary solubilizer.

Unzelman, G.H. and Wolf, C.J., in *Petroleum Processing Handbook*, Bland, W.F. and Davidson, R.L., Eds., McGraw-Hill, New York, 1967, 3–113.

Dualforming A process that enables a petroleum refiner to improve the *Catalytic Reforming step at minimal capital cost. A new reactor with continuous catalyst recirculation is integrated into the existing reactor train. Developed and offered by Axens.

Dual Pressure An ammonia synthesis process, based on the *AMV process but providing more capacity by removing some of the product at an intermediate stage. Developed by Krupp Uhde in 2001–2002.

Eur. Chem. News, 10 Dec 2001, **75**(1985), 26.

Dual-Spectrum Also called Workman. A thermographic copying process. A transparent film base, coated with 4-methoxy-1-naphthol and a photoreducible dye such as erythrosine, receives the image, which is then transferred to a paper sensitized with sodium behenate. Invented in 1961 by W.R. Workman at the Minnesota Mining and Manufacturing Company, MN, and commercialized for office copying but later superseded by various electrophotographic processes.

U.S. Patent 3,094,417.

Dubbs One of a number of thermal (i.e., noncatalytic) processes for cracking petroleum, widely used in the 1920s and 1930s. Invented by J.A. Dubbs and C.P. Dubbs and promoted by the Universal Oil Products Company (now UOP). One of the original patents (U.S. Patent 1,123,502), filed in 1909 but not granted until 1915, was mainly concerned with methods for breaking an emulsion by subjecting it to heat and pressure, but subsequent amendments and filings extended it to cover continuous distillation and condensation under pressure. Widely adopted from 1923 and used until the introduction of *catalytic cracking.

U.S. Patents 1,100,717; 1,123,502.
Enos, J.L., *Petroleum Progress and Profits*, MIT Press, Cambridge, MA, 1962, Chap. 2.
Stanley, H.M., in *Propylene and Its Industrial Derivatives*, Hancock, E.G., Ed., Ernest Benn, London, U.K., 1973, 14.
Achillades, B., *Chem. Ind. (London)*, 1975, (8), 337.

Dubrovai An early thermal process for cracking petroleum, deriving its heat from partial combustion. Invented by K.K. Dubrovai and operated in the USSR in the 1930s.

Asinger, F., *Paraffins: Chemistry and Technology*, translated by B.J. Hazzard, Pergamon Press, Oxford, U.K., 1968, 588.
Skarchenko, V.K., *Russ. Chem. Revs.*, 1968, **37**(1), 1.

Duftschmid A variation of the *Fischer–Tropsch process in which *synthesis gas and an oil are circulated over a fixed bed of iron catalyst in order to increase the yield of olefins from the gas.

Asinger, F., *Paraffins: Chemistry and Technology*, translated by B.J. Hazzard, Pergamon Press, Oxford, U.K., 1968, 151.

Dunlop A process for making foam rubber that uses sodium fluorosilicate to coagulate the rubber particles and deactivate the surfactants. *See also* Talalay.

Madge, E.W., *Latex Foam Rubber*, John Wiley & Sons, New York, 1962.

Dunn *See* Wendell Dunn.

Duo-Sol A process for separating aromatic from aliphatic hydrocarbons by partition between two solvents. The first solvent (Selecto or Selectox) is a mixture of phenol and cresylic acids; the second is liquid propane. Developed by the Max B. Miller Company and licensed by the Milwhite Company.

Pet. Refin., 1960, **39**(9), 234.
Unzelman, G.H. and Wolf, C.J., in *Petroleum Processing Handbook*, Bland, W.F. and Davidson, R.L., Eds., McGraw-Hill, New York, 1967, 3–90.

Duplex (1) An integrated steelmaking process in which iron is converted to steel in a *Bessemer furnace with a basic lining, and the molten product is transferred to a basic, lined arc furnace in which the remaining impurities are oxidized. Developed in Germany and widely used there and elsewhere around 1900.

Dennis, W.H., *A Hundred Years of Metallurgy*, Gerald Duckworth, London, U.K., 1963, 111.

Duplex (2) A process for making sodium perborate from both sodium peroxide and hydrogen peroxide. Developed in the United States from the 1920s and used until the development of the present process, which uses hydrogen peroxide as the only source of the peroxygen. *See also* Acid.

DuraTherm A process for treating waste products from oil refineries by thermal desorption. The wastes are passed through a rotating, externally heated metal drum containing a rotating helix. Volatile waste products are swept from the drum in a stream of nitrogen and condensed.

Oil Gas J., 1996, **94**(38), 49.

Dutch An obsolete process for making basic lead carbonate pigment (white lead) by exposing metallic lead to vinegar and carbon dioxide. Reportedly first described by Theophrastos circa 300 BC. Also known as the Stack process, because the metal ingots were arranged in vertical stacks. *See also* Thompson–Stewart.

Sherwood Taylor, F., *A History of Industrial Chemistry*, Heinemann, London, U.K., 1957, 83.

Duval *See* CLEAR.

Dwight–Lloyd A pyrometallurgical process for simultaneously roasting and smelting a ground ore. The ore is contained in a series of shallow iron boxes with perforated bases, which are drawn continuously through a furnace having a downward draft of air. Developed by A.S. Dwight and R.L. Lloyd at the Greene Consolidated Smelter in Cananéia, Brazil, in 1906; later acquired by Lurgi and now widely used for ferrous and nonferrous ores.

Dennis, W.H., *A Hundred Years of Metallurgy*, Gerald Duckworth, London, U.K., 1963, 60.

DWN [**D**ruck**W**echsel **N**itrogen] A proprietary *PSA system for separating nitrogen from air. Developed and offered by Linde. *See also* DWO.

DWO [**D**ruck**W**echsel **O**xygen] A proprietary *PSA system for separating oxygen from air. Developed and offered by Linde. *See also* DWN.

Dynacracking A petroleum cracking process that combines the best features of the *catalytic cracking and *thermal cracking processes. Dynacracking converts heavy oil feedstocks to fuel gas, gasoline, and fuel oil. No catalyst is used. Developed in the 1950s by Hydrocarbon Research, but not commercialized.

Rakow, M.S. and Calderon, M., *Chem. Eng. Prog.*, 1981, **77**(2), 31.

Dynagas A noncatalytic process for hydrogenating coal to produce gas or light oil fuels. Developed by Hydrocarbon Research.

Dynaphe A process for converting mixed alkyl phenols (from coal liquids or lignin) to benzene, phenol, and fuel gas, by noncatalytic hydrogenation at high temperature. Developed and offered by Hydrocarbon Research.

Dynatec A pressure oxidation process for leaching copper from sulfide ores. Low-grade coal is added to disperse the molten sulfur produced. Being piloted by Dynatec Corporation at Fort Saskatchewan, Canada. Dynatec merged with Sherritt International in 1997. *See also* *Sherritt-Gordon.

Dynatol A continuous process for making sorbitol. Developed by Hydrocarbon Research before 1982.

E

EA [Ethyl Acetate] A one-step process for converting ethanol to ethyl acetate. Acetaldehyde is an intermediate, and the catalyst is proprietary. Developed by Davy Process Technology and Sasol from 2000, following initial work by Davy in 1983. The first commercial plant was built by Sasol in Secunda, South Africa, in 2001. The process won the Kirkpatrick Honor Award for Chemical Engineering Achievement in 2003.

> *Eur. Chem. News*, 2001, **75**(1980), 19.

EARS [Enhanced Acid Regeneration System] A process for recovering hydrochloric acid from the *ERMS ilmenite beneficiation process. It may also be used for recovering waste pickle liquor. The acid liquor, containing ferrous chloride, is evaporated at low temperature to form iron chloride pellets, which are fed to a pyrohydrolysis reactor. This generates hydrochloric acid and iron oxide pellets, which can be used for steel production or disposed of as inert landfill. Developed by E.A. Walpole at the University of Newcastle, Australia, from the early 1990s and piloted by Austpac Gold (now Austpac Resources). Construction of a demonstration plant was announced in 2013.

> *Chem. Eng. (N.Y.)*, 2013, **120**(6), 12.

Eastman Also known as the Tennessee Eastman Challenge Process. Not an actual chemical process but a theoretical chemical engineering exercise, based on a proprietary process.

> Downs, J.J. and Vogel, E.F., *Comput. Chem. Eng.*, 1993, **17**, 245.
> Luyben, W.L., *Ind. Eng. Chem. Res.*, 1996, **35**(10), 3280.
> Wu, K.-L. and Yu, C.-C., *Ind. Chem. Eng. Res.*, 1997, **36**(6), 2239 (23 refs.).

Eastman–Halcon A process for making acetic anhydride from *syngas. The basic process is the carbonylation of methyl acetate. Methanol is made directly from the carbon monoxide and hydrogen of *syngas. Acetic acid is a by-product of the cellulose acetate manufacture for which the acetic anhydride is needed. The carbonylation is catalyzed by rhodium chloride and chromium hexacarbonyl.

> Wittcoff, H.A., Reuben, B.G., and Plotkin, J.S., *Industrial Organic Chemicals*, 2nd ed., Wiley-Interscience, Hoboken, NJ, 2004, 368.

EB *See* Ethylbenzene.

Ebara [Electron beam ammonia reaction] A dry process for removing sulfur and nitrogen oxides from flue gas. A beam of high-energy electrons is injected into the gas, to which a stoichiometric quantity of ammonia has been added. The product, a mixture of ammonium sulfate and ammonium nitrate, is collected downstream by an electrostatic precipitator or a bag filter. Developed by Ebara Corporation, Japan, and piloted in Indianapolis in 1986.

> IEA Coal Research, *Flue-Gas Desulfurization Handbook*, Butterworths, London, U.K., 1987.

Ebara Ube A process for making *syngas from mixed plastic waste including PVC. It uses a two-stage gasification system, pyrolysis followed by partial oxidation. The first stage operates at atmospheric pressure at 600°C–800°C. The second stage operates at 1300°C–1500°C at high pressure. Developed by Ebara Corp and Ube Industries and awarded a prize by the Japan Chemical Industry Association.

> *Jpn. Chem. Week*, 2007, **48**(2422), 5.

Ebex [**E**thylbenzene **ex**traction] A version of the *Sorbex process, for extracting ethylbenzene from mixtures of aromatic C_8 isomers. The adsorbent is a zeolite. It has not been commercialized.

de Rossett, A.J., Neuzil, R.W., Tajbl, D.G., and Braband, J.M., *Sep. Sci. Technol.*, 1980, **15**, 637.
Ruthven, D.M., *Principles of Adsorption and Adsorption Processes*, John Wiley & Sons, New York, 1984, 400.

EBMax A continuous, liquid-phase process for making ethylbenzene from ethylene and benzene. Generally similar to the *Mobil/Badger process, but the improved catalyst (zeolite MCM-22) permits the reactor size to be reduced by two-thirds. Developed by Raytheon Engineers and Constructors and Mobil Oil Corporation and first installed at Chiba Styrene Monomer in Japan in 1995. Licensed by Badger Licensing; over 40 units had been licensed by 2005.

Chem. Eng. (N.Y.), 1995, **109**(9), 21.
Degnan, T.F., Jr., Smith, C.M., and Venkat, C.R., *Appl. Catal. A: Gen.*, 2001, **221**(1–2), 286.

EBV *See* carbonization.

ECF [**E**lemental **C**hlorine **F**ree] A generic term for pulp-bleaching processes that use chlorine dioxide and other oxidants in place of elemental chlorine. *See also* TCF.

Chem. Eng. (N.Y.), 1997, **104**(4), 33.
Nelson, P.J., in *Environmentally Friendly Technologies for the Pulp and Paper Industries*, Young, R.A. and Akhar, M., Eds., John Wiley & Sons, New York, 1998, 215.

Echelon An ion-exchange process for removing perfluoro-octanoic acid (PFOA) from aqueous suspensions of fluoropolymers. The resin containing the fluorine is subsequently calcined at a high temperature. PFOA is not added deliberately in the manufacture of fluoropolymers but is an unwanted by-product. Developed by DuPont from 2006.

ICIS Chem. Bus., 2007, **2**(53), 8.
Chem. Ind. (London), 2007, (17), 24.

Echeverria *See* DR.

ECO₂ A process for absorbing carbon dioxide from power plants using an aqueous ammonia solution. The resulting ammonium bicarbonate solution is regenerated by heating to 55°C, releasing both carbon dioxide and ammonia, which are separated. The ammonia is recycled, and the carbon dioxide is stored. Developed by Powerspan, Portsmouth, NH, and planned to be piloted in Shadyside, OH, in 2007.

Chem. Eng. (N.Y.), 2006, **113**(7), 14.

Ecoclear A continuous process for destroying organic contaminants in wastewater by treatment with ozone in the presence of a proprietary catalyst. Developed in 1995 by Eco Purification Systems, the Netherlands.

Ecofining A catalytic *hydroprocessing process for making diesel fuel from vegetable oils. Developed by ENI and UOP. The first plant was planned for Livorno, Italy, in 2009. A second plant was planned to be used by Galp Energia in Portugal. Valero was building a plant in Norco, LA, in 2012. ENI plans to use it in Venice in 2014. The product is called Green Diesel. The *Renewable Jet process is an improved version.

Chem. Eng. (N.Y.), May 2007, 18.
Chem. Eng. News, 2007, **85**(24), 25.
Chem. Eng. (N.Y.), May 2010, 18.
Perego, C. and Ricci, M., *Catal. Sci. Technol.*, 2012, published online 29 Jun 2012.
Kalnes, T.M., Marker, T., Shonnard, D.R., and Koers, K.P., *Biofuels Technol.*, 2008, (4), 7.

Econ-Abator A process for oxidizing hydrogen sulfide and other sulfur compounds in waste gases by fluid-bed combustion in the presence of an oxide catalyst. Licensed by ARI Technologies. In 1992, there were 90 installations.

> *Hydrocarbon Process. Int. Ed.*, 1992, **71**(4), 99.

Econamine A process for removing acid gases from natural gas by selective absorption in Diglycolamine (also called 2-(2-aminoethoxy) ethanol and DGA). Developed by the Fluor Corporation, the El Paso Natural Gas Company, and the Jefferson Chemical Company and widely used. Later versions, developed by Fluor Daniel International, include the Fluor Daniel Econamine and the *Econamine FG processes. More than 30 units were operating in 1996. *See also* Aromex.

> Dingman, J.C. and Moore, T.F., *Hydrocarbon Process. Int. Ed.*, 1968, **47**(7), 138.
> *Hydrocarbon Process. Int. Ed.*, 1971, **50**(4), 101; 1996, **75**(4), 116.
> Kohl, H.L. and Nielsen, R.B., *Gas Purification*, 5th ed., Gulf Publishing Co., Houston, TX, 1997, 51.

Econamine FG and Econamine FG Plus [Flue Gas] A process for removing carbon dioxide from flue gases by dissolution in an aqueous solution of monoethanolamine and a proprietary corrosion inhibitor. Originally developed by the Dow Chemical Company under the designation Gas/Spec FT, the process was acquired in 1989 by the Fluor Corporation and is now licensed by that company.

Econocat A process for deodorizing gases by catalytic oxidation over manganese dioxide activated with other metals. Developed and sold by Courtaulds Engineering.

> *Chem. Eng. (N.Y.)*, 1984, **91**(13), 156.

Ecoteg A process for drying natural gas and also removing all the aromatic hydrocarbons present. Triethylene glycol is used as the absorbent.

> *Hydrocarbon Process. Int. Ed.*, 2000, **79**(4), 68.

Edeleanu (1) A process for extracting aromatic hydrocarbons from kerosene using liquid sulfur dioxide. Developed in Romania in 1908 by L. Edeleanu to improve the burning properties of lamp kerosene; in 1990, the original process was still in use in commercial plants. The company founded by the inventor in 1910, Edeleanu Gesellschaft mbH, now a part of the RWE Group, Germany, is today better known for its range of processes for making lubricating oils, waxes, and certain organic chemicals.

> British Patent 11,140 (1908).
> Edeleanu, L., *J. Inst. Petr. Technol.*, 1932, **18**, 900.
> *The Petroleum Handbook*, 3rd ed., Shell Petroleum, London, U.K., 1948, 183.

Edeleanu (2) A process for extracting higher aliphatic hydrocarbons from hydrocarbon mixtures by extraction with aqueous urea.

> Weissermel, K. and Arpe, H.-J., *Industrial Organic Chemistry*, 4th ed., translated by C.R. Hindley and S. Hawkins, Wiley-VCH, Weinheim, Germany, 2003, 78.

EDS *See* Exxon Donor Solvent.

EDV A wet-scrubbing process for removing sulfur dioxide, sulfuric acid mist, and catalyst particles from the effluent gases from *FCC units. Developed by Belco Technologies, which had installed 20 units by 2003.

> Weaver, E.H. and Confuorto, N., in *Handbook of Petroleum Refining Processes*, 3rd ed., Meyers, R.A., Ed., McGraw-Hill, New York, 2003, 11.15.

Efflox A process for destroying cyanide wastes in hydrometallurgical effluents by the use of Caro's acid. The cyanide ion is oxidized to cyanate ion.

EFFOL A process for making nylon salt (hexamethylenediamine di-adipate). Developed by Rhône-Poulenc in the 1980s and now used in its plant at Chalampé, France.

E-flex A process for making polypropylene from C_2 hydrocarbons or bioethanol. Developed by Asahi Kasei Chemicals Corp. from 2010 and planned for commercialization in 2012.

There are several unrelated uses of this name and similar ones.

> *Chem. Eng. (N.Y.)*, 2011, **118**(4), 16.

E-Gas Formerly known as the Destec gasification process. A coal gasification process that produces a hydrogen-rich *syngas and an inert, vitreous slag. The coal is fed as an aqueous slurry and reacted with oxygen. First developed by Destec Energy, the process was acquired by ConocoPhillips in 2003. Demonstrated in the 1990s at Wabash River, West Terre Haute, IN. ConocoPhillips plans to build a larger plant in East Dubuque, IL, in 2007.

> Rezaiyan, J. and Cheremisinoff, N.P., *Gasification Technologies*, Taylor & Francis, Boca Raton, FL, 2005, 51.

EHD [ElectroHydroDimerization] Also known as electrodimerization. An electrolytic process for converting acrylonitrile to adiponitrile. *See* Monsanto. The name has also been used for the electrodimerization of formaldehyde to ethylene glycol, which was being developed by Electrosynthesis, Amherst, NY, in 1984.

EIC *See* Cuprasol.

Eichner An early thermographic copying process.

ELCOX A *flue-gas desulfurization process in which the sulfur dioxide is oxidized electrochemically to sulfuric acid, using an organometallic catalyst (e.g., cobalt phthalocyanine) adsorbed on activated carbon. Developed by the Central Laboratory of Electrochemical Power Sources, Sofia, Bulgaria.

> European Patent Application 302,224.
> Vitanov, T., Budevski, E., Nikolov, I., Petrov, V., Naidenov, V., and Christov, Ch., in *Effluent Treatment and Waste Disposal*, Institution of Chemical Engineers, Rugby, England, 1990, 251.

Electrocracking A general name for processes that decompose organic liquids or gases by passage through an electric arc. Thus, methane is converted to acetylene and di-acetylene. Developed in Russia in the early twentieth century but probably never commercialized.

Electrodimerization *See* EHD.

Electrofining A process for purifying petroleum fractions by extracting them with various liquid reagents and then assisting their separation by means of an electric field. Developed by the Petreco Division of Petrolite Corporation and first operated in California in 1951.

> Unzelman, G.H. and Wolf, C.J., in *Petroleum Processing Handbook*, Bland, W.F. and Davidson, R.L., Eds., McGraw-Hill, New York, 1967, 3–137.

Electrohydrodimerization *See* EHD.

Electropox [**Electro**chemical **p**artial **ox**idation] Also called *POX. An electrochemical process for oxidizing methane to *syngas. It combines the partial oxidation and steam reforming of methane with oxygen separation in a single stage. Invented in 1988 by T.J. Mazanec at BP Chemicals. An industrial–academic consortium to develop the process was formed in 1997.

U.S. Patents 4,802,958; 4,793,904.
European Patent Application 399,833.
Mazanec, T.J., Cable, T.L., and Frye, J.G., Jr., *Solid State Ionics*, 1992, **53–56**, 111.
Mazanec, T.J., Cable, T.L., and Frye, J.G., Jr., in *The Role of Oxygen in Improving Chemical Processes*, Felizon, M. and Thomas, W.J., Eds., Royal Society of Chemistry, Cambridge, U.K., 1993, 212.
Mazanec, T.J., in *The Activation of Dioxygen and Homogeneous Catalytic Oxidation*, Barton, D.H.R., Martell, A.E., and Sawyer, D.T., Eds., Plenum Press, New York, 1993, 85.
Mazenac, T.J., *Interface*, 1996, **5**(4), 46.
Chem. Ind. (London), 1997, (22), 911.

Electroslag Also called ESR. A general term for any electrolytic metal extraction process in which the metal is produced in the molten state beneath a layer of molten slag. Used mainly for ferrous alloys.

Duckworth, W.E. and Hoyle, G., *Electroslag Refining*, Chapman & Hall, London, U.K., 1969.

ElectroSlurry An electrolytic process for extracting copper from sulfide ores, liberating elemental sulfur. Developed by the Envirotech Research Center, Salt Lake City, UT.

Chem. Eng. (N.Y.), 1980, **87**(25), 35.

Electrox A process for removing atmospheric pollutants by passing the air through a pulsed corona discharge between concentric metal cylinders. Developed by AEA Technology from 1998.

Ind. Environ. Management, 1999, **10**(1), 15.

Elektrion Also called volto. A method of increasing the molecular weight, and thus the viscosity, of a mixture of light mineral oil and a fatty oil, by subjecting it to an electric discharge in a hydrogen atmosphere.

Elektron A family of processes, operated by Magnesium Elektron, United Kingdom, for making magnesium, magnesium–zirconium alloys, and zirconium chemicals. In the 1920s and 1930s, the names *elektron* and *elektronmetall* were used colloquially in Germany for magnesium metal.

ELEX [**EL**ectrochemical **EX**change] A process for separating lithium isotopes for use in making lithium deuteride for use in nuclear weapons. Operated at Oak Ridge National Laboratory, TN, from 1951 to 1959, until this process was replaced by the *COLEX process.

Elf Activated MDEA An improved *MDEA process for removing carbon dioxide from natural gas rich in hydrogen sulfide. The activator is a proprietary secondary amine that accelerates the reaction between the MDEA and the carbon dioxide. Developed by TotalFinaElf and IFP in the 1990s for use in the Lacq gas field.

Lallemand, F. and Minkkinenm, A., *Oil Gas J.*, 2002, **100**(3), 38.

Elf-SRTI An industrial chromatographic process developed by Elf Aquitaine Développement and the Société de Recherches Techniques et Industrielles. Multiple beds are used. Used for separating perfume ingredients and proposed for separating aliphatic hydrocarbons.

U.S. Patent 4,230,464.
Bonmati, R.G., Chapelet-Letourneux, G., and Margulis, J.R., *Chem. Eng. (N.Y.)*, 1980, **87**(6), 70.
Anal. Chem., 1980, **52**(4), 481A.
Yang, R.T., *Gas Separation by Adsorption Processes*, Butterworths, Guildford, U.K., 1987, 212.

Elix [**El**ectrically regenerated **i**on e**x**change] A water purification process in which an ion-exchange resin is regenerated electrically. Developed by Millipore.

Elkem *See* DR.

Elkington An early electrolytic process for plating gold and silver from cyanide solutions. Invented by J. Wright and then patented by G.R. and H. Elkington in 1840. First commercialized in Birmingham, United Kingdom, in 1841.

> British Patent 8447 (1840).
> *Wikipedia, The Free Encyclopedia, George Richards Elkington.* Revised 11 Mar 2013. Retrieved 20 Jul 2013.

Ellis A process for making isopropyl alcohol from light olefin mixtures by treatment with concentrated sulfuric acid. Operated in World War I by the Melco Chemical Company, as an intermediate for the production of acetone for airplane "dope."

> Ellis, C., *The Chemistry of Petroleum Derivatives*, Chemical Catalog Co., New York, 1934, 349.

Eloxal [**El**ectrolytic **ox**idation of **al**uminum] An electrolytic process for applying an oxide film to the surface of aluminum in order to protect it from further oxidation or to make it capable of adsorbing a dyestuff.

> Institut Fresenius, *Waste Water Technology*, Springer-Verlag, Berlin, Germany, 1989, 70.

ELSE [**E**xtremely **L**ow **S**ulfur **E**mission] A *flue-gas desulfurization process in which the sulfur dioxide is absorbed by zinc oxide. Developed by Amoco, United States.

> Wieckowska, J., *Catal. Today*, 1995, **24**(4), 445.

ELSORB A process for recovering sulfur from the sulfur dioxide in flue gases. The gases are scrubbed with aqueous sodium phosphate, and the sulfur dioxide is recovered by heating the scrubbing liquor. Developed in the 1980s by the Norwegian Institute of Technology, Trondheim, and piloted at the Vitkovice Steel Works, Czechoslovakia. The first commercial plant was built for Esso in Norway.

> Kohl, H.L. and Nielsen, R.B., *Gas Purification*, 5th ed., Gulf Publishing Co., Houston, TX, 1997, 559.

Eluex An early process for extracting uranium from its ores, using both ion-exchange and solvent extraction. Developed by the National Lead Company, United States.

> Merritt, R.C., *The Extractive Metallurgy of Uranium*, U.S. Atomic Energy Authority, Colorado School of Mines, Golden, CO, 1971, 209.
> *Eng. Min. J.*, 1978, **179**(12), 84.
> Eccles, H. and Naylor, A., *Chem. Ind. (London)*, 1987, (6), 174.

Eluxyl A process for separating *p*-xylene from its isomers, using an adsorbent–solvent technique. The process is based on simulated countercurrent adsorption, where the selective adsorbent is held stationary in the adsorption column. The feed mixture to be separated is introduced at various levels in the middle of the column, as in the *Sorbex process. The *p*-xylene product can be more than 99.9% pure. Developed by IFP and Chevron Chemical. A large pilot plant was built in Chevron's site at Pascagoula, MS, in 1994, and a commercial plant on the site was announced in 1996. By 2005, eight plants had been licensed by Axens, of which three were operating.

> *Eur. Chem. News*, 1994, **62**(1648), 18.

Emersol A process for separating stearic and palmitic acids from oleic acid by fractional crystallization from aqueous methanol.

Emert Also known as the Gulf process, the University of Arkansas process, and SSF. A process for converting cellulose to ethanol by simultaneous saccharification and fermentation. Invented by G.H. Emert.

> Emert, G.H., Katzen, R., Fredrickson, R.E., and Kaupisch, K.F., *Chem. Eng. Prog.*, 1980, **76**, 47.

Emert, G.H. and Katzen, R., *Chemtech*, 1980, **10**, 610.
Worthy, W., *Chem. Eng. News*, 1981, **59**(49), 36.
Chem. Eng. News, 23 Apr 2007, **85**(17), 5.

EMICT [**E**xxon**M**obil **I**somerization **C**atalyst **T**echnology] A naphtha isomerization process developed by
ExxonMobil that uses sulfated zirconia with tungsten oxide.

Busca, G., *Chem. Rev.*, 2007, **107**, 5399.

Empty cell *See* Rüping.

ENSEL [The name is probably derived from NCL, the National Chemical Laboratory in Pune, India, whose
former director, Paul Ratnasamy, was the prime inventor of the catalyst for this process] A transesterification
process for making biodiesel from vegetable oils, using a solid double metal cyanide catalyst. Developed
in 2006 by the National Chemical Laboratory, India, and commercialized by Benefuel, a joint venture of
Seymour Biofuels with Süd-Chemie. The first plant was built in Seymour, IN, from 2008. The novel catalyst
was provided by Süd-Chemie India.

Hydrocarbon Process., 2007, **86**(12), 33.
Chem. Eng. (N.Y.), 2008, **115**(2), 12.
U.S. Patent 7,754,643.

Enciforming [**N**ational **C**hemical **reform**ing] A petroleum *reforming process that converts pyrolysis
gasoline to mixtures of propane, butane, and aromatic hydrocarbons, thereby obviating the usual hydrogenation and solvent extraction processes. The catalyst is a ZSM-5-type zeolite containing both iron and
a platinum metal. Developed by the National Chemical Laboratory, Pune, India, since 1988 but not yet
commercialized.

Indian Patent Application 526/DEL/88.

ENCOAL A process for making liquid fuel from low-grade coals. A solid fuel (process-derived coal) is
also produced. Developed on a large scale from 1992 by Encoal Corporation with funding from the U.S.
Department of Energy.

Rezaiyan, J. and Cheremisinoff, N.P., *Gasification Technologies*, Taylor & Francis, Boca Raton, FL, 2005, 154.

Endewax [**N**ational **C**hemical **dewax**ing] A process for dewaxing heavy petroleum fractions by treatment
with a catalyst, which converts the long-chain hydrocarbons to shorter ones. The catalyst is a ZSM-5-type
zeolite in which some of the aluminum has been replaced by iron. Developed by the National Chemical
Laboratory, Pune, India, and piloted in 1991.

Indian Patent Applications 904/DEL/89; 905/DEL/89.

Enefit 140 An improved version of the *Galoter process for extracting crude oil from shale. The ground
shale is mixed with hot ash made by burning spent shale in a separate furnace. Developed by Outotec GmbH
(Germany) and Eesti Energia (Estonia) and scheduled for commissioning in 2012. An earlier version, not
commercialized, was Enefit 280.

Chem. Eng. (N.Y.), 2009, **116**(8), 11.

Energx A process for making linear low-density polyethylene (LLDPE). Developed by Eastman Chemical in
the 1990s and used at its plant in Longview, TX. Licensed to Chevron Chemical in 1999 for use at its plant in
Baytown, TX. By 2002, licenses had been granted in Europe, North America, and Asia. A variation, Energx
DCX, uses a supported catalyst (Sylopol DCX) made by W.R. Grace. The polyethylene products have the trade
name Hifor.

Eur. Chem. News, 1999, **70**(1860), 28; 2000, **73**(1920), 36.
Chem. Mark. Rep., 16 Sep 2002.

Enerkem A process for making a liquid fuel from municipal waste. The waste is first shredded and gasified in a bubbling fluidized bed. The resulting gases are converted to *syngas, which is processed by established methods to liquid fuels. Developed by Enerkem Inc., Montreal, in 2010–2011. A large plant was under construction in Edmonton, Alberta, and another in Mississippi in 2012. A large production facility, to be built in Varennes, Quebec, in association with GreenField Ethanol Inc., was announced in 2012.

Chem. Eng. (N.Y.), 2011, **117**(10), 14.

Enersludge An *OFS process for making fuel oil from sewage sludge. The dried sludge is heated to 450°C in the absence of oxygen, thereby vaporizing about half of it. The vapors are then contacted with hot char from the reaction zone, where catalyzed reactions and thermal cracking convert the lipids and proteins to hydrocarbons. The oil yield is variable, typically 25% of the sludge mass. Developed in the 1980s at the Wastewater Technology Center, Burlington, Canada, based on earlier work by E. Bayer at Tübingen University, Germany, and piloted in Australia and Canada. Marketed by Enersludge, Australia.

U.S. Patents 4,618,735 (process); 4,781,796 (equipment).
European Patent 52,334.
Canadian Patent 1,225,062.
Fernandes, X., *Water Waste Treat.*, 1991, **34**(9), 114.

Engeclor A process for making ammonium chloride by passing gaseous ammonia and hydrogen chloride into an aqueous suspension of the product. Developed by the Brazilian company of the same name.

Bamforth, A.W. and Sastry, S.R.S., *Chem. Process Eng. (London)*, 1972, **53**(2), 72.
Ullmann's Encyclopedia of Industrial Chemistry, 5th ed., Vol. A2, VCH Publishers, Weinheim, Germany, 1989, 259.

Engel A process for making potassium carbonate from potassium chloride obtained from the salt deposits at Stassfurt, Germany. The basis of the process is the formation of the sparingly soluble double salt, $MgKH(CO_3)_2 \cdot 4H_2O$, when carbon dioxide is passed into a suspension of magnesium carbonate in aqueous potassium chloride:

$$3MgCO_3 + 2KCl + CO_2 + H_2O = 2MgKH(CO_3)_2 \cdot 4H_2O + MgCl_2$$

The double salt is decomposed by hot water and magnesia, forming potassium carbonate and insoluble hydrated magnesium carbonate:

$$2MgKH(CO_3)_2 \cdot 4H_2O + MgO = 3MgCO_3 \cdot 3H_2O + K_2CO_3$$

Invented by C.R. Engel in France in 1881.

German Patent 15,218.
Mellor, J.W., *Comprehensive Treatise on Inorganic and Theoretical Chemistry*, Vol. 2, Longmans Green, London, U.K., 1922, 716.

Engel–Precht An improved version of the *Engel process, developed in 1889 by H. Precht at the Salzbergwerk Neu Stassfurt, Germany, and operated there until 1938.

German Patent 50,786.
Thorpe's Dictionary of Applied Chemistry, 4th ed., Vol. 10, Longmans Green, London, U.K., 1950, 150.

English A process for extracting zinc from its ores by reduction with charcoal, invented by the *Champion brothers in Bristol in the eighteenth century.

Cocks, E.J. and Walters, B., *A History of the Zinc Smelting Industry in Britain*, George G. Harrap, London, U.K., 1968, 8.

Ensaco A process for making carbon black by the incomplete combustion of used car tires. Developed and commercialized by IMM, Belgium.

Ensio-Fenox A process for removing chlorinated phenols from pulp-bleaching effluents. It combines anaerobic and aerobic fermentation processes.

Hakulinen, R. and Salkinoja-Salonen, M., *Proc. Biochem.*, 1982, **17**, 18.
Speece, R.E., in *Toxicity Reduction in Industrial Effluents*, Lankford, P.W. and Eckenfelder, W.W., Jr., Eds., Van Nostrand Reinhold, New York, 1990, 146.

ENSOL A combined process for converting *syngas to methanol and then to ethanol. Acetic acid is an intermediate. Developed by Humphreys & Glasgow, in conjunction with BASF and Monsanto.

Chem. Ind. (London), 1985, (8), 240.
Winter, C.L., *Hydrocarbon Process. Int. Ed.*, 1986, **65**(4), 71.

ENSORB [**E**xxo**N** ad**SORB**] A process for separating linear from branched hydrocarbons, using a zeolite molecular sieve. The adsorbed gases are desorbed using ammonia. The process operates in a cyclic, not a continuous, mode. Developed by Exxon Research and Engineering Company and used by that company on a large scale at the Exxon refinery in Baytown, TX.

Asher, W.J., Campbell, M.L., Epperly, W.R., and Robertson, J.L., *Hydrocarbon Process. Int. Ed.*, 1969, **48**(1), 134.
Yang, R T , *Gas Separation by Adsorption Processes*, Butterworths, Guildford, U.K., 1987, 262.

ENsulf A process for desulfurizing gas streams containing hydrogen sulfide, yielding elemental sulfur. Air containing a stoichiometric quantity of oxygen is added to the gas stream, which is passed over a catalyst at about 200°C.

Hydrocarbon Process. Int. Ed., 2000, **79**(4), 69.

EnviNOx Two related processes for removing N_2O and NO_x from the tail gases from nitric acid manufacture by the *Ostwald process. Both variants use two beds of two zeolite catalysts modified with iron. The first variant is for gases hotter than 425°C and involves injecting ammonia between the beds. The second variant, for gases cooler than 425°C, injects ammonia before the first bed and a volatile hydrocarbon between the beds. Developed by Uhde in association with Agrolinz Melamine International. The first variant was installed in Linz, Austria, in 2003. The second variant was installed in Egypt in 2007. Installations were planned for Nitrogenmuvek Zrt, Hungary, in 2007; ENAEX SA, Chile, in 2007; and Bayer, Dormagen, in 2011. The catalysts are supplied by Süd-Chemie.

Groves, M., Schwefer, M., and Siefert, R., *Chem. Eng. (Rugby, Engl.)*, 2006, (778), 30.
Oygarden, A.H. and Pérez-Ramírez, J., *Appl. Catal. B: Environ.*, 2006, **65**(1–2), 163.
Chem. Eng. World, 2006, **41**(3), 18.
Chem. Eng. Prog., 2007, **103**(3), 14.
Hevia, M.A.G. and Perez-Ramirez, J., *Appl. Catal. B: Environ.*, 2008, **77**(3–4), 248.
Chem. Ind. (London), 2010, (23), 13.

Enzink [**Enz**yme de**ink**ing] A paper-deinking process using cellulase enzymes. Invented by K.-E.L. Eriksson and J.L. Yang at the University of Georgia and commercialized in 1994.

Chem. Eng. News, 1998, **76**(12), 42.

EnZone [**En**Z**yme oz**one] A pulp-bleaching process using the enzyme xylanase in combination with oxygen, ozone, and hydrogen peroxide. Invented by K.-E.L. Eriksson at the University of Georgia and piloted in 1998.

Chem. Eng. News, 1998, **76**(12), 42.
Vicuna, R., Yeber, M.C., and Osses, M., *J. Biotechnol.*, 1995, **42**(1), 69.

EnzyMax A process for degumming vegetable oils by use of the enzyme phospholipase A2, which converts water-insoluble phospholipids into hydratable phospholipids that can be extracted with water. Developed by Lurgi and first installed at Shenzhen, China.

> *Eur. Chem. News*, 29 Jun 1998, **69**(1819), 28.

EOF [**E**nergy-**O**ptimizing **F**urnace] An oxygen steelmaking process in which part of the heat is provided by the combustion of carbon powder blown beneath the surface of the molten iron. Developed by the KORF group and being considered for use in India in 1987.

> Bose, P.K., *Gas Sep. Purif.*, 1987, **1**, 30.

Eolys A process for removing 80%–90% of the particulate carbon from diesel exhaust gases, which uses a catalytic fuel additive containing cerium. Developed by Rhône-Poulenc in 1995 and licensed to Sumitomo Metal Mining Company in Japan in 1996.

EP2 A process for polymerizing olefins in the slurry phase. Developed by Borealis.

Epal A process for making linear aliphatic alcohols by reacting ethylene with triethyl aluminum and oxidizing the products. Similar to *Alfol but incorporating a transalkylation stage that permits a wider range of products to be made. Developed by Ethyl Corporation (now Albemarle Corporation) and operated in the United States since 1964.

> *Ullmann's Encyclopedia of Industrial Chemistry*, 5th ed., Vol. A10, VCH Publishers, Weinheim, Germany, 1989, 284.

Epicerol A process for converting glycerol to epichlorohydrin. Glycerol is reacted with hydrogen chloride, over a novel catalyst, to produce mixed dichloropropanols. These are then dehydrochlorinated. Developed by Solvay from 2004 in Tavaux, France, and piloted from 2007, using glycerol derived from rapeseed oil. A large plant was proposed for construction in Thailand for completion at end 2011 and one in China for 2013 or 2014. Solvay took legal action against a Chinese epichlorohydrin producer for infringing its patents on this process.

> U.S. Patent 8,067,645.
> *Chem. Eng. News*, 2006, **84**(6), 7.
> *Chem. Eng. (N.Y.)*, 2006, **113**(3), 14; 2007, **114**(9), 36.

Erco A group of processes for making chlorine dioxide by reducing sodium chlorate with a chloride:

$$NaClO_3 + NaCl + H_2SO_4 = ClO_2 + \tfrac{1}{2}Cl_2 + Na_2SO_4 + H_2O$$

Developed by W.H. Rapson at the Electric Reduction Company of Canada, now Erco Industries, United States. Erco R-3 uses chloride in sulfuric acid as the reductant; Erco R-3H uses mixed hydrochloric and sulfuric acids; Erco R-5 uses chloride in hydrochloric acid.

> Canadian Patents 825,084; 826,577.

Erdmann *See* Normann.

Erdölchemie A process for treating the waste from the *ammoxidation process for making acrylonitrile, yielding ammonium sulfate. Developed by the eponymous German company, a joint venture of Bayer and BP Chemicals.

Erifon A process for making cellulose textiles flame resistant. Titanium and antimony oxychlorides are applied from acid solution, and the cloth is then neutralized with sodium carbonate solution. Invented by DuPont in 1947. *See also* Titanox FR.

> U.S. Patent 2,570,566.
> Gulledge, H.C. and Seidel, G.R., *Ind. Eng. Chem.*, 1950, **42**, 440.

ERIX [**E**lectrochemically **R**egenerated **I**on-e**X**change system] A process for removing hydrofluoric acid from aqueous wastes from the electronics industry. The fluoride ion is trapped in an ion-exchange resin, which is continuously regenerated electrochemically. Developed by BOC Edwards and first installed in the University at Albany, State University of New York, in 2006.

Chem. Eng. (N.Y.), 2006, **113**(10), 13.

Erlenmeyer An early process for making potassium cyanide from potassium ferrocyanide by heating it with sodium:

$$K_4Fe(CN)_6 + 2Na = 4KCN + 2NaCN + Fe$$

First commercialized in 1876. *See also* Rodgers.

ERMS [**E**nhanced **R**oasting **M**agnetic **S**eparation; also known as **E**rnie's **R**eductive **M**agnetic **S**eparation] A process for beneficiating ilmenite, yielding an off-white product containing more than 99% TiO_2. The ilmenite is roasted in a fluid bed, leached with hydrochloric acid, and beneficiated by dry magnetic separation. The acid is recycled by the *EARS process. The product can be used as a *Chloride process feed material or as a low-grade pigment. Developed by E.A. Walpole at the University of Newcastle, Australia, from the early 1990s and piloted by Austpac Gold (now Austpac Resources). Licensed to Iscor, South Africa (later Ticor South Africa), in July 1998 but not exploited there. A 3000 tpa demonstration plant was built in Newcastle, NSW, in 2008. *See also* EARS.

World Patent Application WO 92/04121.
Ind. Miner. (London), 1996, (349), 10.

Escambia (1) A process for oxidizing isobutene to *a*-hydroxy-isobutyric acid (HIBA), for use as an intermediate in the manufacture of methacrylates. The oxidant was dinitrogen tetroxide, N_2O_4. Operated by the Escambia Chemical Corporation, FL, in 1965 before its destruction by an explosion in 1967. It has not been used again.

British Patent 954,035.

Escambia (2) A process for oxidizing propylene to propylene oxide. A transition metal catalyst is used in an organic solvent at high temperature and pressure. Developed by the Escambia Corporation.

Dumas, T. and Bulani, W., *Oxidation of Petrochemicals: Chemistry and Technology*, Applied Science Publishers, London, U.K., 1974, 34.

ESEP A process for removing ethylene from gas streams by absorption in a solution of $CuAlCl_4$ in an aromatic solvent. Developed from the *Cosorb process by Tenneco in 1978.

Safaric, D.J. and Eldridge, R.B., *Ind. Eng. Chem. Res.*, 1998, **37**(7), 2571.

E-SOX A *flue-gas desulfurization process. The gas is contacted with calcium hydroxide or sodium hydroxide solution in a spray drier, and the resulting particles are trapped in an electrostatic precipitator. Developed and piloted by the U.S. Environmental Protection Agency.

Espig A process for making synthetic emeralds by the flux reaction technique. Beryllia and alumina are dissolved in molten lithium molybdate, and silica is floated on the melt. The emerald crystals form at the base of the melt, but because they tend to float and mix with the silica, a platinum screen is suspended in the middle of the melt. Invented by H. Espig.

Espig, H., *Chem. Technol. (Berlin)*, 1960, **12**, 327.
Elwell, D., *Man-Made Gemstones*, Ellis Horwood, Chichester, U.K., 1979, 59.

ESPLIM [**E**lectro**S**tatic **P**seudo-**LI**quid **M**embrane] A solvent extraction system that accomplishes extraction and stripping in a single extractor, using a membrane and an electric field. Invented in China in the 1980s and further developed by BNFL, United Kingdom, and the University of Western Australia.

> *Chem. Eng. (Rugby, Engl.)*, 10 Dec 1998, 15.

ESR *See* Electroslag.

Esso Fior *See* Fior.

Estasolvan A process for removing acid gases from liquefied petroleum gases by absorption in tributyl phosphate and separation by fractional distillation. Developed by the Institut Français du Pétrole and Friedrich Uhde. No commercial plants were operating in 1985, and in 1992 it was unavailable.

> Franckowiak, S. and Nitschke, E., *Hydrocarbon Process. Int. Ed.*, 1970, **49**(5), 145.
> Kohl, H.L. and Nielsen, R.B., *Gas Purification*, 5th ed., Gulf Publishing Co., Houston, TX, 1997, 1224.

ESTER A batch process for immobilizing nuclear waste in a borosilicate glass for long-term disposal. Developed in Italy in the 1970s and installed at the Euratom Research Centre, Ispra, in 1981. Intended for use in the radioactive pilot plant (Impianto Vetrificatione Eurex, IVEX) at the European Extraction Plant (UREX) at Saluggia, Italy.

> Lutze, W., in *Radioactive Waste Forms for the Future*, Lutze, W. and Ewing, R. C., Eds., North-Holland, Amsterdam, the Netherlands, 1988, 12.

ESTEREX A process for extracting neutral and acid esters from the products of sulfuric acid–catalyzed alkylation processes, using sulfuric acid passed through a bundle of hollow fibers immersed in the reaction mixture. Developed by the Merichem Company, Houston, TX.

> *Hydrocarbon Process. Int. Ed.*, 1996, **75**(4), 126.

Esterfip-H A process for converting vegetable oils to methyl esters for use as diesel fuel ("biodiesel"). A heterogeneous transesterification catalyst is used. Developed by IFP and Sofiproteol, France; licensed by Axens. Axens was awarded the Kirkpatrick Chemical Engineering Achievement Award for this in 2007. First commercialized at Sete, France, in 2006. Several more units were announced in 2007. The latest version uses two fixed-bed reactors and a zinc aluminate catalyst developed by Axens. The glycerol by-product is 98% pure.

> *Eur. Chem. News*, 1994, **62**(1648), 18.
> *Chem. Eng. (N.Y.)*, 2004, **111**(11), 13.
> *Eur. Chem. News*, 2006, **168**(12), 20.
> Bournay, L., Casanave, D., Delfort, B., Hilton, G., and Chodorge, J.A., *Catal. Today*, 2005, **106**, 190.
> *Chem. Eng. (N.Y.)*, 2007, **114**(13), 19.
> Santacesaria, E., Vicente, G.M., Di Serio, M., and Tesser, R., *Catal. Today*, 2012, **195**(1), 2.
> Luque, R. and Melero, J.A., Eds., *Advances in Biodiesel Production*, Woodhead Publishing, Cambridge, U.K., 2012, 138.

Ethalk [**Eth**ylene **alk**ylation] A catalytic process for combining ethylene with isobutane, to make a gasoline blending component, using a dilute ethylene stream. Developed by Technip Benelux (formerly KTI) from 1995, but not piloted by mid-1999.

> *Eur. Chem. News*, 1999, **70**(1863), 35.

Ethermax A process for making ethers (e.g., methyl *t*-butyl ether) by reacting tertiary olefins with alcohols. Conversion levels are increased by using reaction with distillation technology (also called *RWD), in which the reaction takes place in a distillation column containing the catalyst, a sulfonic acid ion-exchange resin. Equilibrium limitations are overcome by continuously removing the products in the RWD section as the reaction occurs. Developed jointly by Hüls, UOP, and Koch Engineering Company and licensed by UOP. Eleven units were operating in 1996.

Chem. Eng. (Int. Ed.), 1991, **98**(7), 44.

DeGarmo, J.L., Parulekar, V.N., and Pinjala, V., *Chem. Eng. Prog.*, 1992, **88**(3), 43.

Hydrocarbon Process. Int. Ed., 1996, **75**(11), 113.

Krupa, S., Meister, J., and Luebke, C., in *Handbook of Petroleum Refining Processes*, 3rd ed., Meyers, R.A., Ed., McGraw-Hill, New York, 2003, 3.2.

Etherol A process for making oxygenated fuels (e.g., methyl *t*-butyl ether) from C_4 to C_6 hydrocarbons by reacting them with methanol over an acid resin catalyst in a fixed-bed reactor under mild conditions. Developed by BP with Erdoel Chemie and first used in a refinery at Vohburg, Germany, in 1986. Four units were operating and one was under construction in 1988.

Hydrocarbon Process. Int. Ed., 1988, **67**(9), 83.

Ethoxene A process for making a mixture of ethylene and acetic acid by the catalytic oxidation of a mixture of ethane and ethylene. Developed and piloted by Union Carbide in the 1980s, but not commercialized because of inflexibility in the ratio of acetic acid to ethylene produced.

Hu, Y.C., in *Chemical Processing Handbook*, McKetta, J.J., Ed., Marcel Dekker, New York, 1993, 810.

Plotkin, J., *Eur. Chem. News*, 2003, **78**(2047), 23.

Ethylbenzene Also called UOP Ethylbenzene. A liquid-phase process for making ethylbenzene by reacting ethylene with benzene, catalyzed by a zeolite. The process is usually coupled with one for converting ethylbenzene to styrene. Developed by Unocal Corporation and now licensed by UOP and ABB Lummus Global.

Eureka A process for upgrading bitumen and heavy oils by delayed coking, which yields a heavy pitch rather than a coke. Developed by Chiyoda Corp and Fuji Oil Co and operated in Sodegaura, Japan, since 1976.

Takahashi, R. and Washimi, K., *Hydrocarbon Process. Int. Ed.*, 1976, **55**(11), 93.

Aiba, T., Kaji, H., Suzuki, T., and Wakamatsu, T., *Chem. Eng. Prog.*, 1981, **77**(2), 37.

EUROCLAUS An improved version of the *SUPERCLAUS process for converting hydrogen sulfide to elemental sulfur. Developed by Comprimo and engineered by Jacobs Engineering. First demonstrated in 2000. By 2012, 40 plants were operating or under construction.

EVA-ADAM [Einzelrohr VerzugsAnlage/ADA Methanator] Also called ADAM-EVA. A cyclic process for transporting energy, produced in a nuclear reactor, by a gas pipeline. Heat from the reactor is used for *steam reforming methane. The gaseous products are passed along a pipe, at the end of which the reaction is catalytically reversed. Developed by Kernforschungsanlage Jülich, Germany, and piloted there from 1979 to 1981.

Parmon, V.N., *Catal. Today*, 1997, **35**(1–2), 153.

Weissermel, K. and Arpe, H.-J., *Industrial Organic Chemistry*, 4th ed., translated by C.R. Hindley and S. Hawkins, Wiley-VCH, Weinheim, Germany, 2003, 23.

Evolue Also written Evolué. A gas-phase process for making linear low-density polyethylene using higher α-olefin comonomers and a metallocene catalyst. Developed by Mitsui Chemicals and Idemitsu Kosan, and now manufactured in Japan by Prime Polymer, a joint venture of these two companies.

Jpn. Chem. Week, 2006, **47**(2378), 2.

Excer A process for making uranium tetrafluoride by electrolytic reduction of a uranyl fluoride solution, precipitation of a uranium tetrafluoride hydrate, and ignition of the hydrate.

EXOL N Also called EXOL N Extraction. A solvent extraction process for purifying feedstocks for making lubricating oil. The solvent is N-methyl pyrrolidone. Developed and licensed by Exxon Research and Engineering Company. Seventeen units had been installed by 1994.

Bushnell, J.D. and Fiocco, R.J., *Hydrocarbon Process. Int. Ed.*, 1980, **59**(5), 119.
Hydrocarbon Process. Int. Ed., 1994, **73**(11), 140.

ExSact An *alkylation process catalyzed by a proprietary solid acid catalyst in a fixed-bed reactor. It converts C_3–C_5 hydrocarbons to a gasoline component. Developed by Exelus, NJ, and piloted from 2006 to 2012.

Chem. Eng. (N.Y.), 2007, **114**(1), 11.

ExSyM [**Ex**elus **Sty**rene **M**onomer] A one-step process for making styrene from toluene and methanol. A special zeolite catalyst is used at 400°C–425°C, under atmospheric pressure. Developed by Exelus, NJ, and piloted from 2008 to 2012.

ICIS Chem. Bus., 2008, **273**(3), 28.
Chem. Eng. (N.Y.), 2010, **117**(12), 11.

ExSyte A process for making syngas by steam reforming methane under novel conditions. Partially developed by Exelus, NJ, but since abandoned.

Extafax An early thermographic copying process.

Exxon Donor Solvent Also known as EDS. A coal liquefaction process in which coal in solution in tetrahydronaphthalene is hydrogenated, using a cobalt–molybdenum–alumina catalyst. So-called because the hydrogen is "donated" by the tetrahydronaphthalene to the coal. Developed from the *Pott-Broche process. Piloted by Exxon Research and Engineering Company in the 1970s and operated at 250 tons per day in the Exxon refinery in Baytown, TX, from 1980 to 1982.

Furlong, L.E., Effron, E., Vernon, L.W., and Wilson, E.L., *Chem. Eng. Prog.*, 1976, **72**(8), 69.
Eur. Chem. News, 1982, **39**(1047), 11.
Maa, P.S., Trachte, K.L., and Williams, R.D., *The Chemistry of Coal Conversion*, Schlosberg, R.H., Ed., Plenum Publishing, New York, 1985.
Davies, G.O., in *Chemicals from Coal: New Developments*, Payne, K.R., Ed., Blackwell Scientific Publications, Oxford, U.K., 1985, 109.

Exxpol [**Exx**on **pol**ymerization] A gas-phase process for making polyethylene from ethylene. The process uses single-site catalysis (SSC), based on a zirconium metallocene catalyst. Developed by Exxon Chemical Company in 1990 and now widely used.

Eur. Chem. News, 1992, **57**(1514), 24, 27.

Ezinex A multistage hydrometallurgical process for recovering zinc from electric arc furnace dust. It involves leaching, residue separation, purification, electrowinning, and crystallization. Developed by Engitec Impianti (Milan) in the 1990s and operated at Ferriere Nord, Italy.

EZ-Oil Generator A process for making liquid fuels by pyrolyzing waste plastics. Piloted in China by the Beijing Roy Environment Technology Company, also called ROYCO.

Scheirs, J., in *Feedstock Recycling and Pyrolysis of Waste Plastics*, Scheirs, J. and Kaminsky, W., Eds., John Wiley & Sons, Chichester, U.K., 2006, 422.

F

FAeSTER An enzymic transesterification process for making biodiesel, developed by Piedmont Biofuels and operated in North Carolina since 2012.

Biodiesel Magazine, Mar/Apr 2012.

Falconbridge Also called the matte leach process. A process for extracting copper and nickel from matte (a sulfide ore that has been roasted to remove most of the sulfur). Most of the nickel is leached out with hydrochloric acid and recovered as nickel chloride crystals. The leach residue is roasted and leached with sulfuric acid to dissolve the copper. The process has been operated in Canada and Norway since 1970.

Thornhill, P.G., Wigstol, E., and Van Weert, G., *J. Met.*, 1971, **23**(7), 13.
Burkin, A.R., *Extractive Metallurgy of Nickel*, John Wiley & Sons, Chichester, U.K., 1987, 121.
Mukherjee, C.K. and Mukherjee, T.K., *Hydrometallurgy in Extraction Processes*, Vol. 1, CRC Press, Boca Raton, FL, 1990, 15, 117.
Hill, I., in *Insights into Speciality Inorganic Chemicals*, Thompson, D., Ed., Royal Society of Chemistry, Cambridge, U.K., 1995, 18.

Fan steel A process for extracting tungsten from wolframite, $FeWO_4$. The ore is mixed with sodium carbonate and heated to 800°C, forming sodium tungstate. This is leached out and treated with calcium chloride, precipitating calcium tungstate. The metal is produced via tungstic acid, ammonium tungstate, and tungstic oxide.

FAST (1) A *BAF process.

Stephenson, T., Mann, A., and Upton, J., *Chem. Ind. (London)*, 1993, (14), 533.

FAST (2) [Fast-Acting Selective Thermolysis] A process for converting biopolymers such as poly-4-hydroxybutyrate into gamma butyrolactone. The fermentation broth is dried and heated. Developed by Metabolix.

Chem. Eng. (N.Y.), 2011, **118**(9), 11.

FASTMELT *See* FASTMET.

FASTMET A *DR process, using pulverized coal and iron-ore fines, heated in a rotary hearth furnace. Under development by Midrex Corporation and Kobe Steel from 1991; a pilot plant was operated by Kobe Steel in Japan in 1996. The first commercial installation was at Kobe Steel's Kakogawa plant in Japan in 2000. Further development of the process took place under the name *ITmk3. A variation, known as FASTMELT, conveys the hot iron powder to an adjacent melter. *See also* MIDREX.

Fauser An early process for making ammonia. Developed by G. Fauser in Italy in 1924.

Vancini, C.A., *Synthesis of Ammonia*, translated by L. Pirt, Macmillan Press, Basingstoke, U.K., 1971, 230.

Fauser–Montecatini A *reforming process for making *syngas from heavy hydrocarbons by gasifying with preheated steam and oxygen. Widely operated in Europe and Asia in the 1960s.

FBA [**F**ixed-**B**ed **A**lkylation] An *alkylation process developed by Amoco Corporation and Haldor Topsoe. To be demonstrated at Amoco's Yorktown, VA, refinery from 1997.

Oil Gas J., 1996, **94**(14), 69.

FBD [**F**luidized-**B**ed **D**ehydrogenation] A catalytic process for converting alkanes to alkenes. Two fluidized beds are used: a reactor and a regenerator. The catalyst beads are recirculated from the regenerator to the reactor, providing heat to the reactor. The catalyst is chromia on alumina.

Miracca, I. and Piovesan, L., *Catal. Today*, 1999, **52**, 259.
Sanfillipo, D. and Miracca, I., *Catal. Today*, 2006, **111**(1–2), 136.

FBD-4 [**F**luid-**B**ed **D**ehydrogenation] A process for converting isobutane to isobutene, used in making methyl *t*-butyl ether (MTBE, a gasoline additive). The reactor is a fluidized bed, and the catalyst is regenerated in another fluidized bed. The regeneration system was developed in Russia in the 1960s, based on work by Esso Standard Oil in the 1950s. The catalyst is chromia–silica, stabilized with silica and promoted by alkali metals. The process is now operated on a large scale, using a catalyst made by Haldor Topsøe.

Sanfilippo, D., *Catal. Today*, 2011, **178**(1), 142.

FCC [**F**luid **C**atalytic **C**racking] A process for converting various heavy liquid petroleum fractions into high-octane gasoline and other fuels. The original inventors were W.K. Lewis and E.R. Gilliland at the Massachusetts Institute of Technology. Their patent (US 2,498,088) was filed in 1940, but it was not published until 1950 because of its importance to the U.S. war effort. Developed by Universal Oil Products (now UOP) and several oil companies and first commercialized by Standard Oil of New Jersey (now Exxon) at Baton Rouge, LA, in 1942. The subsequent four inventors at Standard Oil in 1942 were D.L. Campbell, H.Z. Martin, E.V. Murphree, and C.W. Tyson, who are honored in the U.S. National Inventors' Hall of Fame, Akron, OH. In addition to providing high-octane gasoline, the process gave butenes, which were used in the manufacture of synthetic rubber.

In recent years, the process has been modified to increase the yield of lower olefins, too. Continually improved since then, especially in the mid-1960s with the replacement of the original silica–alumina catalyst by a zeolite. The catalyst is now typically a zeolite Y, bound in a clay matrix. The feed is vaporized and contacted in a pipeline reactor with concurrently flowing microspheroidal catalyst particles. The catalyst is then separated from the hydrocarbon products and is continuously regenerated by burning off the coke in a fluidized bed. The process is licensed by UOP; several hundred units are in operation worldwide. *See also* HS-FCC.

U.S. Patents 2,451,804; 2,498,088.
Enos, J.L., *Petroleum Progress and Profits*, MIT Press, Cambridge, MA, 1962, Chap. 6.
Venuto, P.B. and Habib, E.T., Jr., *Fluid Catalytic Cracking with Zeolite Catalysts*, Marcel Dekker, New York, 1979.
Dwyer, J. and Rawlence, D.J., *Catal. Today*, 1993, **18**(4), 487.
Magee, J.S. and Mitchell, M.M., Jr., *Fluid Catalytic Cracking: Science and Technology*, Elsevier, Amsterdam, the Netherlands, 1993.
Sadeghbeigi, R., *Fluid Catalytic Cracking Handbook*, Gulf Publishing Co., Houston, TX, 1995.
Davis, B.H., *Appl. Catal. A: Gen.*, 1999, **189**(2), N8.
Harding, R.H., Peters, A.W., and Nee, J.R.D., *Appl. Catal. A: Gen.*, 2001, **221**(1–2), 389.
O'Donnell, J.P., *Oil Gas J.*, 2002, **100**(35), 127.
Niccum, P.K. and Santner, C.R., in *Handbook of Petroleum Refining Processes*, 3rd ed., Meyers, R.A., Ed., McGraw-Hill, New York, 2003, 3.3.
Flank, W.H., Abraham, M.A., and M.A. Matthews, M.A., Eds., *Innovations in Industrial and Engineering Chemistry*, 2009, American Chemical Society, Washington DC, Chap. 6.
Chem. Eng. (Rugby, Engl.), Apr 2010, (826), 52.

FEAST [**F**urther **E**xploitation of **A**dvanced **S**hell **T**echnology] Not a single process, but a range of processes for converting cyclic diolefins into alpha–omega dienes. The catalyst is based on rhenium on alumina. Operated in France since 1986.

Chem. Eng. (N.Y.), 1987, **94**(11), 22.
Chaumont, P. and John, C.S., *J. Mol. Catal.*, 1988, **46**, 317.

Feld Also called thionite. An early process proposed for removing hydrogen sulfide and ammonia from coal gas by absorption in an aqueous solution of ammonium thionates. Investigated by W. Feld in Germany in the early 1900s; operated at Königsberg, but never fully developed.

German Patent 237,607.
Kohl, H.L. and Nielsen, R.B., *Gas Purification*, 5th ed., Gulf Publishing Co., Houston, TX, 1997, 734.

Fernbach–Strange–Weizmann *See* Weizmann.

Ferrite *See* Löwig.

Ferrofining A mild *hydrotreating process for purifying lubricating oils. The catalyst contained cobalt, molybdenum, and iron (hence the name). Developed by the British Petroleum Company and first operated in Dunkirk, England, in 1961.

Dare, H.F. and Demeester, J., *Pet. Refin.*, 1960, **39**(11), 251.
Hydrocarbon Process. Int. Ed., 1964, **43**(9), 187.
Unzelman, G.H. and Wolf, C.J., in *Petroleum Processing Handbook*, Bland, W.F. and Davidson, R.L., Eds., McGraw-Hill, New York, 1967, 3–45.

FERROSEP A magnetic process for removing iron contaminants from petroleum residues before desulfurization. Developed by Nippon Oil Company and Nippon Petroleum Refining Company and operated in Japan since 1992.

Japanese Patent H5 35754.

Ferrox A process for removing hydrogen sulfide from petroleum refining streams by absorption in an aqueous solution of sodium carbonate containing suspended ferric hydroxide. The absorbent is regenerated by blowing air through it, producing elementary sulfur. The process was invented in 1921 by F.D. Mann, Jr., at the Standard Development Company and subsequently developed by the Koppers Company and widely used. Later, it was replaced mainly by the *Thylox process. *See also* Gluud.

U.S. Patents 1,525,140; 1,841,419.
Kohl, A.L. and Riesenfeld, F.C., *Chem. Eng. (N.Y.)*, 1959, **66**, 152.
Kohl, H.L. and Nielsen, R.B., *Gas Purification*, 5th ed., Gulf Publishing Co., Houston, TX, 1997, 738.

Fersona A process for stabilizing the calcium sulfite–sulfate waste from *FGD processes, so that it may be used for landfill. The waste is mixed with ferric sulfate waste from another process (e.g., metallurgical leaching) to form sparingly soluble basic sodium ferric sulfates. Developed in the 1970s at the Battelle Columbus Laboratories, OH, under contract with Industrial Resources. *See also* Sinterna.

U.S. Patents 3,876,537; 3,984,312; 4,034,063.
Dulin, J.M., in *Toxic and Hazardous Waste Disposal*, Pojasek, R.J., Ed., Ann Arbor Science, Ann Arbor, MI, 1979, Chap. 18.

FFC *See* Fray–Farthing–Chen.

FFGC [Final Flue-Gas Cleaning system] Not a single process, but a combination of processes to remove sulfur oxides, nitrogen oxides, particulates, mercury, other heavy metals, and volatile organic compounds from flue gases. Developed by WOW Energy and demonstrated in two plants in Texas in 2006. Planned to be operated in conjunction with the *CCLC (cascading closed-loop cycle) heat recovery system.

Chem. Eng. (N.Y.), 2006, **113**(1), 14.

FGD [Flue-Gas Desulfurization] *See* Flue-Gas Desulfurization.

Fina/Badger A process for making styrene by dehydrogenating ethylbenzene. The reaction takes place at high temperature and low pressure and in the presence of steam and a proprietary heterogeneous catalyst. The hydrogen produced is used to provide process heat. Developed by the Badger Company in the 1960s, first operated on a large scale by Union Carbide Corporation at Seadrift, TX, and now widely used.

FINGAL [Fixation IN Glass of Active Liquors] A batch process for immobilizing nuclear waste in a borosilicate glass. Developed by the U.K. Atomic Energy Authority at Harwell from 1958 and piloted at its Windscale Works from 1962 to 1966. After a lapse of several years, the project was resumed in 1972 under the acronym *HARVEST.

> Lutze, W., in *Radioactive Waste Forms for the Future*, Lutze, W. and Ewing, R.C., Eds., North-Holland, Amsterdam, the Netherlands, 1988, 14.
> Ojovan, M.I. and Lee, W.E., *An Introduction to Nuclear Waste Immobilisation*, Elsevier, Amsterdam, the Netherlands, 2005, 233.

FINMET An ironmaking process now used by BHP in Australia and under construction in Venezuela.

Finsider *See* DR.

Fior Also called Esso Fior. A direct-reduction ironmaking process, using natural gas as the reductant, in a fluidized bed. Operated in Venezuela since 1976. Licensed by Davy Corporation. *See* DR.

FIPS [FIssion Product Solidification] A process for immobilizing the radioactive waste products from the thorium fuel cycle in a borosilicate glass for long-term storage. Developed at Kernforschungsanlage Jülich, Germany, from 1968, until abandoned in favor of *PAMELA in 1977.

> Lutze, W., in *Radioactive Waste Forms for the Future*, Lutze, W. and Ewing, R.C., Eds., North-Holland, Amsterdam, the Netherlands, 1988, 8.

Fischer A process for removing hydrogen sulfide from coal gas by absorption in an aqueous solution of potassium ferrocyanide and bicarbonate; the solution is regenerated electrochemically with the production of elemental sulfur. Operated at the Hamburg gasworks in the 1930s.

> Müller, H., *Gas Wasserfach.*, 1931, **74**, 653.
> Kohl, H.L. and Nielsen, R.B., *Gas Purification*, 5th ed., Gulf Publishing Co., Houston, TX, 1997, 745.

Fischer–Tropsch A process for converting synthesis gas (a mixture of carbon monoxide and hydrogen) to liquid fuels. Modified versions were known as the *Synol and *Synthol processes. The process is operated under pressure at 200°C to 350°C, over a catalyst. Several different catalyst systems have been used at different periods, notably iron–zinc oxide, nickel–thoria on kieselguhr, cobalt–thoria on kieselguhr, and cemented iron oxide. The main products are C_5 to C_{11} aliphatic hydrocarbons; the aromatics content can be varied by varying the process conditions.

The basic reaction was discovered in 1923 by F. Fischer and H. Tropsch, working at the Kaiser Wilhelm Institute for Coal Research in Mülheim, Germany. In 1984, Mako and Samuel wrote, "The quantity of patents and literature that has appeared on the subject in the past 60 years makes it virtually impossible to retrace step-wise the developments of the Fischer–Tropsch synthesis." The first full-scale plant was built by Ruhrchemie at Holten, from 1934 to 1936; by 1939, eight more plants had been built in Germany. All of these plants were disabled by air attacks in 1944. Used in the *SASOL coal gasification plant in South Africa. The first modern Fischer–Tropsch plant outside Africa was built by Shell in Malaysia in 1993, based on natural gas.

In the 1990s, many companies developed related processes for making liquid fuels from *syngas. *See also* Synthine.

> German Patent 484,337.
> Fischer, F. and Tropsch, H., *Ber. Dtsch. Chem. Ges.*, 1923, **56**, 2428.
> Fischer, F. and Tropsch, H., *Brennstoff-Chemie*, 1923, **4**, 193; 1932, **13**, 62; 1935, **16**, 1.
> Storch, H.H., Golumbic, N., and Anderson, R.B., *The Fischer-Tropsch and Related Syntheses*, John Wiley & Sons, New York, 1951.

Dry, M.E., *Hydrocarbon Process. Int. Ed.*, 1982, **61**(8), 121.

Anderson, R.B., *The Fischer-Tropsch Synthesis*, Academic Press, Orlando, FL, 1984.

Mako, P.F. and Samuel, W.A., in *Handbook of Synfuels Technology*, Meyers, R.A., Ed., McGraw-Hill, New York, 1984, Chap. 2.

Dry, M.E., in *Chemicals from Coal: New Processes*, Payne, K.R., Ed., John Wiley & Sons, Chichester, U.K., 1987.

Bartholomew, C.H., in *New Trends in CO Activation*, Guczi, L., Ed., Elsevier, Amsterdam, the Netherlands, 1991, 159.

Chem. Eng. (N.Y.), 1997, **104**(4), 39.

Davis, B.H., *Catal. Today*, 2002, **71**(3–4), 249.

Weissermel, K. and Arpe, H.-J., *Industrial Organic Chemistry*, 4th ed., translated by C.R. Hindley and S. Hawkins, Wiley-VCH, Weinheim, Germany, 2003, 22.

Davis, B.H., *Catal. Today*, 2009, **141**(1–2), 25.

Glasser, D., Hildebrand, D., Liu, X., Lu, X., and Masuku, C.M., *Curr. Opin. Chem. Eng.*, 2012, **1**(3), 296.

Flakt–Boliden A variation on the *Citrate process for *flue-gas desulfurization, in which the sulfur dioxide is removed from the citrate solution by vacuum. Developed by Flakt, United States, and piloted in 1980 at the TVA Electric Power Research Institute, Muscle Shoals, AL.

Fläkt–Hydro A *flue-gas desulfurization process that uses seawater as the scrubbing liquor. The pH of the effluent is raised with calcium hydroxide before it is discharged to sea. Developed by Norsk Viftefabrikk. Now owned by ABB Fläkt Industri. As of 1996, 16 plants had been installed worldwide.

Bafy, R., Coughlan, J., and Reynolds, S.K., *Environ. Protect. Bull.*, 1991, **12**, 21.

Radojevic, M., *Chem. Br.*, 1996, **32**(11), 47.

Kohl, H.L. and Nielsen, R.B., *Gas Purification*, 5th ed., Gulf Publishing Co., Houston, TX, 1997, 599.

Flame-Smelting *See* DR.

FLC [Flame Chamber] A high-temperature process for pyrolyzing solid waste. The granulated waste is passed down the axial space of a vertical, coaxial reactor and heated by a central combustion chamber.

Martinez, D., in *Chemical Waste Handling and Treatment*, Muller, K.R., Ed., Springer-Verlag, Berlin, Germany, 1986, 150.

Fleming An early liquid-phase, thermal process for cracking petroleum. *See also* Dubbs.

Flesch–Winkler *See* Winkler.

Fletcher A new process proposed for making titanium dioxide pigment from ilmenite, based on its dissolution in hydrochloric acid. Developed by Fletcher Titanium Products, a subsidiary of Fletcher Challenge (a large paper company in New Zealand), based on original work performed in 1979 at the New Zealand Department of Scientific and Industrial Research and the Victoria University of Wellington. The hydrochloric acid was available in the paper mills, and the intended product was anatase, suitable for use in pigmenting paper. In 1989, a pilot unit at Gracefield on the North Island was said to be producing at the rate of 3 tons per year, and a unit for producing 3000 tons had been designed. The process was sold to Sherwin-Williams Company, which continued the development until abandoning it in the mid-1990s.

Eur. Chem. News, 1988, **51**(1354), 24.

Ind. Miner. (London), 1989, (257), 13.

Chem. Week, 1989, **114**(16), 9.

Flex DS An oxidative desulfurization process for decontaminating and upgrading crude oil. The oxidant is an alkyl hydroperoxide and the catalyst an organic titanium complex. Developed from 2007 by Auterra, Malta, NY, formerly called Applied NanoWorks. Funded by a Canadian Oil Sands company, piloted but not yet commercialized.

U.S. Patent Applications 2011-0031164; 2012-0067777; 2012-0022272.

Chem. Eng. (N.Y.), 2013, **120**(5), 21.

FLEXICOKING A continuous process for thermally cracking the residues from petroleum distillation and other heavy oils. It combines fluidized-bed (or fixed-bed) coking of residuum with coke gasification. The products are mainly C_4 hydrocarbons. Developed from the *FLUID COKING process by Esso Research and Engineering Company in the early 1970s and licensed by that company and the Union Oil Company of California. Five units were operating in 1988.

Oil Gas J., 1975, **73**(10), 53.
Allan, D.E., Metrailer, W.J., and King, R.C., *Chem. Eng. Prog.*, 1981, **77**(12), 40.
Hydrocarbon Process. Int. Ed., 1994, **73**(11), 96.
Roundtree, E.M., in *Handbook of Petroleum Refining Processes*, 2nd ed., Meyers, R.A., Ed., McGraw-Hill, New York, 1996, 122.3.

Flexicracking A version of the *FCC process developed by Exxon Research and Engineering Company. Seventeen units were operating in 1996.

Hydrocarbon Process. Int. Ed., 1996, **75**(11), 121.

Flexomer A gas-phase process for making ethylene–propylene copolymers. Developed by Union Carbide Corporation and first commercialized in 1989.

FLEXSORB A group of gas-treating processes using proprietary hindered amines, developed by Exxon Research and Engineering Company and announced in 1983. FLEXSORB SE removes hydrogen sulfide, FLEXSORB HP removes carbon dioxide, and FLEXSORB PS removes both gases. Thirty-five plants were operating in 2000.

Goldstein, A.M., Edelman, A.M., and Ruziska, P.A., in *Acid and Sour Gas Treating Processes*, Newman, S.A., Ed., Gulf Publishing, Houston, TX, 1985, 319.
Hydrocarbon Process. Int. Ed., 1992, **71**(4), 100; 1996, **75**(4), 118; 2000, **79**(4), 69.

Flintshire An early lead-smelting process in which galena was roasted in a reverberatory furnace.

flue-gas desulfurization Often abbreviated to FGD. A general term for the removal of sulfur dioxide from the off-gases from power stations and smelters. Many processes for accomplishing this have been developed; those with special names that are described in this dictionary are Abgas-Turbo-Wascher von Kroll, ADVACATE, ASARCO, Battersea, BF/Uhde, Bischoff, CEC, Citrate, Citrex, CONOSOX, CT-121, CZD, DAP-Mn, DESONOX, DeSOx, Dowa, DRYPAC, Dual Alkali, Ebara, ELSE, E-SOX, Flakt–Boliden, Fläkt–Hydro, Formate (2), FW-BF, HALT, Holter, Howden, Ispra Mark 13A, Kranz MWS, LIFAC, LIMB, Molten carbonate, Nahcolite, NEUTREC, NOXSO, Pox-O-Tec, RCE, Reinluft, Saarburg–Holter, SDA, SGFD, SHU, S-Magyp, SNOX, SOLINOX, Stackpol, Stone & Webster/Ionics, Sulfidine, SULF-X, Sultrol, Thoroughbred, Walther, Wellman–Lord, WSA, WSA-SNOX, and WUK.

Ashley, M.J. and Greaves, R.A., *Chem. Ind. (London)*, 1989, (3), 60.
Wieckowska, J., *Catal. Today*, 1995, **24**(4), 405.

FLUID COKING A noncatalytic, thermal process for converting bitumen, petroleum residues, and coal liquids to lighter hydrocarbon fluids and gases. Developed by the Exxon Research and Engineering Company and used commercially since 1954. *See also* FLEXICOKING.

Allan, D.E., Metrailer, W.J., and King, R.C., *Chem. Eng. Prog.*, 1981, **77**(12), 40.
Massenzio, S.F., in *Handbook of Synfuels Technology*, Meyers, R.A., Ed., McGraw-Hill, New York, 1984, 6–5.

Fluid Hydroforming An early *Catalytic Reforming process in which the catalyst was used in a continuously regenerated fluidized bed. Developed by the M.W. Kellogg Company.

Ciapetta, C.F., Dobres, R.M., and Baker, R.W., in *Catalysis*, Vol. 6, Emmett, P.H., Ed., Reinhold Publishing, New York, 1958, 495.

Fluohmic Also called Shawinigan. A process for making hydrogen cyanide by passing a mixture of ammonia and methane through a fluidized bed of coke. The bed is heated by passing an electric current through it, hence the name. The process is economic only where cheap electricity is available. Operated by Shawinigan Chemicals at Shawinigan Falls, Québec, from 1960 to 1968.

Dowell, A.M., III, Tucker, D.H., Merritt, R.F., and Teich, C.I., in *Encyclopedia of Chemical Processing and Design*, McKetta, J.J. and Cunningham, W.A., Eds., Marcel Dekker, New York, 1988, **27**, 10.

Fluor Econamine *See* Econamine.

Fluorodec [**Fluor**ine on **d**emand by **electrolysis**] An electrolytic process and apparatus for generating fluorine. The electrolyte is molten KHF_2; the fluorine is liberated at a nickel anode. Offered by Fluorogas, United Kingdom.

Fluor Solvent A process for removing carbon dioxide from natural gas and various industrial gas streams by dissolution in propylene carbonate. Carbon dioxide is much more soluble than other common gases in this solvent at low temperatures. The process cannot be used when hydrogen sulfide is present. The process was invented in 1958 by A.L. Kohl and F.E. Miller at the Fluor Corporation, Los Angeles. It is now licensed by Fluor Daniel. The first plant was built for the Terrell County Treating Plant, El Paso, TX, in 1960; by 1985, 13 plants were operating.

U.S. Patents 2,926,751; 2,926,752; 2,926,753.
Kohl, A.L. and Buckingham, P.A., *Pet. Refin.*, 1960, **39**, 193.
Kohl, H.L. and Nielsen, R.B., *Gas Purification*, 5th ed., Gulf Publishing Co., Houston, TX, 1997, 1198.

FMC *See* carbonization.

Folkins A process for making carbon disulfide from methane and sulfur at elevated temperature and pressure. A complex separation system removes the hydrogen sulfide from the products so that this sulfur can be reused. The process can be operated catalytically or noncatalytically. Developed in 1948 by H.O. Folkins and others at the Pure Oil Company, Chicago.

U.S. Patent 2,568,121.
Folkins, H.O., Miller, E., and Hennig, H., *Ind. Eng. Chem.*, 1950, **42**, 2202.

Fondon *See* Cazo.

Footing A process for displacing platinum metals from their chloride solutions by adding metallic iron. *See* Cementation.

Footner *See* metal surface treatment.

Formacell (1) An *Organocell process.

Villa, C., Santos, V., and Parajo, J.C., *Ind. Eng. Chem. Res.*, 2003, **42**(2), 349.

Formacell (2) An *Organosolv process using formic acid.

Pye, E.K., in *Biorefineries: Industrial Processes and Products*, Kamm, B., Gruber, P.R., and Kamm, M., Eds., Wiley-VCH, Weinheim, Germany, 2005, **II**, 179.

Formaldehyde SR [**S**eries **R**eactor] A process for making aqueous formaldehyde from methanol. Developed by Haldor Topsoe. Three plants had been built by 2005, and three more were under construction.

Formate (1) A process for making sodium dithionite by reacting sodium formate with sulfur dioxide in aqueous methanol.

Bostian, L.C., in *Speciality Inorganic Chemicals*, Thompson, R., Ed., Royal Society of Chemistry, London, U.K., 1981, 65.

Formate (2) A *flue-gas desulfurization process. Potassium formate solution reduces the sulfur dioxide to thiosulfate and then to hydrosulfide.

Buckingham, P.A. and Homan, H.R., *Hydrocarbon Process. Int. Ed.*, 1971, **50**(8), 121.
Speight, J.G., *Gas Processing*, Butterworth Heinemann, Oxford, U.K., 1993, 289.

Formcoke *See* carbonization.

Formex A process for extracting aromatic hydrocarbons from petroleum reformate, using N-formyl morpholine at 40°C. Developed by Snamprogetti.

Cinelli, E., Noe, S., and Paret, G., *Hydrocarbon Process. Int. Ed.*, 1971, **51**(4), 141.
Bailes, P.J., in *Handbook of Solvent Extraction*, Lo, C.C., Baird, M.H.I., and Hanson, C., Eds., John Wiley & Sons, Chichester, U.K., 1983, 18.2.4.

Formox [**Form**aldehyde by **ox**idation] A process for oxidizing methanol to formaldehyde, using a ferric molybdate catalyst. Based on the *Adkins–Peterson reaction. Developed jointly by Reichhold Chemicals and Perstorp (a company in the Swedish town of Perstorp). Perstorp has been making formaldehyde since the 1900s and the first Formox plant was built in 1959. The process is now owned by Perstorp Holding AB and has been licensed to more than 100 plants worldwide. Some of these use a continuous Formox process to make urea–formaldehyde resins continuously. Several other companies operate similar processes.

Chem. Week, 7 Mar 1990, 7.
Weissermel, K. and Arpe, H.-J., *Industrial Organic Chemistry*, 4th ed., translated by C.R. Hindley and S. Hawkins, Wiley-VCH, Weinheim, Germany, 2003, 40.

Fosbel [**Fos**eco **Bel**ret] A ceramic welding process for repairing refractories. A dry mixture of a refractory and a metallic powder is projected, in a stream of oxygen, onto the surface to be repaired. Oxidation of the metal produces the necessary high temperature. Developed by Foseco, United Kingdom, and Belret, Belgium, hence the name.

Foster Wheeler–Stoic *See* Stoic.

Foulis–Holmes A process for removing hydrogen cyanide from coal gas by scrubbing with a suspension of freshly precipitated ferrous carbonate. Invented by W. Foulis and P. Holmes in England in the 1890s.

British Patents 9,474 (1893); 15,168 (1896).
Hill, W.H., in *Chemistry of Coal Utilization*, Vol. 2, Lowry, H.H., Ed., John Wiley & Sons, New York, 1945, 1097.

Four-step *See* Aldol.

FR *See* Titanox FR.

Frank–Caro Also called the cyanamide process. An early process for fixing atmospheric nitrogen. Lime and carbon were heated to produce calcium carbide; this was reacted with nitrogen to give calcium cyanamide, which was hydrolyzed with steam to yield ammonia and calcium carbonate. Developed by A. Frank and N. Caro from 1895 at Dynamit, Germany, and used in Germany, Norway, Italy, United States, and Canada until it was replaced by the *Haber–Bosch process after World War I.

German Patents 88,363; 108,971.
Chilton, T.H., *Strong Water: Nitric Acid, Its Sources, Methods of Manufacture, and Uses*, MIT Press, Cambridge, MA, 1968, 61.
Vancini, C.A., *Synthesis of Ammonia*, translated by L. Pirt, Macmillan Press, Basingstoke, U.K., 1971, 2.
Morris, P.J.T., *Chem. Ind. (London)*, 1983, (18), 710.
Leigh, G.J., *The World's Greatest Fix: A History of Nitrogen and Agriculture*, Oxford University Press, Oxford, U.K., 2004, 125–129, 139–140.

Frasch (1) A process for extracting sulfur from underground deposits, developed by H. Frasch between 1890 and 1902 at Sulphur Mine, LA. Three concentric pipes are inserted into a hole drilled into the deposit. The outermost pipe carries water superheated to 140°C–165°C, which melts the sulfur; hot air is forced down the central pipe, which forces the molten sulfur up through the intermediate annular space. Only a small proportion of sulfur deposits have the appropriate geology for extraction in this way. Because of this invention, sulfur came to be exported from the United States to Europe, instead of from Sicily to the United States. In 1991, the process was operated in the United States, Mexico, Poland, and Iraq.

> U.S. Patents 461,429; 461,430; 461,431.
> *J. Soc. Chem. Ind.*, 1914, **33**, 539.
> Haynes, W., *The Stone That Burns*, D. Van Nostrand, New York, 1942.
> Shearon, W.H., Jr. and Pollard, J.H., *Ind. Eng. Chem.*, 1950, **42**, 2188.
> Loughbrough, R., *Ind. Miner. (London)*, 1991, Jul, 19.

Frasch (2) A process for removing sulfur compounds from petroleum fractions by distillation from copper oxide.

Fray–Farthing–Chen Also called FFC Cambridge. A process for converting metal oxides to metals by molten salt electrolysis (electrodeoxidation). Invented by D.J. Fray, T.W. Farthing, and Z. Chen in Cambridge University, United Kingdom, in the 1990s and further developed by QinetiQ, the former U.K. government research organization, for making titanium sponge and powder from titanium dioxide. The process should also be useful for making other reactive metals from their oxides. The process was acquired by Metalysis in 2006; QinetiQ received a share in the company. A version for extracting niobium was announced in 2008. A pilot plant in Rotherham, Yorkshire, was due to be commissioned in 2010. Metalysis has strategic partnerships with BHP Billiton and Rolls Royce.

> U.S. Patent 6,712,952.
> *Nature*, 2000, **407**, 361.
> *Chem. Eng. (Rugby, Engl.)*, 2003, 750/1, 10
> *Chem. World*, 2008, **5**(4), 38.

Freeman *See* DR.

French Also known as the indirect process. A process for making zinc oxide, in the form of a white pigment, from metallic zinc. The zinc is melted and vaporized in a current of carbon monoxide. The vapor is oxidized with air in a second chamber, forming zinc oxide and carbon dioxide. The fume passes through a settling chamber, where oversized particles settle out; from there, it goes to a bag house, where the product is collected. Confusingly, the French process has been operated by North American Oxide Company at Clarksville, TN. *See also* American.

> Buxbaum, G. and Pfaff, G., *Industrial Inorganic Pigments*, 3rd ed., 2005, 91.

Fresnel A predecessor of the *Solvay process, proposed by A.J. Fresnel in France in 1811.

> Cohen, J.M., *The Life of Ludwig Mond*, Methuen, London, U.K., 1956, 267.

Fricker A process for producing zinc oxide by direct oxidation of zinc vapor. Operated by the Fricker's Metal Company at Luton and Burry Port, United Kingdom, in the 1920s and 1930s, subsequently acquired by the Imperial Smelting Corporation. Also operated by the Anglo American Corporation, South Africa, after World War II.

> Cocks, E.J. and Walters, B., *A History of the Zinc Smelting Industry in Britain*, George G. Harrap, London, U.K., 1968, 28, 93, 95, 140.

Fritz Winkler *See* Winkler.

Fröhler *See* Munich.

F-S [Ferrous Sulfate] A process for removing ammonia, hydrogen sulfide, and hydrogen cyanide from coke-oven gas by scrubbing with aqueous ferrous sulfate solution obtained from steel pickling. A complex series of reactions in various parts of the absorption tower yield ammonium sulfate crystals and hydrogen sulfide (for conversion to sulfur or sulfuric acid) as the end products. Developed in Germany by F.J. Collin AG.

Dixon, T.E., *Iron Age*, 1955, **175**(12), 91.
Kohl, A.L. and Riesenfeld, F.C., *Gas Purification*, 4th ed., Gulf Publishing Co., Houston, TX, 1985, 563.

F–T *See* Fischer–Tropsch.

FTC [Fluid-bed Thermal Cracking] A continuous thermal cracking process for making synthesis gas from heavier petroleum fractions.

Full cell *See* Bethell.

Fumaks A process for removing hydrogen sulfide from coke-oven gas by oxidation with picric acid. Developed by Osaka Gas and then commercialized by Sumitomo Metal Industries. Fumaks was used in 11 units in Japan from 1972 to 1997.

Kohl, H.L. and Nielsen, R.B., *Gas Purification*, 5th ed., Gulf Publishing Co., Houston, TX, 1997, 850.

Fumaks–Rhodaks A combination of the *Fumaks and *Rhodaks processes for removing both hydrogen sulfide and hydrogen cyanide from coke-oven gas. Developed by the Osaka Gas Company and marketed by Sumitomo Chemical Engineering Company.

Furacell [**Fura**ne **cell**ulose] A continuous process for making levoglucosenone (an unsaturated ketone derived from glucose) from waste cellulose. The cellulose is heated to about 450° in a vacuum in the presence of a catalyst and an organic solvent. The chiral product has uses in synthetic organic chemistry and can be converted to a solvent ("Cyrene") and a fuel ("Furafuel"). Developed by W.D. Raverty, now at Circa Group, Melbourne, from 2009.

U.S. Patent Application 2012-0111714.
Chem. Eng. (N.Y.), 2013, **120**(5), 12.

Furnace Also called the Readman process. A process for making elementary phosphorus in an electric furnace. The raw materials are phosphate rock, coke, and silica. Large amounts of electric power are consumed, so the process is economic only where cheap power is available. The overall reaction is

$$2Ca_3(PO_4)_2 + 6SiO_2 + 10C = 6CaSiO_3 + 4P + 10CO$$

Invented by J.B. Readman in Edinburgh in 1888 and now operated on a very large scale in at least 10 countries.

British Patent 14,962 (1888).
Hartlapp, G., in *Phosphoric Acid*, Vol. 1, Part II, Slack, A.N., Ed., Marcel Dekker, New York, 1968, 927.
Childs, A.F., in *The Modern Inorganic Chemicals Industry*, Thompson, R., Ed., The Royal Society of Chemistry, London, U.K., 1977, 378.
Hocking, M.B., *Modern Chemical Technology and Emission Control*, Springer-Verlag, Berlin, Germany, 1985, 184.

Furnace Black One of the principal processes used for making carbon black. Aromatic fuel oils and residues are injected into a high-velocity stream of combustion gases from the complete burning of an auxiliary fuel with an excess of air. Some of the feedstock is burned, but most of it is cracked to yield carbon black and hydrogen. The products are quenched with water. Developed in the United States in the 1920s and still of major importance. In 2006, more than 95% of the world production of carbon black was made by this process.

Franck, H.-G. and Stadelhofer, J.W., *Industrial Aromatic Chemistry*, Springer-Verlag, Berlin, Germany, 1988, 384.
Kühner, G. and Voll, M., in *Carbon Black Science and Technology*, Donnet, J.-B., Bansai, R.C., and Wang, M.J., Eds., Marcel Dekker, New York, 1993, 14.
Buxbaum, G. and Pfaff, G., *Industrial Inorganic Pigments*, 3rd ed., 2005, 170.
Chem Week, 2007, **169**(4), 24.

Fuse-quench Also called the Kjellgren–Sawyer process. A process for extracting beryllium from beryl. The beryl is fused at 1600°C and then rapidly quenched by pouring through a water jet of high velocity. The glassy product is heated to 900°C to precipitate beryllia from its solid solution in silica and then extracted with sulfuric acid. Operated by the Brush Beryllium Company, Cleveland, OH.

Schwenzfrier, C.W., Jr., in *The Metal Beryllium*, White, D.W., Jr. and Burke, J.E., Eds., American Society for Metals, Cleveland, OH, 1955.
Everest, D.A., *The Chemistry of Beryllium*, Elsevier, Amsterdam, the Netherlands, 1964, 109.

FW-BF [**F**oster **W**heeler–**B**ergbau-**F**orschung] A dry *flue-gas desulfurization process that combines the sulfur removal system of *Bergbau-Forschung, which uses a carbon adsorbent, with the Foster Wheeler process for oxidizing adsorbed sulfur to sulfuric acid.

Bischoff, W.F. and Habib, Y., *Chem. Eng. Prog.*, 1975, **71**(5), 59.
Habib, Y. and Bischoff, W.F., *Oil Gas J.*, 1975, **75**(8), 53.
Sulphur, 1975, (119), 24.
Environ. Sci. Technol., 1975, **9**, 712.

G

Galoter A process for extracting oil and gas from shale, using a vertical retort. Operated in Estonia since 1964. An improved version is *Enefit 140.

Smith, J.W., in *Handbook of Synfuels Technology*, Meyers, R.A., Ed., McGraw-Hill, New York, 1984, 4–177.

Gardinier A process for making monoammonium phosphate from gaseous ammonia and phosphoric acid by mixing them in a helical reactor. *See also* Swift.

Gardner A process for making magnesium metal by reducing magnesium sulfide, made from magnesium oxide, with calcium carbide in the presence of a flux of calcium and aluminum chlorides. Invented by D. Gardner in 1895 and operated by Murex, United Kingdom, in the 1930s.

British Patent 465, 421.

Garrett *See* ORC.

Garrigue A process for recovering glycerol from the residual liquor from the *kettle soapmaking process. After the solid soap is separated, the liquor is treated with aluminum sulfate, thereby precipitating the residual carboxylic acids as their insoluble aluminum salts. After removing these by filtration, the liquor is concentrated by vacuum evaporation and the glycerol distilled out under vacuum. *See also* Van Ruymbeke (1).

Martin, G. and Cooke, E.I., in *Industrial and Manufacturing Chemistry*, Cooke, E.I., Ed., Technical Press, Kingston Hill, U.K., 1952, 126.

Gas [Gas antisolvent] A process for separating dissolved materials by selective precipitation with added supercritical carbon dioxide. First used for recrystallizing the explosive RDX, subsequently used for recrystallizing other explosives, pharmaceuticals, fine chemicals, and food products. Another use has been for precipitating insulin particles from dimethylsulfoxide solution.

Yeo, S., Lim, G., Debendetti, P.G., and Bernstein, H., *Biotechnol. Bioeng.*, 1993, **41**, 341.
McHugh, M.A. and Krukonis, V.J., in *Supercritical Fluid Extraction: Principles and Practice*, 2nd ed., Butterworth-Heinemann, Boston, MA, 1994, 342.
Brennecke, J.F., *Chem. Ind. (London)*, 1996, (21), 831.
Thiering, R., Dehghani, F., Dillow, A., and Foster, N.R., *J. Chem. Technol. Biotechnol.*, 2000, **75**(1), 29.

Gas Black A process for making carbon black. Similar to the other carbon black processes but uses a coal tar oil as the feedstock. Developed in the 1930s and still in use in 2005.

Buxbaum, G. and Pfaff, G., *Industrial Inorganic Pigments*, 3rd ed., 2005, 173.

GasCat A process for converting natural gas to liquid fuels. Essentially an updated *Fischer–Tropsch process, GasCat is being developed by Energy International, a division of Williams Companies, Tulsa, OK. The process first produces *syngas by *reforming natural gas and oxygen and then passes this gas into a slurry bubble column reactor containing a cobalt oxide catalyst. Not commercialized as of 1997.

Singleton, A.H., *Oil Gas J.*, 1997, **95** (31), 68.

Gasmaco [Gas Machinery Company] A process for making a substitute natural gas from petroleum fractions and residues by thermal cracking. Developed from the *Hall (2) process in the 1940s by the American Gas Association. *See also* Petrogas.

Crane, K.J., *J. Inst. Fuel*, 1957, **30**, 661.
Claxton, G., *Benzoles: Production and Uses*, National Benzole & Allied Products Association, London, U.K., 1961, 97.
Gasmaking, British Petroleum Co., London, U.K., 1965, 70.

GAS/SPEC CS-Plus A process for removing carbon dioxide and hydrogen sulfide from natural gas by washing with a solution of a special amine. Developed and offered by Dow Chemical Company. Operated since 1988.

Oil Gas J., 1996, **94**(8), 38.

GAS/SPEC FT *See* Econamine FG.

GASTAK *See* Purasiv HR.

Gastechnik A process for removing hydrogen sulfide and organic sulfur compounds from coal gas by absorption on formulated iron oxide pellets, which flow by gravity down large absorption towers. The pellets are regenerated by removing their sulfur, either by atmospheric oxidation or by extraction with tetrachloroethylene. Developed by Gastechnik, Germany, in the 1950s and widely operated there and in England.

British Patents 433,823; 683,432.
Moore, D.B., *Gas World*, 1956, **143**, 153.
Egan, P.C., *Gas World*, 1957, **145**, 136.
Claxton, G., *Benzoles: Production and Uses*, National Benzole & Allied Products Association, London, U.K., 1961, 209.
Kohl, H.L. and Nielsen, R.B., *Gas Purification*, 5th ed., Gulf Publishing Co., Houston, TX, 1997, 1300.

Gasynthan A process for making synthetic natural gas from naphtha by a two-stage *steam reforming process. Developed by Lurgi and BASF in the 1960s. Over 30 units were operating in 1975.

Jockel, H. and Triebskorn, B.E., *Hydrocarbon Process. Int. Ed.*, 1973, **52**(1), 93.
Hydrocarbon Process. Int. Ed., 1975, **54**(4), 114.

GDR Parex *See* Parex (2).

GEESI [General Electric Environmental ServIces] A process for making ammonium sulfate from the sulfur dioxide present in flue gas.

Eur. Chem. News, 1996, **65**(1720), 23.

GEGas [General Electricity Gas] A coal gasification process intended for producing gas for combined-cycle power generation. Powdered coal, with steam and air, is fed to a mechanically stirred gasifier. Piloted in the 1970s by the General Electric Company, United States.

Geigy A one-stage process for making ethylenediaminetetraacetic acid (EDTA) from ethylenediamine, hydrogen cyanide, and formaldehyde.

Anderson, E.V. and Gaunt, J.A., *Ind. Eng. Chem.*, 1960, **52**, 191.

GEMINI A family of gas separation processes based on selective adsorption. GEMINI 5 is for upgrading raw landfill gas to methane. GEMINI 9 is for producing hydrogen and carbon dioxide from *steam reformer off-gas and is used in Butler, PA. GEMINI Hyco separates carbon monoxide and hydrogen from *syngas. Developed by S. Sircar at Air Products and Chemicals from 1978.

U.S. Patents 4,171,206; 5,073,356.
Kratz, W.C., Rarig, D.L., and Pietrantonio, J.M., in *Adsorption and Ion Exchange: Fundamentals and Applications*, LeVan, M.D., Ed., American Institute of Chemical Engineers, New York, 1998, 36.
Kumar, R. and Van Sloun, J.K., *Chem. Eng. Prog.*, 1989, **85**(1), 34.
Hydrocarbon Process. Int. Ed., 1996, **75**(4), 118.

Geminox A direct process for converting butane to 1,4-butanediol. The butane is first oxidized in the gas phase to maleic anhydride, using BP's fluidized bed technology. The maleic anhydride is scrubbed with water and then catalytically dehydrogenated to butanediol. Developed in 1994 by BP Chemicals and Lurgi. Modifications of the process can be used to make tetrahydrofuran and γ-butyrolactone. The first plant, on BP's site at Lima, OH, was completed in 2000.

Chem. Eng. (N.Y.), 1995, **102**(8), 17.
Chem. Eng. (Rugby, Engl.), 1997, (638), 26.
Eur. Chem. News, 2000, **73**(1927), 20.

Generon Not a process, but a trademark used by BOC and Dow to designate their processes for separating nitrogen from air by either the *PSA process or a membrane process. The PSA process is based on the *Bergbau-Forschung process. The membrane process uses the GENERON HP membrane developed by the Dow Chemical Company. *See also* NOVOX.

GEOCOAT A biological oxidation process for heap-leaching sulfide mineral concentrates. The concentrate is coated onto a support rock, and the support is made into a heap and inoculated with thermophilic bacteria and a special leach solution. Air is provided through perforated pipes at the base of the heap. Developed by GeoBiotics LLC.

GEODE [General Electric Organic DEstruction] A development of the *Ultrox process in which a combination of ozone and ultraviolet radiation is used to oxidize traces of organic compounds in water. Developed by the General Electric Company and demonstrated at the Commonwealth Edison nuclear power plant at Dresden, IL, in 1989. The requirement was to reduce the concentration of total organic carbon in the process and make up waters to the low parts-per-billion range.

Head, R.A., Alexander, J.E., and Lezon, R.J., *Nucl. Eng. Int.*, 1989, **4**, 40.

GFETC A coal gasification process based on a slagging gasifier, developed for the U.S. Department of Energy, at Grand Forks, ND.

Ghaem A *DR process developed by the Esfahan Steel Company, Iran, in 1996.

GHR [Gas Heated Reformer] A process for making *syngas from natural gas, designed for making methanol. Developed by ICI, Davy, and Aka Kvaerner. *See* LCM.

GHU-1 and GHU-2 Processes for upgrading C_5 hydrocarbon streams, developed by Axens.

GI *See* carbonization.

Giammarco-Vetrocoke Also known as G-V. Two processes are known by this name, both using an aqueous solution of sodium or potassium carbonate and arsenite to absorb acid gases. In one process, the solution is used to extract carbon dioxide from natural gas or synthesis gas. In the other, hydrogen sulfide is extracted from coke-oven or synthesis gas, yielding elemental sulfur from a complex sequence of reactions. The process was invented by G. Giammarco and developed by Giammarco-Vetrocoke, Italy. In 1992, more than 200 plants were operating. *See also* Benfield, Carsol, CATACARB, HiPure.

U.S. Patents 2,840,450; 3,086,838.
Giammarco, G., in *Ammonia*, Part 2, Slack, A.V. and James, G.R., Eds., Marcel Dekker, New York, 1974, 171.
Hydrocarbon Process. Int. Ed., 1975, **54**(4), 90; 1992, **71**(4), 103.
Kohl, H.L. and Nielsen, R.B., *Gas Purification*, 5th ed., Gulf Publishing Co., Houston, TX, 1997, 371, 754.

Gibbs A process for oxidizing naphthalene to phthalic anhydride, using air as the oxidant and catalyzed by vanadium pentoxide. Invented in 1917 by H.D. Gibbs and C. Conover.

U.S. Patent 1,285,117.

Gilchrist *See* Thomas.

Gilson A process for making synthetic opals, invented in France in 1974 by P. Gilson, Sr.

Chem. Eng. News, 2003, **81**(4), 58.

Girbotol [**Gir**dler **Bott**oms] Also spelled **Girbitol**. A gas-scrubbing process using an alkanolamine as the absorbent. Used for removing hydrogen sulfide from refinery and natural gases, and carbon dioxide from hydrogen and combustion products. The gases are subsequently removed by steam stripping. Monoethanolamine, diethanolamine, and triethanolamine have all been used. The process depends on the reversible formation of an amine carbonate:

$$2HO\text{-}C_2H_4\text{-}NH_2 + H_2O + CO_2 \rightleftharpoons (HO\text{-}C_2H_4\text{-}NH_3)_2CO_3$$

Invented in 1930 by R.G. Bottoms at the Girdler Corporation, Louisville, KY. In 1950, it was the most commonly used process for removing hydrogen sulfide from refinery and natural gases.

U.S. Patent 1,783,901.
Reed, R.H. and Updegraff, N.C., *Ind. Eng. Chem.*, 1950, **42**, 2269.
Unzelman, G.H. and Wolf, C.J., in *Petroleum Processing Handbook*, Bland, W.F. and Davidson, R.L., Eds., McGraw-Hill, New York, 1967, 3–132.

Girdler (1) A process for removing acid gases from gas streams by scrubbing with aqueous triethanolamine. Developed by the Girdler Corporation, Louisville, KY, in the 1920s. *See also* Girbotol.

German Patent 549,556.

Girdler (2) An improved *Claus process developed by the Girdler Corporation in 1948.

Sands, A.E. and Schmidt, L.D., *Ind. Eng. Chem.*, 1950, **42**, 2277.

GKT A development of the *Koppers–Totzek coal gasification process.

Glacido A process for making acetic acid from methanol and carbon monoxide.
Various catalysts have been used, but now, the preferred one is believed to be a supported palladium/gold alloy catalyst. Developed in 1980 by Lyondell Chemical Technology, formerly Millennium Petrochemicals, and used in their plant in La Porte, TX. *See also* Vacido. Also a trademark used by LyondellBasell Industries.

U.S. Patents 6,022,823; 7,811,968.

Glanor A *Chlor-Alkali process using a bipolar diaphragm cell. Developed by PPG Industries and Oronzio de Nora Impianti Elettrochimica in the early 1970s.

Gluud A process for removing hydrogen sulfide from gas streams by scrubbing with an aqueous solution containing ammonia and a nickel salt. Invented by W. Gluud in 1921.

U.S. Patent 1,597,964.
Gluud, W. and Schonfelder, R., *Chem. Metall. Eng.*, 1927, **34**(12), 742.
Kohl, A.L. and Riesenfeld, F.C., *Gas Purification*, 4th ed., Gulf Publishing Co., Houston, TX, 1985, 493.
Kohl, H.L. and Nielsen, R.B., *Gas Purification*, 5th ed., Gulf Publishing Co., Houston, TX, 1997, 735.

GMD [**Ga**s to **M**iddle **D**istillate] A process for converting natural gas to diesel fuel or synthetic crude oil. The catalyst is cobalt and rhenium on alumina, used in a slurry reactor. Developed by Statoil in the 1980s.

Jens, K.-J., *Appl. Catal. A: Gen.*, 1997, **152**(1), 161.

GM-IX [**G**as **M**embrane-**I**on e**X**change] A process for treating metal cyanide waste solutions, with recovery of both the metal and the cyanide. The solution is first passed through an anion-exchange column, removing metal cyanide complexes and cyanide ion. Treatment of the column with 10% sulfuric acid releases the metal ions and hydrogen cyanide into solution. Passage of this solution through a gas membrane module, comprising thousands of hollow fibers of microporous polypropylene, strips the hydrogen cyanide gas from the solution. The process not only removes the free hydrogen cyanide, it also forces the hydrolysis of metal cyanide complexes until all the cyanide has been removed from the system. The hydrogen cyanide is scrubbed by sodium hydroxide solution, and the resulting sodium cyanide solution is reused. The metal is recovered from solution by electrowinning. Developed in 1988 at the University of Minnesota's Department of Civil and Mineral Engineering, where a pilot plant was under construction in May 1989.

Semmens, M.J. and Chang, Y., in *Membrane Separation Processes*, Green, A., Ed., BHRA (Information Services), Cranfield, Bedford, U.K., 1989, 167.

GO [**G**lycol **O**xalate] Also called UBE/UCC. A process for making ethylene glycol from carbon monoxide in three stages, involving methyl nitrite and dimethyl oxalate:

$$4CH_3OH + 4NO + O_2 = 4CH_3ONO + 2H_2O$$

$$2CH_3ONO + 2CO = CH_3OOC = COOCH_3 + 2NO$$

$$CH_3OOC=COOCH_3 + 4H_2 = HOCH_2=CH_2OH + 2CH_3OH$$

All the methanol and nitric oxide are recycled. The process was developed jointly by UBE and Union Carbide Corporation in 1983 and piloted by the latter company.

Saunby, J.B., in *Oxygen and the Conversion of Future Feedstocks*, Royal Society of Chemistry, London, U.K., 1984, 235.

Goethite A process for removing iron from leach liquors from hydrometallurgical leaching operations. Used in recovering zinc from the residues of the electrolytic zinc process. *See also* Jarosite, Hematite.

Morgan, S.W.K., *Zinc and Its Alloys and Compounds*, Ellis Horwood, Chichester, U.K., 1985, 121.

GO-fining [**Gas-Oi**l re**f**ining] A *hydrodesulfurization process adapted for gas oil. The proprietary catalyst is regenerable. Developed by Esso Research & Engineering Company and the Union Oil Company of California and jointly licensed by them. First commercialized at Wakayama, Japan, in 1968; by 1972, nine units had been built. In 1998, Esso started to collaborate with M.W. Kellogg in licensing this technology.

Hydrocarbon Process. Int. Ed., 1970, **49**(9), 210; 1994, **73**(11), 135.

GOLDOX [**GOLD OX**idation] A process for improving the extraction of gold from its ores by injecting oxygen into the cyanide solution by the *VITOX process. Developed by Afrox, a subsidiary of BOC, and now used widely in South Africa, Zimbabwe, Australia, Canada, and United States.

Downie, N.A., *Industrial Gases*, Blackie Academic, London, U.K., 1997, 308.

Goldschmidt The German company Th. Goldschmidt AG, founded in 1847, has developed many processes, including the *Thermit process, and many for extracting and recovering tin. The process with which its name is usually associated is one for purifying tin metal by adding aluminum in order to precipitate out arsenic and antimony. Air is then blown into the molten metal, forming a surface scum containing these impurities, which is removed after cooling. A process for removing tin from tinplate by chlorination has also been known by this name.

German Patent 411,477.
Mantell, C.L., *Tin: Its Mining, Production, Technology, and Application*, Reinhold Publishing, New York, 1949, 147. Published in facsimile by Hafner Publishing, New York, 1970.

Gorham *See* Paralene.

GPB [**G**eon **P**rocess **B**utadiene] A process for extracting butadiene from C_4 petroleum cracking fractions, developed by Nippon Zeon. It was in use by more than 30 plants in 1985.

Weissermel, K. and Arpe, H.-J., *Industrial Organic Chemistry*, 4th ed., translated by C.R. Hindley and S. Hawkins, Wiley-VCH, Weinheim, Germany, 2003, 111.

GPP [**G**as-phase **P**oly**P**ropylene] A gas-phase process for making polypropylene. Developed by Sumitomo and first commercialized in 1990.

Grainer Originally called the Michigan process because it was widely used in Michigan for using the waste heat generated at sawmills. This process is used for evaporating brine to yield salt. The brine is indirectly heated to below its boiling point, and the crystals grow at the surface of the liquid.

Richards, R.B., in *Sodium Chloride*, Kaufmann, D.W., Ed., Reinhold Publishing, New York, 1960, Chap. 12.

Grande Paroisse A process for making nitric acid by oxidizing ammonia; claimed to be self-sufficient in power.

Chilton, T.H., *Strong Water: Nitric Acid, Its Sources, Methods of Manufacture, and Uses*, MIT Press, Cambridge, MA, 1968, 110.

Granodizing *See* metal surface treatment.

Grätzel An early process for making aluminum by electrolyzing a solution of alumina in molten calcium chloride. The process was also used for making beryllium, using a mixed fluoride melt.

Gravimelt A coal-cleaning process in which coal is heated with molten sodium hydroxide and then washed with acid. The process removes 90% of the sulfur and 95% of the ash. The process was piloted by TRW between 1988 and 1993.

U.S. Patents 4,497,636; 5,312,462.

Gray An early petroleum desulfurization process that used a fixed bed of an absorbent catalyst, such as fuller's earth. A related process, Gray clay treating, removed diolefins and other gum-forming constituents of thermally cracked gasolines by passing their vapors over hot fuller's earth. These processes, developed by T.T. Gray at the Gray Processing Corporation, were first used in the 1920s and abandoned in the 1950s with the advent of *catalytic cracking.

British Patent 222,481.
Unzelman, G H. and Wolf, C.J., in *Petroleum Processing Handbook*, Bland, W.F. and Davidson, R.L., Eds., McGraw-Hill, New York, 1967, 3–61.

GREENCAT A process for making oxide catalysts that minimizes the production of wastewater and uses no nitrates. Developed by Süd-Chemie and piloted in Louisville, KY, from 2001. The process received the Presidential Green Chemistry Challenge Award in 2003.

Chem. Week, 2003, **165** (37), S4.

Green liquor A papermaking process that uses a mixture of sodium hydroxide and sodium carbonate.

GreenOx A pulp-bleaching process developed by Kemira in Finland.

GREENOX A process for purifying carbon dioxide from combustion gases so that it may be used in greenhouses for enhancing plant growth. Nitric oxide is removed by the *SCR process, using urea as the reductant. Carbon monoxide and ethylene are catalytically oxidized to carbon dioxide. Developed by Haldor Topsoe in 1997.

GRH A process for increasing the calorific value of a lean gas to make it suitable for use as a town gas, by thermal hydrogasification. Developed by British Gas.

Griesheim (1) An early process for producing chlorine by electrolysis, developed by Chemische Fabrik Griesheim-Elektron, Germany, and commercialized in 1890. The electrolyte was saturated potassium chloride solution, heated to 80°C–90°C. The by-product, potassium hydroxide, was recovered. The process was superseded in the United States by several similar electrolytic processes before being ousted by the mercury cell, invented by H.Y. Castner and K. Kellner in 1892. *See* Castner–Kellner.

> Kircher, M.S., in *Chlorine: Its Manufacture, Properties and Uses*, Sconce, J.S., Ed., Reinhold Publishing, New York, 1962, 85.
> Hocking, M.B., *Modern Chemical Technology and Emission Control*, Springer-Verlag, Berlin, Germany, 1985.

Griesheim (2) A process for making potassium by reducing potassium fluoride with calcium carbide:

$$2KF + CaC_2 = CaF_2 + 2C + 2K$$

Grillo-Schröder *See* Schröder-Grillo.

Grosvenor–Miller A two-stage process proposed for making chlorine from hydrogen chloride. In the first stage, the hydrogen chloride reacts with ferric oxide to give ferric chloride:

$$Fe_2O_3 + 6HCl = 2FeCl_3 + 3H_2O$$

In the second, the ferric chloride is oxidized to ferric oxide and chlorine:

$$4FeCl_3 + 3O_2 = 2Fe_2O_3 + 6Cl_2$$

The iron catalyst, on an inert support, is contained in two fixed beds, which alternate in function. Invented by W.M. Grosvenor, Jr. and I. Miller; extensively studied in the 1940s but not commercialized.

> U.S. Patent 2,206,399.
> Redniss, A., in *Chlorine: Its Manufacture, Properties and Uses*, Sconce, J.S., Ed., Reinhold Publishing, New York, 1962, 259.

GRS A process for making a synthetic rubber by copolymerizing styrene and butadiene.

GRT [Gas Reaction Technology] A process for converting low molecular weight hydrocarbons to higher molecular weight ones by bromination, followed by treatment with a "cataloreactant" such as a zeolite exchanged with an alkaline earth cation. The cataloreactant is regenerated by heating in air or oxygen. Developed at the University of California at Santa Barbara, latterly with the Marathon Oil Co. Piloted but not yet commercialized.

> U.S. Patents 7,964,764; 8,278,493.

Grünstein A process for hydrating acetylene to acetaldehyde, invented by N. Grünstein in 1910 and developed by Griesheim-Elektron at Rheinfelden, Germany.

Miller, S.A., *Acetylene: Its Properties, Manufacture and Uses*, Vol. 1, Academic Press, New York, 1965, 135.
Morris, P.J.T., *Chem. Ind. (London)*, 1983, (18), 711.

Grzymek Also known as the sintering/self-disintegration process. A process for making both alumina and cement from aluminous ores and wastes. The ore is mixed with limestone and calcined in a coal-fired rotary kiln. The product spontaneously breaks up into small lumps, suitable for leaching by aqueous sodium carbonate. Alumina is precipitated from the leachate by carbon dioxide. The residues are mixed with more limestone and calcined to produce cement clinker. Developed by J. Grzymek and his colleagues in Poland from the 1930s to the 1980s. A plant at Groszowice has been operating since 1966; by 1985, it had produced 70,000 tons of alumina and 600,000 tons of cement.

Grzymek, J., *Process Eng. (London)*, 1974, Feb, 43.
Grzymek, J., *Light Met. Mat. Ind.*, 1976, **2**, 29.
O'Connor, D.J., *Alumina Extraction from Non-Bauxitic Materials*, Aluminium-Verlag, Düsseldorf, Germany, 1988, 234.

G-S [**G**irdler **S**ulfide] A process for separating hydrogen isotopes, using the equilibrium between water and hydrogen sulfide:

$$HDO + H_2S \; \rightleftharpoons \; H_2O + HDS$$

Several versions of this reaction have been used for separating deuterium and tritium; the G-S version operates in the liquid phase, without a catalyst, at two temperatures. Used at the Savannah River plant of the U.S. Atomic Energy Commission and at several Canadian plants.

Rae, H.K., in *Separation of Hydrogen Isotopes*, Rae, H.K., Ed., American Chemical Society, Washington, DC, 1978, Chap. 1.
Benedict, M., Pigford, T.H., and Levi, H.W., *Nuclear Chemical Engineering*, 2nd ed., McGraw-Hill, New York, 1980, 767.

GSP *See* Noell (2).

GT-Cryst PX A process for separating *p*-xylene from its isomers, by crystallization followed by centrifugation. Developed by GTC.

Eur. Chem. News, 2001, **74**(1949), 25.

GT-DeSulf A process for removing sulfur from cracked gasoline without *hydrotreating the entire stream. The sulfur-containing compounds are separated by solvent extraction and hydrogenated separately. Developed by GTC Technology in 2000.

Eur. Chem. News, Jul 2001, Proc. Tech. Suppl., 23.

GTE [**G**as **T**o **E**thylene] A process for converting methane to liquid fuels. High-temperature cracking first produces acetylene, which is converted to ethylene, which is then oligomerized. Piloted at Texas A&M University in 2005 and licensed to Synfuels International.

Hall, K.R., *Catal. Today*, 2005, **106**(1–4), 243.

GTG [**G**as **T**o **G**asoline] A process for converting natural gas to gasoline by oxidative coupling. Ethylene, formed initially, is oligomerized to a gasoline-like distillate over a zeolite catalyst. Under development by the Arco Chemical Company in 1988.

Parkyns, N.D., *Chem. Br.*, 1990, **26**(9), 841.

GTL [**G**as **T**o **L**iquids] A general term for processes that convert natural gas to liquid fuels. Syngas is usually an intermediate. The first such plant was that of *Sasol in South Africa. Another version was developed by Synfuels International, based on work at Texas A&M University. Piloted in Robertson County, TX, and expected to be commercialized in Kuwait in 2010. *See also* GTSC, SPD, Syntroleum.

U.S. Patent 7,183,451.
Oil Gas J., 1997, **95**(25), 16; 1997, **95**(30), 35; 1998, **96**(24), 34; 1999, **97**(49), 48.
Bakkerud, P.K., *Catal. Today*, 2005, **106**(1–4), 30.
Chem. Eng. (N.Y.), 2008, **115**(3), 13.

GTPG [**G**lycerin **T**o **P**ropylene **G**lycol] A two-stage process operated in one reactor. The glycerol is first dehydrated to acetol and water, and the acetol is then hydrogenated over a copper-based catalyst. A related process has been developed at the University of Missouri in Columbia, MS. Developed by Davy Process Technology. Planned to be installed in Ashland/Cargill's plant, scheduled for startup in early 2009.

Chem. Eng. (N.Y.), 2007, **114**(9), 31.

GTSC [**G**as **T**o **S**yn**C**rude] A process for converting natural gas to a synthetic crude oil, which may be mixed with natural crude oil and used in conventional oil refineries. Based on *Fischer–Tropsch technology but using a proprietary slurry bubble column reactor with a promoted cobalt catalyst. Developed by Syncrude Technology, Pittsburgh, PA, in the 1990s.

GT-STDP A process for disproportionating toluene into *p*-xylene and benzene, using a high-silica zeolite catalyst. Operated in the vapor phase at 390°C–400°C. Developed by Indian Petrochemical and licensed to GT Technology.

Eur. Chem. News, 2000, **72**(1904), 47.

GT-Styrene Recovery A process for recovering styrene from hydrocarbon fractions obtained by the steam cracking of petroleum. It uses extractive distillation with a proprietary selective extractant. Developed by GTC Technology (formerly Glitsch Technology Corporation) and planned for installation at a Chinese oil refinery in 2007.

ICIS Chem. Bus., 9 Jan 2006, 8.

GT-TolAlk A process for making *p*-xylene from toluene and methanol, using a high-silica zeolite catalyst. Operated in a fixed bed at 400°C–450°C in the presence of hydrogen and water in a specific molar ratio. Developed by Indian Petrochemical and licensed to GTC Technology.

Eur. Chem. News, 2000, **72**(1904), 47.

GUD [**G**as **U**nd **D**ampf; German, meaning *gas and steam*] *See* ICG-GUD.

Guggenheim A process for extracting sodium nitrate from caliche, a native sodium nitrate found in Chile. The ore is leached at 40°C with water containing controlled concentrations of magnesium and calcium sulfates. Operated on a large scale in Chile. *See also* Shanks.

Guillini A process for making gypsum from the waste product from the *wet process for making phosphoric acid. The waste is heated with water in an autoclave; this removes impurities and converts the calcium sulfate dihydrate to the hemihydrate.

Guimet A process for making ultramarine, a blue pigment. A mixture of clay, sodium carbonate, and sulfur is heated in the absence of air and then slowly cooled while air is admitted over a long period. Invented in 1826, independently, in France by J.B. Guimet and in Germany by C.B. Gmelin. Guimet won a prize for this invention, which changed the world's pigment industry and was soon operating in 22 European factories. Guimet's son Émile, who inherited the business, collected Asian art and founded the Musée Guimet, now in Paris.

Guimet, J.B., *Ann. Chim. Phys.*, 1831, **46**, 431.
Mellor, J.W., *Comprehensive Treatise on Inorganic and Theoretical Chemistry*, Vol. 6, Longmans Green, London, U.K., 1925, 588.
Buxbaum, G. and Pfaff, G., *Industrial Inorganic Pigments*, 3rd ed., VCH, Weinheim, 2005, 136.

Gulf A process for making ethanol by the simultaneous hydrolysis and fermentation of cellulose. *See* Emert.

Gulf HDS A process for *hydrorefining and *hydrocracking petroleum residues in order to make fuels and feeds for *catalytic cracking. Developed by the Gulf Research & Development Company. *See also* Hydrodesulfurization.

> Unzelman, G.H. and Wolf, C.J., in *Petroleum Processing Handbook*, Bland, W.F. and Davidson, R.L., Eds., McGraw-Hill, New York, 1967, 3–23.

Gulfining [Gulf refining] A *hydrodesulfurization process adapted for heavy gas oils. Developed by Gulf Research & Development Company in the early 1950s.

> Unzelman, G.H. and Wolf, C.J., in *Petroleum Processing Handbook*, Bland, W.F. and Davidson, R.L., Eds., McGraw-Hill, New York, 1967, 3–45.
> *Hydrocarbon Process. Int. Ed.*, 1970, **49**(9), 211.

Gulfinishing A process for *hydrofining lubricating oils. Developed by Gulf Oil Corporation.

> *Hydrocarbon Process.*, 1964, **43**(9), 190.

Gulf Resid A process for desulfurizing petroleum residues, developed by Gulf Oil Corporation.

> Speight, J.G., *The Desulfurization of Heavy Oils and Residua*, Marcel Dekker, New York, 1981, 175.

Gulf SRC *See* SRC.

Gutehoffnungshütte [Named after the company with the same name in Oberhausen, Germany, now a subsidiary of MAN AG, Munich] A process for oxidizing methane to formaldehyde, using nitrogen oxides as the oxidant and sodium borate as the catalyst. Operated at atmospheric pressure and 620°C. Developed in Germany during World War II and improved by P. Nashan in 1951.

> U.S. Patent 2,757,201.
> Sittig, M., *Combining Oxygen and Hydrocarbons for Profit*, Gulf Publishing, Houston, TX, 1962, 130.
> Asinger, F., *Paraffins: Chemistry and Technology*, translated by B.J. Hazzard, Pergamon Press, Oxford, U.K., 1968, 582.

Guyot A continuous sulfonation process for making phenol. Sodium sulfite is a by-product and must be sold if the process is to be economic.

> Molyneux, F., *Chem. Trade J.*, 1960, **147**(497), 518.

G-V *See* Giammarco-Vetrocoke.

Gyro Also called Gyro-cracking. An early vapor-phase *Thermal Cracking process for refining petroleum.

> *The Petroleum Handbook*, 3rd ed., Shell Petroleum, London, U.K., 1948,170.
> Asinger, F., *Mono-olefins: Chemistry and Technology*, translated by B.J. Hazzard, Pergamon Press, Oxford, U.K., 1968, 359.

Gyttorp [Named after a small Swedish town] A continuous process for nitrating polyols to form nitrate esters, used as explosives. Similar to the *Biazzi process.

H

H *See* Pechiney H.

Haber Also called **Haber–Bosch** and **Haber–Bosch–Mittasch**. A process for synthesizing ammonia from the elements, using high temperatures and pressures and an iron-containing catalyst. Invented by F. Haber at BASF in 1908. In 1909, C. Bosch of BASF built a pilot plant using an osmium-based catalyst; in 1913, a larger plant was built at Oppau, Germany. The process has been continually improved and is still of major importance worldwide. Haber was awarded the Nobel Prize for this work in 1918 but was infamous for his introduction of poison gases in World War I. Bosch was awarded the Nobel Prize in 1931.

German Patents 235,421; 293,787.

Haber, F. and van Oordt, G., *Z. Anorg. Allg. Chem.*, 1905, **43**, 111.

Harding, A.J., *Ammonia: Manufacture and Uses*, Oxford University Press, London, U.K., 1959.

Chilton, T.H., *Strong Water: Nitric Acid, Its Sources, Methods of Manufacture, and Uses*, MIT Press, Cambridge, MA, 1968, 64.

Haber, L.F., *The Chemical Industry 1900–1930*, Clarendon Press, Oxford, U.K., 1971, 187.

Vancini, C.A., *Synthesis of Ammonia*, translated by L. Pirt, Macmillan Press, Basingstoke, U.K., 1971, 234.

Jennings, J.R. and Ward, S.A., in *Catalyst Handbook*, 2nd ed., Twigg, M.V., Ed., Wolfe Publishing, London, U.K., 1989, 384.

Travis, T., *Chem. Ind. (London)*, 1993, (15), 581.

Leigh, G.J., *The World's Greatest Fix: A History of Nitrogen and Agriculture*, Oxford University Press, Oxford, U.K., 2004, 129–134.

Stoltzenerg, D., *Fritz Haber: Chemist, Nobel Laureate, German, Jew*, Chemical Heritage Press, Philadelphia, PA, 2004, Chap. 5.

Haber Gold A hydrometallurgical process for extracting gold from its ores. It uses a proprietary extracting agent that is specific for gold, instead of cyanide. Developed by Haber Inc. and proposed for installation by Gold City Inc. at its Winnemucca, NV, plant in 2004.

Chem. Eng. (N.Y.), 2004, **111**(2), 14.

Haifa Also called IMI. One of the two *wet processes for producing phosphoric acid by the acidulation of phosphate rock; the other is the *Dorr process. The Haifa process uses hydrochloric acid for the acidulation and solvent extraction for the purification. It is economic only where by-product hydrochloric acid is available. The overall reaction is

$$Ca_5(PO_4)_3F + 10HCl = 3H_3PO_4 + 5CaCl_2 + HF$$

Various organic extractants may be used, including butyl and amyl alcohols, diisopropyl ether, tri-*n*-butyl phosphate, and tri-2-ethylhexyl phosphate. The fluoride remains with the calcium chloride solution, from which it may be recovered for sale or disposal. The process was invented in 1957 by A. Baniel and R. Blumberg at Israel Mining Industries, Haifa.

U.S. Patent 2,880,063.

British Patents 805,517; 1,051,521.

Baniel, A., Blumberg, R., Alon, A., El-Roy, M., and Goniadski, C., *Chem. Eng. Prog.*, 1962, **58**(11), 100.

Baniel, A. and Blumberg, R., in *Phosphoric Acid*, Vol. 1, Part 2, Slack, A.V., Ed., Marcel Dekker, New York, 1968, 889.

Blumberg, R., Gonen, D., and Meyer, D., in *Recent Advances in Liquid–Liquid Extraction*, Hanson, C., Ed., Pergamon Press, Oxford, U.K., 1971, 93.

Haines A process for recovering sulfur from natural gas, using a zeolite adsorbent. The hydrogen sulfide in the gas is adsorbed on the zeolite; when the bed is saturated, hot sulfur dioxide is passed through it. The zeolite catalyzes the reaction between hydrogen sulfide and sulfur dioxide to form elemental sulfur, which sublimes out and is condensed. The process was invented by H.W. Haines in 1960; it was developed by Krell Associates and piloted in Canada from 1961 to 1962 but not commercialized because of problems caused by fouling of the zeolite with heavy hydrocarbons.

> *Pet. Refin.*, 1960, **39**(12), 208.
> Haines, H.W., Van Wielingen, G.A., and Palmer, G.H., *Pet. Refin.*, 1961, **40**(4), 123.
> *Hydrocarbon Process.*, 1961, **40**(11), 291.

HAL [**H**ot **A**cid **L**eaching] A process for purifying silica sand or zircon by leaching out surface iron compounds with hot sulfuric acid. Derived from an earlier process, invented in 1955 by British Industrial Sand, in which silica sand was treated with hot, gaseous hydrogen chloride. The process for cleaning zircon sand was developed jointly by Hepworth Minerals and Chemicals, United Kingdom, and Metallurgical Services Pty, Australia, in 1991.

> British Patents 845,745; 1,223,177.
> *Ind. Miner. (London)*, 1992, (296), 101.

Halcon (1) Halcon International (later The Halcon SD Group) designed many organic chemical processes but is perhaps best known for its process for making phenol from cyclohexane. Cyclohexane is first oxidized to cyclohexanol, using air as the oxidant and boric acid as the catalyst; the cyclohexanol is then dehydrogenated to phenol. Invented in 1961 by S.N. Fox and J.W. Colton, the process was operated by Monsanto in Australia for several years.

> U.S. Patents 3,109,864; 3,239,552; 3,256,348; 3,932,513.

Halcon (2) A process for oxidizing ethylene to ethylene oxide, using atmospheric oxygen, and catalyzed by silver. Developed by Halcon International in the late 1940s and early 1950s and first commercialized at Lavera, France. *See* Halcon (1).

> Landau, R., *Pet. Refin.*, 1953, **32**(9), 146.
> Porcelli, J.V., *Catal. Rev. Sci. Eng.*, 1981, **23**, 151.

Halex [**Hal**ogen **ex**change] A process for making fluoroaromatic compounds by reacting the corresponding chloro- or bromoaromatic compounds with an inorganic fluoride, usually potassium fluoride. Widely used for the manufacture of fluorointermediates.

> Dolby-Glover, L., *Chem. Ind. (London)*, 1986, 518.
> Siegemund, G., in *Ullmann's Encyclopedia of Industrial Chemistry*, 5th ed., Vol. A11, VCH Publishers, Weinheim, Germany, 1989, 379.

Hall (1) Also called the dry puddling process. An early ironmaking process, invented by J. Hall in 1830.

Hall (2) A process for making fuel gas from petroleum fractions and residues by thermal cracking. Developed by W.A. Hall in 1913, later forming the basis of the *Gasmaco process. *See also* Petrogas.

> U.S. Patents 105,772; 1,175,909; 1,175,910.

Hall (3) An early process for making aromatic hydrocarbons by thermally cracking petroleum naphtha. *See also* Rittman.

> Ellis, C., *The Chemistry of Petroleum Derivatives*, Chemical Catalog Co., New York, 1934, 164.
> Asinger, F., *Mono-olefins: Chemistry and Technology*, translated by B.J. Hazzard, Pergamon Press, Oxford, U.K., 1968, 137.

Hall (4) A process for making alumina by reducing bauxite with coke in an electric furnace. The coproduct is an alloy of iron–silicon–aluminum–titanium. Invented by C.M. Hall in 1901 and later developed and commercialized by ALCOA.

O'Connor, D.J., *Alumina Extraction from Non-bauxitic Materials*, Aluminium-Verlag, Düsseldorf, Germany, 1988, 304.

Hall–Héroult An electrolytic process for making aluminum metal from alumina, invented in 1886 independently by C.M. Hall in the United States and by P.L. Héroult in France. The alumina, made by the *Bayer process, is dissolved in fused cryolite, Na_3AlF_6, and electrolyzed at approximately 1000°C. Because of the large requirement for electricity, the process is operated only where hydroelectric power is available. The cryolite was originally obtained from a deposit in Greenland but is now made synthetically from alumina, hydrofluoric acid, and sodium hydroxide:

$$Al_2O_3 + 6NaOH + 12HF = 2Na_3AlF_6 + 9H_2O$$

Hall's process was first operated by the Pittsburgh Reduction Company, a predecessor of ALCOA, in 1889. Héroult's process was first operated by the Société Metallurgique Suisse at Neuhausen am Rheinfall, Switzerland, in 1887, using electric power generated at the Rhine Falls. Hall has been honored by a National Historic Chemical Landmark of the American Chemical Society in Oberlin College, OH.

Grjotheim, K., Krohn, C., Malinovsky, M., Matiasovsky, K., and Thonstad, J., *Aluminium Electrolysis; Fundamentals of the Hall-Héroult Process*. Aluminium-Verlag GmbH, Düsseldorf, Germany, 1982.
Palmear, I.J., *The Chemistry of Aluminium, Gallium, Indium, and Thallium*, Downs, A.J., Ed., Blackie, London, U.K., 1993, 85.
Chem. Eng. News, 1997, **75**(39), 42.

Haloclean A pyrolytic process for destroying scrap plastics, especially those derived from scrap electronic equipment. The material is passed through a horizontal rotary kiln containing metal spheres that aid the heat transfer. Developed in Karlsruhe, Germany, and piloted from 2006.

Hornung, A. and Seifert, H., in *Feedstock Recycling and Pyrolysis of Waste Plastics*, Scheirs, J. and Kaminsky, W., Eds., John Wiley & Sons, Chichester, U.K., 2006, Chap. 20.
TCE (Rugby), 2007, (793), 46.

Halomet [**Halo**gen **met**al] A process for reducing halides to metals by reaction with metallic aluminum or magnesium in a closed vessel. Invented in 1968 by R. Nowak and W. Schuster at Halomet, Basel, Switzerland.

U.S. Patents 3,244,509; 3,466,169.
Canadian Patent 899,631.

HALT [**H**ydrate **A**ddition at **L**ow **T**emperature] A *flue-gas desulfurization process in which lime slurry is injected into the combustion gases after they have been cooled in the heat exchanger. Developed in Canada in the mid-1970s.

Ashley, M.J. and Greaves, R.A., *Chem. Ind. (London)*, 6 Feb 1989, 62.

Hamburg A process for pyrolyzing waste plastics and rubber in an externally heated, fluidized-bed reactor containing sand. Developed by W. Kaminsky (famous for inventing metallocene polymerization catalysts) at the University of Hamburg and demonstrated in Ebenhausen, Germany, from 1983. A larger plant, fed by whole tires, was built in Grimma, Germany, but was closed when Germany was reunified in 1989. Olefins are major products. The process has not been commercialized. Kaminsky and H. Sinn were awarded the European Research Prize for this development in 1987.

Kaminsky, W., in *Feedstock Recycling and Pyrolysis of Waste Plastics*, Scheirs, J. and Kaminsky, W., Eds., John Wiley & Sons, Chichester, U.K., 2006, Chap. 17.
Aguado, J., Serrano, D.P., and Esola, J.M., *Ind. Eng. Chem. Res.*, 2008, **47**(21), 7982.

Hänisch–Schröder A process for scrubbing sulfur dioxide from smelter gases using water. Developed in 1884 and still in use in 1950, although it was probably obsolete by 1990. It required 100–200 tons of water for each ton of sulfur recovered.

Hänisch, E. and Schröder, M., *J. Soc. Chem. Ind.*, 1884, **3**, 570.
Katz, Z. and Cole, R.J., *Ind. Eng. Chem.*, 1950, **42**, 2263.

Hansgirg Also called Radenthein. A process for making magnesium metal by reducing magnesium oxide with carbon in an electric arc furnace at >2000°C and shock-chilling the vapor with hydrogen. The product was a fine dust that had to be converted to ingot. Operated on a pilot scale at Radenthein, Austria, by Österr.-Amerik. Magnesit in the 1930s. Made obsolete by the invention of the electrolytic process for making magnesium. *See* Elektron.

Hanson-Van Winkle-Munning *See* metal surface treatment.

Hargreaves–Bird An early process for electrolyzing brine, which used a vertical diaphragm, unlike earlier designs. Developed by J. Hargreaves and T. Bird around 1890.

Hardie, D.W.F., *A History of the Chemical Industry in Widnes*, Imperial Chemical Industries, Widnes, U.K., 1950, 193.

Hargreaves–Robinson A process for making sodium sulfate and hydrochloric acid by passing a mixture of wet sulfur dioxide and air through a series of vertical chambers containing briquettes of salt lying on a perforated floor, the temperature being maintained at approximately 500°C:

$$4NaCl + 2SO_2 + O_2 + 2H_2O = 2Na_2SO_4 + 4HCl$$

The addition of a small amount of nitric acid vapor or nitric oxide accelerates the process. Invented by J. Hargreaves and T. Robinson in Widnes in 1870, in order to provide sodium sulfate for the *Leblanc process, circumventing the need for the sulfuric acid used in the salt cake section of that process. Reportedly still in use in 1984.

British Patents 46 (1870); 3,045 (1870); 1,733 (1871); 3,052 (1872).
Hardie, D.W.F., *A History of the Chemical Industry in Widnes*, Imperial Chemical Industries, Widnes, U.K., 1950, 80, 130.

Harloff A process for purifying sugar syrup by the addition of lime and sulfur dioxide. The precipitated calcium sulfite carries down many of the impurities in the syrup.

Spencer, G.L. and Meade, G.P., *Cane Sugar Handbook*, 8th ed., John Wiley & Sons, New York, 1945, 110.

HARP [**H**ybrid **A**rgon **R**ecovery **P**rocess] A process for extracting argon from the hydrogen recycle stream in ammonia synthesis. Both *PSA and a cryogenic process are used.

Krishnamurthy, R., Lerner, S.L., and MacLean, D.M., *Gas Sep. Purif.*, 1987, **1**, 16.

Harris A softening process for removing antimony, arsenic, and tin from lead. The mixed metals are heated with a molten mixture of sodium hydroxide and sodium nitrate. Invented by H. Harris at H.J. Enthoven & Sons.

HARVEST [**H**ighly **A**ctive **R**esidue **V**itrification **E**xperimental **St**udies] A process for immobilizing nuclear waste by incorporation in a borosilicate glass. Developed from *FINGAL. Piloted by the U.K. Atomic Energy Authority at Sellafield in the late 1970s but abandoned in 1981 in favor of the French vitrification process *AVM.

Lutze, W., in *Radioactive Waste Forms for the Future*, Lutze, W., and Ewing, R.C., Eds., North-Holland, Amsterdam, the Netherlands, 1988, 15.
Ojovan, M.I. and Lee, W.E., *An Introduction to Nuclear Waste Immobilisation*, Elsevier, Amsterdam, the Netherlands, 2005, 233.

Hasenclever An improvement to the *Deacon process for oxidizing hydrogen chloride to chlorine, in which the hydrogen chloride is first dried with concentrated sulfuric acid.

> Hardie, D.W.F., *A History of the Chemical Industry in Widnes*, Imperial Chemical Industries, Widnes, U.K., 1950, 69.

Hass (1) A process for making aliphatic nitro-compounds from aliphatic hydrocarbons. The hydrocarbon vapor, mixed with nitric acid vapor, is passed through a narrow tube at 420°C. Invented by H.B. Hass in 1933.

> U.S. Patents 1,967,667; 2,071,122.
> Hass, H.B. and Riley, E.F., *Chem. Rev.*, 1943, **32**, 373.

Hass (2) A process for making cyclopropane by reacting 1,3-dichloropropane with zinc.

> Gustavson, G., *J. Prakt. Chem.*, 1887, **36**, 300.
> Hass, H.B., McBee, E.T., Hinds, G.E., and Gluesenkamp, E.W., *Ind. Eng. Chem.*, 1936, **28**, 1178.

Hass–McBee A thermal, vapor-phase process for chlorinating aliphatic hydrocarbons. The chlorine and the hydrocarbon vapor are separately heated to >250°C and then mixed. Propane is thus converted to 1,3-dichloropropane. Invented in 1934 by H.B. Hass and E.T. McBee at Purdue University, Indiana.

> U.S. Patent 2,004,073.
> Hass, H.B., McBee, E.T., and Weber, P., *Ind. Eng. Chem.*, 1935, **27**, 1192.
> McBee, E.T., Hass, H.B., Burt, W.E., and Neher, C.M., *Ind. Eng. Chem.*, 1949, **41**, 799.
> Asinger, F., *Paraffins: Chemistry and Technology*, translated by B.J. Hazzard, Pergamon Press, Oxford, U.K., 1968, 219.

HAT [Homogeneous Alkylation Technology] An *alkylation process using an alkyl aluminum catalyst. Developed by Kerr-McGee in 1994.

> Rhodes, A.K., *Oil Gas J.*, 1994, **92**(34), 52.

Häusser A process for making nitric acid from air using an internal combustion engine fueled by coal gas. Piloted in Germany before World War II but never commercialized.

> *J. Soc. Chem. Ind. (London)*, 1922, 41, 253.
> Chilton, T.H., *Strong Water: Nitric Acid, Its Sources, Methods of Manufacture, and Uses*, MIT Press, Cambridge, MA, 1968, 116.

HBN *See* carbonization.

HBNPC *See* carbonization.

HB Unibon [Hydrogenation of Benzene] The revised name for *Hydrar.

(HC)₃ A process for *hydrocracking heavy petroleum oils, developed by Headwaters Inc. Conventional catalysts are used but in the form of nanoparticles.

> *Chem. Eng. (N.Y.)*, 2005, **112**(13), 13.

HCAT A process for upgrading heavy oil, developed by Headwaters Heavy Oil and first installed at Neste Oil Corp.'s Porvoo Refinery in 2011. The catalyst is distributed among the feed asphaltene molecules.

HCM [High Carbon Monoxide] Also called HICOM. A process for making methane from coal, based on the British Gas/Lurgi slagging gasifier.

> Tart, K.R. and Rampling, T.W.A, *Hydrocarbon Process. Int. Ed.*, 1981, **60**(4), 114.

H-Coal A coal gasification process. Crushed coal is mixed with process-derived oil and catalytically hydrogenated in an ebullated bed under pressure at 455°C. The catalyst is a mixture of cobalt and molybdenum oxides on alumina. Developed by Hydrocarbon Research from the 1960s and piloted in Catlettsburg, KY, from 1980 to 1982. *See also* CSF, H-Oil, CSF, Synthoil.

U.S. Patent 3,321,393.
Alpert, S.B., Johanson, E.S., and Schuman, S.C., *Hydrocarbon Process.*, 1964, **43**(11), 193.
Eccles, R.M., DeVaux, G.R., and Dutkiewicz, B., in *The Emerging Synthetic Fuel Industry*, Thumann, A., Ed., Fairmont Press, Atlanta, GA, 1981, Chap. 4.
Papso, J.E., in *Handbook of Synfuels Technology*, Meyers, R.A., Ed., McGraw-Hill, New York, 1984.
Davies, G.O., in *Chemicals from Coal: New Developments*, Payne, K.R., Ed., Blackwell Scientific Publications, Oxford, U.K., 1985, 104.

HC Platforming [**H**ydro**C**racking] A version of the *Platforming process that uses different catalyst systems before the reforming catalyst in order to partially hydrocrack the feed before converting it to aromatic hydrocarbons.

HCR (1) [**H**igh **C**apacity **R**eactor] *See* Otto Aqua-Tech HCR.

HCR (2) [**H**igh **C**laus **R**atio] A variation of the *RAR process developed by KTI.

HC Unibon [**H**ydro**C**racking] A version of the *hydrocracking process for simultaneously hydrogenating and cracking various liquid petroleum fractions to form branched-chain hydrocarbon mixtures of lower molecular weight. The catalyst is dual-functional, typically silica and alumina with a base metal, in a fixed bed. Developed by UOP. By 1988, 46 licenses had been granted. Currently offered under the name UOP Unicracking.

Tajbl, D.G., in *Handbook of Petroleum Refining Processes*, 1st ed., Meyers, R.A., Ed., McGraw-Hill, New York, 1986.
Hydrocarbon Process. Int. Ed., 1988, **67**(9), 73.
Ackelson, D., in *Handbook of Petroleum Refining Processes*, 3rd ed., Meyers, R.A., Ed., McGraw-Hill, New York, 2003, 7.23.

H-D (1) *See* DR.

H-D (2) *See* Huron-Dow.

HDA [**H**ydro**D**e**A**lkylation] A proprietary *dealkylation process for making benzene from toluene, xylenes, pyrolysis naphtha, and other petroleum refinery intermediates. The catalyst, typically chromium oxide or molybdenum oxide, together with hydrogen gas, removes the methyl groups from the aromatic hydrocarbons, converting them to methane. The process also converts cresols to phenol. Developed by Hydrocarbon Research with Atlantic Richfield Corporation and widely licensed worldwide.

HDH (1) [**H**emi-**D**i**H**ydrate] A *wet process for making phosphoric acid. The calcium sulfate is first produced as the hemihydrate; then, in another stage, this is converted to the dihydrate. Developed by Fisons, United Kingdom, and operated in Yugoslavia and the United Kingdom.

Blumrich, W.E., Koening, H.J., and Schwer, E.W., *Chem. Eng. Prog.*, 1978, **74**, 58.

HDH (2) [**H**ydrocracking–**D**istillation–**H**ydrotreatment] A process for *hydrotreating and hydroconverting petroleum residues.

Marcos, F. and Rosa-Brussin, D., *Catal. Rev. Sci. Eng.*, 1995, **37**(1), 3.

HDM *See* hydrodemetallation.

HDN *See* hydrodenitrogenation.

HDO [HydroDeOxygenation] A general name for processes that remove oxygenates from various biomass-derived products before they can be used as motor fuels. A variety of processes and catalysts will accomplish this.

Mortensen, P.M., Grunwaldt, J.-D., Jensen, P.A., Knudsen, K.G., and Jensen, A.D., *Appl. Catal. A: Gen.*, 2011, **407**(1–2), 1.

HDS *See* hydrodesulfurization. Not to be confused with *TETRA HDS.

heap leaching A general name for processes for leaching ores with various aqueous solutions in order to extract metals. It is an inexpensive process, normally used with low-grade ores. The leachate must be well contained by use of impervious membranes in order to prevent contamination of groundwater. The first large-scale heap leaching process using sulfuric acid to extract vanadium from a unique ore was being installed by American Vanadium in Nevada in 2013.

Watling, H., *TCE (Rugby)*, 2007, (274), 30.
Chem. Eng., 2013, **120**(4), 12.
Wikipedia, The Free Encyclopedia. Heap leaching. Revised 6 Jun 2013. Retrieved 20 Jul 2013.

Heat-Fast *See* DR.

Heinrich Koppers A fractional distillation process for separating the constituents of coal tar. In 1981, this process was in use in two plants in the United Kingdom.

McNeil, D., *Chemistry of Coal Utilization*, Vol. 2., Elliott, M.A., Ed., John Wiley & Sons, New York, 1981, 1070.

Heiskenskjold A process for making an animal feed by growing the yeast *Saccharomyces cerevisiae* on the waste liquor from paper manufacture by sulfite pulping. Developed in Finland in 1936.

Prescott, S.C. and Dunn, G.G., *Industrial Microbiology*, 3rd ed., McGraw-Hill, New York, 1959, Chap. 3.
Litchfield, J.H., *ChemTech*, 1978, **8**, 218.

Hematite Also spelled haematite. A process for removing iron from leach liquors from hydrometallurgical leaching operations. Used in recovering zinc from the residues of the electrolytic zinc process. *See also* Goethite, Jarosite.

Morgan, S.W.K., *Zinc and Its Alloys and Compounds*, Ellis Horwood, Chichester, U.K., 1985, 122.

Henderson A wet process for extracting copper from its ores. Operated in Widnes, England, in 1860.

Hardie, D.W.F., *A History of the Chemical Industry in Widnes*, Imperial Chemical Industries, Widnes, U.K., 1950, 78.

Henkel Also named Raecke, after the inventor. A process for making terephthalic acid from potassium benzoate by disproportionation:

$$2C_6H_5(COOK) = C_6H_4(COOK)_2 + C_6H_6$$

The reaction is conducted in the melt, in the presence of carbon dioxide under pressure, catalyzed by zinc or cadmium compounds. Invented by B. Raecke at Henkel, Germany. Improved variations of this process, known as Henkel I and Henkel II, were operated by several other companies but, by 1975, had been abandoned in favor of various other methods of oxidizing *p*-xylene.

German Patents 936,036; 958,920.
U.S. Patents 2,794,830; 2,823,229; 2,891,992; 2,905,709.
Raecke, B., *Angew. Chem.*, 1958, **70**, 1.
Weissermel, K. and Arpe, H.-J., *Industrial Organic Chemistry*, 4th ed., translated by C.R. Hindley and S. Hawkins, Wiley-VCH, Weinheim, Germany, 2003, 399.

Heraeus A process for making clear, fused quartz by passing powdered quartz crystals through an oxyhydrogen flame and collecting the product on a rotating tube of fused quartz. Invented by H.R. Heraeus in 1952 and developed by Heraeus Quarzschmelze, Germany.

U.S. Patent 2,904,713.

Hercosett *See* Chlorine/Hercosett.

Hercules An organic nitration process, similar to the *Bofors process but using a tee joint for mixing. Developed by the Hercules Powder Company, Wilmington, DE, from 1957.

U.S. Patents 2,951,746; 2,951,877.

Hercules–BP Also called BP–Hercules–Kellogg. A process for making phenol from cumene, based on processes first developed by Hercules and BP. Engineered by Kellogg International Corporation and first installed in Montreal, PQ, in 1953. By 1993, more than half of the world's production of phenol was made by this process.

Weissermel, K. and Arpe, H.-J., *Industrial Organic Chemistry*, 4th ed., translated by C.R. Hindley and S. Hawkins, Wiley-VCH, Weinheim, Germany, 2003, 356.

Hercules–Witten *See* Katzschmann.

Herz A process for making *o*-aminothiophenols by heating the hydrochlorides of aromatic amines with sulfur monochloride at 50°C–75°C. The products are used as intermediates in the manufacture of thioindigo dyestuffs. Invented by R. Hertz in Germany in 1914.

German Patent 360,690.
U.S. Patents 1,637,023; 1,699,432.
Warburton, W.K., *Chem. Rev.*, 1957, **57**, 1011.

HETACAT An alkylation process using a solid acid catalyst. Not commercialized as of 1997.

Lerner, B.A., *Chem. Ind. (London)*, 1996, (1), 16.

Hexall A process for making hexane by dimerizing propylene. The reaction takes place in the liquid phase in a fixed bed of catalyst. Developed by UOP, but not commercialized as of 1992.

Ward, D.J., Friedlander, R.H., Frame, R., and Imai, T., *Hydrocarbon Process. Int. Ed.*, 1985, **64**(5), 81.
Hydrocarbon Process. Int. Ed., 1987, **66**(11), 77.

Hexone *See* Redox.

Heyden-Wacker A process for making phthalic anhydride by the catalytic oxidation of naphthalene or *o*-xylene. Offered by Lurgi.

HF A predecessor of the *DHD petroleum reforming process, operated in Germany in 1939. The catalyst was a mixed $MoO_3/ZnO/Al_2O_3$ system.

Weisser, O. and Landa, S., *Sulphide Catalysts: Their Properties and Applications*, Pergamon Press, Oxford, U.K., 1973, 379.

HFC [**H**ydrocracking with **F**ine **C**atalyst] A Japanese process for *hydrotreating and hydroconverting petroleum residues.

Marcos, F. and Rosa-Brussin, D., *Catal. Rev. Sci. Eng.*, 1995, **37**(1), 3.

HIB An ironmaking process in which the powdered ore is reduced in a fluidized bed; the reducing gas is made by reforming natural gas. *See* DR.

Hibernia A process for making formaldehyde by the partial oxidation of methane by ozonized oxygen. The catalyst is barium peroxide activated with silver oxide. Developed in Germany during World War II but not commercialized.

> Hann, V., *Chem. Ind. (London)*, 1950, **67**, 386.
> Asinger, F., *Paraffins: Chemistry and Technology*, translated by B.J. Hazzard, Pergamon Press, Oxford, U.K., 1968, 582.

Hichlor A family of processes for making aluminum trichloride and other volatile metal chlorides by chlorinating fly ash and other coal wastes in fixed and fluidized beds. Developed by Ames Laboratory and Iowa State University from the mid-1970s to the mid-1980s. Not commercialized.

> O'Connor, D.J., *Alumina Extraction from Non-bauxitic Materials*, Aluminium-Verlag, Düsseldorf, Germany, 1988, 297.

Hi-Chloroff A thermal (noncatalytic) process for removing chlorine from chlorinated hydrocarbon wastes containing either low or high concentrations of chlorine. Developed by Kinetics Technology International. *See also* Chloroff.

HICOM [**HI**gh **C**arb**O**n **M**onoxide] *See* HCM.

High Productivity Also called HP. An improved method of operating gas-phase plants for making polyethylene. It involves removing gas and liquid from the fluidized bed, separating them, and returning the liquid to the bed via a patented nozzle system. Developed by BP in 1994 and operated in its Grangemouth refinery since 1995. *See also* Supercondensed Mode.

> *Chem. Br.*, 1995, **31**(4), 278; 1997, **33**(5), 31.

Hilgenstock *See* Hoerde.

Hiperion A process for removing hydrogen sulfide and organic sulfur compounds from hydrocarbons. Similar to the *Takahax process but using a solution of chelated iron and naphthaquinone. The elemental sulfur produced is removed by filtration. Licensed by Ultrasystems, CA. Two commercial systems were operating in California in the 1990s.

> *Chem. Eng. (N.Y.)*, 1987, **94**(2), 159.
> Dalrymple, D.A., Trofe, T.W., and Evans, J.M., *Chem. Eng. Prog.*, 1989, **85**(3), 43.
> Kohl, H.L. and Nielsen, R.B., *Gas Purification*, 5th ed., Gulf Publishing Co., Houston, TX, 1997, 794.

HiPOx A process for destroying organic pollutants in drinking water by oxidation with a mixture of ozone and hydrogen peroxide. Developed by Applied Process Technology, CA, and used to destroy MTBE in water from Lake Tahoe. Also used to destroy 1,4-dioxan at several Californian locations and proposed for use in South Korea.

> Foster, J., *Water Wastewater Int.*, 2002, **17**(6), 48.
> Borg, C., *Water Wastewater Int.*, 2004, **19**(7), 26.
> *Water Wastewater Int.*, 2005, **20**(3), 5.

HiPure A variation on the *Benfield process, using two stages of scrubbing by hot potassium carbonate solution in order to reduce the carbon dioxide contents of gases to very low levels. *See also* Carsol, CATACARB, Giammarco-Vetrocoke.

> Benson, H.E. and Parrish, R.W., *Hydrocarbon Process. Int. Ed.*, 1974, **53**(4), 81.
> Kohl, H.L. and Nielsen, R.B., *Gas Purification*, 5th ed., Gulf Publishing Co., Houston, TX, 1997, 339.

Hirohax A process for removing hydrogen cyanide from gas streams. It is first converted to thiocyanate, and this is oxidized at 350°C with oxygen under pressure. Developed in 1972 by Nippon Steel, Tokyo.

U.S. Patent 3,855,390.

H-Iron [**Hydrogen Iron**] A process for making iron by reducing powdered iron oxides from ores or scrap, using hydrogen. A variation on the process will convert iron–titanium ores to a titanium concentrate and metallic iron. Developed by Hydrocarbon Research and the U.S. Steel Corporation and used in Pennsylvania and California. *See also* DR.

HIsmelt A direct iron-smelting process in which noncoking coal, fine iron ore, fluxes, and gases are injected into a molten iron bath; the carbon monoxide produced is used to prereduce the ore in a fluidized bed. Under development by CRA, Australia, since the early 1980s, joined by Midrex Corporation in 1988. Their joint venture company, Hismelt Corporation, commissioned a pilot plant at Kwinana, near Perth, Australia, in 1993.

Hitachi Wet A *flue-gas desulfurization process using activated carbon.

Speight, J.G., *Gas Processing*, Butterworth Heinemann, Oxford, U.K., 1993, 293.

Hitachi Zosen A pyrolytic process for converting waste plastics into liquid fuels. Developed in the 1990s by Hitachi Zosen Corporation.

U.S. Patent 5,584,969.
Scheirs, J., in *Feedstock Recycling and Pyrolysis of Waste Plastics*, Scheirs, J. and Kaminsky, W., John Wiley & Sons, Chichester, U.K., 2006, 427.

Hivalloy A process for grafting styrenic polymers on to polyolefins, using a *Ziegler–Natta catalyst. The products combine the physical properties of both polymer types. Developed by Montell and commercialized in the United States in 1997. *See also* Catalloy.

Oxley, D.F., *Chem. Ind. (London)*, 1998, (8), 307.

Hi-Zex A process for making high-density polyethylene having good processing properties. The properties result from the use of two catalysts and multiple reactors. Developed by Mitsui Chemicals in 1958 and widely used; the latest plant is being built for TVK at Tiszaujvaros, Hungary. Mitsui Chemicals has now merged with Idemitsu Kosan, and the merged companies are named Prime Polymer Co. Ltd., based in Tokyo. The name Hi-Zex is also the trade name for the polymer product.

Chem. Week, 2002, **164**(20), 17.

HOCC [**High-O**lefin Catalytic Cracking] A general name for such processes, including CPP, DCC, and PetroFCC.

Chem. Eng. News, 2007, **85**(17), 27.

Hock Also known as the Hock–Lang process and the cumene peroxidation process. A process for converting isopropyl benzene (cumene) to a mixture of phenol and acetone; *m*-diisopropyl benzene likewise yields resorcinol, and *p*-diisopropyl benzene yields hydroquinone. The basis of the process is the liquid-phase air oxidation of cumene to cumene hydroperoxide:

$$C_6H_5CH(CH_3)_2 + O_2 \rightarrow C_6H_5C(CH_3)_2OOH$$

followed by the cleavage of the hydroperoxide by hot acid:

$$C_6H_5C(CH_3)_2OOH \rightarrow C_6H_5OH + (CH_3)_2CO$$

The second stage can be carried out in the liquid phase by sulfuric acid or in the gas phase by phosphoric acid adsorbed on a carrier.

Invented in Germany during World War II by H. Hock and S. Lang in the course of developing cumene hydroperoxide for initiating the polymerization of butadiene–styrene mixtures. After the war, the process was developed by the Distillers Company in England and Allied Chemical Corporation in the United States. Since 1954, Hock has been the main commercial process for the production of phenol and acetone. By 1987, 97% of the phenol made in the United States was produced via this route. In 1990, both resorcinol and hydroquinone were produced commercially by this route as well. *See also* Cumox.

Hock, H. and Lang, S., *Ber. Dtsch. Chem. Ges.*, 1944, **77**, 257.
Armstrong, G.P., Hall, R.H., and Quin, D.C., *J. Chem. Soc.*, 1950, 666.
Weissermel, K. and Arpe, H.-J., *Industrial Organic Chemistry*, 4th ed., translated by C.R. Hindley and S. Hawkins, Wiley-VCH, Weinheim, Germany, 2003, 355.

Hock–Lang *See* Hock.

Hoechst coker A continuous process for pyrolyzing various oils to yield olefins. The heat is carried by coke spheres. First operated in 1956, later supplanted by *Hoechst HTP.

Asinger, F., *Mono-olefins: Chemistry and Technology*, translated by B.J. Hazzard, Pergamon Press, Oxford, U.K., 1968, 128.
Stokes, R.G., *Opting for Oil*, Cambridge University Press, Cambridge, U.K., 1994, 187.

Hoechst HTP [**H**igh-**T**emperature **P**yrolysis] Also called Hoechst–Uhde. A two-stage process for making a mixture of acetylene and ethylene by cracking higher hydrocarbons. In the first stage, a fuel is burnt with the stoichiometric quantity of oxygen. In the second stage, the hot combustion products from the first stage are contacted with the vapor of the hydrocarbon to be cracked. The process was developed and operated by Farbwerke Hoechst in Germany in the 1950s until 1975. In Czechoslovakia, it was operated until 1985. *See also* Wulff.

Kamptner, H.K., Krause, W.R., and Schilken, H.P., *Hydrocarbon Process.*, 1966, **45**(4), 187.
Asinger, F., *Mono-olefins: Chemistry and Technology*, translated by B.J. Hazzard, Pergamon Press, Oxford, U.K., 1968, 178.
Stokes, R.G., *Opting for Oil*, Cambridge University Press, Cambridge, U.K., 1994, 190.
Weissermel, K. and Arpe, H.-J., *Industrial Organic Chemistry*, 4th ed., translated by C.R. Hindley and S. Hawkins, Wiley-VCH, Weinheim, Germany, 2003, 97.

Hoechst–Shawinigan A catalytic process for oxidizing acetaldehyde directly to acetic anhydride, using oxygen. *See* Shawinigan.

Hoechst–Uhde (1) An electrolytic process for generating hydrogen and chlorine from hydrochloric acid. Widely used.

Hoechst–Uhde (2) A variation of the *Wacker process, which makes vinyl acetate from ethylene and acetic acid. The catalyst is an aqueous solution of palladium and copper chlorides.

Dumas, T. and Bulani, W., *Oxidation of Petrochemicals: Chemistry and Technology*, Applied Science Publishers, London, U.K., 1974, 20.

Hoechst–Uhde (3) *See* Hoechst HTP.

Hoechst–Wacker *See* Wacker (1).

Hoechst–WLP A process for making acetylene by injecting a liquid hydrocarbon into a hydrogen plasma. It is an improvement on the *Hüls process in its ability to use heavy hydrocarbons without forming carbon.

Miller, S.A., *Acetylene: Its Properties, Manufacture and Uses*, Vol. 1, Academic Press, New York, 1965, 407.

Hoerde Also called Massener and Hilgenstock. A process for removing sulfur from molten iron or steel by adding manganese. This produces manganese sulfide, which forms an upper phase that can be skimmed off. Invented in 1891.

Hoesch *See* Bertrand Thiel.

Hoffman A process for making hydrazine by reacting urea and sodium hypochlorite in water:

$$OC(NH_2)_2 + NaOCl + 2NaOH = N_2H_4 + NaCl + Na_2CO_3 + H_2O$$

> *Riegel's Handbook of Industrial Chemistry*, Kent, J.A., Ed., 9th ed., Van Nostrand Reinhold, New York, 1992, 1123.

H-Oil A process for demetallizing, desulfurizing, and hydrocracking heavy fuel oil with the production of lighter distillates and gas. H-Oil is intended for heavy hydrocarbons including residues, tar, and shale oil. The oil is hydrogenated in an "ebullated," fluidized bed containing catalyst granules. Invented by Hydrocarbon Research in the early 1950s and developed jointly with Cities Service Research & Development Company. First demonstrated at Lake Charles, LA, in 1963; commercialized in Kuwait in 1968; and later installed in Mexico and United States. Six units were built between 1963 and 1996. The ebullated, fluidized bed was later used in the *LC-Fining process. *See also* H-Coal, LC-Fining.

> Unzelman, G.H. and Wolf, C.J., in *Petroleum Processing Handbook*, Bland, W.F. and Davidson, R.L., Eds., McGraw-Hill, New York, 1967, 3–19.
> Mounce, W. and Rubin, R.S., *Chem. Eng. Prog.*, 1971, **67**(8), 81.
> Papso, J.E., in *Handbook of Synfuels Technology*, Meyers, R.A., Ed., McGraw-Hill, New York, 1984, 6–19.
> *Hydrocarbon Process. Int. Ed.*, 1996, **75**(11), 126.

Hojanas Also called Siurin. An iron extraction process. Magnetite, mixed with carbon-coke breeze and limestone, is heated in a ceramic retort by passage through a tunnel kiln at 1200°C. Used commercially in Sweden since 1911. *See also* DR.

> Dennis, W.H., *Metallurgy of the Ferrous Metals*, Sir Isaac Pitman & Sons, London, U.K., 1963, 102.

HOKO A process for making nitric acid simultaneously at two concentrations.

> Hellmer, L., *Chem. Eng. Prog.*, 1972, **68**(4), 67.

Holmes–Manley An early mixed-phase petroleum cracking process.

Holmes–Maxted A process for removing organic sulfur compounds from coal gas. The gas, mixed with hydrogen, is passed over a metal thiomolybdate catalyst at 300°C–380°C, which converts the sulfur compounds to hydrogen sulfide, which is then absorbed by iron oxide. Developed by E.B. Maxted at W.C. Holmes & Company, United Kingdom, based on an invention made in 1937. More than 50 units had been used by 1985.

> British Patent 490,775.
> Wedgewood, W., *Inst. Gas Eng.*, 1958, 525; *Chem. Abstr.*, **52**, 14138.
> Kohl, H.L. and Nielsen, R.B., *Gas Purification*, 5th ed., Gulf Publishing Co., Houston, TX, 1997, 1168.

Holmes–Stretford A version of the *Stretford process, developed by W.C. Holmes & Company, Huddersfield, England.

> Moyes, A.J. and Vasan, S., *Oil Gas J.*, 1974, **72**(35), 56.
> Speight, J.G., *Gas Processing*, Butterworth Heinemann, Oxford, U.K., 1993, 229.

Holst A batch process for making chlorine dioxide by reducing sodium chlorate with sulfur dioxide in the presence of sulfuric acid. The sodium hydrogen sulfate by-product can be used in the *Kraft papermaking process. Developed by Moch Domsjo in Sweden in 1946.

> Partridge, H. de V., in *Chlorine: Its Manufacture, Properties, and Uses*, Sconce, J.S., Ed., Reinhold Publishing, New York, 1962, 275.

Holter A *flue-gas desulfurization process in which the sulfur dioxide is absorbed in an aqueous suspension of calcium hydroxide and calcium chloride, yielding gypsum. Operated in an experimental plant at the Weiherr III power station in Quierschied, Germany, in 1988.

Hondo HS-100 A process for removing acid gases from hydrocarbon gas streams by absorption in aqueous potassium nitrite. Developed by Hondo Chemicals. *See also* Sulfa-Check.

> Kohl, H.L. and Nielsen, R.B., *Gas Purification*, 5th ed., Gulf Publishing Co., Houston, TX, 1997, 1312.

Hooker–Raschig *See* Raschig (2).

Hoopes An electrolytic process for refining aluminum metal. The electrolyte is a mixture of fluoride salts. The cell is constructed of graphite. The electrolyte in contact with the sidewalls of the cell is frozen, thus preventing short-circuiting of electricity through the walls. Developed by W. Hoopes and others at Aluminum Company of America in the 1920s.

> U.S. Patents 1,534,321; 1,534,322.
> Frary, F.C., *Trans. Am. Electrochem. Soc.*, 1925, **47**, 275.

Horizone A gas-phase process for making polypropylene, incorporating two horizontal reactors. Developed by Japan Polypropylene Group.

> Soares, J.B.P. and McKenna, T.F.L., *Polyolefin Reaction Engineering*, Wiley-VCH, Weinheim, Germany, 2012, 124.

Hornsey *See* Steelmaking.

Hostalen A slurry process for making high-density polyethylene using a *Ziegler catalyst. Two reactors are used, in series or in parallel, which permits the production of a variety of molecular weight distributions. Developed and licensed by Basell Polyolefins. In 2005, 33 plants were operating or under construction. Hostalen is also a trade name for the polymer products. *See also* next entry.

Hostalen Advanced Cascade Process (ACP) A process for making polyethylene with mixed molecular weight polymers. Developed by LyondellBasell Industries. Operated in Germany, Poland, and Saudi Arabia in 2013.

> *Kunststoffe*, Jul 2006, **96**(7), 10.
> Soares, J.B.P. and McKenna, T.F.L., *Polyolefin Reaction Engineering*, Wiley-VCH, Weinheim, Germany, 2012, 112.

Hot Acid An improved version of the *Cold Acid process, which increases the yield of octanes. It operates at 60°C–90°C.

> *The Petroleum Handbook*, 3rd ed., Shell Petroleum, London, U.K., 1948, 229.

Hot Water A process for separating bitumen from tar sands. Developed by Suncor for treating the sands from Athabasca, Alberta, based on an invention made by K.A. Clarke in 1932. Two large plants were operating in Canada in 1984.

> Clark, K.A. and Pasternack, D.S., *Ind. Eng. Chem.*, 1932, **24**, 1410.
> Camp, F.W., *The Tar Sands of Alberta*, 2nd ed., Cameron Engineers, Denver, CO, 1974.
> Rider, D.K., *Energy: Hydrocarbon Fuels and Chemical Resources*, John Wiley & Sons, New York, 1981.
> Erskine, H.L., in *Handbook of Synfuels Technology*, Meyers, R.A., Ed., McGraw-Hill, New York, 1984, 5–13.

Hou A process for making sodium carbonate, developed by Hou Debang in China in the 1930s. It partially resembles the *Solvay process but generates by-product ammonium chloride instead of calcium chloride. Operated only in China, which has an agricultural market for the large quantities of ammonium chloride produced.

Wikipedia, The Free Encyclopedia. Sodium carbonate. Revised 10 Jul 2013. Retrieved 20 Jul 2013.

Houdresid A catalytic petroleum cracking process, similar to *Houdriflow, adapted for processing residues.

Unzelman, G.H. and Wolf, C.J., in *Petroleum Processing Handbook*, Bland, W.F. and Davidson, R.L., Eds., McGraw-Hill, New York, 1967, 3–10.

Houdriflow A catalytic petroleum cracking process in which the beads of catalyst move continuously through the reactor and the catalyst regenerator.

Enos, J.L., *Petroleum Progress and Profits*, MIT Press, Cambridge, MA, 1962, 177.
Unzelman, G.H. and Wolf, C.J., in *Petroleum Processing Handbook*, Bland, W.F. and Davidson, R.L., Eds., McGraw-Hill, New York, 1967, 3–9.

Houdriforming A continuous *Catalytic Reforming process for producing aromatic concentrates and high-octane gasoline. It used a fixed bed of a platinum catalyst. Developed in the 1950s by the Houdry Process Corporation.

Unzelman, G.H. and Wolf, C.J., in *Petroleum Processing Handbook*, Bland, W.F. and Davidson, R.L., Eds., McGraw-Hill, New York, 1967, 3–28.
Asinger, F., *Mono-olefins: Chemistry and Technology*, translated by B.J. Hazzard, Pergamon Press, Oxford, U.K., 1968, 389.
Oil Gas J., 1971, **69**(51), 45.

Houdry The first catalytic petroleum cracking process, based on an invention by E.J. Houdry in 1927, which was developed and commercialized by the Houdry Process Corporation. The process was piloted by the Vacuum Oil Company, Paulsboro, NJ, in the early 1930s. The catalyst was contained in a fixed bed. The first successful catalyst was an aluminosilicate mineral. Subsequently, other related catalysts were developed by Houdry in the United States, by IG Farbenindustrie in Germany, and by Imperial Chemical Industries in England. After World War II, the clay-based catalysts were replaced by a variety of synthetic catalysts, many based on aluminosilicates. Later, these too were replaced by zeolites.

U.S. Patents 1,837,963; 1,957,648; 1,957,649.
Houdry, E.J., Burt, W.F., Pew, A.E., and Peters, W.A., *Oil Gas J.*, 1938, **37**(28), 40.
Haensel, V. and Sterba, M.J., *Ind. Eng. Chem.*, 1948, **40**, 1662.
Enos, J.L., *Petroleum Progress and Profits*, MIT Press, Cambridge, MA, 1962, Chap. 4.
The Petroleum Handbook, 6th ed., Elsevier, Amsterdam, the Netherlands, 1963, 284.
Oil Gas J., 2002, **100**(35), 118.
Flank, W.H., Abraham, M.A., and Matthews, M.A., Eds., *Innovations in Industrial and Engineering Chemistry*, 2009, American Chemical Society, Washington, DC, 196X.

Houdry–Litol *See* Litol.

Howard A process for separating lignosulfonates from sulfite liquor from papermaking. The lignosulfonates are precipitated as their calcium salts. The product is used in the manufacture of vanillin.

Sarkanen, K.V. and Ludwig, C.H., *Lignins*, Wiley-Interscience, New York, 1971, 800.

Howden An early *flue-gas desulfurization process using a lime or chalk slurry in wooden grid-packed towers. The calcium sulfate–sulfite waste product was intended for use in cement manufacture, but this was never commercialized. The key to the process was the use of a large excess of calcium sulfate in suspension in the scrubbing circuit, which minimized the deposition of scale on the equipment. The process was developed by Imperial Chemical Industries and James Howden & Company in the 1930s and operated for several years at power stations at Fulham, London, and Tir John, South Wales, being finally abandoned during World War II.

British Patents 420,539; 433,039.
Pearson, J.L., Nonhebel, G., and Ulander, P.H.N., *J. Inst. Fuel*, 1935, **8**, 119.

Katz, M. and Cole, R.J., *Ind. Eng. Chem.*, 1950, **42**, 2266.
Achillades, B., *Chem. Ind. (London)*, 19 Apr 1975, 337.

Howe-Baker A process for removing sulfur compounds from petroleum fractions by treatment with sulfuric acid.

HP *See* High Productivity.

HPC [**H**ot **P**otassium **C**arbonate] A generic name for processes for absorbing acid gases by the use of hot aqueous potassium carbonate. Developed by H.E. Benson and J.H. Field at the U.S. Bureau of Mines in the 1950s. Further developed by other organizations, HPC became the basis for the *Benfield, *CATACARB, and other gas purification processes.

U.S. Patent 2,886,405.

HPH [**H**ydrogen **P**urification in ***H**ydroprocessing] A process for improving the efficiency of *hydroprocessing by increasing the partial pressure of the hydrogen to be recycled. The excess of hydrogen from hydro processing typically contains up to 16 mol% of C_1–C_5 hydrocarbons. These are removed by contacting the hydrogen with a liquid hydrocarbon mixture containing C_5 and higher hydrocarbons (this is similar to the *Mehra (2) process). Developed by the Saudi Arabian Oil Co. from 2001 and demonstrated by Hydrocarbon Technologies in New Jersey in 2002–2003. Offered for license in 2005.

U.S. Patent 6,740,226.

H-Plus A process for extracting aluminum from clays by successive treatment with sulfuric and hydrochloric acids. The product is aluminum trichloride hexahydrate.

HPM [**H**ydrochloric acid hydrogen **P**eroxide **M**ix] *See* RCA.

HPO [**H**ydroxylamine **P**hosphate **O**xime] A process for making caprolactam, an intermediate in the manufacture of polyamides, it differs from related processes, such as *HSO, in producing less of the ammonium sulfate by-product. Developed by DSM Research, the Netherlands, operated by DSM Polymers and Hydrocarbons, and offered for license by Stamicarbon.

Dahlhoff, G., Niederer, J.P.M., and Hölderich, W.P., *Catal. Rev.*, 2001, **43**(4), 389.

HPO-plus A process for making caprolactone, an intermediate in the manufacture of polyamides. An improved version of the cyclohexanone oximation process, which produces only a third as much of the by-product ammonium sulfate. Developed by DSM and operated in Nanjing, China, in 2001. *See also* Altam.

Chem. Week, 2000, **162**(32), 17.

HPPO (1) [**H**ydrogen **P**eroxide **P**ropylene **O**xide] A process for making propylene oxide by epoxidizing propylene with hydrogen peroxide, catalyzed by titanium silicalite. The reaction takes place in methanol solution in a proprietary tubular reactor. In a variation of the process, the propylene is made by catalytically dehydrogenating propane; this hydrogen is then used in the manufacture of the hydrogen peroxide. Developed by Dow Chemical Company and BASF and installed at BASF's Antwerp site in 2008. In a further variation, from Degussa, the hydrogen is converted to hydrogen peroxide by a direct process developed by Headwaters Technologies. The titanium zeolite epoxidation catalyst is regenerated by washing with methanol at 150°C. A joint venture between DegussaHeadwaters and SKC Chemicals Group (Seoul) built a propylene oxide plant in Ulsan, South Korea, which operated from 2008.

World Patent Application WO 04020423A1.
U.S. Patent 6,878,836.
Chem. Eng. (N.Y.), 2004, **111**(11), 16; 2006, **113**(6), 14.
U.S. Patent 7,144,565 (describes the H_2O_2 production used in this).
Chimie Pharma Hebdo, 16 Mar 2009, (458), 4.

HPPO (2) A similar process to HPPO (1), developed by the Instituto de Catalisis y Petroquimica, with Repsol YPF.

Blanco-Brieud, G., Capel-Sanchez, M.C., de Fruitas, M.P., Padilla-Palo, A., Campos-Martin, J.M., and Fierro, L.L.G., *Ind. Eng. Chem. Res.*, 2008, **47**(21), 8011.

HR [**H**eat **R**ecycle] *See* UTI.

HS A family of gas purification processes developed by Union Carbide Corporation, based on the use of proprietary solvents known as UCARSOLs. UCARSOL HS-101 is based on methyl diethanolamine and is used for removing hydrogen sulfide and carbon dioxide from other gases. UCARSOL LH-101 is used in its *Cansolv system for *flue-gas desulfurization.

Kohl, A.L. and Riesenfeld, F.C., *Gas Purification*, 4th ed., Gulf Publishing Co., Houston, TX, 1985, 119.

HSC [**H**igh-conversion **S**oaker **C**racking] A continuous *Visbreaking process, developed and offered by Toyo Engineering Corporation, Japan. Demonstrated from 1988 to 1989 in the Schwedt oil refinery, Germany.

Washimi, K. and Limmer, H., *Hydrocarbon Process. Int. Ed.*, 1989, **68**(9), 69.

HS-FCC [**H**igh-**S**everity **F**luid **C**atalytic **C**racking] An oil refinery conversion process, suitable for an integrated refinery–petrochemical complex. It produces high-octane gasoline and propylene from heavy oils. The main operating regimen is a special downflow reactor system, high reaction temperature, short contact time, and high catalyst–oil ratio. HS-FCC was developed from 1996, jointly by Saudi Aramco, King Fahd University (KFUPM), Nippon Oil, and Japan Cooperation Center, Petroleum (JCCP). Demonstration plants have been built in Saudi Arabia and Japan; commercialization was expected in 2008. Licenses are offered by JCCP and KFUPM.

U.S. Patents 5,904,837; 5,951,850; 6,045,690; 6,146,597; 6,656,346.
Fujyama, Y., Redhwi, H., Aitani, A., Saeed, R., and Dean, C., *Oil Gas J.*, 2005, **103**(36), 62.
Chem. Eng. (N.Y.), 2007, **114**(9), 13.

HSO [**H**ydroxylamine **S**ulfate **O**xime] A process for making caprolactam, the intermediate for making polyamide 6. Developed by DSM Research, the Netherlands, and used by DSM Chemicals at Geleen, the Netherlands, and in Augusta, GA. Offered for license by Stamicarbon. *See also* HPO.

HSR *See* Biobor HSR.

HSSX An ion-exchange process for removing salts from alkanolamines used in oil refineries for removing hydrogen sulfide. Developed by Conoco and commercialized by MPR Services (part of the Tessenderlo Group) from 1991.

HTC (1) [**H**igh-**T**emperature **C**hlorination] A general term for the process for making 1,2-dichloroethane from ethylene and chlorine by processes operated above the boiling point of the product (83°C). *See also* CER.

HTC (2) [**H**ydrogen **T**ransfer **C**atalysis] A catalytic process for reducing aromatic nitro-compounds. Developed by Rohner in 1993.

HTP *See* Hoechst HTP. Also used as an abbreviation for **H**igh-**T**est **P**eroxide (i.e., concentrated hydrogen peroxide).

HTR A catalytic process for making gas from oil, designed for meeting peak loads. Developed by North Western Gas, United Kingdom, in the 1960s.

HTS Carter *See* Carter.

HTU [**H**ydro**T**hermal **U**pgrading] A process for making liquid fuels from biomass. The biomass is made into an aqueous slurry and heated to 300°C under 200 bar. The resulting crude "oil" is catalytically hydrogenated to give a high-quality naphtha or diesel fuel. Developed by Royal Dutch Shell from 1982 to 1993, but not commercialized.

Thoen, J. and Busch, R., in *Biorefineries: Industrial Processes and Products*, Kamm, B., Gruber, P.R., and Kamm, M., Eds., Wiley-VCH, Weinheim, Germany, 2005, **II**, 347.

HTW [**H**igh-**T**emperature **W**inkler] A process for gasifying coal, using oxygen and steam in a fluidized bed. An improved version of the *Winkler process, differing from it by being operated under pressure, at a higher temperature, and with dust recycle. Developed first at Aachen Technical University and then by Reinbraun AG. Two pilot plants and then two large demonstration plants were built in Germany between 1974 and 1997. The overall purpose was to demonstrate how to convert lignite to *syngas for producing chemicals and electricity. Since 1997, the process has been investigated using other fuels such as municipal solid waste and waste plastics.

Schrader, L., Teggers, H., and Theis, K.-A., *Chem. Ing. Tech.*, 1980, **52**, 794.
Theis, K.-A. and Nitschke, E., *Hydrocarbon Process. Int. Ed.*, 1982, **61**(9), 233.
Thomas, G. and Nitschke, E., in *Handbook of Petroleum Refining Processes*, 1st ed., Meyers, R.A., Ed., McGraw-Hill, New York, 1986, 11–3.
Cornils, B., in *Chemicals from Coal: New Processes*, Payne, K.R., Ed., John Wiley & Sons, Chichester, U.K., 1987, 17.
Rezaiyan, J. and Cheremisinoff, N.P., *Gasification Technologies*, Taylor & Francis, Boca Raton, FL, 2005, 54.

Hüls Also called the Hüls flaming arc process. A process for making acetylene by passing hydrocarbon gases through a dc electric arc. Electrode erosion is minimized by rotating the arc roots by swirling the gas. Operated on a large scale by Chemische Werke Hüls in Germany from 1939 to 1993, when a smaller plant was built, and in Romania.

Gladisch, H., *Hydrocarbon Process. Int. Ed.*, 1962, **41**(6), 159.
Miller, S.A., *Acetylene: Its Properties, Manufacture and Uses*, Vol. 1, Ernest Benn, London, U.K., 1965, 394.
Weissermel, K. and Arpe, H.-J., *Industrial Organic Chemistry*, 4th ed., translated by C.R. Hindley and S. Hawkins, Wiley-VCH, Weinheim, Germany, 2003, 95.

Humphrey A catalytic process for hydrogenating rosin acids.

Hunter A process for making titanium metal by reducing titanium tetrachloride with sodium:

$$TiCl_4 + 4Na = Ti + 4NaCl$$

The process is operated in heated batch reactors under an inert atmosphere. Two companies (Deeside Titanium, North Wales, and New Metals Industries, Nihongi, Japan) operate a one-stage process. Reactive Metals Industries Company, Ashtabula, OH, operates a two-stage process: In the first stage, at 230°C, the trichloride and dichloride are formed. In the second, more sodium is added and the temperature is raised to 1000°C. The sponge product is mixed with sodium chloride, which is leached out with dilute hydrochloric acid. Based on the work by M.A. Hunter at Rensselaer Polytechnic Institute, New York, in 1910. *See also* Kroll.

Hunter, M.A., *J. Am. Chem. Soc.*, 1910, **32**, 330.
Eveson, G.F., in *Speciality Inorganic Chemicals*, Thompson, R., Ed., Royal Society of Chemistry, London, U.K., 1981, 231.
Minkler, W.W., *The Production of Titanium, Zirconium and Hafnium*, Metal. Treatises, 1981, 171 (*Chem. Abstr.*, **96**, 184821).

Huntington-Heberlein A lead smelting process, succeeded by the *Dwight–Lloyd process.

Cocks, E.J. and Walters, B., *A History of the Zinc Smelting Industry in Britain*, George G. Harrap, London, U.K., 1968, 59.

Huntsman Also called the Crucible process. A method for purifying blister steel, made by *cementation, by melting it in a closed clay or graphite crucible and pouring the melt into a cast iron mold. Developed by B. Huntsman, a clockmaker, around 1740, initially in Doncaster and later in Sheffield, United Kingdom.

The process was not patented but was operated in secrecy for some years. It was subsequently widely operated in the United Kingdom (where the combination of this with the cementation process was known as "Les Procédés Anglais") and in the United States (from 1830), until the introduction of the *Bessemer process. Thereafter, use of the Huntsman process declined, but it continued in use in some countries until the 1920s. The growth of the town of Sheffield in the nineteenth century was largely due to this development.

Barraclough, K.C., *Steelmaking before Bessemer, Vol. 2. Crucible Steel*, The Metals Society, London, U.K., 1984.

Huron–Dow Also called H–D. A process for making hydrogen peroxide by electrolyzing alkaline water. Developed in the 1980s, initially by Huron and Dow Chemical, subsequently as H-D Technologies, and commercialized in 1991. Intended for on-site production of dilute alkaline hydrogen for direct use in the pulp and paper industries. The electrolytic cell has a porous cathode through which oxygen is passed, a platinum-coated titanium anode, and a separator containing an ion-exchange membrane.

Hess, W.T., in *Kirk-Othmer's Encyclopedia of Chemical Technology*, 4th ed., Vol. 13, John Wiley & Sons, New York, 1991–1998, 978.

Hurter *See* Deacon.

HYAN [**HY**brid **AN**aerobic] A process for treating the supernatant liquor from sewage sludge treatment plants by anaerobic digestion. The methane produced is burnt to provide heat for the treatment plant and to destroy odors. Developed by a Canadian consortium and operated at the Lakeview Water Pollution Control Plant, Mississauga, Ontario, from 1991.

Hybas A biological process for removing nitrogen and organic material from wastewaters. It combines the *Activated Sludge process with the *MBBR process. Developed by AnoxKaldnes AS, Norway, and offered by Veolia Water.

Hybinette A process for extracting nickel from sulfide ores. The nickel ore that occurs in Canada is a mixture of the sulfides of nickel, copper, and iron. Several methods have been used to separate these metals. In the Hybinette process, the ore is first smelted in a blast furnace, yielding a nickel–copper matte (i.e., a mixture of their lower sulfides). This matte is roasted to remove sulfur and leached with dilute sulfuric acid to remove copper. The resulting crude nickel oxide is used as the anode of an electrochemical cell. The nickel deposits on the cathode, which is contained in a cloth bag. Precious metals collect in the anode slime. The process was invented by N.V. Hybinette in 1904 and operated at the Kristiansand refinery, Norway, from 1910.

U.S. Patents 805,555; 805,969.
Archibald, F.R., *J. Met.*, 1962, **14**, 648.
Dennis, W.H., *A Hundred Years of Metallurgy*, Gerald Duckworth, London, U.K., 1963, 204.
Burkin, A.R., *Extractive Metallurgy of Nickel*, John Wiley & Sons, Chichester, U.K., 1987, 121.

Hybrisol [**Hybri**d **sol**vent] A process for removing hydrogen sulfide from natural gases. Based on the *Elf Activated MDEA process but using added methanol to improve the removal of mercaptans and COS.

Lallemand, F. and Minkkinenm, A., *Oil Gas J.*, 2002, **100**(3), 38.

Hycar (1) A *reforming process for making *syngas from light hydrocarbons, differing from the standard process in using two reactors. The second reactor (a convective reformer), operated in parallel with the primary reformer, preheats the feedstock. Developed by Uhde.

Chem. Eng. (N.Y.), 1992, **99**(5), 33.

Hycar (2) A hydrovisbreaking process for oil sands. *See* Visbreaking.

Marcos, F. and Rosa-Brussin, D., *Catal. Rev. Sci. Eng.*, 1995, **37**(1), 3.

Hy-C Cracking A *hydrocracking process. The catalyst is nickel–tungsten on alumina. Developed by Cities Service Research and Development Company and Hydrocarbon Research.

Johnson, A.R. and Rapp, L.M., *Hydrocarbon Process.*, 1964, **43**(5), 165.
Unzelman, G.H. and Wolf, C.J., in *Petroleum Processing Handbook*, Bland, W.F. and Davidson, R.L., Eds., McGraw-Hill, New York, 1967, 3–20.

HyChlor Formerly called DHC. A catalytic *hydrodechlorination process that converts organic chlorides to hydrogen chloride and saturated hydrocarbons. The UOP HyChlor process recovers and recycles both the organic and inorganic reaction products in order to minimize waste disposal requirements and maximize yield. Used for treating wastes from the production of chlorinated petrochemicals such as vinyl chloride. Developed by UOP but not commercialized as of 1992.

Johnson, R.W., Youtsey, K.J., Hifman, L., and Kalnes, T., in *Management of Hazardous and Toxic Wastes in the Process Industries*, Kolaczkowsti, S.T. and Cuttenden, B.D., Eds., Elsevier Applied Science, Amsterdam, the Netherlands, 1987.
Kalnes, T.N. and James, R.B., *Environ. Prog.*, 1988, **7**(3), 185.
Kalnes, T.N. and James, R.B., *Effluent Treatment and Waste Disposal*, Institution of Chemical Engineers, Rugby, U.K., 1990, 179.

Hyco *See* GEMINI Hyco.

HYCO A process for separating *syngas into carbon monoxide and hydrogen, using *PSA. Developed by Mitsui Engineering & Shipbuilding Company from 2000.

Chem. Eng. (N.Y.), Dec 2000, **107**(13), 17.

Hycon (1) [**Hy**drogen **con**version] A hydrogenation process for converting high-sulfur petroleum residues to clean transport fuels. Sulfur is removed as hydrogen sulfide. Two catalytic stages are used. The first achieves demetallization with a regenerable catalyst. The second achieves *HDS, *HDN, and partial *hydrocracking. Developed by Shell and used in a semi-industrial plant at Pernis, the Netherlands, from 1990.

Chem. Eng. (Rugby, Engl.), 1989, Nov, 11.
Chauvel, A., Delmon, B., and Hölderich, W.F., *Appl. Catal. A: Gen.*, 1994, **115**, 178.

Hycon (2) A process for making nitric acid, which can provide a range of concentrations. Developed in 1968 by the Chemical Construction Corporation.

U.S. Patent 3,542,510.
Newman, D.J. and Klein, L.A., *Chem. Eng. Prog.*, 1972, **68**(4), 62.

HyCycle An advanced *hydrocracking process that maximizes the yield of diesel fuel. Several engineering modifications are involved. Also known as HyCycle Unicracking. Developed by UOP in 2000.

Chem. Mark. Rep., 4 Dec 2000.
Eur. Chem. News, 4 Dec 2000, 31.
Ackelson, D., in *Handbook of Petroleum Refining Processes*, 3rd ed., Meyers, R.A., Ed., McGraw-Hill, New York, 2003, 7.29.

Hydeal [**Hy**dro**deal**kylation] A process for making benzene by dealkylating other aromatic hydrocarbons. Generally similar to the *Litol process. Developed in the 1950s by UOP and Ashland Oil Company but abandoned in favor of UOP's *THDA process. *See* Dealkylation.

Hydrocarbon Process., 1963, **42**(11), 180.

HYDECAT [**HY**pochlorite **DE**struction **CAT**alyst] A continuous process for destroying unwanted hypochlorite streams. A heterogeneous catalyst containing nickel converts the hypochlorite ion to chloride ion and oxygen gas:

$$2NaOCl = 2NaCl + O_2$$

Developed by ICI Katalco; first demonstrated at Billingham, England, in 1989; and first operated on a large scale at Yarwun, Queensland, Australia, in 1991. Three versions were subsequently developed: HYDECAT TD (total destruction), HYDECAT PD (partial destruction), and HYDECAT ID (integrated destruction). *See also* ODORGARD.

European Patent 397,342.
WPCT Patent WO92/18235.
Chem. Eng. (Rugby, Engl.), 1991, (493), 22.
Chem. Br., 1991, **27**(5), 402.
Eur. Chem. News, 18 Mar 1991, 24.
King, F. and Hancock, F.E., *Catal. Today*, 1996, **27**(1–2), 203.

Hydrane A coal gasification process in which coal is hydrogenated directly to methane, using hydrogen gas under pressure, preheated to a high temperature. Developed by the U.S. Bureau of Mines; as of 1980, it had not been fully piloted.

Hydrocarbon Process. Int. Ed., 1975, **54**(4), 122.

Hydrar A catalytic process for hydrogenating benzene to cyclohexane. It is conducted in the vapor phase with a fixed-bed reactor. The catalysts are based on platinum and modified by lithium; an alternative nickel-based catalyst is also used. Developed in 1960 by UOP, subsequently renamed HB Unibon.

Hydrocarbon Process., 1963, **42**(11), 181.
Franck, H.-G. and Stadelhofer, J.W., *Industrial Aromatic Chemistry*, Springer-Verlag, Berlin, Germany, 1988, 192.
Bertucco, A. and Vetter, G., Eds., *High Pressure Process Technology*, Elsevier, Amsterdam, the Netherlands, 2001.

Hydrisom A selective hydrogenation process offered by Phillips Petroleum. It is used in Argentina to upgrade C_4 olefins before *alkylation.

Oil Gas J., 1998, **96**(11), 66.

Hydrobon A catalytic petroleum hydrotreating process, developed by UOP and formerly licensed jointly with the Union Oil Company of California under the name *Unifining.

Davidson, R.L., *Pet. Process.*, 1956, **11**(11), 115.
Unzelman, G.H. and Wolf, C.J., in *Petroleum Processing Handbook*, Bland, W.F. and Davidson, R.L., Eds., McGraw-Hill, New York, 1967, 3–44.

Hydrocarb A hypothetical process for making methanol from biomass, which simultaneously makes elemental carbon for storage or use as a fuel. Developed in the 1990s at the Brookhaven National Laboratory, New York. *See also* Carnol, Hynol.

Hydrocarb is also the trade name for a range of ground calcium carbonates made by Omya.

Steinberg, M., Yuang, D., and Borgwardt, R.H., in *Carbon Dioxide Chemistry: Environmental Issues*, Paul, J. and Prodier, C.-H., Eds., Royal Society of Chemistry, Cambridge, U.K., 1994, 189–199.
Steinberg, M. and Dong, Y., *Int. J. Power Energy Syst.*, 2004, **24**(3), 194.

Hydrocol A process for making gasoline from natural gas. Partial combustion with oxygen yields *syngas, which is catalytically converted to gasoline in a fluidized bed containing an iron catalyst. Developed by Hydrocarbon Research and commercialized by Carthage Hydrocol in Brownsville, TX, in 1950.

Keith, P.C., *Am. Gas J.*, 1946, **164**(6), 11 (*Chem. Abstr*, **40**, 4873).
Asinger, F., *Paraffins: Chemistry and Technology*, translated by B.J. Hazzard, Pergamon Press, Oxford, U.K., 1968, 161.

HydroCopper A process for leaching copper from sulfide ores, using dilute aqueous cupric chloride. The copper is precipitated from the leach solution by sodium hydroxide, and the precipitated cuprous oxide is reduced to the metal by hydrogen. An integrated *Chlor-Alkali cell provides the sodium hydroxide and hydrogen. Planned for demonstration in Finland in 2003.

Chem. Eng. (N.Y.), 2002, **109**(6), 21.

hydrocracking A general term for any catalytic hydrogenation process for upgrading heavy petroleum fractions to produce fractions of lower molecular weight, which can be used as fuels. The vaporized feed is mixed with hydrogen at high temperature and pressure and allowed to react in a fixed- or fluidized-catalyst bed. The forerunner of today's hydrocracking processes was the *Bergius process. Those hydrocracking processes with special names that are described in this dictionary are ABC, Chiyoda ABC, CANMET Hydrocracking, HC-Unibon, HDDV, H-G, H-Oil, Hy-C Cracking, Isocracking, Isomax, LC-Fining, Lomax, Unicracking, and Veba.

The Petroleum Handbook, 6th ed., Elsevier, Amsterdam, the Netherlands, 1983, 294.
Froment, G.F. and Marin, G.B., *Hydrocracking: Science and Technology*, Elsevier, Amsterdam, the Netherlands, 1987.
Bridge, A.G. and Hamilton, G.L., in *Handbook of Petroleum Refining Processes*, 3rd ed., Meyers, R.A., Ed., McGraw-Hill, New York, 2003, 14.5.

hydrocyanation A complex process for making adiponitrile by adding hydrogen cyanide to butadiene. The homogeneous catalyst used is $Ni[P(OC_6H_5)_3]_4$. Developed by DuPont in 1971.

Chem. Eng. News, 1971, **49**(17), 30.

hydrodealkylation The use of hydrogen to convert an alkyl benzene (typically toluene) to benzene. The reaction takes place at high temperatures and pressures and may or not be catalyzed. Named processes described in this dictionary are DETOL, HDA, Hydeal, Litol, MBE, THD, and THDA.

Weissermel, K. and Arpe, H.-J., *Industrial Organic Chemistry*, 4th ed., translated by C.R. Hindley and S. Hawkins, Wiley-VCH, Weinheim, Germany, 2003, 331.

hydrodechlorination A general term for processes that convert organic chlorine compounds to hydrogen chloride and saturated hydrocarbons. *See* DCII (2), HyChlor.

hydrodemetallation Often abbreviated to HDM. A general term for processes for removing metal compounds from petroleum fractions by catalytic reduction with hydrogen.

Reynolds, J.G., *Chem. Ind. (London)*, 1991, 570.

hydrodenitrogenation Often abbreviated to HDN; also called hydrodenitrification. An essential stage in the *hydrocracking of petroleum fractions in which organic nitrogen compounds are removed by reaction with hydrogen to produce ammonia. The compounds are removed in order to prevent them from poisoning the catalyst used in the following stage.

Reynolds, J.G., *Chem. Ind. (London)*, 1991, 570.

hydrodesulfurization Often abbreviated to HDS. A general term for processes that convert sulfur compounds in petroleum fractions to hydrogen sulfide and simultaneously convert high molecular weight hydrocarbons to more volatile ones. The process operates in the liquid phase under hydrogen pressure, in a trickle flow reactor containing a heterogeneous catalyst. The catalyst is typically a mixture of cobalt and molybdenum oxides on alumina. They are converted to their sulfides prior to use. More recently, transition metal phosphides have been proposed as catalysts. Such processes with special names that are described in this dictionary are

Alkacid, Alkazid, Autofining, Cycloversion, Diesulforming, GO-fining, Gulfining, Hycon, Hyperforming, IsoTherming, RDS Isomax, RESID-fining, Trickel, Ultrafining, VGO Isomax, and VRDS Isomax.

Reynolds, J.G., *Chem. Ind. (London)*, 1991, 570.
Startsev, A.N., *Catal. Rev. Sci. Eng.*, 1995, **37**(3), 353.
Nagai, M., Fukiage, T., and Kurata, S., *Catal. Today*, 2005, **106**(1–4), 201.

Hydrofining A process for desulfurizing and *hydrotreating a wide range of petroleum fractions. Licensed by the Esso Research and Engineering Company in collaboration with M.W. Kellogg.

Unzelman, G.H. and Wolf, C.J., in *Petroleum Processing Handbook*, Bland, W.F. and Davidson, R.L., Eds., McGraw-Hill, New York, 1967, 3–40.
Hydrocarbon Process. Int. Ed., 1994, **73**(11), 135.

Hydrofinishing The final stage in a number of petroleum hydrorefining processes, commonly used in the manufacture of lubricating oil.

Hydroforming [**Hydro**gen re**forming**] A *Catalytic Reforming process, operated in an atmosphere of hydrogen. The catalyst was molybdena–alumina in a fixed bed. Developed jointly by Standard Oil of New Jersey, Standard Oil of Indiana, and M.W. Kellogg; first operated in 1940 in Texas City. An improved version was developed by Standard Oil of California in 1943. The name has also been used as a general term for the catalytic dehydrogenation of aliphatic hydrocarbons to aromatic hydrocarbons.

Ciapetta, F.G., Dobres, R.M., and Baker, R.W., in *Catalysis*, Vol. 6, Emmett, P.H., Ed., Reinhold, New York, 1958, 495.
Asinger, F., *Mono-Olefins: Chemistry and Technology*, translated by B.J. Hazzard, Pergamon Press, Oxford, U.K., 1968, 388.
Spitz, P.H., *Petrochemicals: The Rise of an Industry*, John Wiley & Sons, New York, 1988, 159.

hydroformylation A general name for processes that extend the chain lengths of aliphatic compounds by the catalytic addition of carbon monoxide and hydrogen. An olefin is thus converted to an aldehyde, whose molecule contains one more carbon atom than the original olefin. The *OXO process was the first such process to be developed. Other related processes described in this dictionary are Aldox and RCH/RP.

Weissermel, K. and Arpe, H.-J., *Industrial Organic Chemistry*, 4th ed., translated by C.R. Hindley and S. Hawkins, Wiley-VCH, Weinheim, Germany, 2003, 127.

hydrogenation, catalytic *See* Catalytic Hydrogenation.

Hydrogen Polybed PSA A version of the *Polybed process, for purifying hydrogen from various industrial processes. Developed by the Union Carbide Corporation in 1975 and now licensed by UOP. Over 660 units were operating or under construction in 2000.

Hydrocarbon Process. Int. Ed., 1996, **75**(4), 120; 2000, **79**(4), 72.

hydroisomerization A general name for processes that isomerize aromatic hydrocarbons, operated in a hydrogen atmosphere. Exemplified by *Hysomer, *Isarom, *Isoforming, *Isomar, *Isopol, and *Octafining. The name is used also for processes that convert less-branched hydrocarbons to more-branched ones to increase their octane numbers.

Ravishankar, R. and Sivasenker, S., *Appl. Catal. A: Gen.*, 1996, **142**(1), 47.

Hydromag A process for removing sulfur dioxide from industrial gas streams by absorption in magnesia. Developed in Japan by Nissan Chemical Industries.

HydroMax A process for making *syngas from wood fines at pulp mills. Developed by Arizona-based Diversified Energy and California-based Evergreen Pulp and supported by the California Energy Commission.

The process uses a molten alloy of iron and tin. The resulting syngas may be used either as fuel or in *Fischer–Tropsch processing.

> *Nitrogen + Syngas*, May/Jun 2007, (287), 11–12.

Hydropol A process for co-hydrogenating *n*-butenes with olefinic gasoline fractions. Developed by the Institut Français du Pétrole as part of its "polymer gasoline" process.

> *Hydrocarbon Process. Int. Ed.*, 1980, **59**(9), 219.

Hydroprocessing A general term for a family of petrochemical processes, including *hydrotreating, *hydrorefining, and *hydrocracking. All are heterogeneous catalytic processes for removing sulfur- and nitrogen-containing compounds from crude oil fractions, and they may or may not induce *cracking. *Oil and Gas Journal* classifies these processes according to the amount of cracking induced: Hydrotreating induces no cracking; hydrorefining converts approximately 10% of the feed to lower molecular weight products; and hydrocracking converts approximately 50% to lower molecular weight products. *See also* *HPH.

> *Oil Gas J.*, 1998, **96**(2), 13.
> Bridge, A.G. and Hamilton, G.L., in *Handbook of Petroleum Refining Processes*, 3rd ed., Meyers, R.A., Ed., McGraw-Hill, New York, 2003, 14.5.

Hydropyrolysis A catalytic process for converting coal into a mixture of liquid and gaseous products. It is operated at high temperatures and pressures, with a residence time in the pyrolysis zone of only a few seconds.

Hydrorefining A general name for petroleum refining processes using hydrogen gas. *See also* Hydroprocessing.

HYDROSULFREEN A process for removing sulfur compounds from the tail gas from the *Claus process. It combines the *Sulfreen process with an upstream hydrolysis–oxidation stage, which improves efficiency and optimizes the emission control. Developed jointly by Lurgi and Société National Elf Aquitaine and installed in 1990 in the Mazovian Refinery and Petrochemical Works, near Warsaw, Poland. *See also* Oxysulfreen.

> *Ind. Miner. (London)*, 1990, (274), 14.

Hydrothermolysis (1) A general name for processes that decompose organic materials in hot-compressed, subcritical water. Typical conditions are 370°C and 220 bar. Catalysts are optional. *See* ISOCONVERSION.

Hydrothermolysis (2) A process for generating hydrogen by decomposing ammonia borane. Developed in the laboratory at Purdue University by A. Varma and colleagues in 2010. It has not been commercialized.

> U.S. Patent Application 2010-0047159.

hydrotreating A general name for a family of catalytic petroleum refining processes that use hydrogen. They may remove organic sulfur, nitrogen, metal, and oxygen compounds or hydrogenate olefins. *See also* Hydroprocessing, OCR.

> *The Petroleum Handbook*, 6th ed., Elsevier, Amsterdam, the Netherlands, 1983, 306.
> *Hydrotreating Catalysts*, Occelli, M.L. and Anthony, R.G., Eds., Elsevier, Amsterdam, the Netherlands, 1989.
> Topsoe, H., Clausen, B.S., and Massoth, F.E., *Hydrotreating Catalysis: Science and Technology*, Springer-Verlag, Berlin, Germany, 1996.
> Bridge, A.G. and Hamilton, G.L., in *Handbook of Petroleum Refining Processes*, 3rd ed., Meyers, R.A., Ed., McGraw-Hill, New York, 2003, 14.5.

HydroZinc A biological process for leaching zinc from heaps of zinc sulfide ore. Naturally occurring bacteria attack the ore, and the leachate is then neutralized and extracted with a solvent. The metal is produced from this liquor by electrowinning. The process was developed by Teck Cominco in British Columbia and piloted between 2000 and 2002.

Eur. Chem. News, 2003, **79**(2067), 29.

HYDROZONE [**HYDR**ogen peroxide **OZONE**] A process for destroying organic wastes in aqueous streams by oxidation with a combination of hydrogen peroxide, ozone, and UV illumination. Developed by EA Technology, Chester, United Kingdom, in the 1980s. *See also* Ultrox.

HYFLEX A process for making fuel gases and liquids by hydrogenating coal. Powdered coal is reacted with hydrogen at high temperature and pressure, in a reactor with a reaction time of only 2 s. The products are mostly methane, ethane, and aromatic hydrocarbons. An advantage over other coal gasification processes is that no heavy tars are produced. The process will also accept other plant-derived raw materials, such as eucalyptus wood, sugar cane, and pineapple wastes. Developed and piloted since 1976 by the Institute of Gas Technology, Chicago.

HYGAS [**HY**dro**GAS**ification] Also known as **IGT HYGAS**, after the Institute of Gas Technology where it was developed. A coal gasification process intended to maximize the production of methane for use as substitute natural gas (SNG). Crushed coal is slurried in light oil and dried in a fluidized bed. It is then hydrogenated by one of the three processes: steam treatment over an iron catalyst (S–I), steam and oxygen treatment, or electrothermal treatment. Laboratory work began in 1946 under sponsorship from AGA. The U.S. Office of Coal Research took over sponsorship in 1964 and continued until 1976. Recent funding has come from the U.S. Department of Energy and the American Gas Association. Several pilot units were operated in the 1960s and 1970s. The electrothermal method was abandoned in 1974 as uneconomic; the steam–oxygen process was still being tested in 1976; and sections of the steam–iron process were being piloted in 1976. One version was piloted in Chicago in 1981, and a commercial plant was designed.

Blair, W.G., Leppin, D., and Lee, A.L., *Methanation of Synthesis Gas*, American Chemical Society, Washington, DC, 1975, 123.

Dainton, A.D., in *Coal and Modern Coal Processing*, Pitt, G.J. and Millward, G.R., Eds., Academic Press, London, U.K., 1979, 150.

Hebden, D. and Stroud, H.J.F., in *Chemistry of Coal Utilization*, 2nd Suppl. Vol., Elliott, M.A., Ed., John Wiley & Sons, New York, 1981, 1679.

HyK A hydrocracking process for middle distillates developed by Axens.

HyL [**H**ojalata **y** **L**amina] A direct reduction ironmaking process in which pellets or lumps of ore are reduced in a batch reactor using a mixture of hydrogen and carbon monoxide. Used in countries that have natural gas and cannot afford to invest in blast furnaces. Developed in the 1950s in Mexico by the Hojalata y Lamina Steel Company (now Hylsa) and the M.W. Kellogg Company and now operated in nine other countries, too. *See* DR.

HyLube A process for making high-quality lubricating oils from recycled oils. The oils are vaporized in a hot hydrogen stream and catalytically converted at 300°C–350°C and 870–1200 psig. Developed by UOP from 2000 and licensed to Puralube, Germany. The first plant was built in Elsteraue, Germany, from 2008. Planned for use in Ramadan City, Egypt, from 2012.

Chem. Eng. (N.Y.), 2007, **114**(5), 22.

Hynol A hypothetical process for making methanol from biomass, especially wood. It has three stages:

1. Gasification with hydrogen-rich gas
2. Steam reforming
3. Methanol synthesis from syngas

Developed in the 1990s at the Brookhaven National Laboratory, New York. *See also* Carnol, Hydrocarb.

U.S. Patent 5,344,848.
Dong, Y. and Borgwardt, R.H., *Energy Fuels*, 1998, **12**(3), 479.
Steinberg, M. and Dong, Y., *Int. J. Power Energy Syst.*, 2004, **24**(3), 195.

Hyperforming A *hydrodesulfurization process in which the catalyst moves by gravity down the reactor and is returned to the top by a solids-conveying technique known as hyperflow. Developed by the Union Oil Company of California in 1952 and first operated commercially in 1955.

Berg, C., *Pet. Refin.*, 1952, **31**(12), 131 (*Chem. Abstr.*, **49**, 15223).
Unzelman, G.H. and Wolf, C.J., in *Petroleum Processing Handbook*, Bland, W.F. and Davidson, R.L., Eds., McGraw-Hill, New York, 1967, 3–33.
Little, D.M., *Catalytic Reforming*, PennWell Publishing, Tulsa, OK, 1985, xv.

Hypersorption A continuous chromatographic separation process using a moving bed. Invented at Oxford in 1919 by F.D. Soddy (famed for his work on isotopes) and developed commercially for petroleum refinery separations by the Union Oil Company of California in 1946. Six plants were built in the late 1940s, using activated carbon as the adsorbent. The process was abandoned because attrition of the bed particles proved uneconomic.

U.S. Patents 1,422,007; 1,422,008.
Berg, C., *Trans. Inst. Chem. Eng.*, 1946, **42**, 665.
Berg, C., *Chem. Eng. Prog.*, 1951, **47**(11), 585.
Yang, R.T., *Gas Separation by Adsorption Processes*, Butterworths, Guildford, U.K., 1987, 217.

Hypol [**Hy**osung **pol**ypropylene] Also called Hypol II and Mitsui Hypol. A process for making polypropylene, generally similar to *Spheripol. Developed by Mitsui Petrochemical Company, Japan, and Hyosung Corporation, a South Korean conglomerate. Operated in Japan and Korea.

Soares, J.B.P. and McKenna, T.F.L., *Polyolefin Reaction Engineering*, Wiley-VCH, Weinheim, Germany, 2012, 126.

Hypotreating A process for desulfurizing and hydrogenating petroleum fractions. Developed by the Houdry Process and Chemical Company.

Unzelman, G.H. and Wolf, C.J., in *Petroleum Processing Handbook*, Bland, W.F. and Davidson, R.L., Eds., McGraw-Hill, New York, 1967, 3–42.

Hypro A process for making hydrogen by catalytically decomposing hydrocarbons to carbon and hydrogen. The carbon is burned to provide the heat for the reaction. Developed by UOP.

Pohlenz, J.B. and Stine, L.O., *Hydrocarbon Process.*, 1962, **41**(5), 191.
Hydrocarbon Process., 1964, **43**(9), 232.
Unzelman, G.H. and Wolf, C., in *Petroleum Processing Handbook*, Bland, W.F. and Davidson, R.L., Eds., McGraw-Hill, New York, 1967, 3–147.

Hy-Pro A process for recovering propane and heavier components from natural gas or refinery off-gases. Licensed by ABB Randall Corporation. A version called **Super Hy-Pro** is designed for removing liquid hydrocarbons from LPG.

Hydrocarbon Process. Int. Ed., 1996, **75**(4), 124; 2000, **79**(4), 91.

HySCOP A catalytic process for making a sweet, extra-light synthetic crude oil from heavy crude oil by hydroconversion. It combines Axens' *HYVAHL hydroprocessing technology with Idemitsu's hydroconversion catalyst technology. The process was designed by Axens and licensed to Idemitsu Kosan in 2007.

Jpn. Chem. Week, 5 Apr 2007, **48**(2411), 12.
Hydrocarbon Process., 2007, **86**(5), 27, 29.

HYSEC A process for purifying hydrogen from coke-oven gas by *PSA, developed by Mitsubishi Kakoki Kaisha and the Kansai Coke & Chemicals Company.

> Suzuki, M., in *Adsorption and Ion Exchange: Fundamentals and Applications*, LeVan, M.D., Ed., American Institute of Chemical Engineers, New York, 1998, 122.

Hysomer [**Hy**droi**somer**ization] A process for converting *n*-pentane and *n*-hexane into branched-chain hydrocarbons. Operated in the vapor phase, in the presence of hydrogen, in a fixed bed of a mordenite catalyst loaded with a platinum. Developed by Shell Oil Company and licensed worldwide through UOP. Used in the *Total Isomerization process. Now called the UOP once-through (OT) zeolitic isomerization process.

> Kouwenhoven, H.W., Langhout, W.C., and Van Zijll, *Chem. Eng. Prog.* 1971, **67**(4), 65.
> Symoniac, M.F., *Hydrocarbon Process. Int. Ed.*, 1980, **59**(5), 110.
> Cusher, N.A., in *Handbook of Petroleum Refining Processes*, 1st ed., Meyers, R.A., Ed., McGraw-Hill, New York, 1986, 5–15.
> Cusher, N.A., in *Handbook of Petroleum Refining Processes*, 3rd ed., Meyers, R.A., Ed., McGraw-Hill, New York, 2003, 9.29.

HySorb XP A design for a process for separating *p*-xylene from its isomers. Prepared in 1997 by three companies with long experience in xylene processing—UOP, Washington Group International, and Niro Processing Technology—with the objective of combining the best features of all their technologies. Never constructed; UOP believes that the process offers no cost or performance advantage versus its own *Parex (1) process.

> Commissaris, S.E., in *Handbook of Petroleum Refining Processes*, 3rd ed., Meyers, R.A., Ed., McGraw-Hill, New York, 2003, 2.49.

HySpec A process for reducing the H_2S content of molten sulfur. Gas–liquid contactors remove most of the H_2S, and then a catalyst is used to decompose residual polysulfides. Developed by Enersul Technologies, Alberta.

Hysulf A process for converting hydrogen sulfide to hydrogen and sulfur. Its two steps are conducted in a polar organic solvent. In the first, the H_2S reacts with a quinone to form sulfur and the corresponding hydroquinone. In the second, the hydroquinone solution is catalytically processed to yield hydrogen and the original quinone. Developed by Marathon Oil, with funding from the Institute of Gas Technology, Chicago.

> *Chem. Eng. (N.Y.)*, 1994, **101**(11), 19.
> Quinlan, M.P., Echterhof, L.W., Leppin, D., and Meyer, H.S., *Oil Gas J.*, 1997, **95**(29), 54.

HyTail A *hydrocracking process for making low-sulfur diesel fuel, developed by Axens.

Hytanol A *peak shaving process for making fuel gas from methanol. Offered by Lurgi.

HyTex [**Hy**drogen **Tex**aco] A process for making pure hydrogen from waste gases in oil refineries. There are three stages. In the first, partial combustion with oxygen in a noncatalytic reactor yields a mixture of carbon monoxide and hydrogen (*syngas). In the second, the shift reaction with steam converts the carbon monoxide to carbon dioxide and more hydrogen. The third stage uses *PSA to separate the hydrogen from the carbon dioxide and various impurities. Developed by Texaco Development Corporation and announced in 1991; the first unit, in Anacortes, WA, was scheduled for completion in 1993.

> *Chem. Eng. News*, 1991, **69**(19), 29.
> *Hydrocarbon Process. Int. Ed.*, 1992, **71**(4), 112.

Hytoray [**Hy**drogenation **Toray**] A process for hydrogenating benzene to cyclohexane. Developed by Toray.

> Weissermel, K. and Arpe, H.-J., *Industrial Organic Chemistry*, 4th ed., translated by C.R. Hindley and S. Hawkins, Wiley-VCH, Weinheim, Germany, 2003, 348.

HYTORT A process for making gaseous and liquid fuels from oil shale. Developed by the Institute of Gas Technology, Chicago, in 1959. The process uses high-pressure hydrogenation, which recovers more of the carbon from shale than does pyrolysis. In 1981, a joint venture of IGT with the Phillips Petroleum Company was formed in order to make a feasibility study.

Hydrocarbon Process. Int. Ed., 1986, **65**(4), 99.

HYVAHL A *hydrotreating process for upgrading petroleum residues. Developed in the 1990s by the Institut Français du Pétrole (now Axens) and Total Oil Company. Three units were operating in 1996. Improved versions are HYVAHL F and HYVAHL S. The latter uses two reactors containing two different catalysts, operated in a swing mode. A mathematical model of the process that takes numerous process variables into account is known as **THERMIDOR** (**THER**mal **M**onitoring for **I**soperformance **D**esulfurization of **O**il **R**esidues). *See also* *HySCOP, *TERVAHL.

Hydrocarbon Process. Int. Ed., 1988, **67**(9), 78.
Appl. Catal. A: Gen., 1993, **102**(2), N19.
Hydrocarbon Process. Int. Ed., 1994, **73**(11), 130; 1996, **75**(11), 130.
Kressmann, S., Morel, F., Harlé, V., and Kasztclan, S., *Catal. Today*, 1998, **43**, 203.
Toulhoat, H., Hubedine, D., Raybaudm P., Guillaume, D., and Kressmann, S., *Catal. Today*, 2005, **109**, 135.

I

Ibuk A process for removing free sulfur and sulfur dioxide from "benzole" (a mixture of aromatic hydrocarbons used as a motor fuel) by passing the vapor through sulfuric acid and then hot aqueous sodium hydroxide. Invented by T. Troniseck and used in Germany in the 1930s.

> British Patent 364,778.
> Claxton, G., *Benzoles: Production and Uses*, National Benzole & Allied Products Assoc., London, U.K., 1961, 417.

ICAR [**I**ntermediate **C**atalytic **A**ccumulation of Ionizing **R**adiation Energy] A hypothetical process for converting nuclear energy to chemical energy. Energy from a nuclear reactor is used to promote the catalytic reforming of methane to *syngas. Proposed by Yu. A. Aristov in 1993. *See also* EVA-ADAM.

> Parmon, V.N., *Catal. Today*, 1997, **35**(1–2), 153.

ICEM *See* DR.

ICG-GUD [**I**ntegrated **C**oal **G**asification–**G**as **Und D**ampf] An integrated coal gasification and power generation system developed by Siemens, Germany. The coal is gasified by exposure to steam and oxygen. The gases are then burned in a gas turbine, producing electric power, and the hot combustion products then generate more power in a steam turbine. Removal of sulfur dioxide, nitrogen oxides, and dusts is easier than in conventional coal-fired power stations. A demonstration plant was scheduled to be built in the Netherlands in 1990.

> U.S. Patent 5,451,160.

IChPW *See* carbonization.

ICI Low-Pressure Methanol A process for making methanol from methane and steam. The methanol is first converted to *syngas by *steam reforming at a relatively low pressure. The syngas is then converted to methanol over a copper-based catalyst:

$$CO + 2H_2 = CH_3OH$$

$$CO_2 + 3H_2 = CH_3OH + H_2O$$

$$CO + H_2O = CO_2 + H_2O$$

Conversion per pass is limited by reaction equilibrium; after cooling to condense the product methanol, the unreacted gas is recycled to the reactor. Developed by Imperial Chemical Industries in the late 1960s, since when it has been the leading process. As of 1991, 41 plants had been commissioned and a further seven were under contract or construction.

> *Hydrocarbon Process. Int. Ed.*, 1991, **70**(3), 164.
> Weissermel, K. and Arpe, H.-J., *Industrial Organic Chemistry*, 4th ed., translated by C.R. Hindley and S. Hawkins, Wiley-VCH, Weinheim, Germany, 2003, 31.

ICI Steam Naphtha Reforming A *steam reforming process, adapted for processing high-boiling naphtha. A highly selective nickel-based catalyst is used. Developed by Imperial Chemical Industries in the 1960s, based on its early experience of the steam reforming of methane. Following the development, more than 400 reformers were built in over 30 countries. When North Sea gas and other natural gas reserves around the world were discovered, the use of naphtha as a feedstock declined; nevertheless, the process is still used in many countries.

Weissermel, K. and Arpe, H.-J., *Industrial Organic Chemistry*, 4th ed., translated by C.R. Hindley and S. Hawkins, Wiley-VCH, Weinheim, Germany, 2003, 20.

ICON [Integrated Chlorination and OxidatioN] An improved version of the *Chloride process for making titanium dioxide pigment. ICON operates at above atmospheric pressure and is claimed to be cheaper than previous designs to build. Chlorine from the oxidation section, under pressure, is introduced directly to the chlorinator. Developed by Tioxide Group (now Huntsman Tioxide) and first operated at its plant at Greatham, United Kingdom, in 1990. A new plant was opened in Greatham in 2002, and another is planned for 2008. The name is also a trade name for a range of insecticides produced by Syngenta.

Chem. Eng. (Rugby, Engl.), 1991, (497), 13.
Ind. Miner. (London), 1991, (286), 17.
Chem. Week, 2002, **164**(48), 49.

IDAS

A process for making isoprene (for the manufacture of a synthetic rubber) by the oxidative dehydrogenation of isopentene. Iodine is the initial reactant; the hydrogen iodide produced in the reaction is reconverted to elemental iodine via nickel iodide.

Dumas, T. and Bulani, W., *Oxidation of Petrochemicals: Chemistry and Technology*, Applied Science Publishers, London, U.K., 1974, 139.

Idemitsu A process for making C_6 to C_{18} α-olefins from ethylene, catalyzed by a modified homogeneous *Ziegler–Natta catalyst containing a zirconium chloride. Developed by the Idemitsu Petroleum Company in 1988.

European Patent 328,728.
Al-Sa'doun, A.W., *Appl. Catal. A: Gen.*, 1993, **105**(1), 3.

IDGCC [Integrated Drying Gasification-Combined Cycle] A coal gasification process developed by Herman Research Pty in Australia for processing very wet coals.

A large demonstration plant was being built in Latrobe Valley, Victoria, from 2011, funded by HRL, Harbin Power, the Federal Government, and the Victorian Government.

Rezaiyan, J. and Cheremisinoff, N.P., *Gasification Technologies*, Taylor & Francis, Boca Raton, FL, 2005, 56.
Chem. Eng. (N.Y.), 2007, **114**(7), 10.

IDR (1) [Isobaric Double Recycle] A process for making urea from ammonia and carbon dioxide, via ammonium carbamate:

$$NH_3CO_2 \rightarrow NH_2COONH_4 \rightarrow NH_2CONH_2$$

Developed by Fertimont (a subsidiary of Montedison) and operated in Italy since 1988 and in China since 1989.

Eur. Chem. News, 1982, **39**(1044), 15.
Zardi, U., *Nitrogen*, 1982, (135), 33.

IDR (2) [Integrated Dry Route] A process for making ceramic-grade uranium dioxide from uranium hexafluoride. Uranium hexafluoride vapor is first reacted with superheated steam to form uranyl fluoride;

this is then reduced with hydrogen to form uranium dioxide. Developed at Springfields, United Kingdom, by British Nuclear Fuels in the 1970s. By 1989, more than 7500 tons of uranium had been processed there in this way. Licenses have been granted to Westinghouse Electric Corporation, United States (now BNFL), and FBFC, France.

ifawol A process for removing phenol from aqueous wastes by solvent extraction into a high-boiling organic solvent mixture made by the *OXO process.

> Martinetz, D., *Chemical Waste Handling and Treatment*, Müller, K.R., Ed., Springer-Verlag, Berlin, Germany, 1986, 182.

I-Forming [**I**sobutane re**Forming**] A process for selectively converting heavy naphtha to isobutane. Developed by UOP.

IFP The **I**nstitut **F**rançais du **P**étrole has developed many processes, but the one most associated with its name is that for removing residual sulfur dioxide and hydrogen sulfide from the tail gases from the *Claus process. This process was invented by P. Renault in 1966. The gases are passed through a solvent, such as tributyl phosphate, at a temperature higher than that of the melting point of sulfur. A soluble catalyst promotes a Claus-type reaction, and the resulting liquid sulfur is separated from the base of the vessel. In 1976, 27 plants were either in operation or in design or construction. *See* Clauspol 1500. The process development and licensing activities of IFP are now conducted by Axens.

> U.S. Patent 3,441,379.
> Barthel, Y., Bistri, Y., Deschamps, A., Renault, P., Simadoux, J.C., and Dutriau, R., *Hydrocarbon Process. Int. Ed.*, 1971, **50**(5), 89.
> Kohl, H.L. and Nielsen, R.B., *Gas Purification*, 5th ed., Gulf Publishing Co., Houston, TX, 1997, 843.

Ifpexol A process for removing water and acid gases from hydrocarbon gas streams by washing with methanol at a very low temperature. Developed by the Institut Français du Pétrole. Nineteen units were operating in 2000.

> *Hydrocarbon Process. Int. Ed.*, 1996, **75**(4), 124; 2000, **79**(4), 76.
> *Chem. Technol. Eur.*, 1996, **3**(3), 10.
> Hampton, P., Darde, T., James, R., and Wines, T.W., *Oil Gas J.*, 2001, **99**(16), 54.

IFP Oxypyrolysis Also called *NGOP. A process for converting natural gas to gasoline, based on the oxidative coupling of methane to ethane in a fixed-bed reactor. Developed in 1991 by the Institut Français du Pétrole.

> Mimoun, H., Robine, A., Bonnaudet, S., and Cameron, C.J., *Appl. Catal. A: Gen.*, 1990, **58**, 269.
> Hutchings, G.H. and Joyner, R.W., *Chem. Ind. (London)*, 1991, (16), 575.
> Raimbault, C. and Cameron, C.J., in *Natural Gas Conversion*, Holmen, A., Jens, K.-J., and Kolboe, S., Eds., Elsevier, Amsterdam, the Netherlands, 1991, 479.

IFP-SABIC *See* Alphabutol.

IFP Stackpol *See* Stackpol.

IFR [**I**ntegral **F**ast **R**eactor] A pyrochemical process for processing the fuel from a fast nuclear reactor. The uranium metal fuel is dissolved in a fused melt of lithium and potassium chlorides and electrolyzed. The uranium deposits on the solid cathode. The transuranic elements and fission products remain in the salt and are incorporated in a zeolite matrix, which is hot-pressed into a ceramic composite waste. Developed by the Argonne National Laboratory from 1963. *See also* PYRO-A and PYRO-B.

IG IG Farbenindustrie in Germany developed many processes before World War II, but the one most associated with its name is probably the *Aldol process for making butadiene for synthetic rubber. The name has been used also for the *Bergius–Pier process.

IGCC [Integrated Gasification-Combined Cycle] A family of coal gasification processes in which the hot gases from the combustion are used to generate electric power, in either a gas turbine or a high-temperature fuel cell. Two commercial-scale plants were built in the United States, with U.S. government funding, in the mid-1990s. The need to remove hydrogen sulfide from the gases without cooling them has led to the development of special high-temperature adsorbents, such as zinc ferrite and zinc titanate.

Chem. Eng. (Rugby, Engl.), Nov 1989, 35.
Woods, M.C., Gangwal, S.K., Harrison, D.P., and Jothimurugesan, K., *Ind. Eng. Chem. Prod. Res. Dev.*, 1991, **30**, 100.
Bissett, L.A. and Strickland, L.D., *Ind. Eng. Chem. Prod. Res. Dev.*, 1991, **30**, 170.
Chem. Eng. (N.Y.), 1992, **99**(1), 39.
Mills, G.A. and Rostrup-Nielsen, J., *Catal. Today*, 1994, **22**(2), 337.
Konttinen, J.T., Zevenhoven, C.A.P., and Hupa, M.M., *Ind. Eng. Chem. Res.*, 1997, **36**(6), 2332.
Jothimurugesan, K. and Gangwal, S.K., *Ind. Eng. Chem. Res.*, 1998, **37**, 1929.
Hess, G., *Chem. Eng. News*, 2006, **84**(3), 23.

IG-Hydrogenation An advanced version of the *Bergius process, in which the initial product of coal gasification is refined by centrifugation and the slurry residue is carbonized.

IGI *See* carbonization.

IG-NUE A coal liquefaction process developed by Bergbau-Forschung in Germany during World War II. Catalytic metal salts were impregnated in, or precipitated on, the coal. A pilot plant was to have been built in Westphalia in 1977.

Alpert, S.B. and Wolk, R.H., *Chemistry of Coal Utilization*, 2nd Suppl. Vol., Elliott, M.A., Ed., John Wiley & Sons, New York, 1981, 1963.

IGT HYGAS *See* HYGAS.

IH² [Integrated Hydropyrolysis and Hydroconversion] A process for converting biomass to hydrocarbon fuels using hydropyrolysis and hydroconversion. Developed by the Gas Technology Institute, Des Plaines, IL. Licensed to CRI, United States, which expected to commercialize it in 2014.

Chem. Eng. (N.Y.), 2012, **119**(9), 17.

ILPM [Improved Low-Pressure Methanol] A process for making methanol from natural gas, which has two stages. The first stage converts the gas to *CRG. The second stage converts this gas to methanol in a Kvaerner Steam Raising Reactor. Developed by Kvaerner Process Technology in 2000 and proposed for use in Germany from 2003.

Chem. Eng. (N.Y.), Oct 2000, **107**(11), 17.

Imatra Also called the solid lime process. A method for desulfurizing steel made by an electric arc process. Additions are made of "burned lime" (calcium oxide), fluorspar (mineral calcium fluoride), and ferrosilicon.

Imhausen A process for making dimethyl terephthalate. *See* Katzschmann.

IMI [Israel Mining Industries] *See* Haifa.

Impact A continuous process for making long-chain polyether polyols as precursors for polyurethanes. Uses a double metal cyanide catalyst.

Developed by Arco from 1996, acquired by Bayer in 2000. A further development was *CAOS.
Chem. Week, 2000, **72**(1910), 41.

Imperial Smelting A process for simultaneously extracting zinc and lead from sulfide ores, developed and commercialized by the Imperial Smelting Corporation at Avonmouth, United Kingdom, after World War II, and now widely used. Based on an invention by L.J. Derham in which the vapors emerging from a reducing kiln are rapidly quenched in a shower of droplets of molten lead. The first trial was made in 1943, but most of the development work was done from 1945 to 1947. Eleven plants were operating in 1973.

> Cocks, E.J. and Walters, B., *A History of the Zinc Smelting Industry in Britain*, George G. Harrap, London, U.K., 1968, 166.

InAlk [**In**direct **Alk**ylation] A process for converting C_3 and C_4 streams in oil refinery to C_6 to C_8 streams for use in gasoline. The catalyst is either a sulfonic acid resin or "solid phosphoric acid." Developed by UOP from 2000.

> *Oil Gas J.*, 2000, **98**(24), 66; 2001, **108**(1), 27.

INCO [**I**nternational **N**ickel **CO**mpany] An electrolytic process for extracting nickel from nickel sulfide matte. The matte is melted and cast into anodes. Electrolysis with an aqueous electrolyte containing sulfate, chloride, and boric acid dissolves the nickel and leaves the sulfur, together with precious metals, as an anode slime. Operated in Manitoba by International Nickel Company of Canada.

> Spence, W.W. and Cook, W.R., *Trans. Can. Inst. Min. Met.*, 1964, **67**, 257 (*Chem. Abstr.*, **62**, 4951).
> Gupta, C.K. and Mukherjee, T.K., *Hydrometallurgy in Extraction Processes*, Vol. 1, CRC Press, Boca Raton, FL, 1990, 24.
> Hill, J., in *Insights into Speciality Inorganic Chemicals*, Thompson, D., Ed., Royal Society of Chemistry, Cambridge, U.K., 1995, 18.

Indirect An alternative name for the *French process for making zinc oxide.

Indmax A process for maximizing the yield of liquefied petroleum gas (LPG) and light olefins in a modified *FCC process. A proprietary catalyst is used, having three different functional components. Developed by Indian Oil Technologies from 2005 and licensed to ABB LUMMUS Global from 2006.

> *Oil Gas J.*, 2005, **103**(36), 54.
> *ICIS Chem. Bus.*, 20 Mar 2006.
> *Chem. Mark. Rep.*, 3 Apr 2006.

Infuse Not a process but a range of olefin block copolymers developed by Dow Chemical Co from 2006. *Insite metallocene catalysts are used in their manufacture. They won the Best Product Innovation Prize in the 2012 ICIS Innovation Awards.

INICHAR *See* carbonization.

INIEX *See* carbonization.

Innovene A proprietary process for polymerizing lower olefins in the gas phase, incorporating condensed phase technology. Various types of catalysts may be used, but the *Insite family of catalysts is now preferred. The catalyst remains in the final product. Developed by BP in the 1990s and first used commercially in 1998. Thirty-five units were operating or in design and construction in 2005. In 2012, it was licensed in 24 countries; 16 plants were operating and 10 in design or construction. The original process is now called Innovene PP. There are now also Innovene S (a slurry process for making high density polyethylene) and Innovene G (a swing gas-phase process for making LLDPE and HDPE).

> Innovene is also the name of the petrochemicals business that BP established in 2004 and sold to INEOS in 2005.
> Soares, J.B.P. and McKenna, T.F.L., *Polyolefin Reaction Engineering*, Wiley-VCH, Weinheim, Germany, 2012, 124.

INOTEC A process for making cast aluminum articles using a proprietary sand-binding medium, based on silicates, which emits no volatile products. Developed by Ashland-Südchemie-Kernfest and announced by Süd-Chemie in 2009. Now operated in Moosburg, Germany, where it is used in casting engine blocks for BMW cars.

Insite Not a process but a range of constrained-geometry metallocene catalysts for polymerizing olefins. Olefin block copolymers made using these catalysts have the trade name Infuse. Developed by J.C. Stevens at the Dow Chemical Company, for which he received several medals.

> *Eur. Chem. News*, 1997, **67**(1770), 28.
> *Chem. Week*, 2006, **168**(22), 10.
> *Chem. Eng. News*, 2006, **84**(39), 18.

Intec A complex electrochemical process for extracting copper and other metals from sulfide ores. Piloted by Intec Ltd on a large scale in Sydney in the 1990s, but not yet commercialized.

> U.S. Patent 5,487,819.

IntegRex A process for making polyethylene terephthalate, developed by Eastman Chemical, announced in 2004. It integrates the PX-PTA process with the PTA-PET process. It uses an aqueous solution of tere- phthalic acid instead of the solid acid. Johnson Matthey has developed a special catalyst for it. The first plant was built in S. Carolina in 2006. There was a patent dispute between Eastman Chemical and Wellman regarding the resin product, trade named ParaStar. The process was sold to DAK Americas in 2011 who licensed it to Alpek in 2013.

> *Chem. Week*, 2006, **168**(33), 51.
> *Chem. Eng. News*, 2007, **85**(6), 12 (shows an aerial photo of the Columbia plant).
> *ICIS Chem. Bus.*, 2007, **2**(50), 7; 2010, **278**(13), 47; 2013, **283**(14), 7.
> *Chem. Eng. (N.Y.)*, 2007, **114**(13), 22.

InTox A process for destroying toxic wastes in aqueous solution by oxidation with oxygen at high tempera- tures and pressures in a pipe reactor. No catalyst is required. The reactions take place at approximately 300°C and 120 atm. Developed by InTox Corporation, United Kingdom, based on a process for extracting aluminum from bauxite developed by Lurgi in the 1960s. *See also* Zimpro.

Iodide *See* van Arkel and de Boer.

Ionex A continuous ion-exchange process that uses a rotating valve to control liquid flows through the bed of ion exchanger. Developed by PuriTech. There are some similarities to UOP's Sorbex process.

> *Chem. Eng. (Rugby, Engl.)*, Nov 2006, (785), 43.

Ionics A *flue-gas desulfurization process using aqueous sodium hydroxide. The resulting sodium sulfate solution is electrolyzed to yield sodium hydroxide, sodium bisulfate, sulfuric acid, oxygen, and hydrogen.

> Speight, J.G., *Gas Processing*, Butterworth Heinemann, Oxford, U.K., 1993, 296.

Ionikylation A process for alkylating C_4 olefins with *iso*butene, using a composite ionic liquid solvent. Developed and commercialized by PetroChina.

Iotech A process for preparing wood for enzymic digestion. The wood is heated with aqueous alkali under pressure and then rapidly decompressed.

> Jurasek, L., *Dev. Ind. Microbiol.*, 1979, **78**, 177.

IPA [InterPass Absorption] Also called double absorption and double catalysis. An improved version of the *Contact process for making sulfuric acid, by which the efficiency of the conversion of sulfur to sulfuric acid is increased from 98% to over 99.5%.

> Phillips, A., in *The Modern Inorganic Chemicals Industry*, Thompson, R., Ed., Royal Society of Chemistry, London, U.K., 1977, 184.

Iron Sponge Also called Dry Box. An obsolete process for removing hydrogen sulfide from gas streams by reaction with iron oxide monohydrate. The ferric sulfide formed is periodically reoxidized to regenerate ferric oxide and elemental sulfur. When this process becomes inefficient because of pore-blockage, the sulfur is either oxidized to sulfur dioxide, for conversion to sulfuric acid, or is extracted with carbon disulfide.

Speight, J.G., *Gas Processing*, Butterworth Heinemann, Oxford, U.K., 1993, 296.
Kohl, H.L. and Nielsen, R.B., *Gas Purification*, 5th ed., Gulf Publishing Co., Houston, TX, 1997, 1298.

ISAL A *hydrotreating process for removing sulfur and nitrogen compounds from petroleum fractions without reducing their octane values. Developed by Intevep SA, the research and technology arm of Venezuela's state petroleum company PDVSA. A proprietary zeolite catalyst first saturates the olefins and then isomerizes them to higher octane-value compounds. Recent developments have been in cooperation with UOP.

Chem. Eng. (N.Y.), 1997, **104**(4), 19.
Nafis, D.A. and Houde, E.J., in *Handbook of Petroleum Refining Processes*, 3rd ed., Meyers, R.A., Ed., McGraw-Hill, New York, 2003, 11.71.

Isarom A catalytic process for isomerizing the xylene isomers, developed by Institut Français du Pétrole. The catalyst is aluminum trifluoride.

Isasmelt A high-intensity smelting process that uses a top-submerged lance, whose tip is submerged in the melt. Air, oxygen, and fuel are fed down the lance, creating a very turbulent environment that promotes very fast reactions among the raw materials. The process was developed in the 1980s by Mount Isa Mines and the CSIRO, Australia, and is now widely used internationally for a variety of ores. It is now owned and licensed by Xstrata Technology. The lance is provided by Ausmelt.

Chem. Eng. (N.Y.), 1990, **97**(4), 55.

ISCOR *See* carbonization.

ISEP [Ion Separation Exchange Process] A continuous ion-exchange process for purifying large volumes of water. The granular ion-exchange material is contained in a rotating carousel. Developed by the Progress Water Technologies Corporation, St. Petersburg, FL, and demonstrated for the removal of ammonium ion from municipal water supplies at the Florida South Water Reclamation facility in 1989, using clinoptilolite as the ion exchanger.

Chem. Eng. (N.Y.), 1989, **96**(10), 197.

Iso-CDW [Isomerization and Catalytic DeWaxing] A general term for dewaxing processes that include these two processes. Exemplified by Isodewaxing, MSDW.

Oil Gas J., 1997, **95**(35), 64.

ISOCONVERSION A process for converting vegetable oils to liquid fuels. Developed by Applied Research Associates and Chevron Lummus Global. The process uses catalytic *hydrothermolysis (1), followed by hydroprocessing and fractionation. Licensed in 2012 in North America to Aemetis Advanced Fuels.

The name is used also in thermodynamics.

U.S. Patent 7,691,159.
Biodiesel Magazine, Sep/Oct 2012.

Isocracking A *hydrocracking process, developed by Chevron and now licensed by Chevron Lummus Global. The catalyst contains a mixture of hydrous oxides for cracking, plus heavy metal sulfides for hydrogenation. First commercialized in 1962 and now widely licensed worldwide. *See also* Isomax.

U.S. Patents 2,944,005; 2,944,006.

Unzelman, G.H. and Wolf, C.J., in *Petroleum Processing Handbook*, Bland, W.F. and Davidson, R.L., Eds., McGraw-Hill, New York, 1967, 3–18.

Weisser, O. and Landa, S., *Sulphide Catalysts: Their Properties and Applications*, Pergamon Press, Oxford, U.K., 1973, 308.

Hydrocarbon Process. Int. Ed., 1994, **73**(11), 122.

Bridge, A.G. and Mukherjee, U.K., in *Handbook of Petroleum Refining Processes*, 3rd ed., Meyers, R.A., Ed., McGraw-Hill, New York, 2003, 7.1.

Isocure A process for making foundry molds developed by Ashland Chemical Company. In 1990, it was announced that a pilot plant was to be built in cooperation with the USSR and that the process had been licensed in China. *See also* Pep Set.

Isodewaxing A *Catalytic Dewaxing process developed by Chevron Research & Technology, now offered by Chevron Lummus Global. It incorporates catalysts that achieve both wax isomerization and shape-selective cracking.

Oil Gas J., 1997, **95**(35), 64.

Bridge, A.G. and Mukherjee, U.K., in *Handbook of Petroleum Refining Processes*, 3rd ed., Meyers, R.A., Ed., McGraw-Hill, New York, 2003, 7.17.

ISOFIN [**ISO**-ole**FIN**s] A catalytic process for making iso-olefins from normal olefins by skeletal isomerization. The principle example converts *n*-butenes to isobutylene, needed as a feedstock for making methyl *t*-butyl ether. Developed by BP Oil Company, Mobil Corporation, and M.W. Kellogg from 1992.

Chem. Week, 1992, **151**(2), 8; 1994, **155**(6), 35.

Hydrocarbon Process. Int. Ed., 1994, **73**(11), 138; 1996, **75**(11), 140.

Isoforming A process for increasing the octane rating of thermally cracked gasolines by catalytic isomerization over silica–alumina. Terminal alkenes are thus converted to nonterminal alkenes. Developed by Standard Oil Company of Indiana in the 1940s.

The name is used also for a thermomechanical process for treating steel.

U.S. Patent 2,410,908.

Weissermel, K. and Arpe, H.-J., *Industrial Organic Chemistry*, 4th ed., translated by C.R. Hindley and S. Hawkins, Wiley-VCH, Weinheim, Germany, 2003, 333.

Iso-Kel [**Iso**merization-**Kel**logg] A fixed-bed, vapor-phase isomerization process for making high-octane gasoline from aliphatic petroleum fractions. The catalyst is platinum on alumina. Developed by M.W. Kellogg.

Unzelman, G.H. and Wolf, C.J., in *Petroleum Processing Handbook*, Bland, W.F. and Davidson, R.L., Eds., McGraw-Hill, New York, 1967, 3–49.

Isolene II A catalytic process for converting ethylbenzene to mixed xylenes. The catalyst is platinum on an acidic support. Developed by Toray Industries, Japan. *See also* Isomar.

Otani, S., *Chem. Eng. (N.Y.)*, 1973, **80**(21), 106.

Isomar [**Isom**erization of **ar**omatics] A catalytic process for isomerizing xylene isomers and ethylbenzene into equilibrium isomer ratios. Usually combined with an isomer separation process, such as *Parex (1). The catalyst is a zeolite-containing alumina catalyst with platinum. Developed by UOP and widely licensed by that company. Isomar was first commercialized in 1968; by 2003, 59 units had been licensed worldwide. *See also* Isolene II.

Hydrocarbon Process. Int. Ed., 1991, **70**(3), 192.

Silady, P.J., in *Handbook of Petroleum Refining Processes*, 3rd ed., Meyers, R.A., Ed., McGraw-Hill, New York, 2003, 2.39.

Isomate A continuous, nonregenerative process for isomerizing C_5 to C_8 normal paraffins, catalyzed by aluminum trichloride and hydrogen chloride. Developed by Standard Oil of Indiana.

Unzelman, G.H. and Wolf, C.J., in *Petroleum Processing Handbook,* Bland, W.F. and Davidson, R.L., Eds., McGraw-Hill, New York, 1967, 3–49.

Asinger, F., *Paraffins: Chemistry and Technology,* translated by B.J. Hazzard, Pergamon Press, Oxford, U.K., 1968, 712.

Isomax Originally a general trade name for a family of petroleum processes developed jointly by UOP and Chevron Research Company, including RCD Isomax, RDS Isomax, VGO Isomax, and VRDS Isomax. The name, later used by UOP, has been applied to the family of UOP *hydrotreating and *Unibon processes. *See also* Unicracking.

Isomerate A continuous hydrocarbon isomerization process for converting pentanes and hexanes to highly branched isomers. Developed by the Pure Oil Company, a division of the Union Oil Company of California. The catalyst, unlike those used in most such processes, does not contain a noble metal.

Unzelman, G.H. and Wolf, C.J., in *Petroleum Processing Handbook*, Bland, W.F. and Davidson, R.L., Eds., McGraw-Hill, New York, 1967, 3–48.

Isomerization The following named isomerization processes are described elsewhere in this dictionary. Many of them take place in the presence of hydrogen and may therefore also be called hydroisomerization processes, but it is not always clear whether the hydrogen is essential, and no distinction is made here between isomerization and hydroisomerization: Anglo-Jersey, Butamer, Butomerate, Catstill, Chevron (1), Cold Acid, Cold Hydrogenation, Hysomer, Isarom, Isarom, Isoforming, Iso-Kel, Isolene II, Isomar, Isomate, Isomerate, Koch, MHTI, MLPI, MVPI, Octafining, Octol, Penex, Pentafining, T2BX, Tatoray, TDP, TIP, and Xylofining.

Unzelman, G.H. and Wolf, C.J., in *Petroleum Processing Handbook*, Bland, W.F. and Davidson, R.L., Eds., McGraw-Hill, New York, 1967, 3–46.

ISOMPLUS A process for isomerizing *n*-butenes to *iso*butene and *n*-pentane to *iso*-amylene, using a ZSM-5 zeolite catalyst. Developed by CD Tech and Lyondell Petrochemical. One unit was operating in 1996. Used in their Channelview refinery, TX.

Eur. Chem. News, 1993, **60**(1597), 28.

Hydrocarbon Process. Int. Ed., 1996, **75**(11), 138.

Chimie Pharma Hebdo, 8 Dec 2008, (447), 10.

Iso-Plus Not a process but a range of zeolite catalysts for making *iso*-olefins by *FCC. Developed by Englehard.

Benton, S., *Oil Gas J.*, 1995, **93**(18), 98.

Iso-Plus Houdriforming A complex petroleum *reforming process, based on the *Houdriforming process. Developed by the Houdry Process and Chemical Company.

Asinger, F., *Mono-olefins: Chemistry and Technology*, translated by B.J. Hazzard, Pergamon Press, Oxford, U.K., 1968, 391.

Unzelman, G.H. and Wolf, C.J., in *Petroleum Processing Handbook*, Bland, W.F. and Davidson. R.L., Eds., McGraw-Hill, New York, 1967, 3–36.

Isopol A *hydroisomerization process for converting 1-butene to 2-butene. Developed by the Institut Français du Pétrole (now Axens).

Hydrocarbon Process. Int. Ed., 1980, **59**(9), 219.

Isopure A process for making high-purity isobutene, developed by the Institut Français du Pétrole (now Axens), it combines reactive distillation with hydroisomerization.

Chim. Hebdo, 25 Oct 1999, (64), 18.

IsoSiv [Isomer separation by molecular **Siev**es] A process for separating linear hydrocarbons from naphtha and kerosene petroleum fractions. IsoSiv operates in the vapor phase and uses a modified 5A zeolite molecular sieve, which selectively adsorbs linear hydrocarbons, excluding branched ones. Developed by Union Carbide Corporation and widely licensed, now by UOP. The first plant was operated in Texas in 1961. More than 45 units were operating in 2003. *See also* Total Isomerization.

> Avery, W.F. and Lee, M.N.Y., *Oil Gas J.*, 1962, **60**(23), 121.
> Symoniak, M.F., *Hydrocarbon Process. Int. Ed.*, 1980, **59**(5), 110.
> Yang, R.T., *Gas Separation by Adsorption Processes*, Butterworths, Guildford, U.K., 1987, 242.
> Cusher, N.A., in *Handbook of Petroleum Refining Processes*, 3rd ed., Meyers, R.A., Ed., McGraw-Hill, New York, 2003, 10.63.

Isosynthesis A version of the *Fischer–Tropsch process developed in Germany during World War II.

> Lane, J.C., *Pet. Refin.*, 1946, **25**(8), 87; (9), 423; (10), 493; (11), 587 (*Chem. Abstr.*, **42**, 9118).

Isotex A process for isomerizing olefins, catalyzed by a novel zeolite, which is intended for making methyl *t*-butyl ether from C_4 hydrocarbons.

> *Chem. Week Int.*, 1995, **156**(8), 9.

IsoTherming A two-stage dehydrosulfurization process. Hydrogen is introduced into the pretreat reactor operating under liquid-full conditions. This partially removes the sulfur. The product, which contains all the necessary hydrogen, passes to the main reactor, where the remaining sulfur is removed. This process removes a bottleneck in the usual HDS process, which is the transfer of hydrogen from the gas to the liquid phase. The process can reduce the sulfur in diesel fuel to <10 ppm and is claimed to have a lower capital cost than the usual process. Developed in 2000 by Process Dynamics, Arizona, and Linde and first operated by Giant Industries in New Mexico. Acquired by DuPont in 2007. Process Dynamics had licensed nine packages to US refiners in 2010.

> U.S. Patent 6,123,835.
> Ito, E. and van Veen, J.A.R., *Catal. Today*, 2006, **116**(4), 451.
> Stanislaus, A., Marafi, A., and Rana, M.S., *Catal. Today*, 2010, **153**(1–2), 39.
> *Chem. Eng. (N.Y.)*, 2007, **114**(10), 13.

Ispra Mark 13A A *flue-gas desulfurization process developed at the Joint Research Centre of the European Community at Ispra, Italy, from 1979. It uses a novel electrochemical method to regenerate the solution used for absorbing the sulfur dioxide. The products are concentrated sulfuric acid and hydrogen. The absorbent is a dilute aqueous solution of sulfuric and hydrobromic acids, containing a small amount of elemental bromine. Sulfur dioxide reacts with the bromine; thus,

$$SO_2 + 2H_2O + Br_2 = H_2SO_4 + 2HBr$$

The bromine is regenerated by electrolysis:

$$2HBr = H_2 + Br_2$$

A large pilot plant was built at the SARAS oil refinery in Sardinia in 1989.

> Langenkamp, H. and Van Velzen, D., *Flue-Gas Desulphurization by the Ispra Mark 13 Process*, The European Commission, Ispra, Italy, 1988.
> Kohl, H.L. and Nielsen, R.B., *Gas Purification*, 5th ed., Gulf Publishing Co., Houston, TX, 1997, 588.

ITmk3 [Mark 3 indicates that this is a third-generation ironmaking process, marks 1 and 2 being the blast furnace and direct reduction] A modification of the *FASTMET process, for making molten iron. Pelleted iron ore fines are reduced with a solid reductant. The iron in the reduced pellets separates as molten metal, uncontaminated by gangue. Developed in 1996 by Midrex Corporation and Kobe Steel. First commercialized in Hoyt Lakes, MN, in 2010.

> *Chem. Eng. (N.Y.)*, 2010, **117**(3), 12.

J

Jacobs–Dorr An alternative name for the *Dorr-Oliver process for making phosphoric acid, adopted after the technology was acquired by the Jacobs Company in 1974. Twenty-seven plants were operating in 1989.

Gard, D.R., in *Encyclopedia of Chemical Processing and Design*, McKetta, J.J. and Cunningham, W.A., Eds., Marcel Dekker, New York, 1990, **35**, 455.

James An early process for making mixed oxygenated organic compounds by the catalytic oxidation of petroleum fractions. The products were aldehydes, alcohols, and carboxylic acids. Developed by J.H. James at the Carnegie Institute of Technology, Pittsburgh, PA.

James, J.H., *Chem. Metall. Eng.*, 1922, **26**(5), 209.
Asinger, F., *Paraffins, Chemistry and Technology*, translated by B.J. Hazzard, Pergamon Press, Oxford, U.K., 1968, 632.

Jarosite [Named after the mineral; first recognized at Jarosa, Spain] A process for removing iron from the leach liquors from hydrometallurgical operations. First used in 1964 in processing zinc sulfate liquors at Asturiana de Zinc, Spain. Also used for recovering zinc from the residues from the electrolytic zinc process. *See also* Goethite, Hematite.

Steinveit, G., in *Advances in Extractive Metallurgy and Refining*, Jones, M.J., Ed., Institution of Mining & Metallurgy, London, U.K., 1972, 521.
Morgan, S.W.K., *Zinc and Its Alloys and Compounds*, Ellis Horwood, Chichester, U.K., 1985, 117.

Jenkins An early liquid-phase thermal cracking process. *See also* Dubbs.

Jet Smelting *See* DR.

JGCC [Japan Gas-Chemical Company] *See* MGCC.

Jindal A direct-reduction ironmaking process using coal as the reductant. Two plants were operating in India in 1997, and two more were under construction. *See* DR.

Jones A regenerative process for making carbon black by pyrolyzing petroleum fractions. The gaseous coproduct can be added to town gas.

British Petroleum Co., *Gas Making and Natural Gas*, British Petroleum Co., London, 1972, 66.

Joosten Also known as the two-shot system. A chemical grouting system for solidifying permeable sandy masses and masonry composed of sandy materials. Successive injections of sodium silicate and calcium chloride solutions are made through a pipe, which terminates in the ground to be hardened. Calcium silicate precipitates and binds the soil particles together. Invented by H. Joosten in 1928. *See also* Siroc.

U.S. Patent 1,827,238.
British Patent 322,182.

JPL Chlorinolysis [Jet Propulsion Laboratory] A process for desulfurizing coal by oxidation with chlorine. The sulfur becomes converted to sulfur monochloride, S_2Cl_2. Developed by the Jet Propulsion Laboratory of the California Institute of Technology from 1976 to 1981.

IEA Coal Research, *The Problems of Sulphur*, Butterworths, London, U.K., 1989, 30.

Juratka A process for forming a protective oxide film on aluminum. *See also* Alzac, metal surface treatment.

K

KA A process for making the explosive RDX by nitrating hexamethylenetetramine in acetic anhydride. Developed in Germany during World War II.

KAAP [Kellogg Advanced Ammonia Process] The first high-pressure process developed for synthesizing ammonia from its elements that does not use an iron-containing catalyst. The reformer gas for this process is provided by the *KRES process. The catalyst was developed by BP; it contains ruthenium supported on carbon. Developed by M.W. Kellogg Company in 1990 and first installed by the Ocelot Ammonia Company (now Pacific Ammonia) at Kitimat, British Columbia, in 1992. By 2005, over 200 large-scale plants had been contracted worldwide.

> *Chem. Week*, 3 Apr 1991, 13.
> *Eur. Chem. News*, 1992, **58**(1524), 42; 1993, **60**(1592), 27.
> *Oil Gas J.*, 1996, **94**(47), 37.

KAAP plus An improved version of *KAAP, announced by Kellogg Brown & Root in 1999. It combines the features of the KBR Advanced Ammonia Process, the KBR Reforming Exchanger System, and the KBR Purifier technology. The catalyst is the same as that used in KAAP. In 2003, more than 200 large-scale plants were operating or had been contracted.

> *Eur. Chem. News*, 1999, **71**(1885), 40.
> *Hydrocarbon Process. Int. Ed.*, 2003, **82**(3), 76.

Kaldnes A variation of the *Activated Sludge process for sewage treatment, in which the biological matter is immobilized within short lengths of plastic pipe. It is very effective for removing nitrogenous compounds. Developed in 1987 by Trondheim Technical University and commercialized by Kaldnes, a Norwegian engineering company. Tested in a full-scale plant in Oslo in 1990.

> *Pollution Prevention*, 1994, **4**(5), 50.

Kaldo [**Kal**ling **Do**mnarvets] An oxygen steelmaking process, first operated in Domnarvets, Sweden, in 1956. The furnace rotates at approximately 30 rpm around an axis tilted 17° to the vertical. Variation in the rate of oxygen supply and speed of rotation permits close control of the steel composition. Invented by B. Kalling. The name is now used as a general name for both ferrous and nonferrous metallurgical processes using rotating furnaces, developed by the Boliden group of companies in Sweden.

Kalling *See* DR.

Kalthydrierung [German, meaning *cold hydrogenation*] A process for selectively hydrogenating "pyrolysis gasoline," a petroleum refining by-product, at temperatures below 100°C. A palladium catalyst is used.

> Krönig, W., *Erdoel Kohle*, 1965, **18**, 432.

Kalunite [From **K**, potassium, and **alunite**, the ore] A process for extracting aluminum from alunite, a naturally occurring basic sulfate of aluminum and potassium having the idealized formula $KAl_3(SO_4)_2(OH)_6$. Based on an invention made by G.S. Tilley in 1924. The ore is first dehydrated at up to 600°C. It is then leached with a solution of sulfuric acid and potassium sulfate. After clarification of the leachate, potassium alum is crystallized out. Hydrothermal treatment of potassium alum precipitates a basic potassium alum, $K_2SO_4 \cdot 3Al_2O_3 \cdot 4SO_3 \cdot 9H_2O$, simultaneously regenerating potassium sulfate and sulfuric acid. Calcination of this potassium alum yields a mixture of alumina and potassium sulfate, which is leached out. Piloted by Kalunite in Salt Lake City, UT, in 1943 but later abandoned. *See also* Alumet.

U.S. Patent 1,591,798.
O'Connor, D.J., *Alumina Extraction from Non-bauxitic Materials*, Aluminium-Verlag, Düsseldorf, Germany, 1988, 196.

Kalvar A reprographic process in which microscopic gas bubbles provide white pigmentation over a dark background. The photosensitive layer is a solid solution of a blowing agent in a resin. Exposure to ultraviolet radiation liberates nitrogen from the blowing agent, and heat softens the resin, permitting the nitrogen to collect in the form of submicron-sized bubbles. The optical resolution of such a process can approach the ideal value, because the bubbles can be as small as the wavelength of light. Commercialized by the Kalvar Corporation, New Orleans, LA, from 1967 to 1970, but later abandoned because the film was found to emit hydrogen chloride, which corroded adjacent materials.

U.S. Patents 2,911,299; 3,032,414.

Kanigen An "electroless" process for plating metals with nickel, that is, a process not using electrolysis. A nickel solution in contact with the metal was reduced with sodium hypophosphite. Developed by the General American Transportation Company. Operated from 1956, the hypophosphite being provided by Albright & Wilson. The process had a "short and unprofitable life" but was used on parts of the Comet airliner.

Podger, H., *Albright & Wilson: The Last 50 Years*, Berwin Books, Studley, U.K., 2002, 29.

Kastone A process for destroying cyanide ion in solution by oxidizing it with a mixture of hydrogen peroxide and formaldehyde. Invented by DuPont in 1970 and licensed to Degussa.

U.S. Patent 3,617,582.
Lawes, B.C., Fournier, C.B., and Mathre, D.B., *Plating*, 1973, **60**, 902, 909.

KATAPAK Not a process, but a range of catalysts and catalyst supports using the principle of static mixing. Developed by Sulzer Chemtech since 1991. Used in *catalytic distillation processes. Not to be confused with *Catapac.

Katasorbon A process for removing carbonyl sulfide and other organic sulfur compounds from *syngas by combined catalysis and adsorption. Offered by Lurgi.

Katasulf A process for removing hydrogen sulfide and ammonia from coke-oven gas, developed by IG Farbenindustrie in Germany in the 1920s. The basic reaction involved is the catalyzed oxidation of hydrogen sulfide to sulfur dioxide and water at approximately 420°C:

$$H_2S + \tfrac{3}{2}O_2 = H_2O + SO_2$$

Various catalysts have been used, including activated carbon, bauxite, and bimetallic oxides. The sulfur dioxide is then absorbed in a solution of ammonium sulfite and bisulfite; acidulation of this yields ammonium sulfate and elemental sulfur.

British Patent 310,063.
U.S. Patents 1,678,630; 1,889,942; 2,152,454.
Bähr, H., *Chem. Fabrik.*, 1938, **11**(1/2), 10.
Kohl, H.L. and Nielsen, R.B., *Gas Purification*, 5th ed., Gulf Publishing Co., Houston, TX, 1997, 1163.

KATOX A wet oxidation process for destroying organic residues in aqueous effluents. Oxidation takes place at the surface of active carbon granules.

Wysocki, G. and Hoeke, B., *Wasser Luft Betrieb.*, 1974, **18**, 311.
Martinez, D., in *Chemical Waste Handling and Treatment*, Muller, K.R., Ed., Springer-Verlag, Berlin, Germany, 1986, 229.

Katsobashvili A low-pressure, catalytic process for the destructive hydrogenation of petroleum residues. Piloted in the USSR in the 1950s.

Weisser, O. and Landa, S., *Sulphide Catalysts: Their Properties and Applications*, Pergamon Press, Oxford, U.K., 1973, 297.

Katzschmann A process for making dimethyl terephthalate. Invented in 1953 by E. Katzschmann at the chemical company Imhausen in Witten, Germany, hence the alternative names Imhausen, Witten, and Hercules–Witten. Katzschmann won the DECHEMA award for this invention in 1964. It proceeds by the concurrent catalyzed oxidation and esterification of *p*-xylene. The catalyst is a soluble cobalt compound. Further improvements have been made by Glitsch Technology (now GTC Technology). World production capacity for this process in 1993 was 3 million tons.

U.S. Patent 2,894,978.
Katzschmann, E., *Chem. Ing. Tech.*, 1966, **38**, 1.
Landau, R. and Saffer, A., *Chem. Eng. Prog.*, 1968, **64**(10), 20.
Weissermel, K. and Arpe, H.-J., *Industrial Organic Chemistry*, 4th ed., translated by C.R. Hindley and S. Hawkins, Wiley-VCH, Weinheim, Germany, 2003, 395.

Kawasaki *See* DR.

Kawasaki Kasei A process for oxidizing naphthalene to naphthaquinone. It is operated in the gas phase at 400°C, using air as the oxidant, and uses vanadium pentoxide on silica as the catalyst. Phthalic acid is a coproduct.

Kaysam A process for making rubber articles from rubber latex by flocculating it with a mixed electrolyte and then casting in a rotating, porous mold.

KBW [**K**oppers **B**abcock & **W**ilcox] A coal gasification process developed jointly by the Koppers Company and Babcock & Wilcox, intended to supply the synthetic fuel industry. The product is a mixture of carbon monoxide and hydrogen. Dry, powdered coal, oxygen, and steam are injected into the reactor. The reaction temperature is sufficiently high that the ash is molten; it runs down the reactor walls, is tapped out as a molten slag, and is quenched in water before disposal. In 1984, seven commercial synfuel projects planned to use this process, but it is not known whether any was commercialized.

Dokuzoguz, H.Z., Kamody, J.F., Michaels, H.J., James, D.E., and Probert, P.B., in *Handbook of Synfuels Technology*, Meyers, R.A., Ed., McGraw-Hill, New York, 1984, 3–87.

Kel-Chlor [**Kel**logg **Chlor**ine] A noncatalytic version of the *Deacon process for making chlorine by oxidizing hydrochloric acid, in which nitrosyl sulfuric acid and nitrosyl chloride are intermediates and concentrated sulfuric acid is used as a dehydrating agent:

$$HCl + NO \cdot HSO_4 = NOCl + H_2SO_4$$

$$HCl + NOCl + O_2 = Cl_2 + H_2O + NO_2$$

Developed by M.W. Kellogg and by DuPont and operated by the latter company in Texas from 1974. *See also* Deacon, MT-Chlor.

Chem. Eng. News, 1969, **47**(19), 14.
van Dijk, C.P. and Schreiner, W.C., *Chem. Eng. Prog.*, 1973, **69**(4), 57.
Bostwick, L.E., *Chem. Eng. (N.Y.)*, 1976, **83**(21), 86.
van Dijk, C.P. and Schreiner, W.C., in *Inorganic Chemicals Handbook*, Vol. 2., McKetta, J.J., Ed., Marcel Dekker, New York, 1993, 759.
Tozuka, Y., in *Science and Technology in Catalysis*, Izumi, Y., Arai, H., and Iwamoto, M., Eds., Elsevier, Amsterdam, the Netherlands, 1994, 44.

Kellogg-Hydrotreating A two-stage hydrogenation process for converting olefins in petroleum fractions to benzene. Developed by the M.W. Kellogg Company.

Griffiths, D.J., James, J.L., and Luntz, D.M., *Erdoel Kohle Erdgas Petrochem.*, 1968, **21**(2), 83.

Kemgas A process for making precipitated calcium carbonate from waste calcium hydroxide from acetylene production and waste carbon dioxide. A polyhydroxy compound such as sorbitol is used as a promoter. Developed and piloted by CalciTech (Switzerland). A pilot plant was built in Leuna, Sachsen-Anhalt, Germany. The company changed its name from Kemgas to CalciTech in 2000 and named its product SCC (Synthetic Calcium Carbonate). The process name Kemgas is no longer used.

U.S. Patents 6,458,335; 7,378,073.
Kunesh, C.J., *Indust. Miner.*, Mar 2010, 79.

Kemicond A process for reducing the volume and odor of sewage sludge. Sulfuric acid is added before dewatering and then hydrogen peroxide. Developed in Sweden by Kemira Kemi AB Kemwater and operated in northern Sweden from 2006.

Chem. Eng. (N.Y.), 2006, **113**(8), 13.

Kenox A wet air oxidation system for destroying municipal and industrial organic wastes. It uses a complex mixing system to accelerate the reactions. The process conditions are pH 4, 250°C, and 50 bar. Invented by R.P. McCorquodale in 1984, developed by Kenox Corporation, Mississauga, Ontario, and demonstrated at a drum recycling plant in Toronto. First commercialized by Leigh Environmental near Birmingham, England, in 1992. *See also* Zimpro.

Canadian Patent 1,224,891.
U.S. Patents 4,604,215; 4,793,919.
Chem. Eng. (Rugby, Engl.), 1991, (508), 16.

Kepro [**Ke**mira **pro**cess] A process for recovering valuable products from municipal sewage sludge. It makes four products: crude iron phosphate, a biofuel, water treatment chemicals, and a carbon source for denitrification in the sewage plant. Developed by Kemira Chemicals in the 1990s and first installed in Helsingborg, Sweden.

Eur. Chem. News (Finland Suppl.), 1997, **68**, 8.

Kerpely An early coal gasification process.

Kesting A process for making chlorine dioxide by reducing sodium chlorate with hydrochloric acid. Chlorine is also produced and the usual equation given is

$$2NaClO_3 + 4HCl = 2ClO_2 + Cl_2 + 2NaCl + 2H_2O$$

However, the process is more complex than this, and the usual molar ratio of chlorine dioxide to chlorine produced is about 1:1. The process is integrated with an electrolytic process for making the sodium chlorate, such that the liquor from the reduction step is recirculated to the electrolytic step. The product gas, a mixture of chlorine with chlorine dioxide, is washed with water, which preferentially dissolves the chlorine dioxide. The resulting solution is used for pulp bleaching.

Invented by E.E. Kesting in the 1940s at Elektrochemische Werke München, Germany, and first operated commercially by the Brown Company in Berlin, New Hampshire. A similar process was patented at about the same time by G.A. Day and E.F. Fenn in the United States, so the process has also been called the Day–Kesting process. Later, the development of the process by H. Fröhler, who used titanium metal for the construction and closed the recycle loops, led to what is now known as the *Munich process.

German Patents 831,542; 841,565; 924,689; 971,285.
U.S. Patents 2,484,402; 2,736,636.
Kesting, E.E., *Das Papier*, 1952, **6**, 155 (*Chem. Abstr.*, **46**, 10557).

Ketazine A process for making hydrazine by oxidizing ammonia with chlorine in the presence of an aliphatic ketone. A ketazine is an intermediate.

Audrieth, L.F. and Ogg, B.A., *The Chemistry of Hydrazine*, John Wiley & Sons, New York, 1951, 115.

Kettle A simple batch process for making soap by boiling fat with an aqueous alkali solution in a "kettle."

Keyes A process for separating water from ethanol using azeotropic distillation with benzene. Invented in 1922 by D.B. Keyes.

U.S. Patents 1,676,735; 1,830,469.
Keyes, D.B., *Ind. Eng. Chem.*, 1929, **21**, 998.

KHD-Contop [From the German company **KHD** Humboldt Wedag and **Con**tinuous **top** blowing] A steel-making process.

Kiener-Goldshöfe A process for pyrolyzing solid wastes, used in Germany.

Martinez, D., in *Chemical Waste Handling and Treatment*, Muller, K.R., Ed., Springer-Verlag, Berlin, Germany, 1986, 148.

Kiflu A process for making sodium hydroxide from sodium chloride using sodium fluorosilicate.

KILnGAS A coal gasification process, conducted in a rotary kiln. Developed by Allis-Chalmers Coal Gas Corporation in 1971 and piloted on a large scale at the Wood River Power Station, IL, from 1980.

Kinglor-Metor A *DR process. Lump iron ore, mixed with coke or coal and lime, is passed through a heated rectangular chamber, where it is reduced to sponge iron. First operated in Italy in 1973 and now operating in Myanmar (formerly Burma). *See* DR.

KiOR A process for converting biomass, such as wood waste, into crude oil using standard petroleum refining equipment. Piloted in Columbus, MS, from 2010. A demonstration plant was completed in 2012.

U.S. Patent Applications 2012 0216448; 2012 0214216.

KIP *See* Steelmaking.

Kiss An obsolete process for extracting silver from its ores. The ores were roasted with sodium chloride, producing silver chloride, and this was leached out with a solution of calcium thiosulfate. The process was replaced by the *Cyanide process.

Kivcet [A Russian acronym meaning "vortex oxygen electric smelting"] A flash smelting process for sulfide ores using oxygen. In 1990, three plants had been built in Russia, one in Italy, and one in Bolivia.

Warner, N.A., in *Oxygen in the Metal and Gaseous Fuel Industries*, Royal Society of Chemistry, London, U.K., 1977, 227.
Morgan, S.W.K., *Zinc and Its Alloys and Compounds*, Ellis Horwood, Chichester, U.K., 1985, 96.
Chem. Eng. (N.Y.), 1990, **97**(4), 57.

Kiviter A process for extracting oil and gas from shale using a vertical kiln. Operated in Estonia by VKG Oil.

Smith, J.W., in *Handbook of Synfuels Technology*, Meyers, R.A., Ed., McGraw-Hill, New York, 1984, 4–175.
Wikipedia, The Free Encyclopedia. Kiviter process. Revised 21 Jun 2013. Retrieved 20 Jul 2013.

Kjellgren A process for making beryllium by reducing beryllium fluoride with magnesium. Invented in 1941 by B.R.F. Kjellgren at the Brush Beryllium Company, OH, and now the principal commercial method for making beryllium metal.

U.S. Patent 2,381,291.
Kjellgren, B.R.F., *Trans. Electrochem. Soc.*, 1948, **93**(4), 122.

Kjellgren–Sawyer *See* Fuse-quench.

KK [**K**unigi and **K**unii] A process for cracking crude petroleum or heavy oil in a fluidized bed using coke as the heat carrier. Developed originally by Kunigi and Kunii, subsequently improved by the Japanese Agency of Industrial Science with five Japanese companies. Piloted between 1979 and 1982.

Hu, Y.C., in *Chemical Processing Handbook*, Marcel Dekker, New York, 1993, 776.

Klobbie A process for recycling mixed plastics into molded products with woodlike characteristics. Mixed plastic waste is granulated and then heated in molds such that the lowest-melting components melt and fuse all the components together. The process is most suitable for large-dimension products such as posts and planks. First developed in the 1970s and patented by E.J.G. Klobbie at Lankhorst Touwfabrieken in the Netherlands; the process was further developed by the Irish Superwood Group in the 1980s.

U.S. Patent 4,187,352.

KLP [Dow **K** Catalyst **L**iquid **P**hase] A selective hydrogenation process for removing acetylenes from crude C_4 hydrocarbons from ethylene cracking, with no loss of butadiene. The catalyst is based on either copper metal or alumina. Developed by Dow Chemical Company from 1983 and first commercialized at its plant in Terneuzen, the Netherlands. Eight units were operating in 2012. The KLP licensing business was sold to UOP in 1991. It is now used in conjunction with the BASF butadiene extraction process that is an extractive distillation process using N-methyl pyrrolidone.

U.S. Patent 4,440,956.
Eur. Chem. News, 1991, 22 Apr, 37.

KM CDR [Probable derivation: **K**ansai **M**itsubishi **C**arbon **D**ioxide **R**ecovery] A process for capturing carbon dioxide from flue gas and using it for enhanced oil recovery. The carbon dioxide is captured in a proprietary hindered amine using a packed column. It is used under supercritical conditions to reduce the viscosity of crude oil and thus enhance its recovery. Developed in 2004 by Mitsubishi Heavy Industries, Kansai Electric Power Co., and Shell EP International. Now operated in India and Bahrain.

Chem. Eng. (N.Y.), 2008, **115**(1), 12.

Knapsack A process for making acrylonitrile from lactonitrile, itself made from acetaldehyde and hydrogen cyanide. Operated in Germany from 1958. Not industrially significant today for the production of acrylonitrile, although part of the process is still used for making lactic acid in Japan.

Dumas, T. and Bulani, W., *Oxidation of Petrochemicals: Chemistry and Technology*, Applied Science Publishers, London, U.K., 1974, 141.
Weissermel, K. and Arpe, H.-J., *Industrial Organic Chemistry*, 4th ed., translated by C.R. Hindley and S. Hawkins, Wiley-VCH, Weinheim, Germany, 2003, 305.

Knauf A process for making gypsum, suitable for use as plaster, from the waste from the *wet process for making phosphoric acid. Developed by Research-Cottrell.

Knietsch An early version of the *Contact process for making sulfuric acid. Developed by R. Knietsch at BASF, Ludwigshafen, Germany.

Knietsch, R., *Ber. Dtsch. Chem. Ges.*, 1901, **34**, 4069.

Knox An early vapor-phase thermal cracking process for refining petroleum.

Koch A family of processes for making polymethyl benzenes by isomerization, alkylation, and disproportionation in the presence of a Friedel Crafts catalyst. Invented in 1968 by the Sun Oil Corporation and developed and commercialized by the Koch Corporation.

U.S. Patent 3,542,890.

Kohloel A coal liquefaction process used in a demonstration plant in Germany between 1981 and 1987. The coal is mixed with a recycled solvent and an iron catalyst and hydrogenated at 300 bar, 470°C.

Wikipedia, The Free Encyclopedia. Coal liquefaction. Revised 17 Apr 2013. Retrieved 20 Jul 2013.

Koho *See* DR.

Kolbel–Rheinpreussen A process for converting *syngas to gasoline. The gas was passed through a suspension of an iron catalyst in an oil. Developed by H. Kolbel at Rheinpreussen, Germany, from 1936 until the 1950s, when it was supplanted by the *Fischer–Tropsch process.

Asinger, F., *Paraffins: Chemistry and Technology*, translated by B.J. Hazzard, Pergamon Press, Oxford, U.K., 1968, 153.

Kolsterising A process for hardening the surfaces of the cutting edges of steel tools. Invented in Holland by B.H. Kolster, whose company, Hardiff BV, was acquired by Bodycote International in 1998.

Kombi [**Kombinations-Verfahren**] A liquid-phase petroleum hydrogenation process that combined *hydrogenation with *hydrorefining. The catalyst contained molybdenum and tungsten on an aluminosilicate. Developed by BASF.

Urban, W., *Erdoel Kohle*, 1955, **8**, 780.

KOMBISORBON A process for removing toxic vapors from waste gases by adsorption. The adsorbant is a composite of activated carbon with an inert material derived from a volcanic rock. Developed by Lurgi Bamag in the 1990s and used first in a sewage sludge incineration plant.

Kompogas A process for making fuel gas from kitchen and farm waste by anaerobic fermentation. The product contains about 60% methane, together with CO_2 and minor gases. Developed by the eponymous Swiss company and used on 20 sites in Europe and Asia by 2001.

Bauer, A., *Nickel*, 2001, **16**(3), 10.

Konox A process for removing hydrogen sulfide from industrial gases by absorption in aqueous sodium ferrate (Na_2FeO_4) solution. The ferrate is reduced to ferrite ($NaFeO_2$), and the sulfide is oxidized to elemental sulfur. The main reactions are

$$4Na_2FeO_4 + 6H_2S = 4NaFeO_2 + 4NaOH + 6S + 4H_2O$$

$$4NaFeO_2 + 4NaOH + 3O_2 = 4Na_2FeO_4 + 2H_2O$$

Developed in Japan in 1975 and licensed by Sankyo Process Services, Kawasaki.

Kasai, T., *Hydrocarbon Process. Int. Ed.*, 1975, **54**(2), 93.
Kohl, H.L. and Nielsen, R.B., *Gas Purification*, 5th ed., Gulf Publishing Co., Houston, TX, 1997, 851.

Kontisorbon A process for removing and recovering soluble solvents from water. Developed and offered by Lurgi.

Koppers *See* Heinrich Koppers.

Koppers CAS [**C**yanogen, **A**mmonia, **S**ulfur] A scrubbing process for removing hydrogen cyanide, ammonia, and hydrogen sulfide from coal gas using a polythionate solution. Described in 1945 but probably never commercialized.

> Kohl, H.L. and Nielsen, R.B., *Gas Purification*, 5th ed., Gulf Publishing Co., Houston, TX, 1997, 735.

Koppers Hasche A cyclic process for converting methane to *syngas by partial oxidation over an alumina catalyst:

$$2CH_4 + O_2 = 2CO + 4H_2$$

Operated in a pair of horizontal catalyst chambers that alternate their functions at 1 min intervals.

Koppers Kontalyt A gas-making process.

> British Petroleum Co., *Gas Making and Natural Gas*, British Petroleum Co., London, U.K., 1972, 113.

Koppers Phenolate A process for removing hydrogen sulfide from coal gas by absorption in aqueous sodium phenolate. Invented in 1931 by J.A. Shaw at the Koppers Company, Pittsburgh, PA, but possibly never used.

> U.S. Patent 2,028,124.

Koppers–Totzek A coal gasification process using an entrained bed. The coal is finely ground and injected in a jet of steam and oxygen into a circular vessel maintained at 1500°C. The reaction is complete within 1 s. The ash is removed as a molten slag. The process was invented by F. Totzek at Heinrich Koppers, Essen, Germany, and further developed by Koppers Company in Louisiana, MO, under contract with the U.S. Bureau of Mines. The first commercial operation was at Oulu, Finland, in 1952; by 1979, 53 units had been built. Most of the plants are operated to produce a hydrogen-rich gas for use in ammonia synthesis. Developed by Lurgi. *See also* PRENFLO.

> Totzek, F., *Chem. Eng. Prog.*, 1954, **50**(4), 182.
> Dainton, A.D., in *Coal and Modern Coal Processing*, Pitt, G.J. and Millward, G.R., Eds., Academic Press, London, U.K., 1979, 137.
> Firnhaber, B. and Wetzel, R., in *Coal Chem 2000*, Institution of Chemical Engineers, Rugby, U.K., 1980, K1.
> Cornils, B., in *Chemicals from Coal: New Processes*, Payne, K.R., Ed., John Wiley & Sons, Chichester, U.K., 1987, 13.

Korte A process for retting flax or hemp by treatment with hydrochloric acid, then with hypochlorous acid, and then neutralizing.

Kossuth An electrochemical process for extracting bromine from brines. The cell had bipolar electrodes and no diaphragm. It was developed in Germany in 1897 but abandoned in favor of the *Kubierschky process. *See also* Wunsche.

> German Patent 103,644.
> Yaron, F., *Bromine and Its Compounds*, Jolles, Z.E., Ed., Ernest Benn, London, U.K., 1966, 16.

KPA A process for making phosphoric acid from low-grade ores, developed by the Occidental Chemical Company.

> *Eur. Chem. News*, 1983, **41**(1110), 19.

KPEG [**K** (potassium) **P**oly**E**thylene **G**lycol] A process for destroying polychlorinated biphenyls in contaminated soil by heating to 150°C, under pressure, with potassium hydroxide, a polyethylene glycol, and a sulfoxide. Based on an invention made at the Franklin Institute, Philadelphia, in which metallic sodium was used. Later developed by Galson Research Corporation, New York, and first demonstrated in 1988. *See also* CDP.

U.S. Patents 4,337,368; 4,447,541; 4,574,013.
De Fillipis, P., Scarsella, M., and Pochetti, F., *Ind. Eng. Chem. Res.*, 1999, **38**(2), 380–384.

K-Process [**K**alocsai or **K**aljas] A process for extracting gold from ores, concentrates, tailings, and scrap by means of a proprietary solution containing a bromide and an oxidizing agent. Invented in 1983 by G.I.Z. Kalocsai and developed by Kaljas Pty, Australia. A pilot plant was under construction in 1987.

PCT Patent WO 85/00384.
German Patent 3,424,460.

Kraft [German, meaning *strength*] An alkaline papermaking process, also known as sulfate pulping. Wood chips are digested in an aqueous solution of sodium sulfate, which becomes reduced to sodium sulfide by the organic matter, and sodium hydroxide. The overall chemical process is the attack of the bisulfide ion on the lignin molecule, depolymerizing it, with the formation of lignosulfonates. Originally developed by C.F. Dahl in Danzig in the 1870s.

Sawyer, F.G., Beals, C.T., and Neubauer, A.W., in *Modern Chemical Processes*, Vol. 2, Reinhold Publishing, New York, 1952, 267.
Grant, J., *Cellulose Pulp and Allied Products*, Leonard Hill, London, U.K., 1958, Chap. 9.
Chem. Eng. News, 1980, **58**(21), 26.

Kramfors A two-stage variation on the *Sulfite papermaking process. The first stage uses slightly alkaline sodium sulfite; the second uses acid calcium bisulfite. *See also* Stora.

Higham, R.R.A., *A Handbook of Papermaking*, Business Books, London, U.K., 1963, 261.

Kranz MWS A *flue-gas desulfurization system based on activated carbon. One carbon bed removes most of the sulfur dioxide. Ammonia is then injected for the *SCR process to occur in the second bed, which also removes the residual sulfur dioxide. The carbon is regenerated off-site. Developed by Krantz & Company, Germany. Three plants were operating in Germany in 1986.

KRES [**K**ellogg **R**eforming **E**xchanger **S**ystem] A *reforming process for providing *syngas to the *KAAP process. Two plants were operating in 2005.

Eur. Chem. News, 1993, **60**(1592), 27.
Oil Gas J., 1996, **94**(47), 37.

Kroll (1) A process for making a metal by reducing its halide with another metal. Thus, titanium is prepared by reducing titanium tetrachloride with magnesium:

$$TiCl_4 + 2Mg = Ti + 2MgCl_2$$

Also used commercially for making tantalum, niobium, and zirconium. The reduction takes place in a batch reactor under an inert gas atmosphere. Invented by W.J. Kroll in Luxembourg in 1937, first commercialized by DuPont in 1948, and now widely used. *See also* Hunter.

U.S. Patent 2,205,854.
Kroll, W.J., *Trans. Electrochem. Soc.*, 1940, **78**, 35.
McQuillan, A.D. and McQuillan, M.K., *Titanium*, Butterworths, Guildford, U.K., 1956, 57.

Kroll (2) *See* Abgas-Turbo-Wascher von Kroll.

Kroy A variation on the *Hercosett process for making wool fire resistant. The chlorination step is effected by hypochlorous acid. Invented in Toronto in 1975 by F. Mains of Kroy Unshrinkable Wools.

British Patent 1,524,392.

Krupp-Kohlechemie A process for making hard paraffin wax from *water gas by a variant of the *Fischer–Tropsch process. The products were called "Ruhrwachse." Developed by Ruhr Chemie and Lurgi Ges. für Warmetechnie.

Ziesecke, K.H., *Fette, Seifen, Astrichm.*, 1957, **59**(6), 409.
Asinger, F., *Paraffins: Chemistry and Technology*, translated by B.J. Hazzard, Pergamon Press, Oxford, U.K., 1968, 175.

Krupp–Koppers (1) A process for separating *p*-xylene from its isomers by crystallization. Eight plants were operating in 1979.

Hydrocarbon Process. Int. Ed., 1979, **58**(11), 253.

Krupp–Koppers (2) A process for separating butane and butene isomers from their mixtures by extractive distillation. The added solvent (Butenex) is a morpholine derivative, possibly N-formyl morpholine.

Krupp–Lurgi *See* carbonization.

Krupp–Renn *See* DR.

Krupp sponge iron *See* DR.

Krutzsch A vapor-phase process for making hydrogen peroxide from a mixture of hydrogen and oxygen, saturated with water vapor, in a silent electric discharge. Invented and developed by J. Krutzsch at the Elektrochemische Werke München from 1931 to 1944. The electrodes were silica coated with aluminum. The electric discharge was at 12,000 V, 9,500 Hz. The product was an aqueous solution containing 10% hydrogen peroxide. The process was not commercialized; the *Pietzsch and Adolph process continued in use at Munich until the introduction of the *AO process.

British Patent 453,458.
French Patent 790,916.
Wood, W.S., *Hydrogen Peroxide*, Royal Institute of Chemistry Lectures, London, U.K., 1954, 11.
Schumb, W.C., Satterfield, C.N., and Wentworth, R.L., *Hydrogen Peroxide*, Reinhold Publishing, New York, 1955, 52.

KRW [**K**ellogg–**R**ust–**W**estinghouse] A fluidized-bed process for gasifying coal. A mixture of ground coal and limestone is fed, together with air and steam, into a pressure vessel. The sulfur in the coal forms calcium sulfide, which is oxidized to calcium sulfate in another vessel. Initially developed by the Westinghouse Corporation. A large demonstration plant, sponsored by the U.S. Department of Energy, was operated from 1998 at the Pinon Pine Power Plant in Nevada but suffered numerous technical problems.

Rezaiyan, J. and Cheremisinoff, N.P., *Gasification Technologies*, Taylor & Francis, Boca Raton, FL, 2005, 105.

Kryoclean A process for removing volatile organic compounds from effluent gas streams by low-temperature condensation. The refrigerant is liquid nitrogen, used subsequently in various ways. Developed by BOC in the 1990s. A simplified version of the process was announced in 1997.

Chem. Eng. (Rugby, Engl.), 1996, (614), 23.

Kryosol An adsorptive process for purifying methane from landfill gas. Operated at high pressure. The overall methane recovery is 90%–95%.

Kumar, R. and Van Sloun, J.K., *Chem. Eng. Prog.*, 1989, **85**(1), 36.

K-T A coal gasification process, used as a source of synthesis gas for making ammonia.

Chemistry of Coal Utilization, Elliott, M.A., Ed., John Wiley & Sons, New York, 1981, 1759.

Kubierschky A process for extracting bromine from brines. Chlorine gas is passed in, and the liberated bromine is removed by steaming out.

Yarron, F., in *Bromine and Its Compounds*, Yarron, F., Joller, L.E., Ed., Ernest Benn, London, U.K., 1966, 17.

Kubota A process for treating municipal wastes, incorporating a membrane through which the liquor is recycled to a bioreactor. Eight plants were operating in 1996. Developed in Japan.

Brindle, K. and Stephenson, T., *Water Waste Treat.*, 1996, **12**(39), 18.

Kurabo A *flue-gas desulfurization process involving scrubbing with an acidic solution of ammonium sulfate. The product is gypsum. Used in Japan.

Kohl, H.L. and Nielsen, R.B., *Gas Purification*, 5th ed., Gulf Publishing Co., Houston, TX, 1997, 581.

KURASEP [KURAray **SEP**aration**]** A process for separating nitrogen from air by a variant of the *PSA process using a carbon molecular sieve as the adsorbent. Developed by Kuraray Chemical Company.

Suzuki, M., in *Adsorption and Ion Exchange: Fundamentals and Applications*, LeVan, M.D., Ed., American Institute of Chemical Engineers, New York, 1998, 120.

Kureha A process for making di*iso*propyl naphthalene mixtures from naphthalene and propylene by trans-alkylation. It operates at 200°C using a silica–alumina catalyst. Operated in 1988 at the Rutgerswerke plant in Duisburg-Meiderich, Germany. The name has also been used for a process for making acetylene from petroleum.

Franck, H.-G. and Stadelhofer, J.W., *Industrial Aromatic Chemistry*, Springer-Verlag, Berlin, Germany, 1988, 330.

Kureha/Union Carbide A process for cracking crude oil to olefins and aromatic hydrocarbons using steam superheated to 2000°C. The reaction time is only 15–20 ms.

Kurtz A process for making acrylonitrile by reacting hydrogen cyanide with acetylene in the presence of aqueous cuprous chloride. Invented by P. Kurtz at IG Farbenindustrie in the 1940s. The process was widely used but, by 1970, had been abandoned in the United States in favor of the *Ammoxidation processes.

German Patent 728,767.
Kurtz, P., *Pet. Refin.*, 1953, **32**(11), 142.

Kvaerner The engineering company Kvaerner ASA became Aker Kvaerner ASA in 2004. It engineered the *LP OXO process for making *OXO alcohols using technology from Davy Process Technology, and Dow Chemical Co. Plants have been built in several countries.

Chem. Mark. Rep., 1996, **250**(22), 9.

KVT-Sulfox NK A process for removing sulfur compounds from gas streams by two-bed catalytic oxidation. The first bed converts CS_2 and H_2S to SO_2 and much of the SO_2 to SO_3. The second bed converts the residual SO_2 to SO_3. Four plants were operating in 2000.

Hydrocarbon Process. Int. Ed., 2000, **79**(4), 76.

L

LABSORB A process for removing residual sulfur dioxide from the tail gases from the *Claus process by reversible adsorption. Developed by Belco Technologies Corporation, acquired by DuPont in 2006.

Sulphur, Jul–Aug 1999 (263), 53–54, 56.
Hydrocarbon Process. Int. Ed., 2002, **81**(6), 39.

LAC [Linde Ammonia Concept] A simplified process for making ammonia from light hydrocarbons. It comprises a hydrogen generator, using *steam reforming, a cryogenic nitrogen separation unit, and an ammonia-synthesis loop. The first plant was built in India and the later ones in Australia and China.

Nitrogen, Mar–Apr 1994.

Lacell A zinc extraction process in which zinc sulfide is converted to zinc chloride and molten sulfur and the molten zinc chloride is electrolyzed.

Lachmann An early process for refining gasoline by treatment with aqueous zinc chloride.

Lacy–Keller A process for removing hydrogen sulfide and mercaptans from natural gas by absorption in a proprietary solution. Elemental sulfur precipitates as a colloid and is separated from the solution by means of an electrolytic flotation cell. The process does not remove carbon dioxide.

Kohl, A.L. and Riesenfeld, F.C., *Gas Purification*, 4th ed., Gulf Publishing, Houston, TX, 1985, 546.

La–Mar [Named after the inventors R. J. **La**gow and J. L. **Mar**grave] A process for fluorinating organic compounds, using fluorine gas at low partial pressures. Commercialized by the 3M company.

Bedford, C.T., Blair, D., and Stevenson, D.E., *Nature (London)*, 1977, **267**, 35.
Lagow, R.J. and Margrave, J.L., *Prog. Inorg. Chem.*, 1979, **26**, 161.
Lagow, R.J., *J. Fluorine Chem.*, 1986, **33**, 321.

Lampblack The original process for making carbon black by the incomplete combustion of oils. Largely superseded by the *Acetylene Black, *Channel Black, *Furnace Black, and *thermal processes, but still in use in 2005.

Kühner, G. and Voll, M., in *Carbon Black Science and Technology*, Donnet, J.-B., Bansai, R.C., and Wang, M.-J., Eds., Marcel Dekker, New York, 1993, 54.
Buxbaum, G. and Pfaff, G., *Industrial Inorganic Pigments*, 3rd ed., VCH, Weinheim, Germany, 2005, 175.

Lane A process for making hydrogen by passing steam over sponge iron at approximately 650°C. The iron becomes converted to magnetite.

LANFILGAS An integrated process for dealing with municipal waste, which stabilizes the solid residue and generates methane by bacteriological inoculation. Developed by the Institute of Gas Technology, Chicago.

Lanxide A process for making composites of metals with oxides. A molten metal reacts with an adjacent oxidant and is progressively drawn through its own oxidation product so as to yield a ceramic–metal composite. Fibers or other reinforcing materials can be placed in the path of the oxidation reaction and so be incorporated into the final product. The Lanxide Corporation was founded in 1983 in Newark, DE, to exploit this invention. In 1990, it formed a joint venture with DuPont to make electronic components by this process. Variations are directed metal oxidation (DIMOX), for making ceramic–metal composites, and pressureless infiltration by metal (PRIMEX), for making metal matrix composites.

Newkirk, M.S., Urquart, A.W., and Zwicker, H.R., *J. Mater. Res.*, 1986, **1**(1), 81.
Chiang, Y.-M., Haggerty, J.S., Messner, R.P., and Demetry, C., *Am. Ceram. Soc. Bull.*, 1989, **68**(2), 423.

Lanza Tech A process for converting waste gases containing CO and CO_2 into fuels and useful chemicals. The gases may originate in the steel manufacturing, oil refining, and chemical production industries, as well as by gasification of forestry and agricultural residues, municipal waste, and coal. The gases are passed into a special column containing a suspension of proprietary microorganisms. The preferred microorganisms are the Clostridium species. A variety of common organic chemicals have been produced. The company was founded in 2005 in New Zealand where the process was piloted. It is now based in Chicago and has a demonstration plant in Shanghai.

U.S. Patents 8,178,330; 8,263,372.

LAR [**L**ow **A**ir **R**atio] A process for oxidizing *o*-xylene or naphthalene to phthalic anhydride, using a titania–vanadia catalyst containing molybdenum. Developed by Alusuisse Italia in the 1980s. A plant was operated at Valdarno, Italy, in 1984.

Verde, L. and Neri, A., *Hydrocarbon Process. Int. Ed.*, 1984, **11**, 83.

LARAN [**L**inde **A**nae**R**obic meth**AN**e] An anaerobic process for treating industrial wastewaters, generating methane for use as fuel. The process uses a fixed-bed loop reactor. Developed by Linde in the early 1980s and first commercialized in 1987.

European Patent 161,469.

Larkin An early direct process for reducing iron ore to iron metal by heating with carbon. *See also* DR.

Laux Also called the aniline process. A process for making red iron-oxide pigment in the course of making aniline by reducing nitrobenzene with scrap iron:

$$4C_6H_5NO_2 + 9Fe + 4H_2O = 4C_6H_5NH_2 + 3Fe_3O_4$$

Invented in 1926 by J. Laux at Bayer and used commercially thereafter. The Laux process was an improvement on the *Béchamp process.

Kirk-Othmer's Encyclopedia of Chemical Technology, 4th ed., Vol. 19, John Wiley & Sons, New York, 1991–1998, 24.
Buxbaum, G. and Pfaff, G., *Industrial Inorganic Pigments*, 3rd ed., VCH, Weinheim, Germany, 2005, 106.

Laxal *See* metal surface treatment.

LBE [**L**ance **B**ubbling **E**quilibrium] A steelmaking process in which nitrogen or argon is injected at the base of the furnace and oxygen is introduced at the top. Introduced in the 1970s. *See* Steelmaking.

LC[Lummus Crest] A coal gasification process developed by ABB Lummus Crest.

LCA [**L**eading **C**oncept for **A**mmonia, formerly **L**ow-**C**ost **A**mmonia] A process for making ammonia from air and natural gas. Essentially a simplified form of the standard ammonia-synthesis process, more suitable for smaller plants. Thermal economies are achieved in the *steam reforming section. Developed by ICI from 1985 to 1988. Two units began operating at the ICI plant in Severnside, United Kingdom, in 1988. The first non-ICI installation was designed by KTI for Mississippi Chemicals, Yazoo City, MS. The name appears no longer to be used.

Chem. Eng. (N.Y.), 1989, **96**(7), 43.
Chem. Eng. (Rugby, Engl.), 1990, (471), 21.
Hydrocarbon Process. Int. Ed., 1991, **70**(3), 134.

LC-Fining [**L**ummus **C**ities re**Fining**] A *hydrocracking process using an ebullated catalyst bed. Developed by Lummus Crest and Cities Service Research and Development Company since the 1960s, initially for upgrading bitumen from tar sands. Three units were operating in 1996. The process is now licensed through Chevron Lummus Global.

Van Driessen, R.P., Caspers, J., Campbell, A.R., and Lunin, G., *Hydrocarbon Process. Int. Ed.*, 1979, **58**(5), 107.
Chillingworth, R.S., Potts, J.D., Hastings, K.E., and Scott, C.E., in *Handbook of Synfuels Technology*, Meyers, R.A., Ed., McGraw-Hill, New York, 1984, 6–47.
Gupta, A., in *Handbook of Petroleum Refining Processes*, 3rd ed., Meyers, R.A., Ed., McGraw-Hill, New York, 2003, 8.67.

LCM [**L**eading **C**oncept for **M**ethanol, formerly **L**ow-**C**ost **M**ethanol] A process for making methanol, combining the *ICI Low-Pressure Methanol process with the *steam reforming section of the *LCA ammonia process. Developed by ICI in 1990 and piloted with BHP in Melbourne, Australia, from 1994. Envisaged for floating factories in offshore gas fields. The name appears no longer to be used.

Chem. Br., 1991, **27**(12), 1100.

L-D [**L**inz, Austria, and either **D**usenverfahren (nozzle process) or **D**onawitz, the other Austrian town where it was developed] A basic steelmaking process in which oxygen is used instead of air to remove most of the carbon from the molten pig iron. Developed in Austria by the Vereinigte Österreichische Eisen und Stahlwerke of Linz and Österreichisch-Alpine of Donawitz in the 1930s and 1940s; commercialized in 1952; and now widely adopted. The furnace is essentially a Bessemer converter, modified with a water-cooled oxygen injector. *See also* Bessemer.

Dennis, W.H., *A Hundred Years of Metallurgy*, Gerald Duckworth, London, 1963, 121.
Finniston, M., *Chem. Ind. (London)*, 1976, 19 June, 501.

LD/AC [Named after *L-D, **A**RBED (a company in Luxembourg), and **C**NRM (a Belgian metallurgical research laboratory)] Also called the *OCP process. A version of the *L-D steelmaking process in which powdered lime is introduced with the oxygen in order to remove phosphorus from the steel. *See also* OLP.

Dennis, W.H., *A Hundred Years of Metallurgy*, Gerald Duckworth, London, U.K., 1963, 121.
Jackson, A., *Oxygen Steelmaking for Steelmakers*, Newnes-Butterworths, London, U.K., 1969, 165.
Boltz, C.L., *Materials & Technology*, Vol. 3, Longman, London, U.K., and J.H. de Bussy, Amsterdam, the Netherlands, 1970, Chap. 3.

LDF *See* DF.

LEAD An integrated ammonia-synthesis process, developed by Humphreys and Glasgow.

Saviano, F., Lagana, V., and Bisi, P., *Hydrocarbon Process. Int. Ed.*, 1981, **60**(7), 99.

Leap A process for making vinyl acetate monomer. It uses a fluidized bed of a new catalyst in powder form; the reactants are acetic acid, ethylene, and oxygen. Developed by BP Amoco and first operated in Hull, England, in 2001. The catalyst is a gold–palladium alloy supported on silica having a core–shell structure, made by Johnson Matthey. The catalyst is continuously removed and replenished. This was the first large-scale application of gold in catalysis, and the process received the AspenTech Award for Business Innovation in 2002.

Chem. Br., 1998, **34**(12), 5.
Chimie Hebdo, 30 Nov 1998, (26), 19.
Oil Gas J., 1998, **96**(47), 32.
U.S. Patent 6,534,438, filed by both BP and Johnson Matthey, July 2000, granted 2003.
Thompson, D.T., *Plat. Met. Rev.*, 2004, **48**(4), 169.

Lebedev A one-step process for converting ethanol, derived from carbohydrates, to butadiene, using a mixed alumina–zinc oxide catalyst at ca. 400°C:

$$2CH_3CH_2OH \rightarrow CH_3=CH_2-CH=CH_2$$

Invented by S. V. Lebedev in Leningrad in 1929 and used in Germany during World War II. In 1997, the process was still in use in the CIS, Poland, and Brazil.

British Patent 331,482.
Lebedev, S.V., *J. Gen. Chem. USSR*, 1933, **3**, 698.
Egloff, G. and Hulla, G., *Chem. Rev.*, 1945, **36**, 67.
Corson, B.B., Stahly, E.E., Jones, H.E., and Bishop, H.D., *Ind. Eng. Chem.*, 1949, **41**, 1012.
Weissermel, K. and Arpe, H.-J., *Industrial Organic Chemistry*, 4th ed., translated by C.R. Hindley and S. Hawkins, Wiley-VCH, Weinheim, Germany, 2003, 108.

Leblanc (Also written as LeBlanc and Le Blanc) An obsolete, two-stage process for making sodium carbonate from sodium chloride. In the first stage, the "Salt Cake" process, the salt was heated with sulfuric acid, yielding sodium sulfate (salt cake) and gaseous hydrogen chloride:

$$2NaCl + H_2SO_4 = Na_2SO_4 + 2HCl$$

In the second stage, the "Black Ash process," the sodium sulfate was reduced to sodium sulfide and then converted to sodium carbonate by calcining with limestone and coal in a rotating kiln known as a black ash furnace or revolver:

$$Na_2SO_4 + 4C = Na_2S + 4CO$$

$$Na_2S + CaCO_3 = Na_2CO_3 + CaS$$

The black product was extracted with water and the sodium carbonate in it was recovered by concentration and crystallization. The residue, chiefly calcium sulfide, known as "galigu," was dumped on land and created an environmental nuisance for many years because it never hardened. The process was invented by N. Leblanc in France in 1789, in response to a competition organized by the French Academy of Sciences. The operation of the first factory was delayed for several years because of the French Revolution. The process was operated widely until it was progressively superseded by the *ammonia soda process from 1872. But it was still in use in Bolton, United Kingdom, until 1938, and the last plant in Europe closed in 1992. *See also* Black Ash.

Taylor, F.S., *A History of Industrial Chemistry*, Heinemann, London, U.K., 1957, 183.
Hardie, D.W.F. and Pratt, J.D., *A History of the Modern British Chemical Industry*, Pergamon Press, Oxford, U.K., 1966, 21.
Smith, J.G., *The Origins and Early Development of the Heavy Chemical Industry in France*, Clarendon Press, Oxford, U.K., 1979, 209.
Lord Todd, *Chem. Ind. (London)*, 1989, 519.
Brown, A.H., *Chem. Br.*, 1993, **29**(10), 866.
Campbell, W.A., in *Chemistry, Society and Environment*, Russell, C.A., Ed., Royal Society of Chemistry, London, U.K., 2000, 77.
Sutton, M., *Chem. World*, 2006, **3**(11), 54.

Leckie *See* Steelmaking.

LEDA [**L**ow **E**nergy **D**e-**A**sphalting] A process for removing the asphalt fraction from petroleum residues by liquid–liquid extraction in a special rotating disc contactor. The extractant is a C_3 to C_6 aliphatic hydrocarbon or a mixture of such hydrocarbons. Developed in 1955 by Foster Wheeler USA Corporation and still widely used; 42 units were operating in 1996.

Hydrocarbon Process. Int. Ed., 1996, **75**(11), 106.

Ledgemont A process for removing sulfur from coal by an oxidative leach with lime and ammonia. Developed by Hydrocarbon Research.

IEA Coal Research, *The Problems of Sulphur*, Butterworths, London, U.K., 1989, 24.

Lefort A process for making ethylene oxide by oxidizing ethylene in the presence of a silver catalyst. Invented and developed in the 1930s by T. E. Lefort at the Société Française de Catalyse. For many years, refinements of this basic process were operated in competition with the ethylene chlorohydrin process, but by 1980, the Lefort process was the sole process in use.

French Patent 794,751.
U.S. Patent 1,998,878.
Françon, J., *Chim. Ind.*, 1933, **29**, 869.

Leidie A process for extracting the platinum metals from their ores by fusion with sodium peroxide, followed by a complex separation process. Developed by A. Quennessen, a leading French manufacturer of platinum in the nineteenth century, and E. Leidie. The process is still used for extracting precious metals and in chemical analysis.

McDonald, D., *A History of Platinum*, Johnson Matthey, London, U.K., 1960.

Leming A process for removing sulfur compounds from coal gas by the reaction with iron oxide. Invented in 1847

Lentz A process for making a hydrophobic silica gel for use as a filler for silicone rubber. The wet-process silica is treated with a strong mineral acid before being treated with an organosilicon compound. Invented by C. W. Lentz of Dow Corning in 1959.

U.S. Patent 3,122,520.

Lenze A process for removing naphthalene from coal gas by washing with cold, aqueous ammonia. Operated in Germany in the 1930s.

Lenze, F. and Rettenmaier, A., *Gas Wasserfach.*, 1926, **69**, 689 (*Chem. Abstr.*, **20**, 33556).

Leonard (1) A process for making mixed methylamines by reacting ammonia with methanol over a silica–alumina catalyst at elevated temperature and pressure. Developed and licensed by the Leonard Process Company, NY. Licensed to Alkyl Amine Chemicals in Mumbai in 1979. In 1993, the installed worldwide capacity of this process was 270,000 tons per year.

U.S. Patent 3,387,032.
Hydrocarbon Process., 1979, **58**, 194.
Weissermel, K. and Arpe, H.-J., *Industrial Organic Chemistry*, 4th ed., translated by C.R. Hindley and S. Hawkins, Wiley-VCH, Weinheim, Germany, 2003, 51.

Leonard (2) An early version of the *Chloride process, designed by the Leonard Process Company, NY. It was never patented or published. It incorporates a pebble bed heater for heating the oxygen and a quartz heat exchanger (Selas Furnace) for heating the titanium tetrachloride.

7.5 Ton/Day TiO$_2$ Design Manual, J.D. Leonard and Associates, New York, 1961; deposited at the Catalyst Science Discovery, Centre, Widnes, in 2010.

Le Seur An early process for electrolyzing brine. Developed in 1891 by E. A. Le Seur in Ottawa, Canada, and commercialized in Rumford, ME, in 1893.

British Patent 5,983 (1891).

LETS A process for making triple superphosphate (a calcium hydrogen phosphate). Developed by the J. R. Simplot Company in 1976.

Bierman, L.W., in *Sulphuric/Phosphoric Acid Plant Operations*, American Institute of Chemical Engineers, New York, 1982, 81.

Levinstein A process for making mustard gas, $(ClCH_2CH_2)_2S$, by reacting sulfur monochloride with ethylene:

$$2CH_2{=}CH_2 + S_2Cl_2 \rightarrow ClCH_2CH_2SCH_2CH_2Cl + S$$

LF *See* Steelmaking.

LFC [**L**iquids **F**rom **C**oal] A general term for such processes.

Lidov A process for chlorinating cyclopentadiene to octachloropentadiene, which is then thermally dechlorinated to hexachlorocyclopentadiene (HCCP), used as an intermediate in the manufacture of insecticides and flame retardants. The initial chlorination is catalyzed by phosphorus pentachloride or arsenious oxide. Invented by R. E. Lidov in the Netherlands and commercialized by the Shell Chemical Company.

British Patent 703,202.

LIFAC [**L**imestone **I**n-**F**urnace and **A**dded **C**alcium] A dry *flue-gas desulfurization process in which limestone is injected into the furnace and calcium hydroxide is injected after it. Developed by Tampella in 1984 and used in a power station in Finland. Commercialized also in the Soviet Union, Canada, and the United States.

Kenakkala, T. and Valimaki, E., *Desulphurisation in Coal Combustion Systems*, Institution of Chemical Engineers, Rugby, U.K., 1989, 113.
Kohl, H.L. and Nielsen, R.B., *Gas Purification*, 5th ed., Gulf Publishing Co., Houston, TX, 1997, 620.

LIGA [A German acronym meaning lithography, electroplating, and molding] A three-stage process for making microstructures.

Barrow, D., Cefai, J., and Taylor, S., *Chem. Ind. (London)*, 1999, (15), 591.

Lightox A photochemical process for destroying organic materials in aqueous solution by oxidation with chlorine, activated by ultraviolet radiation. Developed by the Taft Water Research Center, United States, in the 1960s.

Martinez, D., in *Chemical Waste Handling and Treatment*, Muller, K.R., Ed., Springer-Verlag, Berlin, Germany, 1986, 253.

Lignite Ash A *flue-gas desulfurization process that uses ash from lignite combustion as the adsorbent.

Speight, J.G., *Gas Processing*, Butterworth Heinemann, Oxford, U.K., 1993, 300.

LignoBoost A process for extracting lignin from the black liquor from the *Kraft process. Developed from 1996 by Innventia and now offered by Metso Corp. Commercialized by Domstar in Plymouth, NC, in 2012.

Lignol [**Lign**in phe**nol**] A catalytic process for hydrogenating lignin to a mixture of phenol, benzene, and fuel gas. Developed by Hydrocarbon Research. *See also* Noguchi.

Lignox [**Lign**in **ox**idation] A pulp-bleaching process using hydrogen peroxide as the oxidant and pretreatment by EDTA. Developed in Aspabruk, Sweden, from 1990 by Eka Nobel. *See also* Acetox, Macrox.

Bajpai, P., *Environmentally Benign Approaches for Pulp Bleaching*, 2nd ed., Elsevier, Amsterdam, the Netherlands, 2012, 113.

LIMB [**L**ime/**L**imestone **I**njection into a **M**ultistage **B**urner] A *flue-gas desulfurization process used in Germany and Finland. Dry, ground limestone is injected directly into the combustion chamber. This reacts with the sulfur dioxide, and the dry particulate product is collected downstream together with the ash. The process is suitable only for those systems that limit the maximum combustion temperature by staging, in order to minimize the production of oxides of nitrogen. Piloted by Ohio Edison.

Kohl, H.L. and Nielsen, R.B., *Gas Purification*, 5th ed., Gulf Publishing Co., Houston, TX, 1997, 618.

lime soda *See* causticization.

Linde Also called Hampson–Linde. A process for separating oxygen and nitrogen from air by liquefaction followed by fractional distillation. Developed by K. P. G. von Linde in Germany and W. Hampson in England at the start of the twentieth century.

Linde–Frank–Caro A process for extracting hydrogen from *water gas by liquefaction.

Linde/Yukong A two-stage, catalytic process for making 1,4-butanediol from acetylene. The first stage, catalyzed by palladium on alumina, makes 1,4-butynediol. The second stage, catalyzed by nickel on a silicate, hydrogenates this to 1,4-butanediol. Developed by Linde, Germany, and Yukong, Korea, in 1993 and then offered for license.

Heidegger, E. and Schädel, N., *Reports on Science and Technology*, Linde AG, Wiesbaden, Germany, 1997, **59**, 17.

LINDOX [**LIND**e **OX**idation] A variation of the *Activated Sludge sewage treatment system, using industrial oxygen (90%–98%) instead of air. The liquor passes through several closed tanks in series, and the oxygen is absorbed through the surface of the liquor. It is particularly suitable for treating effluents from the food processing industry. Developed by Linde, Munich, in the 1970s and first operated at a meat-rendering plant in Oberding in 1974. It was superseded by the *UNOX process in 1980.

Linear-1 A process for making linear C_6 to C_{10} α-olefins from ethylene. Developed by UOP in 1996 but not commercialized as of 1997.

Eur. Chem. News, 1996, **67**(1755), 16; 1997, **68**(1778), 2.
Eur. Chem. News, May 1998 (*Chemscope*), 24, 26.
Eur. Chem. News, 2004, **80**(2102), 17.

LINPOR [**LIN**de **POR**ous medium] A biological wastewater treatment process, using an open-pore plastic foam for retaining the biomass. Its use enables the capacity of an activated sludge plant to be increased without adding extra tanks. Invented at the Technische Universität, Munich, and further developed by Linde, Munich. *See also* CAPTOR.

European Patent 92,159.
Cooper, P.F., in *Topics in Wastewater Treatment*, Sidgewick, J.M., Ed., Blackwell, Oxford, U.K., 1985, 49.

Linz–Donan *See* L-D.

Linz Donawitz *See* L-D.

LIPP-SHAC [**LI**quid **P**olymerization of **P**ropylene with **S**uper **H**igh **A**ctivity **C**atalyst] A process for making polypropylene. Developed by the Shell Chemical Company in 1987 and used at Pernis, Netherlands; Carrington, England; and Geelong, Australia. The process now belongs to LyondellBasell.

Chem. Eng. (Rugby, Engl.), 1990, Nov, 7.

liquation A metallurgical process for separating metals by partial melting. Used for purifying zinc and tin and in conjunction with the *Parkes process for desilvering lead.

Liquicel A liquid–liquid extraction process in which the two liquids are separated by a permeable membrane in the form of hollow plastic fibers. Developed by Hoechst Celanese Corporation.

Chem. Eng. (Rugby, Engl.), 1992, (513), 10.

Liquefin A highly efficient process for liquefying natural gas. The efficiency results from the use of large plate-fin heat exchangers. Developed by Axens but not commercialized by 2013.

Eur. Chem. News, 2002, **77**(2012), 28.
Chem. Week, 17 Jul 2002, **164**(28), 27.
Oil Gas J., 19 Aug 2002, 100.33, 66.

Liritan A leather tanning process, introduced in 1960 and now widely used worldwide. The leather is first pickled in a solution of sodium hexametaphosphate and then soaked in baths of various vegetable products.

Shuttleworth, S.G., *J. Soc., Leather Trades' Chem.*, 1963, **47**, 143.

Litol Also called **Houdry–Litol**. A process for making benzene by dealkylating other aromatic hydrocarbons. This complex process achieves desulfurization, removal of paraffins and naphthenes, and saturation of unsaturated compounds, in addition to dealkylation. The catalyst contains cobalt and molybdenum. Developed by the Houdry Process and Chemical Company and Bethlehem Steel Corporation. First installed by the Bethlehem Steel Corporation in 1964. Subsequently used at British Steel's benzole refinery, Teesside, England.

Tarhan, M.O. and Windsor, L.H., *Chem. Eng. Prog.*, 1966, **62**(2), 67.
Lorz, W., Craig, R.G., and Cross, W.J., *Erdoel Kohle Erdgas Petrochem.*, 1968, **21**, 610.
Dufallo, J.M., Spence, D.C., and Schwartz, W.A., *Chem. Eng. Prog.*, 1981, **77**(1), 56.
Franck, H.-G. and Stadelhofer, J.W., *Industrial Aromatic Chemistry*, Springer-Verlag, Berlin, Germany, 1988, 123.

Llangwell A process for making acetic acid by fermenting the cellulose in corn cobs. Xylose is a coproduct. The microorganism was isolated from the gut of the goat. Piloted on a large scale by the Commercial Solvents Company, Terre Haute, IN, from 1928 to 1930.

LM [Lurgi–Mitterberg] A process for extracting copper from chalcopyrite. The ore is subjected to "activation grinding" and then dissolved in sulfuric acid under oxygen pressure. The copper dissolves as copper sulfate, leaving a residue of elemental sulfur and gangue. Developed in the early 1970s by Lurgi Chemie and Hüttentechnik, Kupferbergbau Mitterberg, and the Technical University of West Berlin. A demonstration plant operated in Mühlbach, Austria, from 1974 until the mine became exhausted.

Locap A process for removing mercaptans from gasoline by catalytic oxidation to disulfides, using a fixed bed of catalyst that is continuously treated with aqueous sodium sulfide. Commercialized by Petrolite Corporation in 1963.

Hydrocarbon Process. Int. Ed., 1964, **43**(9), 210.
O'Brien, G.A., Newell, O., Jr., and DuVon, R.H., Jr., *Hydrocarbon Process.*, 1964, **43**(6), 175.

LO-CAT A process for removing hydrogen sulfide and organic sulfur compounds from petroleum fractions by air oxidation in a cyclic catalytic process similar to the *Stretford process. The aqueous solution contains iron, two proprietary chelating agents, a biocide, and a surfactant; the formulation is known as ARI-310. The sulfur product is removed as a slurry. Developed in 1972 by Air Resources (now ARI Technologies) and first commercialized in 1976. Over 125 units were operating in 1996. An improved version, LO-CAT II, was announced in 1991, used for removing hydrogen sulfide from landfill gas, wellhead casing gas, and geothermal power plants. The process now belongs to U.S. Filter. *See* ARI LO-CAT II.

U.S. Patent 4,189,462.
British Patent 1,538,925.
Hardison, L.C., *Oil Gas J.*, 1984, **82**(23), 60.
Hardison, L.C., in *Acid and Sour Gas Treating Processes*, Newman, S.A., Ed., Gulf Publishing Co., Houston, TX, 1985, 678.

Dalrymple, D.A., Trofe, T.W., and Evans, J.M., *Chem. Eng. Prog.*, 1989, **85**(3), 43.
Hardison, L.C. and Ramshaw, D.E., *Hydrocarbon Process. Int. Ed.*, 1992, **71**(1), 89.
Hydrocarbon Process. Int. Ed., 1996, **75**(4), 106.
Oil Gas J., 2000, **98**(8), 68; **98**(11), 55.

LO-FIN [Last Out-First IN] A version of the *PSA process for separating hydrogen from other gases. It includes a unique gas-retaining vessel that preserves the concentration gradient in one stream before using it to repressurize another bed. Developed jointly by Toyo Engineering Corporation and Essex Corporation.

Suzuki, M., in *Adsorption and Ion Exchange: Fundamentals and Applications*, LeVan, M.D., Ed., American Institute of Chemical Engineers, New York, 1998, 122.

Lomax An obsolete name for a *hydrocracking process now offered under the name *Unicracking.

LOMI [Low Oxidation Metal Ions] A process for decontaminating parts of nuclear reactors by washing with aqueous solutions of low-valency transition metal ions. Developed at the Berkley laboratories of the U.K. Atomic Energy Authority in the early 1980s.

Longmaid–Henderson A process for recovering copper from the residue from the roasting of pyrites to produce sulfur dioxide for the manufacture of sulfuric acid. The residue was roasted with sodium chloride at 500°C–600°C, and the evolved sulfur oxides and hydrochloric acid were scrubbed in water; the resulting solution was used to leach the copper from the solid residue. Copper was recovered from the leachate by adding scrap iron. The process became obsolete with the general adoption of elemental sulfur as the feedstock for sulfuric acid manufacture.

Lonza (1) A process for oxidizing isobutene to α-hydroxyisobutyric acid, a precursor for methacrylic acid, using a solution of dinitrogen tetroxide in acetic acid as the oxidant. *See also* Escambia (1).

Lonza (2) A process for making malononitrile. Acrylonitrile is reacted continuously with cyanogen chloride, in the vapor phase, in a quartz tube at 900°C. Developed by Lonza, Basel, Switzerland.

German Patents 1,921,662; 1,946,429.

Loop A continuous process for polymerizing aqueous emulsions of olefinic compounds such as vinyl acetate. Polymerization takes place in a tubular reactor (the loop) with recycle. Invented by Gulf Oil Canada in 1971 and further developed by several U.K. paint companies. It is now used for making copolymers of vinyl acetate with ethylene, used in solvent-free paints and adhesives.

Canadian Patent 907,795.
Wilkinson, M. and Geddes, K., *Chem. Br.*, 1933, **29**(12), 1050.

Loprox [Low-pressure wet oxidation] Also written LOPROX. A process for partially oxidizing waste organic products, to render them digestible in biological waste treatment systems such as the *Activated Sludge process. The oxidant is oxygen, at a pressure of 3–20 atm. The temperature is 120°C–200°C, and the reaction is catalyzed by quinonoid substances and iron salts. Developed by Bayer in Germany in 1980, for use in its own works, and piloted in several countries from 1991 to 1992. Six units were operating in Europe in 1997. Now engineered and offered by Bertrams Chemical Plants, Switzerland. The first U.K. plant was built for Hickson & Welch, Castleford, in 1999. It was combined with a *VITOX plant, the first time that these two processes had been operated together.

New Sci., 1992, **133**(1811), 24.
Water Waste Treat., 1998, **41**(5), 22.
Luck, F., *Catal. Today*, 1999, **53**(1), 86.
Water Effluent Treat. News, 2000, **6**(6), 4.
Eur. Chem. News, 2000, **72**(1902), 47.
Debellefontaine, H. and Foussard, J.N., *Waste Management*, 2000, **20**, 15.

LOR [**L**iquid-phase **O**xidation **R**eactor] Not a process but a piece of equipment in which to conduct liquid-phase oxidations (e.g., the *Mid-Century process) safely with oxygen rather than with air. The oxygen is introduced into the liquid phase and rapidly dispersed in the form of bubbles 1–5 mm in diameter. Developed by Praxair and ABB Lummus Global in 1996.

> *Chem. Week*, 1996, **158**(15), 28.

LoTOx A process for removing NO_x from *FCC off-gases. Injection of ozone oxidizes the NO_x to N_2O_5, which is easily removed by water scrubbing. Developed by BOC.

Lovacat [**Lo**w **va**lency **cat**alyst] A catalytic process for making ethylene propylene and EPDM rubbers. Developed by DSM Elastomers in 1996.

> *Kunst. en Rubber*, 1996, **49**(8), 41.

Lowe *See* Water Gas.

Löwenstein–Riedel An electrolytic process for making hydrogen peroxide by the electrolysis of a solution of sulfuric acid and ammonium sulfate. Ammonium peroxodisulfate, $(NH_4)_2S_2O_8$, is an intermediate. This and the *Weissenstein process were made obsolete by the invention of the *AO process.

> Schumb, W.C., Satterfield, C.N., and Wentworth, R.L., *Hydrogen Peroxide*, Reinhold Publishing, New York, 1955, 145.

Löwig Also called Ferrite. A causticization process: the conversion of sodium carbonate to sodium hydroxide. The sodium carbonate is mixed with iron oxide and heated for several hours in a rotating kiln. Carbon dioxide is evolved and sodium ferrite remains:

$$Na_2CO_3 + Fe_2O_3 = 2NaFeO_2 + CO_2$$

The product is agitated with water, producing sodium hydroxide and ferric oxide for reuse:

$$2NaFeO_2 + H_2O = 2NaOH + Fe_2O_3$$

The process was invented in Germany by C. Löwig in 1882 and used at Joseph Crosfield and Sons, Warrington, United Kingdom, in the late nineteenth and early twentieth centuries. *See also* causticization.

> British Patent 4,364 (1882).
> Musson, A.E., *Enterprise in Soap and Chemicals*, Manchester University Press, Manchester, U.K., 1965, 82, 203.

LPG Unibon An outdated UOP version of the *hydrocracking process for simultaneously hydrogenating and cracking a naphtha petroleum fraction to form C_3 and C_4 hydrocarbons. In 1992, the technology was offered under the umbrella of *Unicracking.

LPM [**L**ow-**P**ressure **M**ethanol] A process for making methanol, developed by ICI in the 1960s. A copper catalyst permitted operation at low pressures. By 2004, it was used to make more than 50% of the methanol produced in the world. Offered by Johnson Matthey in 2013.

LPMEOH A process for making methanol. Developed by Air Products and Chemicals and Chem Systems in the late 1970s.

> Cybulski, A., *Catal. Rev. Sci. Eng.*, 1994, **36**(4), 558.

LPO Another name for *Celanese LPO.

LP OXO [**L**ow-**P**ressure **OXO**] *See* OXO. In 2013, this was being marketed jointly by Dow Chemical and Davy Process Technology and was used in making 70% of all global propylene LP OXO products.

> *ICIS Chem. Bus.*, 2013, **283**(10), 42.

LPS [**L**ignin **P**recipitation **S**ystem] A process for removing lignin from the black liquor produced by the soda pulping process, fed by annual plant crops. Developed by Granit SA, Switzerland, and first operated at Papeteries du Léman, Thonon, France, in 2000. The liquor is filtered, acidified, aged, and filtered again to separate the lignin. This produces a liquor, which can be processed in municipal sewage treatment plants, and a saleable lignin product.

> Pye, E.K., in *Biorefineries: Industrial Processes and Products*, Kamm, B., Gruber, P.R., and Kamm, M. (Eds.), Wiley-VCH, Weinheim, Germany, 2005, **I**, 171.

LR *See* Lurgi–Ruhrgas.

LSE [**L**iquid **S**olvent **E**xtraction] A coal liquefaction process, under development in 1990 by British Coal, at Point of Ayr, North Wales, now closed. The coal is dissolved in a coal-derived hydrocarbon solvent and then catalytically hydrocracked.

> *Chem. Br.*, 1990, **26**(10), 922.

LTC [**L**ow-**T**emperature **C**hlorination] A general term for processes for making 1,2-dichloroethane from ethylene and chlorine by processes operated below the boiling point of the product (83°C). *See also* HTC.

LT Unibon A two-stage, catalytic *hydrotreating process for removing deleterious components from naphtha without cracking it. Developed by UOP.

Lucas [**Lu**rgi-**C**laus-**A**bgas-**S**chwefelgewinnung] A process for removing residual sulfur compounds from the tail gases from the *Claus process. The gases are incinerated and then passed over a bed of hot coke, which converts all the sulfur to sulfur dioxide. This is absorbed in aqueous sodium phosphate, which releases it on heating:

$$Na_2HPO_4 + H_2O + SO_2 = NaH_2PO_4 + NaHSO_3$$

The sulfur dioxide is returned to the Claus process for reuse. Developed by Lurgi Mineralöltechnik, Germany.

> Doerges, A., Bratzler, K., and Schlauer, J., *Hydrocarbon Process. Int. Ed.*, 1976, **55**(10), 110.
> *Sulphur*, 1977, (128), 41.

Luce-Rozan A variation of the *Pattinson process, in which steam is blown through the molten metal as cold water is sprayed on the surface.

Lupotech A family of processes for making polyethylene, developed in the 1980s by BASF and now offered by LyondellBasell. Lupotech G is a gas-phase process that uses a chromium-containing catalyst to make medium-density polyethylene (MDPE) and high-density polyethylene (HDPE). In 2006, Lupotech G was being superseded by *Spherilene S. Lupotech T uses a high-pressure tubular reactor to make low-density polyethylene (LDPE). The products are sold under the trade name Lupolen.

> *Chem. Ind. (London)*, 21 Aug 2006, (16), 14.

Lurgi [Meta**lurgi**schegesellschaft] Lurgi, previously called Metallurgische Gesellschaft, now a subsidiary of Metallgesellschaft, is a large chemical engineering company that has particularly given its name to two coal gasification processes. The first was a fixed-bed gasifier, in which a bed of coal particles rested on a rotating hearth through which oxygen and steam were injected. The temperature was kept below the slagging temperature of the ash, which was withdrawn from the base of the bed. The process was first used commercially at Zittau, Germany, in 1936; 65 units had been built by 1979. The second design was a slagging gasifier, of which an experimental model was installed in Solihull, United Kingdom, in 1956.

Wilke, G., *Chem. Fabr.*, 1938, **11**, 563.

Chem. Eng. News, 1958, **36**(16), 88.

Dainton, A.D., in *Coal and Modern Coal Processing*, Pitt, G.J. and Millward, G.R., Eds., Academic Press, London, U.K., 1979, 135.

Cornils, B., in *Chemicals from Coal: New Processes*, Payne, K.R., Ed., John Wiley & Sons, Chichester, U.K., 1987, 12.

Rezaiyan, J. and Cheremisinoff, N.P., *Gasification Technologies*, Taylor & Francis, Boca Raton, FL, 2005, 63.

Lurgi–Ruhrgas Also known as **LR**. A process originally intended for pyrolizing fine-grained solids, such as coal, peat, shale, and tar sands, to produce mixed hydrocarbons. The process is based on flash heating in a mixer by means of circulated hot powders, usually obtained from the process. The process was later modified to allow ethylene to be produced from heavy hydrocarbons. Ethylene was first made in this way in Germany in 1958. Developed jointly by the Lurgi and Ruhrgas companies in Germany in 1949 and now offered by Lurgi.

Rammler, R.W. and Weiss, H.-J., in *Handbook of Synfuels Technology*, Meyers, R.A., Ed., McGraw-Hill, New York, 1984, 4–17.

Lurgi Spülgas *See* Spülgas.

Lyocell Also called the **Courtaulds Tencel** process. A process for making cellulose fibers based on dissolving wood pulp in N-methylmorpholine-N-oxide (NMMNO). The name is also a generic name for the fiber product, most commonly known by the trade name Tencel. Developed by Courtaulds Fibres in the late 1970s and first commercialized in the United Kingdom in 1988. Lenzing AG in Austria also made it from 1990, and the process has been the subject of prolonged patent disputes (resolved in 2005) and several changes of ownership. Lenzing acquired the Tencel group of companies in 2004, but some of the engineering technology belongs to Zimmer AG. *See also* Alceru.

Kirk-Othmer's Encyclopedia of Chemical Technology, 4th ed., Vol. 10, John Wiley & Sons, New York, 1991–1998, 717.

Woodings, C., Ed., *Regenerated Cellulose Fibres*, Woodhead Publishing Series in Textiles, No. 18, Woodhead Publishing, Cambridge, U.K., 2001.

Chem. Week, 17 Dec 2003, 9; 14 Dec 2005, **167**(42), 13.

Eur. Chem. News, 2005, **83**(2174), 17.

M2M [**M**ud-to-**M**oney] A process for recovering more alumina from the "red mud" from the *Bayer process. Developed by Alcor Technology and Alfa Laval Benelux. There are several other meanings for this acronym.

Chem. Eng. (N.Y.), 2007, **114**(8), 16.

MAAP [**M**icrobial **A**mino **A**cid **P**roduction] A general name for processes that use microorganisms for making amino acids. The feed material may be a sugar or biomass. Many organisms can be used, but *Corynebacterium glutamicum* is popular. The amino acid products are used as food supplements and animal feeds. The largest tonnage amino acids made in this way are sodium glutamate and L-lysine.

Marx, A., Wendisch, V.F., Kelle, R., and Bucholz, S., in *Biorefineries: Industrial Processes and Products*, Kamm, B., Gruber, P.R., and Kamm, M., Eds., Wiley-VCH, Weinheim, Germany, 2005, **II**, Chap. 6.

MacArthur–Forrest *See* Cyanide. J.S. MacArthur is also famous for establishing the radium extraction industry in the United Kingdom from 1911.

Harvie, D.I., *Deadly Sunshine: The History and Fatal Legacy of Radium*, Tempus Publishing, Stroud, U.K., 2005, 66.

Macrox A pulp-bleaching process using hydrogen peroxide and EDTA at 90°C. Developed in the 1990s by Kvaerner Pulping Technologies, Sweden. *See* Lignox.

There are many unrelated uses of this name.

U.S. Patent 5,571,377.
U.S. Patent 6,019,870 (a similar process developed by Elf Atochem).

Madaras *See* DR.

Madison–Scholler An improved version of the *Scholler *saccharification process for making glucose from wood. Developed at the U.S. Forest Products Laboratory, Madison, WI.

Harris, E.E. and Berlinger, E., *Ind. Eng. Chem.*, 1945, **38**, 890.
Harris, E.E., Berlinger, E., Hajny, G.J., and Sherrard, E.C., *Ind. Eng. Chem.*, 1945, **37**, 12.
Riegel's Handbook of Industrial Chemistry, 9th ed., Kent, J.A., Ed., Van Nostrand Reinhold, New York, 1992, 256.
Katzen, R. and Schell, D.J., in *Biorefineries: Industrial Processes and Products*, Kamm, B., Gruber, P.R., and Kamm, M., Eds., Wiley-VCH, Weinheim, Germany, 2005, **I**, 131.

Madsenell *See* metal surface treatment.

Magchar A process for extracting gold from solution by adsorption on composites of magnetic particles deposited on grains of activated carbon. Invented by E. Herkenhoff and N. Hedley.

Eng. Min. J., Aug 1982, 84.
Yannopoulos, J.C., *The Extractive Metallurgy of Gold*, Van Nostrand Reinhold, New York, 1991, 224.

Magdeburg-P A high-pressure hydrogenation process developed by Lurgi in World War II.

BIOS Report 623.

MagnaCat A process for selectively removing metal-contaminated catalyst particles from an *FCC reactor by magnetic separation. A high-strength magnet made from a neodymium–boron–iron alloy is used. Developed by the Ashland Oil, which sold it to MW Kellogg in 1997.

U.S. Patent 5,147,527.
Hettinger, W.P., Jr., *Catal. Today*, 1992, **13**, 157.
Andersson, S.-I. and Myrstad, T., *Appl. Catal. A: Gen.*, 1997, **159**(1–2), 291.
Chem. Eng. (N.Y.), 1997, **104**(11), 122.
Johnson, T.E., Goolsby, T.L., Silverman, M.A., Kowalczyk, D.C., and Moore, H.F., *Oil Gas J.*, 1998, **96**(24), 65.
Niccum, P.K. and Santner, C.R., in *Handbook of Petroleum Refining Processes*, 3rd ed., Meyers, R.A., Ed., McGraw-Hill, New York, 2003, 3.30.

Magnaforming A *catalytic reforming process developed by the Atlantic Richfield Corporation and Engelhard Corporation. First announced in 1965, it was commercialized in 1967; by 1988, 150 units were operating worldwide. Hydrocarbon Research installed units in Argentina, Algeria, and the USSR.

Magnaforming is also the name of a process for shaping metal parts by means of a magnet.

Nevison, J.A., Obaditch, C.J., and Dalson, M.H., *Hydrocarbon Process. Int. Ed.*, 1974, **53**(6), 111.
Little, D.M., *Catalytic Reforming*, PennWell Publishing, Tulsa, OK, 1985, 158.
Hydrocarbon Process. Int. Ed., 1988, **67**(9), 80.

Magnefite [Magnesium sul**fite]** A process for separating lignosulfonates from the sulfite liquor in papermaking. The lignosulfonates are separated as magnesium salts. Six paper mills in the United States and Canada were using this process as of 1981.

Magnetherm [Magee magnesium **E**isenberg **therm**al] A process for making magnesium by reducing dolomite with ferrosilicon at 1600°C in a vertical electric arc furnace:

$$2CaO \cdot MgO + Fe_xSi = 2Mg + Ca_2SiO_4 + xFe$$

The process takes place in a partial vacuum and uses a flux containing Ca, Al, and Mg oxides or bauxite. Invented in 1965 by E.M. Magee and B. Eisenberg at Esso Research and Engineering Company. An improved version of this process, *MAGRAM, which did not require a vacuum and which used a mixture of waste products to provide the flux, was developed at the University of Manchester Institute of Science and Technology.

U.S. Patent 3,441,402.
New Sci., 1995, **145**(1967), 21.
Chem. Br., 1996, **32**(4), 12.
Trocmé, F., in *Advances in Extractive Metallurgy and Refining*, Jones, M.J., Ed., Institution of Mining and Metallurgy, London, U.K., 1971, 517.

Magnex A process for removing mineral matter from coal by first rendering it magnetic. The coal is treated with iron carbonyl vapor, which deposits a thin skin of magnetic material on the pyrite and other mineral matter, but not on the coal. Conventional magnetic separation is then used. Developed by Hazen Research in 1976. Piloted but not commercialized, probably because of the explosion and toxicity hazards involved.

IEA Coal Research, *The Problems of Sulphur*, Butterworths, London, U.K., 1989, 38.
Kawatra, S.K. and Eisele, T.C., *Coal Desulfurization*, Taylor & Francis, New York, 2001, 253, 332.

Magnicol [Magnetic col**umnar]** A process for making alnico (an iron-based magnetic alloy containing Al, Ni, Co, and Cu) crystallize with a columnar grain structure in order to optimize its magnetic properties. Successive additions of silicon, carbon, and sulfur are made to the initial melt.

Palmer, D.J. and Shaw, S.W.K., *Cobalt*, 1967, **43**, 63 (*Chem. Abstr.*, **71**, 32827).

MAGRAM A process for extracting magnesium metal from dolomite. Similar to *Magnetherm but uses asbestos as the fluxing material and does not require the use of vacuum. Developed by an international consortium financed by the European Union at the University of Manchester Institute of Science and Technology. *See* Magnetherm.

> *Chem. Br.*, 1996, **32**(4), 12.

MagSonic [**Mag**nesium super**Sonic**] A carbothermal process for making magnesium metal from magnesium oxide by reduction with carbon at about 1800° in an inert atmosphere. The essence of this process is to prevent the back reaction of the magnesium with the carbon monoxide produced. This is achieved by very rapid cooling in a special nozzle through which the gases pass at ultrasonic speeds. Developed and piloted by the CSIRO, Australia, from 2008.

The name has a number of other meanings.

> Prentice, L., *TCE*, 2012, (848), 50.

MAGSORB A process for removing carbon dioxide from hot gas streams by reversible absorption on magnesium oxide modified with potassium carbonate. Developed by the Institute of Gas Technology, Chicago, for fuel gas derived from coal.

MAK fining A petroleum refining process that combines *MAK hydrocracking with a cold-flow improvement process. Developed by Nippon Ketjen and first licensed in 1998.

> *Jpn. Chem. Week*, 1998, **39**(1976), 4, 7.
> Stanislaus, A., Marafi, A., and Rana, M.S., *Catal. Today*, 2010, **153**, 38.

MAK hydrocracking [**M**obil **A**kzo **K**ellogg] A process for making high-quality, low-sulfur fuels from a variety of petroleum intermediates. Developed jointly by the three companies named. Two units were operating in 1996.

> *Hydrocarbon Process. Int. Ed.*, 1996, **75**(11), 126.

Malaprop A process for removing carbonyl sulfide from gas streams by scrubbing with diethylene glycolamine (DGA).

> Moore, T.F., Dingman, J.C., and Johnson, F.L., Jr., in *Acid and Sour Gas Treating Processes*, Newman, S.A., Ed., Gulf Publishing, Houston, TX, 1985, 290, 313.

Mallet A process for separating oxygen from air by selective dissolution in water. Oxygen is more soluble than nitrogen in water, so by contacting water with compressed air and desorbing the gases at a lower pressure, it is possible to make air enriched in oxygen. Repetition of the process yields progressively purer oxygen. Invented by J.T.A. Mallet in Paris in 1869; supplanted by liquefaction processes at the start of the twentieth century.

> British Patent 2,137 (1869).
> Aitkin, W., *Chem. Eng. (Rugby, Engl.)*, 1998, (659), 18.

Maloney A system for drying gases with zeolites. Three beds are used, with one of them being regenerated by hot gas at any one time. Developed by Maloney Steel Company, Calgary, Canada.

> Palmer, G.H., *Hydrocarbon Process. Int. Ed.*, 1977, **56**(4), 103.
> Kohl, H.L. and Nielsen, R.B., *Gas Purification*, 5th ed., Gulf Publishing Co., Houston, TX, 1997, 1066.

Manasevit A process for making electronic devices by depositing thin films of elements or simple compounds, such as gallium arsenide, on flat substrates by *CVD from volatile compounds such as trimethyl gallium and arsine. Harold Manasevit died in 2008. He was the first to demonstrate the epitaxial growth of silicon on sapphire, which is important in the production of microelectronic devices.

Manasevit, H.M., *Appl. Phys. Lett.*, 1968, **12**, 156.
Manasevit, H.M. and Simpson, W.I., *J. Electrochem. Soc.*, 1969, **116**, 1725.
Manasevit, H.M., *J. Electrochem. Soc.*, 1971, **118**, 647.
Chem. Eng. News, 2009, **87**(17), 43.

Manchester A variation on the *Ferrox process for removing hydrogen sulfide from industrial gases in which several absorbers are used, and delay stages permit completion of the reaction with the iron oxide absorbent. Developed by the Manchester Corporation Gas Department in the 1940s and installed in several British gasworks.

British Patents 550,272; 611,917.
Kohl, H.L. and Nielsen, R.B., *Gas Purification*, 5th ed., Gulf Publishing Co., Houston, TX, 1997, 742.

Manhès A metallurgical process for removing sulfur from copper matte by blowing air through the molten material. Invented by P. Manhès in France in 1880.

German Patent 15,562.
Manhès, P., *Ber. Dtsch. Chem. Ges.*, 1881, **14**, 2432.

Manley An early thermal process for cracking petroleum.

Mannheim (1) A process for making hydrochloric acid by roasting sulfuric acid and sodium chloride together in a closed cast iron furnace equipped with a plough. The by-product sodium sulfate, known as salt cake, may be recrystallized after neutralization and filtration and used as a detergent ingredient. A potassium variant is used in those locations where native potassium chloride occurs.

Mannheim (2) An early version of the *Contact process for making sulfuric acid. Two catalysts were used: ferric oxide, followed by platinum. The first Mannheim plant was built in Buffalo, NY, in 1903.

Miles, F.D., *The Manufacture of Sulfuric Acid (Contact Process)*, Gurney & Jackson, London, U.K., 1925, Chap. 10.
Levy, S.I., *An Introduction to Industrial Chemistry*, G. Bell & Sons, London, U.K., 1926, 218.

MANOIL A process for converting municipal waste to a liquid fuel. Invented by F.R. Benn and C.A. McAuliffe at the University of Manchester Institute of Science and Technology and piloted there, but not commercialized. Tetralin is used as the solvent; a nickel catalyst is desirable although not essential. The company Manoil Ltd was created to commercialize it, with funding from the City of Manchester and the EU.

U.S. Patent 4,618,736.

Mansfield A process for extracting copper from sulfide ores by roasting with anthracite or coke and a siliceous flux in a special blast furnace.

Mark and Wulff A process for making styrene from benzene and ethylene. Developed in Germany in the 1930s.

Marqueyrol and Loriette A process for making nitroguanidine, an explosive. Cyanamide dimer is converted to guanidinium sulfate by heating with sulfuric acid; this is then nitrated with nitric acid. *See also* Welland.

Aubertein, P., *Mém. Poudres*, 1948, **30**, 143 (*Chem. Abstr.*, **45**, 8250).

Mars [**M**embrane **a**romatic **r**ecovery **s**ystem] A membrane-based system for recovering aromatic compounds from aqueous waste streams. The membrane was initially made from a silicone rubber, but later versions used polypropylene glycol entrapped in a porous polypropylene sheet. Demonstrated first at a cresol-recovery plant in Knottingley, United Kingdom.

Chem. World, 2005, **2**(9), 11.

Maruzen (1) A process for making terephthalic acid from *p*-xylene. Similar to the *Amoco process but yielding a purer product in one stage. Operated in Japan by Matsuyama Chemical Company.

Raghavendrachar, P. and Ramachandran, S., *Ind. Eng. Chem. Res.*, 1992, **31**, 453.

Maruzen (2) A process for purifying *p*-xylene by crystallization, using ethylene as the direct coolant. Developed by Maruzen Gas Oil Company, United States. Now probably superseded by the *Parex (1) process.

Hatanaka, Y. and Nakamura, T., *Oil Gas J.*, 1972, **70**(47), 60.

Maruzen (3) A process for making methyl ethyl ketone by the direct oxidation of *n*-butane, catalyzed by palladium and ferric sulfate. Not commercialized.

MAS [**M**ethanolo **A**lcooli **S**uperiori] A process for making mixtures of methanol with higher alcohols, for use as gasoline extenders, developed by a consortium of Snamprogetti, Haldor Topsoe, and Anic. Piloted in a demonstration plant in Italy.

Asinger, F., *Methanol: Chemie und Energierhostoff*, Springer-Verlag, Berlin, Germany, 1986, 120.

Massener *See* Hoerde.

Mathieson (1) A process for making chlorine dioxide gas by passing sulfur dioxide, diluted with air, into aqueous sodium chlorate and sulfuric acid. The product is absorbed in water. Operated in the United States on a large scale for pulp bleaching.

Sheltmire, W.H., in *Chlorine, Its Manufacture, Properties and Uses*, Sconce, J.S., Ed., Reinhold Publishing, New York, 1962, 303, 539.

Mathieson (2) A process for making calcium hypochlorite dihydrate by mixing sodium hypochlorite and calcium chloride. Invented by A. George and R.B. MacMullin at the Mathieson Alkali Works, New York, in the 1920s. *See also* Perchloron.

U.S. Patents 1,713,650, 1,787,048.
Sheltmire, W.H., in *Chlorine, Its Manufacture, Properties and Uses*, Sconce, J.S., Ed., Reinhold Publishing, New York, 1962, 523.

Matthey A complex sequence of chemical operations for purifying platinum. Developed by G. Matthey and used in his factory from 1879.

McDonald, D., *A History of Platinum*, Johnson Matthey, London, U.K., 1960, 220.

Maunsell A process for converting elemental phosphorus to phosphoric acid by the use of the "Maunsell burner." Invented in the 1940s by M. Maunsell at the Albright & Wilson plant near Melbourne, Australia, and subsequently used in Canada and England.

Podger, H., *Albright & Wilson, The Last 50 Years*, Berwin Books, Studley, U.K., 2002, 120.

MAWR [**M**obil **A**lkanolamine **W**aste **R**ecovery] A process that reduces the quantity of waste generated by alkanolamine processes, which remove acid gases from oil refinery gas streams. Developed by Mobil Oil, Germany, and used commercially there since 1979.

MaxCat A process for reducing coke buildup in catalytic reformers. Developed by Phillips Petroleum and first licensed to LLC, Coffeyville, LA, in 2000. UOP was granted a license in November 2000.

Eur. Chem. News, 2000, **72**(1908), 33.
Oil Gas J., 2000, **98**(21), 8.
Chem. Week, 2000, **162**(43), 18.

MaxEne A process for increasing the yield of propylene from naphtha crackers without increasing that of propylene. A version of the *Sorbex process is used to separate the normal paraffins from the branched paraffins before the cracker. Developed and offered by UOP in 2000, but not reported to have been licensed by 2005.

> *Eur. Chem. News*, 2000, **73**(1927), 39.
> *Chem. Eng. (Rugby, Engl.)*, 2000, (711), 13.

MAXISULF A process for removing sulfur from tail gases from the *Claus process. It combines an adsorption stage with the *Sulfreen process. Developed and offered by Lurgi.

MAXOFIN [**MAX**imize Ole**FIN**] Also called MAXOFIN FCC. A modification to the *FCC process, which maximizes the production of propylene. A proprietary additive containing zeolite ZSM-5 is used (MAXOFIN-3), together with engineering modifications. Developed by ExxonMobil and Kellog Brown & Root and now licensed by KBR. *See also* SUPERFLEX.

> U.S. Patent 7,128,827.
> Niccum, P.K. and Santner, C.R., in *Handbook of Petroleum Refining Processes*, 3rd ed., Meyers, R.A., Ed., McGraw-Hill, New York, 2003, 3.11.
> *Eur. Chem. News*, 2005, **82**(2137), 21.

Mazzoni A family of continuous soapmaking processes.

> Lanteri, A., *Seifen, Oele, Fette, Wachse*, 1958, **84**, 589.

MBBR [**M**oving **B**ed **B**io**R**eactor] A wastewater treatment system in which the biofilm is supported on special plastic support that are circulated through several reactors. Offered by several companies.

MBG [**M**AN **B**ergbau-Forschung **G**asification] A coal gasification process, suitable for all grades of coal, especially those difficult to gasify. Under development by MAN Gutehoffnungshütte, in collaboration with Deutsche Montan Technologie, in 1991.

MBR (1) [**M**embrane **B**io**R**eactor] A system that combines a suspended growth bioreactor with a membrane filtration device. Most systems involve aeration in which the bubbles minimize fouling of the membrane; anaerobic systems are used in special situations. Widely used in water and wastewater treatment since the 1980s. In 2012, it was offered by 13 companies.

> Judd, S., *The MBR Book: Principles and Applications of Membrane Bioreactors for Water and Wastewater Treatment*, Elsevier, Oxford, U.K., 2006.
> *Wikipedia, The Free Encyclopedia*. Membrane bioreactor. Revised 17 Jun 2013. Retrieved 20 Jul 2013.

MBR (2) [**M**obil **B**enzene **R**eduction] A catalytic process for reducing the benzene content of gasoline. It combines features of three earlier processes: benzene alkylation with light olefins, olefin equilibration with aromatization, and selective paraffin cracking. The olefins are obtained from *FCC off-gas. The catalyst is a modified ZSM-5 zeolite. Developed by Mobil Research & Development Corporation in 1993.

> U.S. Patent 5,347,061.
> *Chem. Eng. News*, 1993, **71**(38), 36.
> *Oil Gas J.*, 1993, **91**, 63.
> *Hydrocarbon Process. Int. Ed.*, 1994, **73**(11), 90.

M-C *See* Mid-Century.

Mcgyan [Named after the inventors **Mc**Neff, **Gy**berg, and **Yan**] A catalytic process for making biodiesel. It combines esterification with transesterification. Developed by Augsburg College and SarTec. A demonstration plant in Isanti, MN, started operating in 2009.

McNeff, C.V., McNeff, L.C., Yan, B., Nowlan, D.T., Rasmussen, M., Gyberg, A.E., Krohn, B.J., Fedie, R.L., and Hoye, T.R., *Appl. Catal. A: Gen.*, 2008, **343**(1–2), 40.
Biofuels Int., Mar 2009, **3**(2), 12.
Krown, B.J., McNeff, C.V., Yan, B., and Nowlan, D., *Biosource Technol.*, 2011, **1029**, 94.

McKechnie–Seybolt A process for making vanadium by reducing vanadium pentoxide with calcium in the presence of iodine. It is conducted in a steel bomb at 700°C.

McKenna *See* Menstruum.

M-coke A homogeneous *desulfurization process that uses an oil-soluble molybdenum compound as the catalyst.

Beardon, C.L.A., *Chem. Eng. Prog.*, 1981, 44.
Rueda, N., Bacaud, R., Lanteri, P., and Vrinat, M., *Appl. Catal. A: Gen.*, 2001, **215**(1–2), 81.

MCRC [**M**aximum **C**laus **R**ecovery **C**oncept] A variation on the *CBA sulfur recovery process using multiple Claus converters. Developed by the Delta Engineering Corporation in 1983; 16 plants had been built by 1993.

Davis, G.W., *Oil Gas J.*, 1985, **83**(8), 110.
Sulphur, 1994, (231), 50.
Kohl, H.L. and Nielsen, R.B., *Gas Purification*, 5th ed., Gulf Publishing Co., Houston, TX, 1997, 707.

MDDW [**M**obil **D**istillate **DeW**axing] A process for removing waxes (long-chain normal paraffins) from petroleum fractions by cracking over zeolite ZSM-5. The waxes are converted to liquid hydrocarbon fuels. Twenty-one units were operating in 1990.

Chen, N.Y., Gorring, R.L., Ireland, H.R., and Stein, T.R., *Oil Gas J.*, 1977, **75**(23), 165.
Perry, R.H., Jr., Davis, F.E., and Smith, R.B., *Oil Gas J.*, 1978, **76**(21), 78.
Ireland, H.R., Redini, C., Raff, A.S., and Fava, L., *Oil Gas J.*, 1979, **77**(24), 82.

MDEA [**M**ethyl **Di**Ethanol**A**mine] A general name for processes using methyl diethanolamine for absorbing hydrogen sulfide and carbon dioxide from other gases. *See also* Activated MDEA.

Kohl, H.L. and Nielsen, R.B., *Gas Purification*, 5th ed., Gulf Publishing Co., Houston, TX, 1997, 42.

MECER A process for recovering copper from solutions used for etching printed circuit boards. The process uses solvent extraction with a solution of a proprietary ß-diketone. Developed by Sigma Metallextraktion, Sweden, from 1978 and now used by many companies in many countries.

PCT Patent Application WO/2000/052229.
Cox, M., in *Developments in Solvent Extraction*, Alegret, S., Ed., Ellis Horwood, Chichester, U.K., 1988, 177.

MEDISORBON An adsorptive process for removing mercury and dioxins from flue gas. The adsorbent is a dealuminated zeolite Y manufactured by Degussa. For mercury removal, the zeolite is impregnated with sulfur. Developed in 1994 by Lurgi Energie und Umwelt and piloted in Germany and the Netherlands.

Chem. Eng. (N.Y.), 1994, **101**(10), 19.

MegaDME A process for making dimethyl ether from methanol. Developed by Lurgi, based on their *Mega-Methanol process, but not yet commercialized.

Pontzen, F., Liebner, W., Gronemann, V., Rothaemel, M., and Ahlers, B., *Catal. Today*, 2011, **171**(1), 242.

Mega-Methanol This is a category of methanol-producing plant, not a particular process. Such plants have capacities of up to 10,000 tons per day. Engineering companies that have designed or built such plants include Lurgi and Davy.

> *Chem. Eng. (Rugby, Engl.)*, 2002, (732), Suppl., 5.

Megammonia A process for making ammonia. It uses oxygen, instead of steam, and novel axial–radial reactors that reduce pressure-drop and catalyst quantities. Developed by Lurgi Oel-Gas-Chemie and Ammonia Casale, for which they received the AstraZeneca Award for Green Chemistry in 2003.

MEGOX A process for increasing the rates of microbiological processes by the use of pure oxygen instead of air.

Mehra (1) [Named after the inventor] A process for extracting particular hydrocarbons from natural or synthetic gas streams using solvent extraction into polyalkylene glycol dialkyl ethers. Invented in 1982 by Y.R. Mehra at the El Paso Hydrocarbons Company, Odessa, TX.

> U.S. Patent 4,421,535.
> Mehra, Y.R., in *Encyclopedia of Chemical Processing and Design*, McKetta, J.J. and Cunningham, W.A., Eds., Marcel Dekker, New York, 1990, **31**, 35.

Mehra (2) [Named after the inventor] A gas separation process utilizing absorption in a solvent at moderate pressures. Developed by Advanced Extraction Technologies and applied to hydrogen recovery, nitrogen rejection, and recovery of natural gas liquids.

> U.S. Patent 5,551,972.
> *Hydrocarbon Process. Int. Ed.*, 1997, **76**(5), 15.
> Bell, C.J. and Mehra, Y.R., *Oil Gas J.*, 1997, **95**(39), 86.

Meissner *See* Schmidt.

Mellon A process for re-refining used oil, using solvent extraction and distillation.

> *Oil Gas J.*, 1994, **92**(22), 87.

membrane cell A refinement of the *diaphragm cell process in which the diaphragm is made from a cation-exchange membrane. *See also* Castner–Kellner.

MEMBREL A process for making an aqueous solution of ozone by electrolyzing water, using a solid perfluorinated cation-exchange membrane as the electrolyte. The membrane was invented at Brookhaven National Laboratory, originally for use in fuel cells. It was subsequently developed in the 1980s by Asea Brown Boveri, Switzerland, for making ozone for purifying water. The process is now offered by Ozonia International, Switzerland. Ozonia is a joint venture of Degrémont with L'Air Liquide. Water purified in this way is used in the electronics and pharmaceutical industries.

> Stucki, S., Theis, G., Kötz, R., Devanaty, H., and Christen, H.J., *J. Electrochem. Soc.*, 1985, **132**(2), 367.
> *Water Waste Treat.*, Mar 1996, 26.

Menstruum Also known as the McKenna process. A process for making the carbides of niobium and tantalum from their respective oxides. The oxide is reduced by heating to 2000°C with aluminum in a graphite vessel. Graphite lumps are then added and the heating continued. After cooling, the excess of aluminum and aluminum carbide is dissolved out with hydrochloric acid. Graphite is removed by flotation, leaving the crystalline carbide. Invented in 1937 by P.M. McKenna.

> U.S. Patents 2,113,353; 2,113,354; 2,113,355; 2,113,356.

MEP [**M**ultiple **E**thylene **P**ropylene technology] A process for making these gases from heavy hydrocarbon feedstocks. Invented by Keyuan Petrochemicals, Ningbo, China, in ca. 2010. Currently used in China.

Chem. Week, 2011, **173**(3), 5.

Mercapsol A process for removing mercaptans from petroleum fractions, using aqueous sodium or potassium hydroxide containing cresols and solubility promoters. Developed by the Pure Oil Company, a division of the Union Oil Company of California, and first operated in West Virginia in 1941.

Unzelman, G.H. and Wolf, C.J., in *Petroleum Processing Handbook*, Bland, W.F. and Davidson, R.L., Eds., McGraw-Hill, New York, 1967, 3–116.

Mercerization A process for modifying cotton textiles by treatment with alkali. The alkali is cold, concentrated aqueous sodium hydroxide; it is subsequently removed by washing with acetic acid. The process is generally conducted while the textile is held under tension. The product has improved luster and is easier to dye. Invented by J. Mercer in 1844.

MercOx A process for removing mercury and sulfur dioxide from flue gases. Hydrogen peroxide is first sprayed into the gas, converting metallic mercury to mercuric ions in solution. A water spray removes the sulfur dioxide as sulfuric acid. Mercury is removed from the liquor by ion exchange, and the sulfate is precipitated as gypsum. Developed by Uhde and Gotaverken, with the Institut für Technische Chemie.

Chem. Eng. (N.Y.), 1996, **103**(6), 19.

mercury cell *See* Castner–Kellner.

MERICAT A process for removing mercaptans from petroleum fractions by a combination of catalytic oxidation and extraction with aqueous sodium hydroxide, using a proprietary contactor based on a bundle of hollow fibers. The sulfur products are disulfides, which remain in the hydrocarbon product. Developed by the Merichem Co., Houston, TX, and used in 61 plants as of 1991. MERICAT II is a variation that also includes a carbon bed; there were four installations as of 1991. *See also* Thiolex.

Hydrocarbon Process. Int. Ed., 1996, **75**(4), 126.
Kohl, H.L. and Nielsen, R.B., *Gas Purification*, 5th ed., Gulf Publishing Co., Houston, TX, 1997, 407.

MERICON A process for oxidizing and neutralizing spent alkali solutions from oil refining. Developed by the Merichem Co., Houston, TX. Five units were operating as of 1991.

Hydrocarbon Process. Int. Ed., 1992, **71**(4), 120.

MERIFINING A process for extracting aromatic mercaptans and organic acids from cracked hydrocarbon fractions by aqueous alkali, using a bundle of hollow fibers. Developed by the Merichem Company, Houston, TX. Twelve units were operating as of 1991.

Hydrocarbon Process. Int. Ed., 1996, **75**(4), 126.

Merox [**Mer**captan **ox**idation] A process for removing mercaptans from petroleum fractions by extracting them into aqueous sodium hydroxide and then catalytically oxidizing them to disulfides using air. The catalyst is an organometallic compound: either a vanadium phthalocyanine supported on charcoal or a sulfonated cobalt phthalocyanine. Developed by UOP in 1958 and widely licensed. More than 1550 units had been commissioned by 1998. *See also* Minalk, Sweetfrac.

Unzelman, G.H. and Wolf, C.J., in *Petroleum Processing Handbook*, Bland, W.F. and Davidson, R.L., Eds., McGraw-Hill, New York, 1967, 3–128.
Basu, B., Satapathy, S., and Bhatnagar, A.K., *Catal. Rev. Sci. Eng.*, 1993, **35**(4), 571.

Hydrocarbon Process. Int. Ed., 1996, **75**(4), 128; 2000, **79**(4), 79.
Dziabis, G.A., in *Handbook of Petroleum Refining Processes*, 3rd ed., Meyers, R.A., Ed., McGraw-Hill, New York, 2003, 11.31.

Merrifield A process for synthesizing peptides, using a solid-phase support. The basic process was invented in 1962 by R.B. Merrifield at the Rockefeller Institute for Medical Research, New York, and it was subsequently automated with the assistance of J. Stewart. The solid supports are chloromethylated copolymers of styrene with divinylbenzene, now commercially available. The process is now widely used commercially for making therapeutic medicines. Merrifield received the Nobel Prize for this work in 1984, which he chose not to patent.

Merrifield, R.B., *J. Am. Chem. Soc.*, 1963, **85**, 2149.
Merrifield, R.B., *Protein Sci.*, 1996, **5**, 1947.

Merrill–Crowe An improvement on the *Cyanide process for extracting gold from rock. The solution of gold cyanide is reduced with zinc dust, thereby precipitating the gold as a fine powder, which is filtered off and smelted. Operated in South Africa.

Yannopoulos, J.C., *The Extractive Metallurgy of Gold*, Van Nostrand Reinhold, New York, 1991, 72, 189.

Merseburg A process for making ammonium sulfate fertilizer from gypsum. The gypsum is slurried with water and ammonium carbonate solution added. Calcium carbonate precipitates and is removed, any excess of ammonium carbonate is neutralized with sulfuric acid, and the solution is concentrated until it crystallizes:

$$CaSO_4 + (NH_4)_2CO_3 = CaCO_3 + (NH_4)_2SO_4$$

Developed by IG Farbenindustrie and first installed at Oppau, Germany, in 1913. Subsequently, widely used worldwide.

Gopinath, N.D., in *Phosphoric Acid*, Vol. 1, Part 2, Slack, A.V., Ed., Marcel Dekker, New York, 1968, 541.

Meta-4 A process for converting ethylene and 2-butene into propylene by metathesis. The process operates in the liquid phase at low temperatures in the presence of heterogeneous catalyst based on rhenium oxide on alumina. The catalyst is constantly regenerated by coke combustion. Developed by IFP and the Chinese Petroleum Corporation of Taiwan. A demonstration plant was operated from 1988 to 1990, and the process was demonstrated at Kaohsiung, Taiwan, in 1999. Now offered by Axens.

Torck, B., *Chem. Ind. (London)*, 1993, (19), 742.
Chim. Hebdo, 1999, (57), 18.
Eur. Chem. News, 1999, **71**(1873); **76**(1998), 21.
Chodorge, J.A., Cosyns, J., Commerceuc, D., Debuisschert, Q., and Travers, P., in *Novel Metathesis Chemistry*, Imamoglu, Y. and Bencze, L., Eds., Kluwer Academic Publishers, Dordrecht, the Netherlands, 2003.

Metabolix A fermentation process for making gamma butyrolactone from biomass, using genetically engineered bacteria. Developed by Metabolix in 2011 and piloted in 2012. An auxiliary process is *FAST.

Chem. Eng. (N.Y.), 2011, **118**(9), 11.

Metallocene Not a process but a range of organometallic catalysts for making polyolefins. First commercialized in the early 1990s and now widely used.

Long, N.J., *Metallocenes: An Introduction to Sandwich Complexes*, Blackwell Science, Oxford, U.K., 1998.
Wittcoff, H.A., Reuben, B.G., and Plotkin, J.S., *Industrial Organic Chemicals*, 2nd ed., Wiley-Interscience, New York, 2004, 494–498, 522.
Soares, J.B.P. and McKenna, T.F.L., *Polyolefin Reaction Engineering*, Wiley-VCH, Weinheim, Germany, 2012, 62.

metal surface treatment Many processes have been developed for treating the surfaces of metals in order to protect or decorate them. Some involve chemical reactions. Others are ostensibly physical, although surface chemical reactions doubtlessly occur in all. Such processes fall outside the scope of this dictionary, but for convenience, the major ones with special names are listed here. Descriptions of most of them may be found in the references that follow: Aldip, Alplate, Alrak, Alumilite, Alzak, Angus Smith, Atrament, Banox, Barff, Bengough–Stuart, Bethanizing, Bonderizing, Borchers–Schmidt, Bowers–Barff, Bullard–Dunn, Calorizing, Chromizing, Coslettizing, Footner, Granodizing, Hanson-Van Winkle-Munning, Laxal, Madsenell, MVB (modified Bauer Vogler), Nitralizing, ONERA, Parkerizing, Phosbrite, Protal, Pylumin, Sendzimir, Sheppard, Sherardizing, Shimer, Walterization, Zincote.

Burns, R.M. and Bradley, W.W., *Protective Coatings for Metals*, 3rd ed., Reinhold Publishing, New York, 1967.
Tottle, C.R., *An Encyclopedia of Metallurgy and Materials*, The Metals Society and Macdonald & Evans, London, U.K., 1984.

Metalysis A one-step process for making titanium metal powder from mineral rutile. Developed by Metalysis in Rotherham, Yorkshire, and announced in 2012.

Metasolvan Solvent extraction of phenols from tar oils. Developed by Lurgi during World War II.

BIOS Report 623.

METC [Morgantown Energy Technology Center] A coal gasification process based on a stirred, fixed-bed gasifier. Developed for the U.S. Department of Energy.

METEOR A process for making ethylene glycol from ethylene oxide. Developed by Union Carbide, first operated in 1994, and now offered by Dow Chemical Company and Aker Kvaerner. A plant in Kuwait started operating in 2008. The first Chinese plant was a joint venture of Sinopec and Sabic and started operating in Tianjin in 2010.

Chem. Week, 2006, **18**(23), **6**, 15.
Chem. Eng. World, 2009, **44**(1), 104.

METEX [METal EXtraction] A process for extracting heavy metals from industrial waste waters by adsorption on activated sludge under anaerobic conditions. It is operated in an up-flow, cylindrical reactor with a conical separation zone at the top. Developed by Linde, originally for removing dissolved copper from wine-making wastes. First commercialized in 1987.

Methanation Any process for making methane. Most commonly, from carbon monoxide; less commonly, from carbon dioxide or a mixture of the two. Biomethanation is the production of methane from biomass.

METLCAP A process for encapsulating hazardous heavy metal wastes in a proprietary type of cement. Developed and offered by Environmental Remediation Technology, Cleveland, OH.

Metocene A range of propylene polymerization processes using *metallocene catalysts developed and now licensed by LyondellBasell Industries. The name is used also for polypropylenes made by these processes.

High Perform. Plast., May 2006, 10.
Chem. Eng. Prog., 2006, **102**(5), 14.

Metolachlor A process for making the herbicide Dual Magnum. Developed by Ciba-Geigy from 1982. The catalyst is a chiral iridium ferrocenyl compound (the Josephson ligand). In 2008, more than 10,000 tons of metolachlor per year were being made by this process.

Rothenberg, G., *Catalysis: Concepts and Green Applications*, 2008, Wiley-VCH, Weinheim, Germany, 102.

Metrex A process for recycling spent hydroprocessing catalysts. Developed in 1993 by Metrex BV.

Met-X A continuous process for removing traces of metals from cracking catalysts by ion exchange. Developed by Atlantic Refining Company and first operated in Philadelphia in 1961.

Leum, L.M. and Connor, J.E., Jr., *Ind. Eng. Chem. Prod. Res. Dev.*, 1962, **1**(3), 145.
Unzelman, G.H. and Wolf, C., in *Petroleum Processing Handbook*, Bland, W.F. and Davidson, R.L., Eds., McGraw-Hill, New York, 1967, 3–15.

MeXTH [**Me**thyl halides **t**o **H**ydrocarbons] (X = Cl or Br) A general name for processes for converting methanol (from methane) to hydrocarbons via methyl halides. This has been proposed as an alternative to the *MTO process, requiring less energy, but it has not yet been commercialized. An oxychlorination step is followed by reaction over a zeotype catalyst such as H-SAPO-34.

U.S. Patents 4,769,504; 5,087,786.
Svelle, S., Aravintham, S., Bjorgen, M., Lllirad, K.-P., Kolboe, S., Dahl, I.M., and Olsbye, U., *J. Catal.*, 2006, **241**, 243.
Olsbye, U., Saure, O.V., Muddada, N.B., Bordiga, S., Lamberti, C., Nilsen, M.H., and Lillerud, K.P., *Catal. Today*, 2011, **171**(1), 211.

Meyers *See* TRW Meyers.

M-forming A process for increasing the octane rating of gasoline by cracking and isomerization, catalyzed by the zeolite ZSM-5. Developed in the 1970s by Mobil Corporation but not commercialized. A related process, M2-forming, for aromatizing light aliphatic hydrocarbons over HZSM-5, was not commercialized either. *See also* Cyclar.

Chen, N.Y. and Yan, T.Y., *Ind. Eng. Chem. Proc. Des. Dev.*, 1986, **25**(1), 151.
Chen, N.Y., Garwood, W.E., and Heck, R.H., *Ind. Eng. Chem. Prod. Res. Dev.*, 1987, **26**, 706.
Chen, N.Y. and Degnan, T.F., *Chem. Eng. Prog.*, 1988, **84**(2), 32.

MGCC [**M**itsubishi **G**as **C**hemical **C**ompany] Also called **JGCC**. A process for extracting *m*-xylene from mixed xylene isomers by making the fluoroboric acid complex. All the xylene isomers form such complexes, but that formed by the *m*-isomer is much more stable than the others. Development started in 1962; by 1979, three plants were operating.

Hydrocarbon Process. Int. Ed., 1969, **48**(11), 254.
Herrin, G.R. and Martel, E.H., *Chem. Eng. (Rugby, Engl.)*, 1971, (253), 319.
Masseling, J.J.H., *ChemTech*, 1976, **6**, 714.

MHC [**M**itsubishi **H**ydro**C**racking] A process for making benzene and other aromatic hydrocarbons by hydrogenating cracked petroleum fractions.

MHC Unibon [**M**ild **H**ydro**C**racking] A mild *hydrocracking process for desulfurizing gas oil and converting it to lower molecular weight hydrocarbons, suitable for further processing by catalytic cracking. Developed by UOP.

MHD [**M**itsubishi **H**ydro**D**ealkylation] A thermal process for converting toluene to benzene. Developed by Mitsubishi Chemical.

Weissermel, K. and Arpe, H.-J., *Industrial Organic Chemistry*, translated by C.R. Hindley and S. Hawkins, 4th ed., Wiley-VCH, Weinheim, Germany, 2003, 332.

MHDV [**M**obil **H**ydrogen **D**onor **V**isbreaking] A modified *visbreaking process in which a hydrogen donor stream from the oil refinery is added to the heavy hydrocarbon stream before thermal cracking. Developed by Mobil Corporation.

MHI [**M**itsubishi **H**eavy **I**ndustries] A coal gasification process designed for coals whose ash has a high melting point. Piloted at Nakoso, Japan.

Rezaiyan, J. and Cheremisinoff, N.P., *Gasification Technologies*, Taylor & Francis, Boca Raton, FL, 2005, 46.

MHO [**M**etallurgie **H**oboken-**O**verpelt] A process for extracting manganese and other metals from nodules from the sea bed by extraction with hydrochloric acid:

$$MnO_2 + 4HCl = MnCl_2 + 2H_2O + Cl_2$$

The chlorine formed in this stage is reused in a subsequent stage, where it oxidizes the manganese (II) to manganese (IV), which precipitates as MnO_2. Developed by the Belgian company named.

Gupta, C.K. and Mukherjee, T.K., *Hydrometallurgy in Extraction Processes*, Vol. 1, CRC Press, Boca Raton, FL, 1990, 10.

MHTI [**M**obil **H**igh **T**emperature **I**somerization] A process for converting mixed xylene streams to *p*-xylene. The catalyst is the zeolite ZSM-5. Developed by Mobil Research & Development Corporation and first commercialized in 1981. Eleven units were operating as of 1991. *See also* MLPI and MVPI.

Hydrocarbon Process. Int. Ed., 1991, **70**(3), 166.

MHUG [**M**ild **H**ydrocracking **U**p**G**rading] A single-stage hydrocracking process for converting petroleum intermediates into valuable end products. The catalyst contains nickel and tungsten on a zeolite. Developed by Sinopec (China) in the early 1990s and engineered by KTI.

Donghong, J., Yuying, Z., and Hong, N., *China Pet. Process. Petrochem. Technol.*, 2012, **14**(1), 1.

Michigan *See* Grainer.

Micro-Simplex *See* MS.

Mid-Century Also called **M-C**. A process for oxidizing *p*-xylene to terephthalic acid, using oxygen in acetic acid and catalyzed by a mixture of cobalt and manganese bromides. Developed in the 1950s by Halcon International and commercialized by Standard Oil Company (Indiana). The first plant was built at Joliet, IL, in 1938. The *Amoco and *Maruzen processes are improved versions.

U.S. Patent 2,833,816.
Landau, R. and Saffer, A., *Chem. Eng. Prog.*, 1968, **64**(10), 20.
Landau, R., *Chem. Eng. Prog.*, 1988, **84**(7), 31.
Parteinheimer, W., in *Catalysis of Organic Reactions*, Blackburn, D.W., Ed., Marcel Dekker, New York, 1990, 321.
Raghavendrachar, P. and Ramachandran, S., *Ind. Eng. Chem. Res.*, 1992, **31**, 453.

Middox A process for delignifying wood pulp by the use of oxygen. Developed jointly by Air Products and Chemicals and Black Clawson Company. The process removes half of the lignin from the pulp, thereby halving the chlorine usage.

Chem. Week, 1981, **129**(17), 17.

Midforming [**Mid**dle-range distillate **forming**] A process for converting lower olefins to transport fuels. The catalyst is either a ZSM-5-type zeolite, in which some of the aluminum has been replaced by iron, or a heteropoly acid. Developed in the 1980s by the National Chemical Laboratory, Pune, India. To be piloted by Bharat Petrochemical Corporation, Bombay, and Davy Powergas.

Indian Patent Application 985/DEL/87.
U.S. Patent 4,950,821.

MIDREX [**Mid**land-**R**oss] A process for the direct reduction of iron ore, using a mixture of carbon monoxide and hydrogen obtained by reforming of natural gas with water vapor and carbon dioxide. The keys to the process are a special shaft furnace and a stoichiometric reformer. The shaft furnace has several design features to facilitate gas–solid contact. The reformer is fed by a mixture of natural gas and off-gas from the shaft furnace. The reformer catalyst is nickel oxide supported on alumina. Developed by the Surface Combustion Corporation (a subsidiary of Midland-Ross Corporation), Toledo, OH, in the 1960s, by D. Beggs and C. Sanzenbacher. The prototype plant was completed in Portland, OR, in 1969; the first large-scale plant was built in 1971 by the Georgetown Steel Corporation, Georgetown, SC. As of mid-2007, 60 plants were either operating or under construction worldwide, with a total capacity of 40 million tons. The process is now licensed by Kobe Steel and marketed by Midrex Technologies of Charlotte, NC. *See also* DR.

Dayton, S., *Eng. Min. J.*, 1979, **180**(1), 80.
Metius, G.E., McClelland Jr., J.M., and Anderson, S.H., *Steel Times Int.*, 2006, **30**(2), 32.

MIDW [**M**obil **I**somerization **DeW**axing] A petroleum refining process that improves yield and quality by isomerizing and selectively cracking paraffins in waxy oils. The catalyst is a noble metal, supported on a zeolite. Developed by Mobil Corporation from 1991 to 1996. In 2005, ExxonMobil announced that MIDW units were to be installed at Petro-Canada's refineries in Montreal and Edmonton.

Chem. Eng. (N.Y.), 1996, **103**(3), 19.

Miex [**M**agnetized **i**on **ex**change resin] A process for removing dissolved organic compounds from potable water by adsorption on a magnetic anionic exchange resin. This is a continuous process: the water makes contact with the resin in a stirred tank, followed by a settler. The magnetization promotes aggregation of the resin beads, giving fast settling. The resin was developed by the CSIRO and Orica, Melbourne, Australia. The process was developed by Orica and SA Water Corporation, Adelaide, Australia. First operated in 2002 in Perth by the Water Corporation of Western Australia, for removing high levels of dissolved organic carbon.

Rogella, F., *Water Waste Treat.*, 2002, **45**(7), 30.
Water Wastewater Int., 2002, **16**(4), 37.

MIGAS [**MI**tsubishi **GAS**] A process for making methyl methacrylate. Developed by the Mitsubishi Gas Chemical Company in 1992.

Miller A process for purifying and removing silver from gold by passing chlorine gas through the molten metal, covered with borax. The silver forms silver chloride, which floats to the top. Bismuth, antimony, and arsenic are eliminated as their volatile chlorides. Developed by F.B. Miller at the Sydney Mint in Australia in 1867 and soon used worldwide.

Yannopoulos, J.C., *The Extractive Metallurgy of Gold*, Van Nostrand Reinhold, New York, 1991, 242.

MILOS [**MI**ddle distillates and **L**ower **O**lefins **S**elective process] A modified *FCC process for increasing the yield of diesel and propylene. It uses two risers, which recycle and reactivate the catalyst.

Oil Gas J., 2008, **106**(16), 10.
Chem. Eng. (N.Y.), 2008, **115**(6), 20.
Hydrocarbon Proc., 2009, Suppl., 9.

MILOX [**MIL**ieu **OX**idative] A wood-pulping and bleaching process. Wood chips are treated with hydrogen peroxide and formic acid in a three-stage process. Developed by the Finnish Pulp and Paper Research Institute (KCL) and Kemira. Piloted at Oulu, Finland, in 1991.

Eur. Chem. News, 1990, **55**, 8.
Eur. Chem. News (Finland Suppl.), May 1991, **56**, 4.
Chem. Br., 1991, **27**, 687.
Sundquist, J. and Poppius-Lerlin, K., in *Environmentally Friendly Technologies for the Pulp and Paper Industries*, Young, R.A. and Akhar, M., Eds., John Wiley & Sons, New York, 1998, 157.

Minalk A variation of the *Merox process in which a small quantity of a weak aqueous alkali is continuously added to the catalyst bed in order to promote the oxidation of mercaptans to disulfides.

> Dziabis, G.A., in *Handbook of Petroleum Refining Processes*, 3rd ed., Meyers, R.A., Ed., McGraw-Hill, New York, 2003, 11.34.

Minemet A hydrometallurgical process for extracting metals from sulfide ores by leaching with ferric chloride solution. Developed by Minemet Recherche, France.

> Gupta, C.K. and Mukherjee, T.K., *Hydrometallurgy in Extraction Processes*, Vol. 1, CRC Press, Boca Raton, FL, 1990, 196.

Minex A process for removing hydrogen sulfide and carbon dioxide from gases and light hydrocarbon streams in oil refineries. Developed by the Merichem Company, Houston, TX.

> *Hydrocarbon Process. Int. Ed.*, 1992, **71**(4), 120.

Minifos A process for making monoammonium and diammonium phosphates by causing a reaction between ammonia and phosphoric acid. Offered by Lurgi.

> Ranney, M.W., *Ammonium Phosphates*, Noyes Data Corp. Process Review No. 35, Noyes Publications, Park Ridge, NJ, 1969, 30.

MISHA [Microwave-Induced Sulfur and Hydrogen Autolysis] A process for splitting hydrogen sulfide into its elements, using microwave energy. Developed in 1998 at the Argonne National Laboratory, Argonne, IL.

> *Chem. Eng. (N.Y.)*, 1998, **105**(5), 116.

Mitsui-BF [Bergbau-Forschung] A process for removing sulfur dioxide from flue gas by adsorption on a moving bed of activated coke. A complex process leads to the production of sulfuric acid.

> Kohl, H.L. and Nielsen, R.B., *Gas Purification*, 5th ed., Gulf Publishing Co., Houston, TX, 1997, 639.

Mitsui Hypol *See* *Hypol.

Mitsui Toatsu A high-pressure process for making urea from ammonia and carbon dioxide. Invented in 1967 by Toyo Koatsu Industries.

> U.S. Patent 3,506,710.
> Zardi, U., *Nitrogen*, 1982, (135), 26.

MixAlco [Mixed Alcohol] A fermentation process for making a mixed alcohol fuel from biomass.

> *Fuels and Chemicals from Biomass*, Saha, B.C. and Woodward, J., Eds., American Chemical Society, Washington, DC, 1997.

MKS [Melnichuk-Kelly-Stanko] A process for making biogasoline from wood. The wood is first gasified in a fluidized bed, and the resulting syngas is catalytically converted to a C_5–C_{12} hydrocarbon mixture. Developed by CORE Biofuel. The first plant, announced in 2012, is to be engineered by Technip.

> U.S. Patent Application 2011-0152593.

MLDW [Mobil Lube DeWaxing] A catalytic process for removing waxes (long-chain linear aliphatic hydrocarbons and alkyl aromatic hydrocarbons) from lubricating oil. Developed by Mobil Research & Development Corporation and operated at Mobil Oil refineries since 1981. Eight units were operating in 1991.

MLPI [**M**obil **L**ow-**P**ressure **I**somerization] One of a family of processes developed by Mobil Corporation for isomerizing xylene mixtures, using a zeolite catalyst. This one was developed in 1977. *See also* LTI, MHTI, MVPI.

U.S. Patent 4,101,596.

Mobil The Mobil Corporation has developed many processes, but since the 1980s, the one most associated with its name has been the methanol-to-gasoline process, using a zeolite catalyst. *See* MTG.

Mobil/Badger A process for making ethylbenzene by reacting benzene with ethylene, in the vapor phase, over a ZSM-5 zeolite catalyst containing phosphorus. Diluted ethylene streams, from a variety of industrial sources, may be used. Developed in the 1970s by the Badger Company, using a catalyst developed by Mobil Corporation. First operated on a large scale in 1980 by American Hoechst Company; by 1991, 21 plants had been built. Over 45 units had been licensed by 2005. An improved version of the ZSM-5 catalyst, EBUF-1, was developed by Fina Oil & Chemical Company and United Catalysts from 1981 and used at the Cosmar company's plant in Louisiana from 1994. *See also* EB Max.

U.S. Patent 3,962,364.
Hölderich, W.F. and van Bekkum, H., in *Introduction to Zeolite Science and Practice*, van Bekkum, H., Flanigen, E.M., and Jansen, J.C., Eds., Elsevier, Amsterdam, the Netherlands, 1991, 664.
Fallon, K.J., Wang, H.K.W., and Venkat, C.R., *Oil Gas J.*, 1995, **93**(16), 50.
Degnan, T.F., Jr., Smith, C.M., and Venkat, C.R., *Appl. Catal. A: Gen.*, 2001, **221**(1–2), 284.

Mobil/Badger cumene A process for making cumene by reacting benzene with propylene. Developed from the *Mobil/Badger process for making ethylbenzene. First commercialized at the Georgia Gulf plant in Pasadena, TX, in 1996. A variation of this process, announced in 1993, reduces the benzene content of gasoline by reaction with propylene from an *FCC plant. The process now uses zeolite MCM-22 as the catalyst.

Goelzer, A.R., Hernandez-Robinson, A., Ram, S., Chin, A.A., Harandi, M.H., and Smith, C.M., *Oil Gas J.*, 1993, **91**(937), 63.
Eur. Chem. News, 1997, **68**(1792), 26.

Mobil–Witco–Shell A process for making poly (1-butene) by polymerizing 1-butene with a *Ziegler catalyst in an excess of liquid monomer.

Chem. Week, 7 Dec 1977, 9.

MOCVD [**M**etal **O**xide **C**hemical **V**apor **D**eposition] A general name for a group of processes used to make microelectronic devices by depositing thin films of metal oxides on suitable substrate surfaces by means of chemical vapor deposition. *See* CVD.

MODAR [Named after **M**odell, the inventor, and his original partners **O**'**D**onnell **A**nd **R**ich] An application of *supercritical water oxidation (SCWO). A process for destroying organic wastes, especially hydrocarbons, by oxidation under supercritical aqueous conditions. The wastewater and alkali are heated to 450°C–650°C under a pressure of >200 atm. Hydrocarbons are rapidly oxidized to carbon dioxide; organic compounds of halogens, sulfur, and phosphorus are converted to the respective inorganic acids, which are neutralized. Invented in 1980 by M. Modell in Cambridge, MA, developed by MODAR Inc., and engineered and commercialized with ABB Lummus Crest in 1989. The process was piloted with pharmaceutical wastes at Pfinztal, near Karlsruhe, Germany, in 1994. Modar Inc. was acquired by General Atomics in 1996.

Modar is also the trade name for a range of acrylic resins and the name of a maker of modular architectural components.

U.S. Patents 4,113,446; 4,338,199; 4,543,190; 4,822,497.
Chem. Week, 1986, **139**(14), 40.
Modell, M., 1992, *Chem. Eng. News*, **70**(4), 2.
Bettinger, J.A., *Chem. Eng. News*, 1992, **70**(10), 2.

Water Waste Treat., 1994, **37**(9), 32.
Chem. Eng. (Rugby, Engl.), 1994, (568), 24.

MODOP [**M**obil **O**il **D**irect **O**xidation **P**rocess] A process for removing residual sulfur-containing gases from the tail gas from the *Claus process. The catalyst is titanium dioxide, pelletized with calcium sulfate. Developed in the 1980s by Rhône-Poulenc, Procatalyse, and Mobil Oil. In 1995, three plants were operating in Germany and one in the United States.

European Patents 60,742; 78,690.
Kettner, R. and Liermann, N., *Oil Gas J.*, 1988, **86**(2), 63.
Hydrocarbon Proces. Int. Ed., 1992, **71**(4), 122.
Wieckowska, J., *Catal. Today*, 1995, **24**(4), 445.

Moebius An electrolytic process for removing gold and platinum from silver. The crude metal, known as Doré, is used as the anode. The cathodes are of silver or stainless steel. The electrolyte is a diluted solution of silver nitrate and nitric acid. Gold and other metals collect as anode slimes. Invented in Mexico by B. Moebius, first used there in 1884, and subsequently widely used in Germany and the United States. *See also* Balbach, Thum, Wohlwill.

British Patent 16,554 (1884).
Dennis, W.H., *A Hundred Years of Metallurgy*, Gerald Duckworth, London, U.K., 1963, 288.

Mofex A liquid–liquid extraction process for removing aromatic hydrocarbons from hydrocarbon mixtures. The solvent is a monomethylformamide–water mixture, operated at 20°C–30°C, 0.1–0.4 bar. Developed by Leuna Werke.

Weissermel, K. and Arpe, H.-J., *Industrial Organic Chemistry*, 4th ed., translated by C.R. Hindley and S. Hawkins, Wiley-VCH, Weinheim, Germany, 2003, 322.

Moffat *See* Steelmaking.

MOG [**M**obil **O**lefins to **G**asoline] A process for converting dilute streams of C_2 to C_4 hydrocarbons to gasoline, using a fluidized bed of zeolite ZSM-5 catalyst. Developed by Mobil Research & Development Corporation and piloted in 1990.

Hydrocarbon Process. Int. Ed., 1994, **73**(11), 142.

MOGD [**M**obil **O**lefins to **G**asoline and **D**istillate] A process for converting C_2 to C_{10} olefins to high-octane gasoline and other hydrocarbons using an HZSM-5 catalyst. Developed by Mobil Corporation and first used at its refinery in Paulsboro, NJ, in 1982.

Garwood, W.E., in *Intrazeolite Chemistry*, Stucky, G.D. and Dwyer, F.G., Eds., American Chemical Society, Washington, DC, 1983, 383.
Van Grieken, R., Escola, J.M., Moreno, J., and Rodriguez, R., *Appl. Catal. A: Gen.*, 2008, **337**(2), 173.

Mohawk A process for recovering used automotive oils for reuse. Mohawk uses distillation and catalytic hydrogenation. Developed in Canada. The first plant was built by Evergreen Holdings and Chemical Engineering Partners in Newark, CA.

Hydrocarbon Process. Int. Ed., 1990, **69**(9), 26.

MOI [**M**obil **O**lefin **I**nterconversion] A process for increasing the yield of ethylene and propylene from *steam crackers and *fluid catalytic crackers, using a ZSM-type catalyst. Developed from 1998 by Mobil Technology.

Eur. Chem. News, 1998, **69**(1808), 39.
Eur. Chem. News, Mar 2002, *Petrochemicals Fact File, Suppl.*, 15.
Wittcoff, H.A., Reuben, B.G., and Plotkin, J.S., *Industrial Organic Chemicals*, 2nd ed., Wiley-Interscience, Hoboken, NJ, 2004, 171.

Molex A version of the *Sorbex process, for separating linear aliphatic hydrocarbons from branched-chain and cyclic hydrocarbons in naphtha, kerosene, or gas oil. The process operates in the liquid phase and the adsorbent is a modified 5A zeolite; the pores in this zeolite will admit only the linear hydrocarbons, so the separation factor is very large. First commercialized in 1964; 28 units had been built by 2005. *See also* Parex (2).

Carson, D.B. and Broughton, D.B., *Pet. Refin.*, 1959, **38**(4), 130.
Broughton, D.B., *Chem. Eng. Prog.*, 1968, **64**(8), 60.
Sohn, S.W., in *Handbook of Petroleum Refining Processes*, 3rd ed., Meyers, R.A., Ed., McGraw-Hill, New York, 2003, 10.75.

MOLPSA-nitrogen [**MOL**ecular sieve **P**ressure **S**wing **A**dsorption] A version of the *PSA process for separating nitrogen from air, developed by Kobe Steel. Most PSA processes for nitrogen use molecular sieve carbon as the adsorbent, but this one uses zeolite X. Water and carbon dioxide are first removed in a two-bed PSA system; the nitrogen is then concentrated and purified in a three-bed system.

Suzuki, M., in *Adsorption and Ion Exchange: Fundamentals and Applications*, LeVan, M.D., Ed., American Institute of Chemical Engineers, New York, 1998, 121.

Molten Carbonate A *flue-gas desulfurization process in which the sulfur dioxide contacts a molten mixture of inorganic carbonates. These are converted to sulfates and sulfides and then reduced to hydrogen sulfide, which is treated in a *Claus kiln. The advantage of this process over most others is that it does not cool the flue gases. Developed by Rockwell International, but not commercialized.

Oldenkamp, R.D. and Margolin, E.D., *Chem. Eng. Prog.*, 1969, **65**(11), 73.
Kohl, H.L. and Nielsen, R.B., *Gas Purification*, 5th ed., Gulf Publishing Co., Houston, TX, 1997, 603.

Moltox A process for separating oxygen from air by selective absorption in a molten salt mixture at high temperature. Invented by D.C. Erickson of Energy Concepts and developed by Air Products and Chemicals. The salts are a mixture of the nitrites and nitrates of sodium and potassium. The reaction is

$$NO_2^- + \tfrac{1}{2}O_2 \rightarrow NO_3^-$$

The operating temperature range is 450°C–700°C; either pressure swing or thermal-swing modes can be used.

U.S. Patents 4,132,766; 4,287,170; 4,340,578.
Erickson, D.C., *Chem. Eng. (N.Y.)*, 1983, **90**(21), 28.
Dunbobbin, B.R. and Brown, W.R., *Gas Sep. Purif.*, 1987, **1**, 23.

Monarch A process for saving energy in sulfuric acid plants using the *contact process. Developed by Monsanto.

Mond A process for recovering sulfur from the residues from the *Leblanc process. The sulfur is partially oxidized to thiosulfate and converted to elemental sulfur by adding hydrochloric acid. This process recovers only half the sulfur and was supplanted by the *Chance process. Invented by L. Mond and operated by the Netham Chemical Company at Bristol from 1868 to 1888.

Holland, R., *Chem. Ind. (London)*, 1985, (11), 367.

Mond gas A process for gasifying coal at a relatively low temperature, using a mixture of air and steam. The use of steam increases the yield of ammonia. The process was invented primarily to produce the ammonia needed for the *ammonia soda process. The gas is of low calorific value but can be used for industrial heating. Developed by L. Mond at Brunner Mond, Winnington, England, in 1883. Subsequently commercialized by the South Staffordshire Mond Gas Corporation at Dudley Port, near Birmingham, United Kingdom, which distributed the gas to local industry through the world's first gas grid. The engineering and further commercialization were carried out by the Power Gas Corporation.

British Patent 3,923 (1883).
Hill, W.H., in *Chemistry of Coal Utilization*, Vol. 2, Lowry, H.H., Ed., John Wiley & Sons, New York, 1945, 1028.
Cohen, J.M., *The Life of Ludwig Mond*, Methuen & Co., London, U.K., 1956, 176.

Mond nickel A process for extracting nickel from its ores by the intermediary of the volatile nickel tetracarbonyl. Sulfide ores are first roasted to convert sulfides to oxides and then reduced by heating in hydrogen and carbon monoxide (water gas). The crude metal is caused to react with carbon monoxide at 50°C, producing $Ni(CO)_4$, which is subsequently decomposed at 180°C–200°C. Invented by L. Mond and C. Langer in 1889, piloted at the works of Henry Wiggin & Company in Smethwick, Scotland in 1892, and subsequently commercialized on a large scale in Swansea, South Wales, where it still operates (under the ownership of CVRD Inco). A new plant was built in Canada in 1986.

British Patent 12,626 (1890).
Canadian Patents 35,427; 35,428.
U.S. Patents 455,228; 455,229; 455,230.
Mond, L., Langer, C., and Quincke, F., *J. Chem. Soc.*, 1890, **57**, 749.
Cohen, J.M., *The Life of Ludwig Mond*, Methuen, London, U.K., 1956, 282.
Abel, E., *Chem. Br.*, 1989, **25**, 1014.

Monell *See* Steelmaking.

Monk–Irwin An unsuccessful predecessor of the *Sulfate process for making titanium dioxide pigment from ilmenite. Invented by C.R. Whittemore at McGill University, Montreal, in the early 1920s and subsequently developed by J. Irwin and R.H. Monk in Canada and B. Laporte Limited in Luton, United Kingdom. Ilmenite from the deposit at Ivry, Québec, was reduced by heating with coke, leached with ferric chloride solution, and then roasted with a mixture of sulfuric acid and sodium sulfate. The resulting cake, containing titanyl sulfate, was dissolved in water and hydrolyzed, and the titania hydrate calcined. Some of the product was extended with barium sulfate. The project was abandoned in 1928.

U.S. Patent 1,542,350.

Mono² A process for making silicon cells for collecting solar energy. Developed by BP Solar from 2006 and expected to be commercialized in 2007.

Monsanto (1) A process for making adiponitrile, an intermediate in the manufacture of Nylon 66, by the electrolytic hydrodimerization (EHD) of acrylonitrile:

$$2CH_2 5CHCN + 2H_2O + 2e^- = NC(CH_2)_4CN + 2OH^-$$

The original process used aqueous tetraethylammonium ethylsulfate as the electrolyte, a lead cathode, and a lead–silver alloy anode. The Mark II process, commercialized in the mid-1970s, uses an emulsion of acrylonitrile in aqueous sodium phosphate containing a salt of the hexamethylene-bis-(ethyldibutylammonium) cation. The process was invented in 1959 by M.M. Baizer at Monsanto Corporation, St. Louis, MO. It was commercialized in 1965 and has been continually improved ever since. Asahi Chemical Industry Company also operates the process in Japan. In 1990, the world production of adiponitrile by this process was over 200,000 tons per year.

Prescott, J.H., *Chem. Eng. (N.Y.)*, 1965, **72**(23), 238.
Baizer, M.M. and Danly, D.E., *Chem. Ind. (London)*, 1979, (435), 439.
Pletcher, D. and Walsh, F.C., *Industrial Electrochemistry*, 2nd ed., Chapman & Hall, London, U.K., 1990, 298.

Monsanto (2) A catalytic process for synthesizing the drug L-DOPA. The catalyst is a chiral diphosphine–rhodium complex. Invented in the early 1970s.

Monsanto acetic acid A process for making acetic acid by carbonylation of methanol, catalyzed by rhodium carbonyl iodide. Operated by BP. A variation of this process, the "low-water" process, used added Group 1 metal iodides such as lithium iodide to enhance the productivity; this was practiced by Celenese and by Daicel.

Haynes, A., Mann, B., Morris, G.E., and Maitlis, P.M., *J. Am. Chem. Soc.*, 1993, **115**(10), 4093.
Parkins, A.W., in *Insights into Speciality Inorganic Chemicals*, Thompson, D., Ed., Royal Society of Chemistry, Cambridge, U.K., 1995, 110.
Yoneda, N., Kusano, S., Yasui, M., Pujado, P., and Wilcher, S., *Appl. Catal. A: Gen.*, 2001, **221**(1–2), 253.
Wikipedia, The Free Encyclopedia. Monsanto process. Revised 31 Mar 2013. Retrieved 20 Jul 2013.

Mon Savon A continuous soapmaking process.

Kirk-Othmer's Encyclopedia of Chemical Technology, 3rd ed., Vol. 21, John Wiley & Sons, New York, 1983, 173.

Mont Cenis [Named after a coal mine in the Ruhr, Germany] An early ammonia synthesis process, basically similar to the *Haber–Bosch process but using coke-oven gas. Operated by the Royal Dutch Group at Ijmuiden, the Netherlands, since 1929.

Scholvien, W.F., *Chem. Met. Eng.*, 1931, **38**(2), 82.
Spitz, P.H., *Petrochemicals: The Rise of an Industry*, John Wiley & Sons, New York, 1988, 84.

Montecatini A process for making nitric acid by oxidizing ammonia. It differs from related processes in the equipment used to absorb the gases: connected horizontal steel chambers operated in a countercurrent mode. Widely operated in Europe in the early twentieth century.

Chilton, T.H., *Strong Water: Nitric Acid, Its Sources, Methods of Manufacture, and Uses*, MIT Press, Cambridge, MA, 1968, 109.

Montoro A process for making styrene and propylene oxide. Named after the eponymous company. The process was to be used in Repsol Quimica's plant in Tarragona, Spain.

Chem. Mark. Rep., Apr 1997.

Morex A process for removing mercaptans from petroleum streams. Developed by UOP and intended for installation in Conoco's Lake Charles, LA, refinery in 2003.

Hydrocarbon Process. Int. Ed., 2002, **81**(2), 29.

Morgas [**Mor**gantown **gas**ification] A coal gasification process using a stirred, fixed-bed gasifier. Piloted in the 1970s at the Morgantown Research Center of the U.S. Bureau of Mines, West Virginia.

Hebden, D. and Stroud, H.J.F., in *Chemistry of Coal Utilization*, 2nd Suppl. Vol., Elliott, M.A., Ed., John Wiley & Sons, New York, 1981, 1627.

MORPHYLANE An extractive distillation process for removing aromatic hydrocarbons from hydrocarbon mixtures. The added solvent is N-formylmorpholine. The process was developed by Krupp-Koppers in the 1960s; since 1972, more than 55 plants have been built worldwide. The original design used two columns, one for extractive distillation to remove the nonaromatics and absorb the aromatics into the solvent and another to separate the solvent and aromatics. The current design, now offered by Krupp Uhde, has combined the two columns into one, with savings of energy and capital costs. *See also* MORPHYLEX, OCTENAR, Super MORPHYLANE.

Chem. Eng. (N.Y.), 2005, **112**(10), 33.

MORPHYLEX A liquid–liquid extraction process for removing aromatic hydrocarbons from hydrocarbon mixtures. The solvent is N-formylmorpholine; the operating temperature is 180°C–200°C. The process was developed by Krupp-Koppers in the 1960s. Only one plant had been built as of 1994. *See also* MORPHYLANE, OCTENAR.

Franck, H.-G. and Stadelhofer, J.W., *Industrial Aromatic Chemistry*, Springer-Verlag, Berlin, Germany, 1988, 110.

Morphysorb A process for removing organic sulfur compounds from acid gases using sorption in a solvent provided by BASF. Developed by Uhde.

Morse *See* Steelmaking.

MOSC [**M**obil **O**il **S**ludge **C**oking] A process, used in oil refineries, that converts aqueous sludges to coke, thereby reducing the quantity of waste discharged. Developed by Mobil Corporation.

MOST [**M**obil **O**il SO$_x$ **T**reatment] A catalytic process for removing sulfur-containing gases from the tail gases from the *Claus process and other SO$_x$-containing gases. The gases are first combusted with air, converting all the sulfur-containing species to SO$_2$. The SO$_2$ is adsorbed on a solid sorbent/catalyst such as vanadia-promoted magnesia spinel and then reductively desorbed as a mixture of H$_2$S and SO$_2$ for recycle to the Claus plant.

> Buchanan, J.S., Stern, D.L., Nariman, K.E., Teitman, G.J., Sodomin, J.F., and Johnson, D.L., *Ind. Eng. Chem. Res.*, 1996, **35**, 2495.

MOX [**M**ixed **OX**ides] A process for making mixed uranium and plutonium oxides for use as a nuclear fuel. Developed for the Thorp plant in Cumbria, United Kingdom.

MOXY [**M**ead **OX**ygen] A variation of the *Kraft papermaking process in which the sulfides are oxidized to polysulfides, with some increase in efficiency. Developed by Mead Corporation.

MPC [**M**itsui **P**etro**C**hemical] A continuous process for polymerizing propylene, based on the *Ziegler–Natta process, but using a much more active catalyst, so that deashing (catalyst removal) is not required. The catalyst contains magnesium in addition to titanium; successive versions of it have been known as HY-HS (high yield, high stereospecificity), HY-HS II, and T-catalyst. Developed jointly by Mitsui Petrochemical Industries, Japan, and Montedison, Italy, in 1975, and subsequently licensed in 56 plants worldwide.

MQD Unionfining A process for making ultralow-sulfur diesel fuel, using several different catalysts. Developed by UOP.

> Stanislaus, A., Marafi, A., and Rana, M.S., *Catal. Today*, 2010, **153**, 36.

MRG [**M**ethane **R**ich **G**as] A catalytic steam-reforming system, similar to the classic *syngas reaction of steam with a hydrocarbon mixture, but yielding hydrogen, methane, and carbon monoxide in different proportions. The system is thermodynamically balanced, requiring no heat other than that required to raise the reactants to the operating temperature. Developed by the Japan Gasoline Company.

MRH (1) [**M**ethanol **R**eformer for **H**ydrogen] A process for generating hydrogen from methanol, separating it by *PSA. Developed by the Marutani CPE Company.

> Suzuki, M., in *Adsorption and Ion Exchange: Fundamentals and Applications*, LeVan, M.D., Ed., American Institute of Chemical Engineers, New York, 1998, 121.

MRH (2) A *hydrocracking process for "difficult" petroleum residues, that is, those containing high levels of metals, sulfur, and nitrogen compounds. MRH uses catalytic hydrogenation in a slurry bed. Developed by the M.W. Kellogg Company.

> Marcos, F. and Rosa-Brussin, D., *Catal. Rev. Sci. Eng.*, 1995, **37**(1), 3.

MRU [**M**ethanol **R**ecovery **U**nit] A process for removing methanol from the unreacted components from the synthesis of methyl *t*-butyl ether. It uses selective adsorption on multiple beds of a zeolite such as 4A. Developed by Union Carbide Corporation and now licensed by UOP; as of 1992, eight units had been licensed. *See also* ORU.

> U.S. Patent 4,740,631.

MS [**M**icro-**S**implex] A *steam-reforming process for making town gas from petroleum fractions or LPG. Developed by Gaz de France and Stein & Roubaix.

British Petroleum Co., *Gas Making and Natural Gas*, British Petroleum Co., London, U.K., 1972, 91.

MS-2 A molecular sieving processes for separating branched-chain aliphatic hydrocarbons from unbranched ones by selective adsorption on a zeolite. Developed by the British Petroleum Company in the 1970s, using zeolites provided by Laporte Industries, but not commercialized.

Grebell, J., *Oil Gas J.*, 1975, **73**(15), 85.

MSCC [**M**illi**S**econd **C**atalytic **C**racking] A *fluid catalytic cracking process that uses an ultrashort contact time reaction system. It is claimed that less capital investment and higher liquid yields can be achieved using this process, compared with conventional *FCC units. Developed by Bar-Co and now offered by UOP, MSCC has been operating since 1994.

Eur. Chem. News, 1995, **64**(1682), 28.
Hydrocarbon Process. Int. Ed., 1996, **75**(11), 96.
Oil Gas J., 1998, **96**(25), 53.

MSDW [**M**obil **S**elective **DeW**axing] A catalytic dewaxing process that uses a catalyst containing a shape-selective molecular sieve and a noble metal.

Oil Gas J., 1997, **95**(35), 64.

MSDW-2 An improved version of *MSDW, which uses an improved catalyst. It selectively places side chains on the longer aliphatic hydrocarbons. Announced in 1998.

Helton, T.E., Degnan, T.F., Mazzone, D.N., McGuiness, M.P., Hilbert, T.L., and Dougherty, R.C., *Oil Gas J.*, 20 Jul 1998.
Wuest, R.G., Anthes, R.J., Hanlon, R.T., Loke, L., and Tan, C.T., *Oil Gas J.*, 1999, **27**(29), 70.
Kerby, M.C., Degnan, Jr., T.F., Marler, D.O., and Beck, J.S., *Catal. Today*, 2005, **104**, 58.

MSP3 [**M**icro-**S**uspension **P**rocess] A process for making polyvinyl chloride in suspension. Developed by Atochem, which has granted four licenses since 1977.

Chem. Mark. Rep., 22 Oct 1990, 4.

MS Sorbex A *Sorbex process used in the production of *m*-xylene from C_8 aromatic mixtures. A zeolite is used as the sorbent, and toluene is the desorbent.

Eur. Chem. News, 1995, **64**(1687), 32.

MSTDP [**M**obil **S**elective **T**oluene **D**ispro**P**ortionation] A process for converting toluene to benzene and a xylene mixture rich in *p*-xylene. The catalyst is the zeolite ZSM-5, selectively coked to constrict the pores and thus increase the yield of *p*-xylene produced. Developed and licensed by the Mobil Oil Corporation and first commercialized in Sicily in 1988. *See also* MTDP.

Chen, N.Y., Kaeding, W.W., and Dwyer, F.G., *J. Am. Chem. Soc.*, 1979, **101**, 6783.
Kaeding, W.W., Chu, C., Young, L.B., and Butter, S.A., *J. Catal.*, 1981, **69**, 392.
Hydrocarbon Process. Int. Ed., 1989, **68**(11), 93; 1991, **70**(3), 140.

MTA [**M**ethanol **T**o **A**romatics] A common abbreviation for any process that achieves this conversion, notably a Mobil process.

MT-chlor [**M**itsui **T**oatsu **chlor**ine] A process for recovering chlorine from hydrogen chloride. The hydrogen chloride is mixed with oxygen and passed through a fluidized bed of chromia–silica catalyst. Developed by Mitsui Toatsu and first operated in Japan in 1988. *See also* Deacon, Kel-Chlor.

Tozuka, Y., in *Science and Technology in Catalysis*, Izumi, Y., Aral, H., and Iwamoto, M., Eds., Elsevier, Amsterdam, the Netherlands, 1994, 41.

MTDP [**M**obil **T**oluene **D**is**P**roportionation] A catalytic process that converts 2 mol of toluene to 1 mol of mixed xylenes and 1 mol of benzene. The catalyst is the zeolite ZSM-5. Developed by Mobil Research and Development Corporation and first commercialized in 1975. Superseded by *MSTDP and then by MTDP-3. Now licensed by ExxonMobil Chemical Technology and Axens.

Oil Gas J., 12 Oct 1992, 60.

MTE [**M**ulti-**T**ask **E**ngineering] A process for recovering sulfur from acid gases, based on the *Claus process, but using a circulating, powdered catalyst instead of the usual fixed catalyst bed. Developed by Multi-Task Engineering Ltd. in 1987, but not yet commercialized.

U.S. Patent 4,801,443.
Simek, I.O., *Hydrocarbon Process. Int. Ed.*, 1991, **70**(4), 45.

MTG [**M**ethanol **T**o **G**asoline] A common abbreviation for any process achieving this conversion, notably the Mobil process. This uses as a catalyst, the synthetic zeolite ZSM-5, invented at the Mobil Research Laboratory in 1972. The process was first disclosed in 1976 and commercialized in 1985 by New Zealand Synfuels, a joint venture of Mobil Corporation and Petrocorp. It was subsequently acquired by Methanex. In 1990, this process was providing one-third of New Zealand's gasoline requirements, but in 1997, it was shut down because it was uneconomic.

Chem. Eng. News, 30 Feb 1978, 26.
Kam, A.Y., Schreiner, M., and Yurchak, S., in *Handbook of Synfuels Technology*, Meyers, R.A., Ed., McGraw-Hill, New York, 1984, 2–75.
Pujado, P.R. and Andersen, J.M., in *Handbook of Petroleum Refining Processes*, 3rd ed., Meyers, R.A., Ed., McGraw-Hill, New York, 2003, 15.3.
Weissermel, K. and Arpe, H.-J., *Industrial Organic Chemistry*, 4th ed., translated by C.R Hindley and S. Hawkins, Wiley-VCH, Weinheim, Germany, 2003, 34.
Hamelinck, C.N. and Faaij, A.P.C., in *Alcoholic Fuels*, Minteer, S., Ed., Taylor & Francis, Boca Raton, FL, 2006, 31.
Olah, G.A., Goeppart, A., and Prakash, G.K.S., *Beyond Oil and Gas: The Methanol Economy*, Wiley-VCH, Weinheim, Germany, 2006, 251.

MTH (1) [**M**ittel-**T**emperatur-**H**ydrierung; German, meaning *medium-temperature hydrogenation*] A version of the *TTH process, using different processing conditions, by which a larger proportion of transport fuels could be produced.

Weisser, O. and Landa, S., *Sulphide Catalysts: Their Properties and Applications*, Pergamon Press, Oxford, U.K., 1973, 333.

MTH (2) [**M**ethanol **T**o **H**ydrocarbons] A process for converting methanol to hydrocarbon mixtures suitable for use as gasoline, invented in 1977. Many zeolites have been used to catalyze this reaction, especially zeolite ZSM-5.

Chang, C.D. and Silvestri, A.J., *J. Catal.*, 1977, **47**, 249.
Chang, C.D., *Catal. Rev.*, 1983, **25**, 1.
Olsbye, U., Bjorgen, M., Svelle, S., Lillerud, K.-P., and Kolboe, S., *Catal. Today*, 2005, **106**(1–4), 108.
Olsbye, U., Saure, O.V., Muddada, N.B., Bordiga, S., Lamberti, C., Nilsen, M.H., Lillerud, K.P., and Svelle, S., *Catal. Today*, 2011, **171**(1), 211.

MTO [**M**ethanol **T**o **O**lefins] A catalytic process for converting methanol to olefins, mainly propylenes and butenes. Developed by Mobil Research & Development Corporation and first demonstrated in 1985. Another version of this process was developed by UOP and Norsk Hydro and has been run at a demonstration unit at Porsgrunn, Norway, since Jun 1995. It is based on fluidized bed technology using a SAPO molecular sieve catalyst. It converts 80% of the carbon in the feed to ethylene and propylene. The first commercial plant built in Nigeria in 2008.

U.S. Patent 4,499,327.
Chem. Eng. (N.Y.), 1996, **103**(1), 17.

Picciotti, M., *Oil Gas J.*, 1997, **95**(26), 72.
Stöcker, M., *Micropor. Mesopor. Mater.*, 1999, **29**(1–2), 3.
Barger, P., in *Zeolites for Cleaner Technologies*, Guisnet, M. and Gilson, J.-P., Eds., 2002, Imperial College Press, London, U.K., 239.
Pujado, P.R. and Andersen, J.M., in *Handbook of Petroleum Refining Processes*, 3rd ed., Meyers, R.A., Ed., McGraw-Hill, New York, 2003, 15.3.
Chen, J.Q., Bozzano, A., Glover, B., Fuglerud, T., and Kvisle, S., *Catal. Today*, 2005, **106**(1–4) 103.
Olah, G.A., Goeppart, A., and Prakash, G.K.S., *Beyond Oil and Gas: The Methanol Economy*, Wiley-VCH, Weinheim, Germany, 2006, 249.
ICIS Chem. Bus., 2012, **281**(8), 33.

MTP [**M**ethanol **To P**ropylene] A general name for processes that convert methanol to propylene. One version was developed by Lurgi and offered for license in 2000, the catalyst being provided by Süd-Chemie under the trade name MTProp. UOP developed a version in China through Wilson Nanjing, scheduled for start-up in 2013.

Chem. Week, 2000, **162**(23), 33; 2003, **165**(22), 20; 2006, **168**(7), 22.
Jpn. Chem. Week, 15 Nov 2007, 1.
ICIS Chem. Bus., 2012, **281**(14), 38.

MTPX [**M**obil **T**oluene to **P**-**X**ylene] *See* PxMax.

MTR [**M**elt **to R**esin] A process for making PET developed by Uhde Inventa-Fischer. Also called 2R.

ICIS Chem. Bus., 2007, **2**(55), 21.

MuCell A process for making foamed plastics by the use of supercritical fluids during plastics processing. Developed by Trexel, Germany, and licensed to Ticona.

Eur. Chem. News, 2001, **74**(1949), 27.

Mülheim *See* Alfol.

Müller–Kühne A process for recovering sulfuric acid from phosphogypsum, the waste product from the manufacture of phosphoric acid. The process is economic only if the lime coproduct is converted to cement. Based on the work of W.S. Müller and H.H. Kühne at Bayer, Leverkusen, Germany, from 1915 to 1918. Further developed in Germany in the 1950s and still in operation in Germany and Austria in 1989.

Hull, W.Q., Schon, F., and Zirngibl, H., *Ind. Eng. Chem.*, 1957, **49**(8), 1204.
Becker, P., *Phosphates and Phosphoric Acid*, 2nd ed., Marcel Dekker, New York, 1989, 560.

Multicont A continuous fermentation process for making ethanol, similar to the *Biostil process. Developed by Vogelbusch, Vienna, Austria.

Munich An integrated process for making chlorine dioxide from hydrochloric acid. Sodium chlorate is made electrochemically from sodium chloride, and this is reduced with hydrochloric acid. Developed from the *Kesting process by H. Fröhler and E. Rossberger at the Elektrochemische Werke München, Germany, and first commercialized in 1974. The essential improvement over the Kesting process is the use of titanium electrodes coated with ruthenium oxide for the electrolytic cell and the use of titanium metal for the construction generally. The hydrogen produced in the electrolysis is burned with the chlorine from the reduction stage to produce hydrochloric acid. The process is licensed by Fröhler AG and has been engineered by Uhde. As of 1990, 18 plants had been built, with capacities of between 8 and 40 tons per day.

German Patents 2,407,312; 2,645,121.

Murso *See* Musro.

Musro [**Mu**rphy ores, **CSIRO**] Also written Murso. A process for beneficiating ilmenite by a combination of oxidation, reduction, and pressure leaching with hydrochloric acid. Invented in Australia in 1967 and developed jointly by Murphyores Pty and the Commonwealth Scientific and Industrial Research Organization, but not commercialized then. Further developed in 1992 by Pivot Mining NL, Queensland, Australia.

> British Patent 1,225,826.
> Gupta, C.K. and Mukherjee, T.K., *Hydrometallurgy in Extraction Processes*, Vol. 1, CRC Press, Boca Raton, FL, 1990, 111.

MVB *See* metal surface treatment.

MVPI [**Mo**bil **V**apor **P**hase **I**somerization] A process for converting mixed xylene streams to *p*-xylene, catalyzed by the zeolite ZSM-5. Invented by Mobil Corporation in 1973, later superseded by *MHTI. *See also* LTI, MHTI, MLPI.

> U.S. Patents 3,856,871; 3,856,872; 3,856,873; 3,856,874.

MWB *See* Sulzer MWB.

MWI [**Mo**bil **W**ax **I**somerization] A process for improving the quality of petroleum-based lubricating oils. The undesirable wax constituents are *hydroisomerized to products of lower molecular weight, using a zeolite catalyst, and the resulting product is treated with an organic peroxide to increase its viscosity. Developed by Mobil Oil Corporation in 1990.

> U.S. Patent 5,037,528.
> Helton, T.E., Degnan, T.F., Jr., Mazzone, D.N., McGuiness, M.P., Hilbert, T.L., and Dougherty, R.C., *Oil Gas J.*, 1998, **96**(29), 58.

MXCOL [**M**idrex **Co**al] A version of the *MIDREX process, using syngas made from coal. Announced in 2010; a plant was being built in India in 2012. *See also* TRS.

MX Sorbex A version of the UOP *Sorbex process for removing *m*-xylene from its isomers. First licensed in 1995, six units had been licensed by 2009.

> Johnson, J.A., in *Handbook of Petroleum Refining Processes*, 3rd ed., Meyers, R.A., Ed., McGraw-Hill, New York, 2003, 2.6.
> *Chem. Week*, 23 May 2007.

N

Nahcolite A *flue-gas desulfurization process. Nahcolite is a mineral containing 70%–90% sodium bicarbonate, which is found in Colorado. In this process, the powdered nahcolite is injected into the baghouse and the following reaction occurs:

$$SO_2 + 2NaHCO_3 + \tfrac{1}{2}O_2 = Na_2SO_4 + 2CO_2 + H_2O$$

The solid product is dumped. Piloted by Battelle Columbus Laboratories.

Genco, J.M., Rosenberg, H.S., Anastis, M.Y., Rosar, E.C., and Dulin, J.M., *J. Air Pollut. Control Assoc.*, 1975, **25**(12), 1244.
Genco, J.M. and Rosenberg, H.S., *J. Air Pollut. Control Assoc.*, 1976, **26**(10), 989.

Nalfining [**Nal**co re**fining**] A process for purifying petroleum fractions by treatment with acetic anhydride, followed by extraction with aqueous sodium hydroxide. The first stage converts the O-, N-, and S-containing impurities into oxygenates that are removed in the second stage. Invented by Nalco in 1959.

Chem. Eng. News, 1959, **37**(23), 42.
Unzelman, G.H. and Wolf, C.J., in *Petroleum Processing Handbook*, Bland, W.F. and Davidson, R.L., Eds., McGraw-Hill, New York, 1967, 3–20.

NanoSelect Not a process but a range of nanoparticulate catalysts offered by BASF. The name has several other meanings.

NAPEX [**NAP**htha **EX**traction] A process that upgrades the naphtha fed to an ethylene cracking unit. It combines naphtha *hydrotreating with aromatics extraction. Developed by Ethylene Consultants. There are also several nonchemical uses of the word.

Hydrocarbon Process. Int. Ed., 2005, **84**(4), 29.

NAPFINING A process for removing naphthenic acids from petroleum fractions by extracting with aqueous alkali, using a bundle of hollow fibers. Developed by the Merichem Company, Houston, TX, and used in 19 plants in 1991.

Hydrocarbon Process. Int. Ed., 1996, **75**(4), 126.

Naphtachimie A gas-phase process for making high-density polyethylene in a fluidized bed. Invented by Naphtachimie in 1973 and operated by that company at Lavera, France, since 1975.

U.S. Patent 3,922,322.

Naphthamax-LSG [Low-Sulfur Gasoline] A process for reducing sulfur levels in gasoline. Developed by Engelhard in 2001.

Chem. Mark. Rep., 19 Nov 2001.

Natta A process for polymerizing propylene and other higher olefins, catalyzed by crystalline titanium trichloride and an alkyl aluminum compound, such as triethyl aluminum. The polymer can exhibit various types of stereoregularity, depending on the catalyst and the conditions. Invented in 1954 by G. Natta at the Istituto de Chimica Industriale del Politecnico di Milano, Italy, and commercialized in 1957. Now used widely, worldwide. *See also* Ziegler, Ziegler–Natta.

U.S. Patents 3,112,300; 3,112,301.
Natta, G., Pino, P., Corradini, P., Danusso, F., Mantica, E., Mazzanti, G., and Moraglio, G., *J. Am. Chem. Soc.*, 1955, **77**, 1708.
Natta, G., *J. Polymer Sci.*, 1955, **16**, 143.
Natta, G., *Angew. Chem.*, 1956, **68**, 393.
Natta, G., *Inaugural Lecture, 16th International Conference on Pure and Applied Chemistry, Paris*, Birkhauser Verlag, Basel, Switzerland, 1957, 21.
Raff, R.A.V., in *Ethylene and Its Industrial Derivatives*, Miller, S.A., Ed., Ernest Benn, London, U.K., 1969, 335.

NCB *See* carbonization.

NDA [**N**atural **D**etergent **A**lcohol] A process for making long-chain alcohols, for use in detergent synthesis, from fatty acids from vegetable oils. The fatty acids are esterified with methanol, and the resulting methyl esters are catalytically hydrogenated. Developed by Kvaerner Process Technology in 1988; the first commercial plant began operating in the Philippines in 1998.

Eur. Chem. News, 1998, **69**(1807), 19.

NEC [**N**itrogen **E**ngineering **C**orporation] A modification of the *Haber process for making ammonia.

Vancini, C.A., *Synthesis of Ammonia*, translated by L. Pirt, Macmillan, London, U.K., 1971, 237.

NEDOL [**N**ew **E**nergy **D**evelopment **O**rganization **L**iquefaction] Similar to *Kohloel. A coal liquefaction process in development in Japan by the New Energy and Industrial Technology Development Organization (NEDO), Tokyo. Crushed coal is mixed with a pyrite catalyst and slurried in a hydrogenated heavy oil. Liquefaction takes place at 450°C, 170 bar. The overall oil yield is 59%. The used solvent is hydrogenated and recycled. The product requires intensive upgrading. Piloted in Kashima, Japan, in 1997 to 1998. Two Chinese companies were licensed to build test units in 2006.

Chem. Eng. (N.Y.), 1998, **105**(2), 29.
Onozaki, M., Namiki, Y., Aramaki, T., Takagi, T., Kobayashi, M., Itoh, H., Hiraide, M., Kidoguchi, A., Onozaki, M., Ishibashi, H., Namiki, Y., Ikeda, K., Inokuchi, K., and Morooka, S., *Ind. Eng. Chem. Res.*, 2001, **40**(1), 210.
Jpn. Chem. Week, 2006, **47**(2374), 5.
Wikipedia, The Free Encyclopedia. Coal liquefaction. Revised 12 Apr 2013. Retrieved 20 Jul 2013.

NENATECH Also called **Albion**, after the suburb of Brisbane, Australia, where it was developed. A hydrometallurgical process for extracting copper from sulfide ores. Nena is the local name for a chalcopyrite ore found in Papua New Guinea. The process involves leaching with aqueous ferric sulfate and air or oxygen sparging. It was piloted in Australia in 2002 but has not been licensed.

Balaz, P., *Mechanochemistry in Nanoscience and Minerals Engineering*, Springer, Berlin, Germany, 2008, 321, 322.

NEOCHROME A process for making colored acrylonitrile fibers by dyeing during the spinning process. Developed by Courtaulds.

Eur. Chem. News CHEMSCOPE, 8 May, 1995.

Neostar A process for destroying waste organic chlorides (e.g., polychlorinated biphenyls) by heating with steam and hydrogen at over 1000°C. The products are methane, ethane, other chlorine-free hydrocarbons, and hydrochloric acid. Developed by Cerchar, France.

Nesbitt *See* Steelmaking.

Netto An early process for extracting aluminum from cryolite by reducing it with sodium. Operated in the 1890s.

Neuberg A process for increasing the yield of glycerol from the fermentation of glucose by adding sodium sulfite.

Neuberg, C. and Reinfurth, E., *Biochem. Z.*, 1918, **89**, 365 (*Chem. Abstr.*, **13**, 328).
Baldwin, E., *Dynamic Aspects of Biochemistry*, 5th ed., Cambridge University Press, Cambridge, U.K., 1967, 347.

neutralization A family of processes for making sodium cyanide by neutralizing anhydrous hydrocyanic acid with aqueous sodium hydroxide. These replaced the *Castner (2) process in the 1960s.

Neutralysis A process for converting municipal waste into lightweight aggregate by calcination with clay. Developed by Neutralysis Industries Pty Ltd. in Australia from 1987 but abandoned in 1992 when the company went into receivership. The process was considered for use in Indiana in 1992 but was never practiced on a large scale.

U.S. Patents 4,890,563; 4,993,331.
Krol, A., White, K., and Hodson, B., in *Waste Materials in Construction: Proceedings*, Goumans, J.J.J.M., van der Sloot, H A., and Aalbers, T.G. (eds.), Elsevier, Amsterdam, the Netherlands, 1991.

NEUTREC A *flue-gas desulfurization process, intended for treating the waste gases from incinerators for municipal, hospital, and industrial wastes. Sodium bicarbonate, optionally mixed with active carbon, is injected into the gases after the usual bag filter, and the solid products are removed in a second bag filter. Sodium compounds can be recovered from the product for reuse, and any toxic compounds disposed of separately. Developed by Solvay and operated in Europe since 1991. *See also* Solatech.

Chem. Ind. (London), 1997, (19), 762.
Eur. Chem. News, 1999, **71**(1878), 34.

Newcell A process for making reconstituted cellulose fibers by dissolving cellulose in N-methyl morpholine N-oxide and injecting the solution into water. Invented in 1977 by Akzona, NC.

U.S. Patents 4,142,913; 4,144,080; 4,145,532.

New Jersey A continuous process for extracting zinc from zinc oxide, made by roasting zinc sulfide ore, by reduction with carbon in a vertical retort. First operated by the New Jersey Zinc Company in Palmerton, PA, in 1929, and introduced into the Avonmouth, United Kingdom, works of the Imperial Smelting Company in 1934.

Newton Chambers A process for purifying benzene by fractional solidification; cooling is accomplished by mixing it with refrigerated brine. The process does not remove thiophene.

NExBTL [NEste Biomass-to-Liquids] A process for converting triglycerides (animal or vegetable) into biodiesel without producing glycerol. Developed by Neste Oil and first operated at Porvoo, Finland. Awarded the 2006 Finnish Chemical Industry Innovation Award. Plants in Singapore and Rotterdam were built in 2010 and 2011.

Chem. Ind. (London), 2006, (10), 10.
Chem. Week, 2008, **170**(23), 8.
Wikipedia, The Free Encyclopedia. NExBTL. Revised 26 Jun 2013. Retrieved 21 Jul 2013.

NEXCC [NEste Catalytic Cracking] A *catalytic cracking process developed by Neste Oil, Finland, from 1998. Novel engineering features permit the very short residence time of 0.7–2.2 s. The name was abandoned by Neste, possibly to avoid confusion with the abbreviated name of an unrelated organization.

Hiltunun, J. et al., in Occelli, M.L. and O'Connor, P., *Fluid Catalytic Cracking* V, 2001, Elsevier, Amsterdam, the Netherlands, 114.

NExETHERS A process for converting C_5, C_6, and C_7 olefins to ethers for blending into gasoline to increase its octane rating. The process resembles *NExTAME but uses additional methanol to increase the yield of ethers.

Oil Gas J., 1997, **95**(1), 44.

NExOCTANE A process for making isooctane by dimerizing isobutylene over an acidic ion-exchange catalyst. The product is used as a gasoline additive. Developed from 1997 by Neste Engineering (a subsidiary of Fortum Oil and Gas, Finland) and Kellogg Brown & Root. The first commercial unit started operating in 2002. *See also* SP-Isoether.

Eur. Chem. News, 2000, **77**(1893), 35.
Birkhoff, R. and Nurminen, M., in *Handbook of Petroleum Refining Processes*, 3rd ed., Meyers, R.A., Ed., McGraw-Hill, New York, 2003, 1.3.

NExSELECT A catalytic, selective hydrogenation process developed by Neste Oy. Operated in Porvoo, Finland, since 1996.

Oil Gas J., 1997, **95**(1), 45.

NExTAME [**NE**ste **T**ertiary **A**myl **M**ethyl **E**ther] A catalytic process for converting C_5, C_6, and C_7 tertiary olefins to ethers for blending into gasoline to increase its octane rating. Developed by Neste Oy in 1994 and commercialized at Porvoo, Finland, in 1995.

Hydrocarbon Process. Int. Ed., 1996, **75**(11), 110.
Oil Gas J., 1997, **95**(1), 44.

N-Free A process for removing nitrates from industrial wastewaters. N-Free has two stages: In the first, nitrates are reduced to nitrites using hydrazine and a fixed-bed copper catalyst. In the second, the nitrite is reduced to elemental nitrogen using amidosulfonic acid as the reductant. Developed by Sumitomo Metal Mining Company and piloted in 2002.

Chem. Eng. (N.Y.), 2002, **109**(12), 21.

NGOP [**N**atural **G**as **O**xy**P**yrolysis] *See* IFP Oxypyrolysis.

Nicaro [Named after the Cuban town] A process for extracting nickel from low-grade ores. The ore is reduced by heating with *producer gas and is then leached with aqueous ammonia.

Niers A process for treating the aqueous effluent from dyeworks by a combination of chemical precipitation and biological purification.

Nippon Steel *See* Steelmaking.

NIPR *See* carbonization.

N-ISELF A process for separating linear hydrocarbons from light naphtha by selective adsorption on a zeolite. Developed by Société Nationale Elf Aquitaine, France.

Bernard, J.R., Gourlia, J.-P., and Guttierrez, M.J., *Chem. Eng. (N.Y.)*, 1981, **88**(10), 92.

NITECH A cryogenic process for removing nitrogen from natural gas, mainly methane. The high-pressure gas is liquefied by expansion and then fractionated. The essential feature is the use of an internal reflux condenser within the fractionating column. Developed by BCCK Engineering and demonstrated at a full-scale plant in Oregon in 1994.

U.S. Patent 5,375,422.
Butts, R.C., Chou, K., and Slaton, B., *Oil Gas J.*, 1995, **93**(11), 92.

Nitralizing *See* metal surface treatment.

Nitreat A process for removing nitrates from potable water by continuous countercurrent ion exchange. Developed by Anglian Water and operated at the Keldgate reservoir, Yorkshire, since 2009.

NitRem [**Nit**rate **Rem**oval] A process for removing nitrate from water supplies by electrodialysis through a selective membrane. Developed in the 1980s by Otto Oeko-Tech & Company, Germany.

NITREX A process for removing nitrogen from natural gas by *PSA. Developed by UOP.

Hydrocarbon Process. Int. Ed., 1996, **75**(4), 128.

NitroGEN [**Nitro**gen **GEN**erator] A version of the *VPSA process for separating nitrogen from air by vacuum-pressure swing adsorption. Developed by the Linde Division of the Union Carbide Corporation. The name has been used also for two membrane systems for extracting pure nitrogen from air.

Eur. Chem. News, 1989, **53**(1391), 31.
Chem. Mark. Rep., 29 Oct 1990, 5.

Nitro Nobel A process for making nitrate esters such as nitroglycerol. A special injector is used to mix the liquid polyol with the nitrating acid.

Nittetu A process for destroying waste organic chlorides by submerged combustion.

Santoleri, J.J., *Chem. Eng. Prog.*, 1973, **69**(1), 68.

Nixan [**Ni**trocyclohe**xan**e] A process for making cyclohexane oxime (an intermediate in the manufacture of nylon) from benzene by liquid-phase nitration, followed by hydrogenation of the nitrobenzene. Invented by DuPont and operated from 1963 to 1967.

Dahlhoff, G., Niederer, J.P.M., and Hölderich, W.P., *Catal. Rev.*, 2001, **43**(4), 389.
Weissermel, K. and Arpe, H.-J., *Industrial Organic Chemistry*, 4th ed., translated by C.R. Hindley and S. Hawkins, Wiley-VCH, Weinheim, Germany, 2003, 256.

Noell (1) A pyrolytic process for treating solid wastes. Developed in Germany by Preussag Noell Wassertechnik.

Waste Manag. Environ., 1998, **8**(2), 20.

Noell (2) A coal gasification process using entrained-flow technology. Also called GSP and BBP. Developed in East Germany in 1975 for processing lignite and now operated in Germany and England.

Rezaiyan, J. and Cheremisinoff, N.P., *Gasification Technologies*, Taylor & Francis, Boca Raton, FL, 2005, 50.

Noell-KRC A scrubbing process for removing hydrochloric and sulfuric acids from flue gas from domestic waste disposal plants. It produces saleable hydrochloric and sulfuric acids. Developed by Noell-KRC Umwelttechnik in 1992.

Kohl, H.L. and Nielsen, R.B., *Gas Purification*, 5th ed., Gulf Publishing Co., Houston, TX, 1997, 589.

Noguchi A catalytic process for hydrogenating lignin to a mixture of monophenols. Invented in 1952 at the Noguchi Institute of Japan, but not commercialized because the yields were uneconomic.

Goheen, D.W., in *Lignin Structure and Reactions*, American Chemical Society, Washington, DC, 1966, 205 (*Chem. Abstr.*, **43**, 21091).

NoNOx A two-stage combustion system that does not produce oxides of nitrogen. The first stage operates under reducing conditions, at a controlled, low partial pressure of oxygen. The second stage uses a small excess of oxygen. Developed by Boliden Contech, Sweden, originally for the roasting of arsenical pyrite, but now used principally for incinerating waste.

Noranda A continuous copper smelting process, developed in Canada by Noranda Mines, from 1964.

Canadian Patent 758,020.
Themelis, N.J. and McKerrow, G.C., in *Advances in Extractive Metallurgy and Refining*, Jones, M.J., Ed., Institution of Mining and Metallurgy, London, U.K., 1971, 3.

Nordac A process for concentrating *wet-process phosphoric acid by submerged combustion. Operated in Europe since 1947. *See also* Ozark-Mahoning.

Forster, J.H., in *Phosphoric Acid*, Vol. 1, Part 2, Slack, A.V., Ed., Marcel Dekker, New York, 1968, 594.

Nordac-Aman *See* Woodall–Duckham.

Normann Also called Sabatier–Normann, after P. Sabatier, one of the inventors of *catalytic hydrogenation. The first commercial process for hardening fats by catalytic hydrogenation over nickel. Invented in 1902 by K. P. W. T. Normann at the Herforder Maschinenfett und Ölfabrik, Germany, and initially licensed to Joseph Crosfield and Sons, United Kingdom. The first large-scale plant was built at Warrington in 1909. Competing processes were developed by E. Erdmann, C. Paal, N. Testrup, and M. Wilbuschewitsch, and much patent litigation ensued. Normann's patent was eventually declared invalid in 1913 because of incomplete disclosure.

British Patent 1,515 (1903).
German Patents 139,457; 141,029.
Wilson, C., *The History of Unilever*, Cassell, London, U.K., 1954, 110.
Musson, A.E., *Enterprise in Soap and Chemicals*, Manchester University Press, Manchester, U.K., 1965, 165.

Norsk Hydro This large Norwegian company has given its name to a number of processes based on hydroelectric power. One such process, offered by Lurgi, is for producing ammonium phosphates. Another is for making magnesium by electrolyzing molten magnesium chloride, derived indirectly from seawater; this has been in operation at Porsgrunn, Norway, since 1951.

Höy-Petersen, N., *J. Met.*, 1969, **21**(4), 43.

NORSOLOR A continuous process for making polystyrene. Licensed by Badger Company and operated in France and South Korea.

Hydrocarbon Process. Int. Ed., 1989, **68**(11), 110.

Norsorex A process for making a thermosetting polymer from norbornene by ring-opening metathesis (ROMP). The product contains double bonds and can be cross-linked; it can be made into a porous polymer, useful for absorbing oil. Various organometallic catalysts may be used in the polymerization.

Janiak, C. and Lassahn, P.G., *J. Mol. Catal.*, 2001, **166**(2), 193.

North Thames Gas Board A process for removing organic sulfur compounds from coal gas by catalytic oxidation over nickel sulfide at 380°C. The sulfur dioxide produced is removed by scrubbing with dilute aqueous sodium hydroxide. Operated by the gasworks of the North Thames Gas Board, London, between 1937 and 1953.

Plant, J.H.G. and Newling, W.B.S., *Trans. Inst. Gas Eng.*, 1948, **98**, 308.

Norzink *See* Boliden/Norzink.

NoTICE [**No** **Ti**e **In** **C**laus **E**xpansion] A process for oxidizing sulfur for the manufacture of sulfuric acid. Oxygen is introduced below the surface of a pool of molten sulfur. This permits easy temperature control. Developed by Brown & Root and first used at Port Neches, TX, in 1989.

Schendel, R.L., *Oil Gas J.*, 1993, **91**(39), 63.

Novacon An adsorptive process for removing oxides of sulfur and nitrogen and carbon monoxide, from combustion gases. The adsorbent is an active form of natural marble.

Novalfer *See* DR.

Novolen A process for making polypropylene in the gas phase, using a vertical stirred-bed reactor. Developed by BASF and engineered by Uhde. Licensed by Lummus Technology. Eight plants had been licensed as of 1985. A metallocene-catalyzed version was introduced in 1996. The name is used also for the product.

Bajpai, P., *Environmentally Benign Approaches for Pulp Bleaching*, 2nd ed., Elsevier, Amsterdam, the Netherlands, 2012, 123.

NOVOSOL A process for treating sludge sediments from canal and port dredging. The heavy metals are rendered insoluble, and the organic material is destroyed by calcination. The product may be used in civil engineering works. Developed by Solvay from 2001.

NOVOX Not a process, but a trademark used by BOC to designate its *PSA process for separating oxygen from air.

NO$_x$Out A process for removing oxides of nitrogen from flue gases by the reaction with urea:

$$CO(NH_2)_2 + 2NO + \tfrac{1}{2}O_2 = 2N_2 + CO_2 + 2H_2O$$

The problem with this reaction is that it takes place over a narrow temperature range, between about 930°C and 1030°C. Below this range, ammonia is formed; above it, more nitrogen oxides are formed. In the NO$_x$Out process, proprietary additives are used to widen the usable temperature range. Developed by the Electric Power Research Institute, Palo Alto, CA, from 1976 to 1980, and then further developed by Fuel Tech. It was first commercialized in Germany in 1988. In February 1990, Fuel Tech formed a joint venture with Nalco Chemical Company—Nalco Fuel Tech—to further develop and promote the process. By September 1990, 16 systems had been sold in the United States and 2 in Europe. In November 1991, an improved version was announced: NO$_x$Out Plus.

U.S. Patents 4,208,386; 4,325,924; 4,719,092.
Chem. Mark. Rep., 1990, **238**(12), 31.
Hydrocarbon Process. Int. Ed., 1993, **72**(8), 80.
Lin, M.L., Comparato, J.R., and Sun, W.H., in *Reduction of Nitrogen Oxide Emissions*, Ozkan, U.S., Agarwal, S.K., and Marcelin, G., Eds., American Chemical Society, Washington, DC, 1995, Chap. 17.

NOXSO A process for simultaneously removing sulfur oxides and nitrogen oxides from flue gases. The sorbent is a regenerable mixture of sodium carbonate with alumina in a hot fluidized bed. The gases are retained as sodium sulfate, nitrate, and nitrite. Regeneration is carried out by heating first with air and then with a reducing gas, such as hydrogen or natural gas. Developed and piloted by the Noxso Corporation with MK-Ferguson Company and WR Grace and Company. A demonstration plant was installed in Ohio Edison's power plant in Toronto, OH, in 1991, funded by the U.S. Department of Energy and the Ohio Coal Development Office. Noxso Corporation was declared bankrupt in June 1997, following a dispute with Olin Corporation.

U.S. Patent 4,755,499.
Chem. Eng. (N.Y.), 1989, **96**(6), 21.
Chem. Eng. News, 1990, **68**(38), 35.
Neal, L.G., Woods, M.C., and Bolli, R.E., in *Processing and Utilization of High-sulfur Coals, IV*, Dugan, P.R.,
    Quigley, D.R., and Attia, Y.A., Eds., Elsevier, Amsterdam, the Netherlands, 1991, 651.
Chem. Eng. News, 1997, **75**(23), 15.

NOXSORB A process for removing oxides of sulfur and nitrogen from "dirty" gases. Developed by Tri-Mer Corporation in 1995.

Am. Ceram. Soc. Bull., 1995, **74**(8), 86.

NOxStar A process for removing oxides of nitrogen from combustion product gases. Controlled amounts of hydrocarbon (usually natural gas) and ammonia are injected into the flue gas. Combustion of the hydrocarbon generates a plasma, and the ammonia reacts with the NOx, generating elemental nitrogen. The process operates at a lower temperature than competing *SCR processes and is claimed to have several advantages over them. Developed by Mitsui Babcock Energy Ltd. (MBEL) and first demonstrated in 2001.

NPEX A process for removing neptunium and plutonium during the processing of nuclear fuel by solvent extraction. The solvent is tributyl phosphate diluted with *n*-dodecane. Developed by E. P. Horwitz at the Argonne National Laboratory, Argonne, IL. *See also* SREX, UREX+.

NRS [New Regeneration System] A process for regenerating the ion-exchange resin used for removing calcium from sugar solution. If sodium chloride were used, the waste calcium chloride solution would have to be disposed of; if sodium hydroxide were used, calcium hydroxide would be precipitated in the resin. The NRS process uses sodium hydroxide in the presence of sucrose, which retains the calcium in solution as calcium saccharate. Developed by the IMACTI Division of Duolite International, the Netherlands.

NSC An obsolete direct-reduction ironmaking process, operated in Japan and Malaysia. *See* DR.

NSM A Dutch process for making ammonium nitrate, offered by Uhde. Not to be confused with another NSM (New Smoking Material), a tobacco substitute developed by ICI in the 1970s but later abandoned.

NSSC [Neutral Sulfite SemiChemical pulping] A papermaking process in which wood chips are digested in an aqueous solution of sodium sulfite and sodium carbonate at 140°C–170°C for several hours.

Nu-Iron *See* DR.

Nulite A catalytic, photochemical process for oxidizing toxic organic compounds in water. The catalyst is titanium dioxide supported on a mesh; the light is sunlight. Intended for treating groundwater containing not more than 500 ppm of toxic organic materials. Developed and offered by Nutech Environmental, London, Ontario.

Nurex A process for extracting C_8 to C_{30} linear hydrocarbons from petroleum fractions, using their ability to form urea inclusion complexes. Branched-chain hydrocarbons do not form such complexes. Developed by the Nippon Mining Company, Japan, and operated until 1979.

Scholten, G.G., in *Petroleum Processing Handbook*, McKetta, J.J., Ed., Marcel Dekker, New York, 1992, 587.
Weissermel, K. and Arpe, H.-J., *Industrial Organic Chemistry*, 4th ed., translated by C.R. Hindley and S. Hawkins,
    Wiley-VCH, Weinheim, Germany, 2003, 78.

Nutriox A process for eliminating the odor and septicity of liquid effluent. Nitrate is added to the water, which prevents bacteria from forming hydrogen sulfide. Developed by Norsk Hydro in 1996.

Nuvalon A development of the *Aloton process for extracting aluminum from clay. As in the Aloton process, clay is first heated with ammonium hydrogen sulfate. In the Nuvalon version, the product from this reaction is digested under pressure with 30% nitric acid, producing a solution of basic aluminum nitrate. Iron is removed by hydrolysis or by cooling. Normal aluminum nitrate is crystallized out and calcined to alumina. The process was piloted in Germany in 1951 but not commercialized.

Gewecke, F., *Chem. Fabr.*, 6 June 1934, (21–22), 199.
O'Connor, D.J., *Alumina Extraction from Non-bauxitic Materials*, Aluminium-Verlag, Düsseldorf, Germany, 1988, 163.

O

OASE Not a single process but the trade name for a group of gas-treating processes offered by BASF.

OATS [**O**lefinic **A**lkylation of **T**hiophenic **S**ulfur] A gasoline desulfurization process. Thiophenes and mercaptans are catalytically reacted with olefins to produce higher-boiling compounds that can more easily be removed by distillation prior to hydrodesulfurization. This minimizes hydrogen usage. The process uses a solid acid catalyst in a liquid-phase, fixed-bed reactor. Developed by BP Amoco in 2000 and tested in Bavaria and Texas. First used commercially at the Bayernoil refinery, Neustadt, in 2001. The process won a European Environment Award in 2002.

> *Chem. Eng. (N.Y.)*, 2000, **107**(13), 19.
> *Chem. Eng. (Rugby, Engl.)*, 21 Jun 2001, *Awards Supplement*, 5.
> *Eur. Chem. News*, 19 Nov 2001, 31.
> *Oil Gas J.*, 2001, **99**(49), 8.
> *Proc. Eng.*, 2002, **83**(10), 2.

Oberphos A version of the superphosphate process for making a fertilizer by treating phosphate rock with sulfuric acid, which yields a granular product. Used in the United States and Canada but superseded in the United States by the *Davison process.

> Gray, A.N., *Phosphates and Superphosphate*, Vol. 1, Interscience Publishers, New York, 1947, 124.

OBM *See* Steelmaking.

OCET [**O**pti-**C**rude **E**nhancement **T**echnology] A process for converting residual refinery oil into petroleum distillates and a coal substitute. A pulsed electric field is applied to the oil. Developed by SGI International in 1996 but probably not commercialized.

> *Hydrocarbon Process. Int. Ed.*, 1996, **75**(1), 42.

OCM [**O**xidative **C**oupling of **M**ethane] *See* Oxidative Coupling.

OCP (1) [**O**xygène **C**haux **P**ulverisée] A steelmaking process in which powdered lime is blown into the furnace through the oxygen stream in order to combine with the phosphorus. Developed in the 1950s by CNRM, a Belgian metallurgical research organization. Similar to the *OLP process. *See also* LD/AC.

OCP (2) [**O**lefins **C**onversion (or **C**racking) **P**rocess] A process for making propylene from C_4 to C_8 olefins in catalytic cracking product streams. A proprietary zeolite catalyst is used in fixed beds in a swing reactor system: one bed is regenerated while the other is in use; the roles are then reversed. Developed by Total Petrochemicals from 1995 and subsequently with UOP and piloted in Total's facility in Feluy, Belgium. The first commercial unit was being engineered in 2005.

> *Chem. Eng. (N.Y.)*, 2004, **111**(3), 20.
> Grootjans, J., Vanrysselberghe, V., and Vermeiren, W., *Catal. Today*, 2005, **106**(1–4), 57.
> *Chem. Week*, 23 Feb; 31 May, 2006.

OCP (3) [**O**lefins **C**onversion **P**rocess] A process for making propylene by reacting ethylene with mixed refinery C_4 streams. The catalyst is a mixture of tungsten oxide on silica (the actual metathesis catalyst) with magnesia (an isomerization catalyst). The process is the reverse of the Phillips *Triolefin process. Developed by Phillips and purchased by ABB Lummus Global in 1997. First installed at the Karlsruhe oil refinery of Mineraloelraffinerie Oberrhein (Miro) for startup in 2000. Installed at Port Arthur, TX, in 2003–2004 and planned for installation in Osaka, Japan, in 2004. In 2005, it was operated by Lyondell Petrochemical, BASF Fina Petrochemical, and Mitsui Petrochemical. *See also* Meta-4.

> *Eur. Chem. News*, 1997, **68**(1792), 45; 2002, **76**(1998), 21; **77**(2028), 34.
> *J. Mol. Catal.*, 2004, **213**(1), 40.

OCR [**O**n-stream **C**atalyst **R**eplacement] A technology for continuously replacing spent catalyst used in *hydrotreating. The catalyst flows downward through the reactor, while the petroleum fraction and the hydrogen flow upward. Developed from 1979 and first demonstrated in ChevronTexaco's Richmond refinery in 1985. Now offered by Chevron Lummus Global.

> Earls, D.E., in *Handbook of Petroleum Refining Processes*, 3rd ed., Meyers, R.A., Ed., McGraw-Hill, New York, 2003, 10.3.

OCT *See* **O**lefins **C**onversion **T**echnology.

Octafining A process for isomerizing *m*-xylene to *o*- and *p*-xylene, developed by the Atlantic Richfield Company in 1960. The catalyst was originally platinum on an aluminum silicate base; now, a zeolite base is used. The reaction takes place in a hydrogen atmosphere. Hydrocarbon Research installed units in Argentina and the USSR.

> *Hydrocarbon Process. Int. Ed.*, 1963, **42**(11), 206.
> Uhlig, H.F. and Pfefferle, W.C., in *Refining Petroleum for Chemicals*, Spillane, L.J. and Leftin, H.P., Eds., American Chemical Society, Washington, DC, 1970, 204.

Octamix [**Octa**ne **mix**ture] A process for converting *syngas to a mixture of methanol with higher alcohols by reducing the CO/H_2 ratio below that required for the usual process for making methanol. The process is operated at 270°C–300°C, 50–100 bar, in the presence of a copper-based catalyst. Developed by Haldor Topsoe and Lurgi but not commercialized. The name is also a trade name used by Lurgi to denote a mixture of methanol and higher alcohols made by this process, suitable for blending with gasoline to increase its octane number.

> Weissermel, K. and Arpe, H.-J., *Industrial Organic Chemistry*, 4th ed., translated by C.R. Hindley and S. Hawkins, Wiley-VCH, Weinheim, Germany, 2003, 34.

Octanizing A continuous *reforming process for making high-octane gasoline. Developed by IFP, now Axens, from *Dualforming. Licensed to ANCAP, Uruguay, in 1999. By 2005, more than 35 units had been licensed worldwide. A unit was licensed to Bharat Petrochemical Corp for use in its Mumbai refinery in 2010.

> *Wikipedia, The Free Encyclopedia*. Catalytic reforming.

OCTENAR [**OCT**ane **EN**hancement by removing **AR**omatics] A process for removing aromatic hydrocarbons from petroleum reformate by extractive distillation with N-formylmorpholine. The product can be blended with gasoline to increase its octane number, hence the name. A paraffin mixture is obtained as a side product. Developed by Krupp Koppers from its *MORPHYLANE and *MORPHYLEX processes.

OCTGAIN A *hydrofinishing process that reduces the sulfur and olefin content of gasoline without reducing its octane number. A proprietary zeolite catalyst is used in a fixed-bed, multibed reactor having an intermediate cooling system. Developed by ExxonMobil and first commercialized in 1991.

> *Chem. Eng. (N.Y.)*, 1994, **101**(7), 25.

Octol A process for making mixed linear octenes by the catalytic dimerization of mixed butenes. A proprietary heterogeneous catalyst is used. Developed jointly by Hüls and UOP and now offered for license by UOP.

First operated in 1983 in the Hüls refinery in Marl, Germany. Another installation began production in 1986 at the General Sekiyu Refineries in Japan.

Also the name of an explosive composition.

Friedlander, R.G., Ward, D.J., Obenaus, F., Nierlich, F., and Neumeister, J., *Hydrocarbon Process. Int. Ed.*, 1986, **65**(2), 31.
Nierlich, F., *Erdoel, Erdgas, Kohle*, 1987, **103**(11), 486.
Hydrocarbon Process. Int. Ed., 1991, **70**(3), 166.
Golombek, M. and de Bruijn, J., *Appl. Catal.*, 2001, **208**, 47.

Odda A process for making a fertilizer by treating phosphate rock with nitric acid. Developed by Odda Smelteverk, Norway, in the early 1900s and still in use in 1988. Licensed by BASF and offered by Uhde.

Piepers, R.J., in *Phosphoric Acid*, Vol. 1, Part 2, Slack, A.V., Ed., Marcel Dekker, New York, 1968, 916.

Odex A process for controlling odors from sewage-treatment systems. Amines are destroyed by ozone, and sulfur compounds by UV radiation. Both the ozone and the UV are produced in special lamps designed by Hanovia.

Water Waste Treat., 1999, **42**(12), 14.

ODORGARD A process for removing odorous gases from air streams by scrubbing with an aqueous solution of sodium hypochlorite in the presence of a proprietary heterogeneous catalyst. The catalyst contains nickel and is based on the *HYDECAT catalyst. Developed by ICI Katalco and F.H.H. Valentin. Nine units had been installed in the United Kingdom by 1995. Offered by Enviro-Chem Systems (a Monsanto subsidiary) in 2000.

World Patent WO 94/11091.
Chem. Eng. Prog., 1995, **91**(6), 19.
Hancock, F.E., King, F., Flavell, W.R., and Islam, M.S., *Catal. Today*, 1998, **40**(4), 289.

ODR [Oxidative Dry Reforming] A general term for the catalytic reaction of methane with carbon dioxide to produce *syngas. The word "dry" signifies the absence of steam. Not yet perfected.

Nematollahi, B., Rezaei, M., and Khajenoon, M., *Int. J. Hydrogen Reform.*, 2011, **36**(41), 2969.

ODS *See* Oxydesulfurization (2).

OFS [Oil From Sludge] A generic term for processes for converting sewage sludge into fuel oils. *See* Enersludge.

OGR [Off-Gas Recovery] A solvent-based technology for recovering olefins, hydrogen, or both from *FCC or coker off-gases. Developed by KTI.

oil gasification Processes that convert liquid petroleum fractions into gaseous fuels. Such processes with special names that are described in this dictionary are CRG, HTR, Petrogas, Pintsch, Recatro, SEGAS, and SSC.

Olefining [Olefin refining] A process for converting *syngas or methanol to a mixture of ethylene, propylene, and butenes. The catalyst is a ZSM-5-type zeolite in which some of the aluminum has been replaced by iron. Developed in 1984 by the National Chemical Laboratory, Pune, India; the process was piloted by Bharat Petrochemical Corporation, Bombay, in 1992.

Indian Patents 159,164; 160,038; 160,212.
European Patent 161,360.

Olefins Conversion Technology [OCT] A process for converting mixtures of ethylene and butenes to propylene by metathesis (disproportionation). Based on technology developed by Phillips Petroleum in the mid-1960s. The catalyst contains tungsten oxide and magnesia. First commercialized in 1986 and further developed by Lummus Technology from 1997. Twenty-two plants were operating worldwide in 2013.

Mol, J.C., *J. Mol. Catal. A. Chem.*, 2004, **213**, 39.
Chem. Eng. (N.Y.). 2013, **120**(3), 39.

OlefinSiv A process for isolating isobutene from a mixture of C_4 hydrocarbons by chromatography over a zeolite molecular sieve. Developed by the Linde Division of the Union Carbide Corporation, as one of its *IsoSiv family of processes.

Adler, M.S. and Johnson, D.R., *Chem. Eng. Prog.*, 1979, **75**(1), 77.

Oleflex [**Ole**fin **flex**ibility] A process for converting normal hydrocarbons into the corresponding olefins (e.g., propane to propylene, or isobutane to isobutene) by catalytic dehydrogenation. Similar to the *Pacol process but incorporating a continuous catalyst regeneration unit. The catalyst is platinum and tin on alumina, operating at above 550°C. So called because of its flexibility in the production of a range of monoolefins from a range of C_2 to C_5 hydrocarbons. Developed and offered for license by UOP. The first commercial plant began operation in Thailand in 1989. Eleven units were operating in 2005. By 2012, UOP had licensed 20 plants. *See also* *CCR.

Vora, B.V. and Imai, T., *Hydrocarbon Process. Int. Ed.*, 1982, **61**(4), 171.
Pujado, P.R. and Vora, B.V., *Hydrocarbon Process. Int. Ed.*, 1990, **69**(3), 65.
Gregor, J. and Wei, D., in *Handbook of Petroleum Refining Processes*, 3rd ed., Meyers, R.A., Ed., McGraw-Hill, New York, 2003, 5.3.
ICIS Chem. Bus., 2012, **281**(14), 37.

Olex A version of the *Sorbex process for separating olefins from paraffins in wide-boiling mixtures. Olex can be used for hydrocarbons in the range C_6–C_{20}. Based on the selective adsorption of olefins in a zeolite and their subsequent recovery by displacement with a liquid at a different boiling point. Mainly used for extracting C_1–C_{14} olefins from the *Pacol process. Six units were commissioned since the first one came on stream in 1972.

Broughton, D.B. and Berg, R.C., *Hydrocarbon Process. Int. Ed.*, 1969, **48**(6), 115.
Sohn, S.W., in *Handbook of Petroleum Refining Processes*, 3rd ed., Meyers, R.A., Ed., McGraw-Hill, New York, 2003, 10.79.

Olgone A catalytic process for removing gum-forming olefins from aromatic hydrocarbon mixtures. Developed in 2002 by ExxonMobil, as an alternative to *clay treating. Licensed to Nippon Petrochemical Refining Co. in 2006 for use in its Muroran refinery.

Chem. Eng. (N.Y.), 2006, **113**(12), 19.
Chem. Eng. Prog., 2007, **103**(5), 12.

Olin Raschig A refinement of the basic *Raschig (1) process for making hydroxylamine. *See* Raschig (1).

OLP [**O**xygène **L**ance **P**oudre] A steelmaking process in which powdered lime is blown into the furnace through the oxygen stream in order to combine with the phosphorus. Developed in the late 1950s by Institut de Recherche de la Sidérurgie. *See also* LD/AC.

Jackson, A., *Oxygen Steelmaking for Steelmakers*, Newnes-Butterworths, London, U.K., 1969, 165.

Omega (1) A process for making ethylene and propylene from C_4 and C_5 streams in oil refineries. It uses a proprietary zeolite catalyst in a fixed-bed reactor at about 500°C. Developed in 2004 by Asahi Kasei Chemicals Corporation and first commercialized at Mizushima, Japan, in 2006.

Chem. Eng. (N.Y.), 2004, **111**(7), 13.
Chem. Week, 2005, **167**(10), 21; 28 Jun 2006.

Omega (2) [**O**nly **M**ono-**E**thylene **G**lycol **A**dvanced] A two-stage process for making monoethylene glycol from ethylene. The first stage converts the ethylene to ethylene oxide, using a catalyst developed by CRI Catalyst. The second stage is another catalytic process that converts the ethylene oxide to monoethylene oxide with a 99% selectivity, via ethylene carbonate. The second stage was developed by Mitsubishi Chemical and acquired by Shell in 1992. The combined process was developed by Shell. In 2012, it was in use by Lotte Daesan in South Korea, Petro Rabigh in Saudi Arabia, and Shell in Singapore.

Chem. Ind. (London), 2007, (17), 22.
Chem. World, 2008, **5**(7), 15.
Chemisch2Weekblad, 16 Jun 2008.
Chem. Eng. (N.Y.), 2009, **116**(1), 16.
TCE, 2009, (815), 38.
ICIS Chem. Bus., 2013, **283**(2), 35.
Wikipedia, The Free Encyclopedia. OMEGA Process, 19 Jun 2013. Retrieved 20 Jul 2013.

OmniSulf A process for removing CO_2, H_2S, and mercaptans from acid gases. A joint development of Lurgi and BASF, based on the *aMDEA process.

ONERA *See* metal surface treatment.

One-shot *See* Siroc.

ONIA-GEGI [Office National des Industriel de l'Azote–Gaz a l'Eau et Gaz Industricl] Acyclic catalytic process for producing either town gas by *steam reforming or *syngas from a variety of hydrocarbon feeds by reaction with oxygen. Developed by the companies named, engineered by Humphreys & Glasgow, and used in England since the 1950s.

Claxton, G., *Benzoles: Production and Uses*, National Benzole & Allied Products Association, London, U.K., 1961, 96.
British Petroleum Co., *Gas Making and Natural Gas*, British Petroleum Co., London, U.K., 1972, 85.

ONYX A process for making catalysts, used by Albemarle in its latest GO-ULTRA range of *FCC catalysts announced in 2009. The pore size distribution is optimized, and a proprietary additive provides sites for the aluminum atoms, which enhance their catalytic activity.

Open Hearth Also called the Siemens process and the pig and scrap process. A steelmaking process in which a mixture of pig iron, iron ore, scrap iron, and limestone was heated in a special reverberatory furnace using regenerative heating. It differed from the earlier Siemens–Martin process in that additional iron ore was added to the slag. The regenerative principle was invented by Frederic Siemens (who changed his name to Charles William Siemens) in England in 1856 and applied to steelmaking by him and E. Martin and P. Martin at Sireuil, France, in 1864. *See also* Acid Open Hearth, Basic Open Hearth, Thomas.

British Patent 2,861 (1856).
Barraclough, K.C., *Steelmaking before Bessemer, Vol. 2, Crucible Steel*, The Metals Society, London, U.K., 1984, 106.
Barraclough, K.C., *Steelmaking 1850–1900*, The Institute of Metals, London, U.K., 1990, 137.

OptiCAT plus A process for regenerating *hydrotreating catalysts. It uses a fluidized-bed pretreatment to control the initial exotherms found in regeneration, followed by a moderate heat soak to eliminate sulfur and carbon. Developed in the 1990s by the Criterion Catalyst Company.

Eur. Chem. News, 1996, **66**(1738), 25.

Optisol A process for removing acid gases and sulfur compounds from gas streams by scrubbing with an aqueous solution of an amine and a proprietary physical solvent.

Hydrocarbon Process. Int. Ed., 1986, **65**(4), 82.

ORC [Occidental Research Corporation] Also called the Garrett process. A coal gasification process using flash pyrolysis at approximately 900°C in the absence of oxygen. Piloted by the Occidental Research Corporation in the 1970s.

McMath, H.G., Lumpkin, R.E., Longanbach, J.R., and Sass, A., *Chem. Eng. Prog.*, 1974, **70**(6), 72.
Sass, A., *Chem. Eng. Prog.*, 1974, **70**(1), 72.

ORCAN [ORganic CArbon and Nitrogen] A *CWAO process for destroying organic wastes in water. It uses hydrogen peroxide and ferrous iron (Fenton's reagent), in addition to atmospheric oxygen. First operated in a textile plant in Switzerland in 1996.

Luck, F., *Catal. Today*, 1999, **53**(1), 87.
Pintar, A., *Catal. Today*, 2003, **77**(4), 451.

Orcarb *See* DR.

OREX [ORganic EXchange] A process for separating lithium isotopes for use in making lithium deuteride for use in nuclear weapons. A solution of a lithium salt in propylene diamine was equilibrated with lithium amalgam. Operated at Oak Ridge National Laboratory from 1952 to 1957 until it was replaced by the *COLEX process.

ORF *See* DR.

Orford An obsolete metallurgical process for separating copper and nickel. Ores containing these metals were smelted in a *Bessemer converter, forming matte. This was melted with sodium sulfate and coke, which yielded copper and nickel sulfides, which are immiscible and easily separated on cooling.

Morgan, G.T. and Pratt, D.D., *British Chemical Industry*, Edward Arnold & Co., London, U.K., 1938, 107.

Organocell A combined process for delignifying and bleaching wood pulp. Developed by Technocell, Düsseldorf, Germany.

Chem. Eng. (N.Y.), 1991, **98**(1), 39.

Organosolv A general name for a papermaking process that achieves separation of the lignin from the cellulose by dissolving the lignin in an organic solvent. The first pilot plant, built in Munich in 1984 with support from the European Commission, used aqueous methanol containing a small concentration of anthraquinone. Several variations on this process were tried in 1989, but none was commercialized. A mixture of acetic and hydrochloric acids has been investigated. *See also* Acetosolv, ASAM, Formacell.

Phillips, G.O., *Chem. Br.*, 1989, **25**, 1007.
Stockburger, P., *Tappi J.*, 1993, **76**(6), 71.
Parajo, J.C., Alonso, J.L., and Santos, V., *Ind. Eng. Chem. Res.*, 1995, **34**, 4333.
Hegert, H.L., in *Environmentally Friendly Technologies for the Pulp and Paper Industries*, Young, R.A. and Akhar, M.,
    Eds., John Wiley & Sons, New York, 1998, 5.
Pye, E.K., in *Biorefineries: Industrial Processes and Products*, Kamm, B., Gruber, P.R., and Kamm, M., Eds.,
    Wiley-VCH, Weinheim, Germany, 2005, **II**, 179.
Lee, S. and Shah, Y.T., *Biofuels and Bioenergy: Process and Technologies,* CRC Press, Boca Raton, FL, 2013, 119.

Orkla A complex process for recovering sulfur from pyrite. The ore was smelted with coke, limestone, and quartz, with very little air at 1600°C, and the iron was removed as a slag. The copper and other nonferrous metals formed a "matte" with the sulfur. Pyrolysis of this matte removed half of the sulfur. An air blast removed the other half without oxidizing it. Developed by the Orkla Mining Company, Norway, between 1919 and 1927. First commercialized at Thamshavn, Norway, in 1931, but the plant closed in 1962. The process was used for many years in Spain, Portugal, and Hungary.

Haynes, W., *The Stone That Burns*, D. Van Nostrand, New York, 1942, 271.
Katz, M. and Cole, R.J., *Ind. Eng. Chem.*, 1950, **42**, 2266.
Kaier, T., *Eng. Min. J.*, 1954, **155**(7), 88.

Orthoflow A fluidized-bed *catalytic cracking process in which the reactor and regenerator are combined in a single vessel. Designed by the M.W. Kellogg Company and widely used in the 1950s. First operated in 1951 by the British American Oil Company at Edmonton, Alberta, Canada. By 1994, more than 120 units had been built.

Reidel, J.C., *Oil Gas J.*, 1952, **50**(46), 200.
Unzelman, G.H. and Wolf, C.J., in *Petroleum Processing Handbook*, Bland, W.F. and Davidson, R.L., Eds., McGraw-Hill, New York, 1967, 3–4.
Hydrocarbon Process. Int. Ed., 1994, **73**(11), 114.
Niccum, P.K. and Santner, C.R., in *Handbook of Petroleum Refining Processes*, 3rd ed., Meyers, R.A., Ed., McGraw-Hill, New York, 2003, 3.3.

Orthoforming A fluidized-bed *Catalytic Reforming process. Developed by the M.W. Kellogg Company in 1953 and first operated commercially in 1955.

Oil Gas J., 1955, **53**(46), 162.
Little, D.M., *Catalytic Reforming*, PennWell Publishing, Tulsa, OK, 1985, xv.

ORU [**O**xygenate **R**emoval **U**nit] A fixed-bed adsorption system for removing oxygenated hydrocarbons (e.g., methanol, methyl *t*-butyl ether) from light hydrocarbon liquid streams. Used particularly for removing all alcohols and ethers from streams resulting from the manufacture of methyl *t*-butyl ether. Developed by UOP in the mid-1980s. By 1992, more than 25 process licenses had been granted. *See also* MRU.

U.S. Patents 4,575,566; 4,575,567.

OSA [**O**xic-**S**ettling-**A**nox] A refinement of the *Activated Sludge sewage-treatment process in which the sludge is periodically exposed to anaerobic conditions by means of a sludge holding tank. This reduces the quantity of sludge to be disposed of. Piloted in China.

Ye, F.-X., Zhu, R.-F., and Li, Y., *J. Chem. Technol. Biotechnol.*, 2008, **83**(1), 109.

OSAG A process for making ammonium sulfate from gypsum. Developed by Österreichische Stickstoffwerke, Linz, Austria, and offered by Power Gas Corp. *See also* Merseburg.

Gopinath, N.D., in *Phosphoric Acid*, Vol. 1, Part 2, Slack, A.V., Ed., Marcel Dekker, New York, 1968, 545.

OSCAR [**O**hio **S**tate **C**arbonation **A**sh **R**eactivation] A clean-coal process developed at the Ohio State University from 2005. The coal is burned in a fluidized bed.

Chem. Eng. News, 2006, **84**(6), 34.

OSIL A *DR ironmaking process. Two plants were operating in India in 1997.

Osmond A medieval process for making wrought iron.

Wikipedia, The Free Encyclopedia. Osmond Process. Revised 15 Mar 2013. Retrieved 20 Jul 2013.

Ostromislenski A process for making butadiene by condensing ethanol with acetaldehyde over an oxide catalyst at 360°C–440°C. Invented by I.I. Ostromislensky in Russia in 1915.

Ostromislensky, I.I., *J. Russ. Phys. Chem.*, 1915, **47**, 1472.
Riegel's Handbook of Industrial Chemistry, 7th ed., Kent, J.A., Ed., Van Nostrand Reinhold, New York, 1974, 232.

Ostwald The basis of the modern family of processes for making nitric acid by the oxidation of ammonia over a platinum catalyst. Named after the eminent German physical chemist Friedrich Wilhelm Ostwald (1853–1932). His invention was patented in the United States in 1902, but the patent was not granted in Germany, where the process had to be operated in secret. Ostwald received the Nobel Prize for this work in 1909.

U.S. Patent 858,904.

Chilton, T.H., *Strong Water: Nitric Acid, Its Sources, Methods of Manufacture, and Uses*, MIT Press, Cambridge, MA, 1968, 73.

OSW [Österreichische StickstoffWerke] An *Ammoxidation process for making acrylonitrile from propylene. Operated in Austria by the named company.

Dumas, T. and Bulani, W., *Oxidation of Petrochemicals: Chemistry and Technology*, Applied Science Publishers, London, U.K., 1974, 155.

OSW/Krupp A process for making sulfuric acid and cement from gypsum, developed by Krupp Koppers.

OT [Once Through] *See* Hysomer.

OTO [Oxygenate To Olefin] A generic name for processes that convert an oxygenate (e.g., ethanol) to an olefin (e.g., ethylene). *See* MTO.

U.S. Patent 6,121,504.

OTP [Oxygenates to Propylene] A range of processes for converting oxygenates, such as methyl *t*-butyl ether, to propylene using a zeolite or zeotype catalyst in a moving bed. A commonly used catalyst is SAPO 34. Developed by Mobil from 1992, later UOP.

U.S. Patents 7,405,337; 7,414,167.

Otto Aqua-Tech HCR [High-Capacity Reactor] A high-intensity biological treatment process for purifying effluents from food processing, certain industrial processes, and landfill leachate. The waste is circulated rapidly through a vertical loop reactor, and air is injected at the top. Invented at the Technical University of Clausthal-Zellerfeld and developed and commercialized by Otto Oeko-Tech. Eleven plants had been installed in Germany and Italy by 1991. *See also* Deep Shaft, Biobor HSR.

Otto–Rummel A coal gasification process based on a double-shaft furnace, developed in Germany by Dr. C. Otto & Company.

Outokumpu [Named after a hill in Finland, near Kuusjärvi] A flash-smelting process for sulfide ores. It is an energy-efficient process (also called an autogenous process), using mainly the heat of combustion of the contained sulfur to sulfur dioxide, rather than any external source of heat. Developed by P. Bryk and J. Ryselin at the Harjavalta works of Outokumpu Oy, Finland, in 1946. Used mainly for copper ores but also for nickel, iron, and lead; by 1988, 40 plants were using the process worldwide.

Kuisma, M., *A History of Outokumpu*, Gummerus Kirjapaino Oy, Jyväskylä, Finland, 1989.

Oxco [Oxidative coupling] A process for converting natural gas to transport fuels and chemicals, based on the oxidative coupling of methane to ethane in a fluidized-bed reactor. Developed in Australia by the Division of Coal and Energy Technology, CSIRO, and BHP. *See also* IFP Oxypyrolysis.

Edwards, J.H., Do, K.T., and Tyler, R.J., *Catal. Today*, 1990, **6**, 435.

Edwards, J.H., Do, K.T., and Tyler, R.J., in *Natural Gas Conversion*, Holmen, A., Jens, K.-J., and Kolboe, S., Eds., Elsevier, Amsterdam, the Netherlands, 1991, 489.

Hutchings, G.H. and Joyner, R.W., *Chem. Ind. (London)*, 1991, 575.

OXD *See* oxidative dehydrogenation.

O-X-D [OXidative Dehydrogenation] A process for converting *n*-butane to butadiene by selective atmospheric oxidation over a catalyst. Developed by the Phillips Petroleum Company and used by that company in Texas from 1971 to 1976. *See also* Oxo-D.

Husen, P.C., Deel, K.R., and Peters, W.D., *Oil Gas J.*, 1971, **69**(31), 60.
Weissermel, K. and Arpe, H.-J., *Industrial Organic Chemistry*, 4th ed., translated by C.R. Hindley and S. Hawkins, Wiley-VCH, Weinheim, Germany, 2003, 113.

oxidative coupling A general term for processes that convert methane to ethane, ethylene, and higher hydrocarbons by heterogeneous catalytic oxidation. If the feed is methane, the process is also called *OCM. *See* IFP Oxypyrolysis, Oxco.

Srivastava, R.D., Zhou, P., Stiegel, G.J., Rao, V.U.S., and Cinquegrane, G., in *Catalysis*, Vol. 9, Spivey, J.J., Ed., Royal Society of Chemistry, Cambridge, U.K., 1992, 191.

oxidative dehydrogenation Also called OXD. A general term for processes that convert monoalkenes to dialkenes, or alkanes to alkenes, by partial oxidation. *O-X-D is a commercialized example. The conversion of ethane to ethylene by such a process was studied by several companies in the 1970s but is not yet commercial.

Eastman, A.D., Kolts, J.H., and Kimble, J.B., in *Novel Production Methods for Ethylene, Light Hydrocarbons, and Aromatics*, Albright, L.F., Crynes, B.L., and Nowak, S., Eds., Marcel Dekker, New York, 1992, 21.

Oxirane A general process for oxidizing olefins to olefin oxides by using an organic hydroperoxide, made by autoxidation of a hydrocarbon. Two versions are commercial. The first to be developed oxidizes propylene to propylene oxide, using as the oxidant *t*-butyl hydroperoxide made by the atmospheric oxidation of isobutane. Molybdenum naphthenate is used as a homogeneous catalyst. The *t*-butanol coproduct may be reconverted to isobutane or sold. The second uses ethylbenzene hydroperoxide as the oxidant, the coproduct 2-phenylethanol being converted to styrene. The process was developed by Halcon Corporation, and the first plant began operation in 1968. Halcon formed a joint venture with Atlantic Richfield Corporation (ARCO) under the name Oxirane Chemical Corporation, so the process came to be called the Oxirane process. It is operated on a large scale in Texas, the Netherlands, and France.

Gait, A.J., in *Propylene and Its Industrial Derivatives*, Hancock, E.G., Ed., Ernest Benn, London, U.K., 1973, 282.
Inform. Chim., 1979, (188), 175.
Braithwaite, E.R., in *Speciality Inorganic Chemicals*, Thompson, R., Ed., Royal Society of Chemistry, London, U.K., 1981, 359.
Weissermel, K. and Arpe, H.-J., *Industrial Organic Chemistry*, 4th ed., translated by C.R. Hindley and S. Hawkins, Wiley-VCH, Weinheim, Germany, 2003, 270.

Oxispec A catalytic process for removing trace impurities from effluent gas streams. Developed by ICI Katalco in the 1990s.

Eur. Chem. News Suppl., Dec 1996, 13.

OXITRON A municipal sewage-treatment process designed by Dorr Oliver. Starfish Industries, UK, has adapted it for coastal towns. The reactor is a biological fluidized bed installed on the seabed.

Water Waste Treat., 1995, **38**(6), 42.

OXO [From **OX**ierung, German, meaning ketonization] Also called hydroformylation and Oxoation. A process for converting olefins to aldehydes containing an additional carbon atom, provided by carbon monoxide:

$$RCH{=}CH_2 + CO + H_2 \rightarrow xR{-}CH_2{-}CH_2{-}CHO + (1-x)R{-}CH{-}CHO$$
$$|$$
$$CH_3$$

In recent years, the name has also come to include the production of downstream products (e.g., alcohols and acids) from the aldehydes. Invented by O. Roelen in 1938 at the Chemische Verwertungsgesellschaft Oberhausen. Further developed by Ruhrchemie and IG Farbenindustrie in Germany during World War II, the process was first commercialized in 1948. Originally, the process operated at high pressure, and dicobalt octacarbonyl, $Co_2(CO)_8$, was used as a homogeneous catalyst. The present process, known as the LP Oxo SELECTOR process (also the LPO process), developed in the 1970s jointly by Union Carbide Corporation, Johnson Matthey, and Davy Corporation, operates at a lower pressure and uses the *Wilkinson catalyst (a complex of rhodium carbonyl hydride with triphenylphosphine). The resulting carbonyl compound can be hydrogenated *in situ* to the corresponding alcohol, if an excess of hydrogen is used, or oxidized in a subsequent operation to the corresponding carboxylic acid. In 1990, 11 plants had been licensed worldwide. More than six million tons of chemicals were made by this family of processes in 1999. *See also* RCH/RP.

German Patent 849,548.
U.S. Patent 2,327,066.
Storch, H.H., Golumbic, N., and Anderson, R.B., *The Fischer-Tropsch and Related Syntheses*, John Wiley & Sons, New York, 1951, 441.
Kirch, L. and Orchin, M., *J. Am. Chem. Soc.*, 1959, **81**, 3597.
Allen, P.W., Pruett, R.L., and Wickson, E.J., in *Encyclopedia of Chemical Processing and Design*, McKetta, J.J. and Cunningham, W.A., Eds., Marcel Dekker, New York, 1990, **33**, 46.
Chem. Eng. (Rugby, Engl.), Mar 1990, 65.
Cornils, B., Herrmann, W.A., and Rasch, M., *Angew. Chem. Int. Ed.*, 1994, **33**, 2144.
Weissermel, K. and Arpe, H.-J., *Industrial Organic Chemistry*, 4th ed., translated by C.R. Hindley and S. Hawkins, Wiley-VCH, Weinheim, Germany, 2003, 136.

Oxoation *See* OXO.

Oxo-D A process for converting *n*-butene to butadiene by selective atmospheric oxidation over a catalyst in a fixed-bed reactor. Developed by Petro-Tex Chemical Corporation and operated by that company in Texas since 1965. *See also* O-X-D.

Welch, L.M., Croce, L.J., and Christmann, H.F., *Hydrocarbon Process. Int. Ed.*, 1978, **57**(11), 131.

Oxorbon A process for removing sulfur compounds from *syngas by adsorption on activated carbon. Offered by Lurgi. *See also* Desorex.

OXY+ A process for making a reducing gas mixture by the partial oxidation of natural gas, using oxygen. Developed by Midrex Technologies, Charlotte, NC, and commercialized in 2004 for use in the *MIDREX process.

Oxycat A catalytic oxidation process for removing combustible vapors from air and industrial exhaust gases. The catalyst is platinum on alumina, supported inside a porcelain tube.

Houdry, J.H. and Hayes, C.T., *J. Air Pollut. Control Assoc.*, 1957, **7**(3), 182.
Resen, L., *Oil Gas J.*, 1958, **56**(1), 110.

Oxychloride A process for making a titanium dioxide concentrate from a low-grade titanium ore. The ore is leached with hydrochloric acid and the iron removed by solvent extraction. Developed by Process Research Ortech and Canadium Titanium Ltd.

Canadian Patent 2,289,967.
U.S. Patent 6,699,446.

oxychlorination An adaptation of the *Deacon process, used for converting ethylene to 1,2-dichloroethane. A mixture of ethylene, air, and hydrogen chloride is passed over a catalyst of cupric chloride on potassium chloride. For the manufacture of vinyl chloride, the dichloroethane is cracked and the hydrogen chloride recycled. Several companies developed and commercialized this process in the 1960s.

Spitz, P.H., *Petrochemicals, the Rise of an Industry*, John Wiley & Sons, New York, 1988, 403.

OxyClaus A variation of the *Claus process, using combustion with oxygen to convert a fraction of the sulfur compounds to sulfur dioxide before reaction. Developed by Lurgi Oel Gas Chemie and Pritchard

Corporation. Ten plants were operating in 1995. In 2000, 16 plants were operating, and 14 more were in design or construction.

Kohl, H.L. and Nielsen, R.B., *Gas Purification*, 5th ed., Gulf Publishing Co., Houston, TX, 1997, 693.
Hydrocarbon Process. Int. Ed., 2000, **79**(4), 82.

oxydehydrogenation A general name for the conversion of saturated aliphatic hydrocarbons to olefins, using atmospheric oxygen; exemplified by *Oxo-D, *O-X-D.

OXYDEP An *Activated Sludge process using pure oxygen. Developed by Air Products and Chemicals.

oxydesulfurization (1) A general name for processes that remove sulfur from coal by oxidation. *See also* IGT, PETC, oxydesulfurization (2).

oxydesulfurization (2) Also known as **ODS**. A process for extracting elemental sulfur from carbon dioxide or natural gas, developed by the M.W. Kellogg Company. *See also* oxydesulfurization (1).

OxyGEN [Oxygen GENcrator] A version of the *VPSA process for separating oxygen from air by vacuum pressure swing adsorption. Developed by the Linde Division of the Union Carbide Corporation. Economical for quantities between 10,000 and 150,000 cf/h, at purities between 90% and 95%. Intended primarily for providing air enriched in oxygen for the steel and pulp and paper industries. First operated in 1989 at the Manitoba Rolling Mills in Selkirk, Manitoba, Canada.

Eur. Chem. News, 1989, **53**(1391), 31.
Chem. Mark. Rep., 3 Sep 1989, 11.

Oxyhydrochlorination A two-stage process for making gasoline from lower paraffinic hydrocarbons, especially methane. The methane, mixed with oxygen and hydrogen chloride, is passed over a supported copper chloride catalyst, yielding a mixture of chloromethanes:

$$CH_4 + \tfrac{1}{2}O_2 + HCl = CH_3Cl + H_2O$$

These chloromethanes are converted to gasoline using a zeolite catalyst, and the hydrogen chloride coproduct is recycled. Developed from 1987 by Allied Chemical Corporation and the Pittsburgh Energy Technology Center of the U.S. Department of Energy.

U.S. Patent 4,769,504.
Srivastava, R.D., Zhou, P., Stiegel, G.J., Rao, V.U.S., and Cinquegrane, G., in *Catalysis*, Vol. 9, Spivey, J.J., Ed., Royal Society of Chemistry, Cambridge, U.K., 1992, 205.

OxyNol Also called CES Oxynol. A process for converting the cellulose in domestic waste and sewage sludge to ethanol. Developed by Pencor-Masada Oxynol. There are five stages: feedstock preparation, hydrolysis with sulfuric acid, acid recovery, fermentation, and distillation. Developed by Masada Resource Group and Auburn University and partially demonstrated during World War II at the TVA Muscle Shoals Demonstration Facility. Not yet commercialized. *See also* CESL.

U.S. Patents 5,407,817; 6,419,828.
Katzen, R. and Schell, D.J., in *Biorefineries: Industrial Processes and Products*, Kamm, B., Gruber, P.R., and Kamm, M., Eds., Wiley-VCH, Weinheim, Germany, 2005, **I**, 133.
TCE (formerly The Chemical Engineer), 2007, (793), 14.

Oxypro (1) A process for making di-*iso*propyl ether (DOPE) from a propane–propylene stream from *FCC. The catalyst system is superior to other acid catalysts, such as zeolites, because of its greater activity at low temperatures. The Oxypro catalyst functions at below 175°C, whereas zeolites require temperatures closer to 260°C. DOPE is used as a gasoline additive. Developed by UOP in 1994; the first commercial unit was commissioned in 2002.

Hydrocarbon Process. Int. Ed., 1995, **74**(7), 15; **74**(8), 42.

Krupa, S., Richardson, L., and Meister, J., in *Handbook of Petroleum Refining Processes*, 3rd ed., Meyers, R.A., Ed., McGraw-Hill, New York, 2003, 13.19.

Oxypro (2) A family of pulp-bleaching processes developed by Air Products and Chemicals. Piloted in 1995 at Pittsfield, MA. Oxypro O$_R$ is for processing mixed office waste papers; it uses molecular oxygen, optionally hydrogen peroxide, sodium hydroxide, and a stabilizer. First commercialized in Scotland in 1994.

U.S. Patent 5,211,809.

Chem. Eng. (N.Y.), 1995, **102**(12), 66.

Oxypyrolysis *See* IFP Oxypyrolysis.

Oxy-Rich [**Oxy**gen en**Rich**ment] A version of the *PSA process for producing air enriched with oxygen in the range of 23%–50%. It is for use where higher degrees of enrichment are unnecessary—for enhanced combustion and for improved oxidation in chemical and biochemical reactors. It is more energy-efficient than the standard PSA process, because only some of the air is compressed to the highest level of the cycle. It was announced in 1987 by Air Products and Chemicals.

U.S. Patent 4,685,939.

Sircar, S. and Kratz, W.C., in *Adsorption and Ion Exchange: Fundamentals and Applications*, LeVan, M.D., Ed., American Institute of Chemical Engineers, New York, 1988, 141.

Oxysulfreen A multistage variation of the *Sulfreen process for removing sulfur compounds from the off-gases from the *Claus process. In the first stage, the sulfur in all the organic sulfur compounds is converted to hydrogen sulfide by *hydrodesulfurization over a cobalt–molybdenum–alumina catalyst. After removal of water, some of this hydrogen sulfide is oxidized to sulfur dioxide over a titania catalyst. The final stage is the Sulfreen version of the Claus process, in which the remaining hydrogen sulfide is reacted with the sulfur dioxide to produce elemental sulfur. *See also* Hydrosulfreen, Sulfreen.

OXYWELL A process for separating oxygen from air by *PSA, using vacuum desorption from a zeolite. Used for medical oxygen generators, hence the name. Developed by Nippon Sanso.

Suzuki, M., in *Adsorption and Ion Exchange: Fundamentals and Applications*, LeVan, M.D., Ed., American Institute of Chemical Engineers, New York, 1998, 120.

Ozalid *See* Reprography.

Ozark-Mahoning A process for concentrating *wet-process phosphoric acid by submerged combustion evaporation. Developed by the Ozark-Mahoning Company and first installed in Pasadena, TX, in 1945. *See also* Nordac.

Forster, J.H., *Phosphoric Acid*, Vol. 1, Part 2, Slack, A.V., Ed., Marcel Dekker, New York, 1968, 599.

OZIOLE *See* carbonization.

Ozocarb A process for purifying municipal drinking water by treatment with ozone, hydrogen peroxide, and activated carbon. Developed by Trailigaz, a subsidiary of Compagnie Générale des Eaux.

OZOFLOT A process for treating raw water with ozone. The design of the treatment vessel encourages bubbles of ozonized air to become attached to algae and particulate solids and float to the surface, where they can be skimmed off. Developed in France by OTV.

Ozonix A process for destroying microorganisms in contaminated water. It involves ozone, cavitation, and electrooxidation. Developed by Ecosphere Technologies in 2012.

U.S. Patent Application 2012-0205301.

Chem. Eng. (N.Y.), 2013, **120**(2), 12.

P

Paal *See* Normann.

Pacol [**Pa**raffin **co**nversion, **l**inear] A process for converting C_{10} to C_{20} linear aliphatic hydrocarbons to the corresponding olefins by catalytic dehydrogenation. Used mainly for the production of linear intermediates for detergents. Used also, in combination with *Olex, for making monoolefins. The catalyst is a platinum metal on an alumina support; several generations of the catalyst have been developed, the latest is known as DeH-7. Developed by UOP in the 1960s and first commercialized in 1968. By 2005, 29 plants had been built. *See also* DeFine, Detergent Alkylate, PEP.

Bloch, M.S., *Eur. Chem. News*, 1966, **10**(254), 46.
Broughton, D.B. and Berg, C.R., *Chem. Eng. (N.Y.)*, 1970, **77**(2), 86.
Vora, B., Pujado, P.R., Imai, T., and Fritsch, T., *Chem. Ind. (London)*, 6 Mar 1990, 187.
Pujado, P.R., in *Handbook of Petroleum Refining Processes*, 3rd ed., Meyers, R.A., Ed., McGraw-Hill, New York, 2003, 5.11.

PACT (1) [**P**owdered **A**ctivated **C**arbon **T**reatment] A wastewater treatment process that combines activated carbon treatment with biological treatment, providing a single-stage treatment of toxic liquid wastes. Developed by DuPont in the 1970s at its Chambers Works, Deepwater, NJ, and now licensed by U.S. Filter. More than 50 units were operating in 1990.

Lankford, P.W., in *Toxicity Reduction in Industrial Effluents*, Lankford, P.W. and Eckenfelder, W.W., Jr., Eds., Van Nostrand Reinhold, New York, 1990, 229.
Chem. Eng. (N.Y.), 1990, **97**(2), 44.
Hutton, D.G., Meidl, J.A., and O'Brien, G.J., in *Environmental Chemistry of Dyes and Pigments*, Reife, A. and Freeman, H.S., Eds., John Wiley & Sons, New York, 1996, 105–164.
McIntyre, D.C., in *Environmental Chemistry of Dyes and Pigments*, Reife, A. and Freeman, H.S., Eds., John Wiley & Sons, New York, 1996, 165–190.

PACT (2) [**P**lasma **A**nd **C**atalyst integrated **T**echnology] A process for oxidizing gaseous pollutants in gas streams. The gas is passed through an electrical discharge between metal electrodes having catalytically active surfaces. The electrodes either vibrate or rotate. Invented in 1993 by Fujitsu Ltd., Japan, and further developed at the University of Connecticut.

U.S. Patent 5,474,747.
Chen, X., Rozak, J., Lin, J.-C., Suib, S.L., Hayashi, Y., and Matsumoto, H., *Appl. Catal. A: Gen.*, 2001, **219**(1–2), 25.

PAL [**P**eroxide-**A**ssisted **L**each] An improved *Cyanide process for extracting gold from its ores. Addition of hydrogen peroxide to the system improves the yield of gold and reduces the usage of cyanide. First operated in South Africa in 1987.

PAL II [**P**ressure **A**cid **L**each] A process for extracting nickel from laterite ores. Operated in Australia.

PAMCO [**P**ittsburgh **A**nd **M**idway **CO**al Mining Company] *See* SRC.

PAMELA [**P**rototyp**A**nlage **M**ol zur **E**rzeugung **L**agerfähiger **A**bfälle] A continuous process for immobilizing nuclear waste in a borosilicate glass. Developed by the Deutsche Gesellschaft für Wiederaufarbeitung von Kernbrennstoffen, and Eurochemic, initially in Germany and later in Mol, Belgium. The plant was first operated with radioactive materials in 1985. *See also* VERA.

Lutze, W., in *Radioactive Waste Forms for the Future*, Lutze, W. and Ewing, R.C., Eds., North-Holland, Amsterdam, the Netherlands, 1988, 7, 612.

Ojovan, M.I. and Lee, W.E., *An Introduction to Nuclear Waste Immobilisation*, Elsevier, Amsterdam, the Netherlands, 2005, 235.

Panindco An early entrained-flow coal gasification process.

Paragon A two-stage *hydrocracking process, based on the zeolite ZSM-5, claimed to increase the yield and quality of the gasoline produced. Developed by Chevron Research Company, but not commercialized by 1991.

O'Rear, D.J., *Ind. Eng. Chem. Res.*, 1987, **26**, 2337.

Maxwell, I.E. and Stork, W.H.J., in *Introduction to Zeolite Science and Practice*, van Bekkum, H., Flanigen, E.M., and Jansen, J.C., Eds., Elsevier, Amsterdam, the Netherlands, 1991, 610.

Paraho [**Par**a **ho**mem, Portuguese, for mankind] A process for making oil and gas from oil shale. Development began in 1971 by the Paraho Development Corporation at Grand Junction, CO. Since then, in conjunction with a variety of American companies and agencies, a number of pilot plants have been operated and plants designed.

Jones, J.B., Jr. and Glassett, J.M., in *Handbook of Synfuels Technology*, Meyers, R.A., Ed., McGraw-Hill, New York, 1984, 4–63.

Paralene [**Para**-xylene] Also called Gorham and also spelled parylene. A process for coating articles with poly-p-xylene. The vapor of di-p-xylylene is pyrolyzed at 550°C, yielding p-xylyl free radicals, $\cdot CH_2$–C_6H_4–$CH_2\cdot$, which deposit and polymerize on cooled surfaces. Developed by W. F. Gorham at Union Carbide Corporation.

Gorham, W.F., *J. Polymer Sci., A-1*, 1966, **4**, 3027.

Kirk-Othmer's Encyclopedia of Chemical Technology, 4th ed., Suppl., John Wiley & Sons, New York, 1998, 864.

Fortin, J.B. and Lu, T.-M., *Chem. Mater.*, 2002, **14**(5), 1945.

ParamaX Not one process but a suite of processes offered by Axens from about 2002 to maximize the production of p-xylene from mixtures of benzene, toluene, and xylenes. One installation was planned for installation at the Japan Energy oil refinery in Kashima, Japan, in 2008. Another installation was planned for Korea in 2008.

Chem. Week, 14 Dec 2005, **167**(42), 16.

Chimie Pharma Hebdo, 26 Jun 2006, (345).

PARC A process for making ammonia, developed by KTI.

Parex (1) [**Par**a **ex**traction] A version of the *Sorbex process, for selectively extracting p-xylene from mixtures of xylene isomers, ethylbenzene, and aliphatic hydrocarbons. The feedstock is usually a C_8 stream from a catalytic reformer, mixed with a xylene stream from a xylene isomerization unit. The process is operated at 177°C; the desorbent is usually p-diethylbenzene. The first commercial plant began operation in Germany in 1971; at the end of 2006, 90 units had been licensed, of which 76 were onstream. Not to be confused with *Parex (2).

U.S. Patents 3,524,895; 3,626,020; 3,734,974.

Broughton, D.B., Neuzil, R.W., Pharis, J.M., and Brearley, C.S., *Chem. Eng. Prog.*, 1970, **66**(9), 70.

Ruthven, D.M., *Principles of Adsorption and Adsorption Processes*, John Wiley & Sons, New York, 1984, 400.

Seidel, R. and Staudte, B., *Zeolites*, 1993, **13**, 348.

Commissaris, S.E., in *Handbook of Petroleum Refining Processes*, 3rd ed., Meyers, R.A., Ed., McGraw-Hill, New York, 2003, 2.47.

Parex (2) [**Par**affin **ex**traction] Also called **GDR Parex**. A process for separating linear aliphatic hydrocarbons from branched-chain and cyclic hydrocarbons by means of a zeolite 5A adsorbent. The products are desorbed with a mixture of steam and ammonia. Developed in the mid-1960s by Leuna-Werke and Schwedt in East Germany and operated in East Germany, Bulgaria, and the USSR. Broadly similar to *Molex, and not to be confused with *Parex (1).

East German Patents 49,962; 64,766.
Wehner, K., Welker, J., and Seidel, G., *Chem. Tech. (Leipzig)*, 1969, **21**, 548.
Seidel, G., Welker, J., Ermischer, W., and Wehner, K., *Chem. Tech. (Leipzig)*, 1979, **31**(8), 405.
Schirmer, W., Fiedler, K., Stach, H., and Suckow, M., in *Zeolites as Catalysts, Sorbents and Detergent Builders*, Karge, H.G. and Weitkamp, J., Eds., Elsevier, Amsterdam, the Netherlands, 1989, 439.
Seidel, R. and Staudte, B., *Zeolites*, 1993, **13**, 348.

Par-Isom [**Par**affin **Isom**erization] A process for isomerizing light naphtha in order to improve the octane number. The catalyst, a sulfated zirconia containing platinum, was developed by Cosmo Oil Company and Mitsubishi Heavy Industries, and the process was developed by UOP. The oxide catalyst is claimed to be more efficient than zeolite catalysts currently used for this process.

Cusher, N.A., in *Handbook of Petroleum Refining Processes*, 3rd ed., Meyers, R.A., Ed., McGraw-Hill, New York, 2003, 9.41.
Kimura, T., *Catal. Today*, 2003, **81**, 57.

Parker A hydrometallurgical process for extracting copper from sulfide ores using an aqueous solution containing an organic nitrile. Invented in Australia by A. J. Parker. Extensively studied in the laboratory but never piloted.

U.S. Patent 3,865,744.
Biographical Memoirs of the Australian Academy of Science, 2006.

Parkerizing *See* metal surface treatment.

Parkes A process for removing silver from lead, based on the use of zinc, which forms intermetallic compounds of lower melting point. Developed by A. Parkes in Birmingham, England, in the 1850s. Parkes also invented the first plastic (Parkesine), used for making billiard balls.

British Patents 13,675 (1850); 13,997 (1852).
Barrett, K.R. and Knight, R.P., *Silver: Exploration, Mining and Treatment*, Institute of Mining and Metallurgy, London, U.K., 1988.

parting A general name for the separation of silver, gold, and platinum from each other, practiced since antiquity. Early processes involved dissolution in nitric acid, but only electrochemical processes are used now. *See* Balbach, Moebius, Thum, Wohlwill.

Patera A process for extracting silver from its ores, invented in 1858.

Patio [Spanish, a courtyard] A medieval process for extracting silver from argentite, Ag_2S. The ore was mixed with salt, mercury, and roasted pyrites that contains cupric sulfate. This mixture was crushed by stones dragged by mules walking on the paved floor of a courtyard. The overall reactions are as follows:

$$Ag_2S + 2NaCl + 2Hg + CuSO_4 = 2Ag + Hg_2Cl_2 + CuS + Na_2SO_4$$

$$2Ag + 2Hg = Hg_2Ag_2$$

Invented by Bartolomé de Medina, a Spanish trader, in Mexico in 1554, and used there until the end of the nineteenth century. The invention changed the course of economic history in Hispanic America; 40% of all the silver recorded to have been produced in the world before 1900 was extracted by this process. *See also* Cazo, Washoe.

Probert, A., *J. West*, 1969, **8**, 90 (118 refs).
Jacobsen, R.H. and Murphy, J.W., *Silver: Exploration, Mining and Treatment*, Institute of Mining & Metallurgy, London, U.K., 1988, 283.
Nriago, J.O., *Chem. Ind. (London)*, 1994, **30**(8), 650.

Pattinson (1) A process for extracting silver from lead by selective crystallization. When molten lead is cooled, the first crystals of lead contain less silver than the residual melt. Repetition of this process a number of times yields a silver concentrate, which is further purified by *cupellation. Invented in 1833 by H.L. Pattinson. Largely superseded by the *Parkes process, except for metals containing bismuth, for which Pattinson is the preferred process. *See also* Luce-Rozan.

Dennis, W.H., *A Hundred Years of Metallurgy*, Gerald Duckworth, London, U.K., 1963, 194.

Pattinson (2) A process for making pure magnesium compounds from calcined dolomite, using the high solubility of magnesium hydrogen carbonate, $Mg(HCO_3)_2$, in water. Invented by H.L. Pattinson in Gateshead, England, in 1841.

British Patent 9,102 (1841).

Pauling–Plinke A process for concentrating and purifying waste sulfuric acid by distillation and addition of nitric acid. It was obsolete by 1994.

Büchner, W., Schliebs, R., Winter, G., and Büchel, K.H., *Industrial Inorganic Chemistry*, VCH Publishers, Weinheim, Germany, 1989, 117.

PCA [**P**recipitation with a **C**ompressed **A**ntisolvent] A process for making a solid with unusual morphology by spraying a solution of it into a supercritical fluid. The process resembles spray drying into a supercritical fluid. Used for making microspheres, microporous fibers, and hollow microporous fibers.

Brennecke, J.F., *Chem. Ind. (London)*, 1996, (21), 831.

PCC [**P**artial **C**ombustion **C**racking] Not to be confused with precipitated calcium carbonate. A process for *cracking crude petroleum or heavy oil to a mixture of olefins and aromatic hydrocarbons. The heat carrier is steam, produced by the partial combustion of the feed. Developed by Dow Chemical Company. It was piloted in 1979, and a larger plant was built in Freeport, TX, in 1984.

Kirk, R.O., *Chem. Eng. Prog.*, 1983, **79**(2), 78.
Hu, Y.C., in *Chemical Processing Handbook*, Marcel Dekker, New York, 1993, 780.

PDH [**P**ropane **D**e**H**ydrogenation] A catalytic process for converting propane to propylene. The catalyst was originally chromium on alumina but is now platinum on magnesium aluminate. Developed by a consortium of Linde, BASF, and Statoil.

U.S. Patent 5,378,350.
Sanfillipo, D. and Miracca, I., *Catal. Today*, 2006, **111**(1–2), 136.

Peachy A process for vulcanizing rubber by successive exposure to hydrogen sulfide and sulfur dioxide. Not commercialized.

peak shaving A term used in the gas industry for gas-producing processes that can be started quickly, to satisfy sudden increases in demand. One such process is *Hytanol.

Pearl A process for recovering phosphorus from wastewaters by precipitation as struvite crystals. Developed by Ostara Nutrient Recovery Technologies in Vancouver, BC.

Chem. Eng. (N.Y.), 2013, **120**(2), 17.

PEATGAS A process for converting peat to gaseous fuels. Developed from 1974 to 1980 by the Institute of Gas Technology, Chicago, and the Minnesota Gas Company, to use the peat deposits in Northern Minnesota.

Pechiney (1) A process for making aluminum by electrolyzing a molten mixture of the chlorides and fluorides of Al, Ba, Ca, Na, and alumina. Developed in 1922.

Pechiney (2) A process for making urea from ammonia and carbon dioxide. The ammonium carbamate intermediate is handled and heated as an oil slurry.

> Lowenheim, F.A. and Moran, M.M., *Faith, Keyes, and Clark's Industrial Chemicals*, 4th ed., John Wiley & Sons, New York, 1975, 856.

Pechiney H$^+$ A process for extracting aluminum from clays and other aluminous ores and wastes by hydrochloric acid. The ore is first attacked by sulfuric acid, and a hydrated aluminum chloride sulfate is isolated. Sparging a solution of this with hydrogen chloride precipitates aluminum trichloride hexahydrate, which is pyrohydrolyzed in two stages. Invented in 1977 by J. Cohen and A. Adjemian at Aluminium Pechiney, France, and subsequently developed in association with Alcan. Piloted in France but not yet commercialized.

> French Patent 1,558,347.
> European Patents 5,679; 6,070.
> U.S. Patent 4,124,680.
> Cohen, J. and Mercier, H., *Light Met. Met. Ind.*, 1976, **2**, 3.
> O'Connor, D.J., *Alumina Extraction from Non-bauxitic Materials*, Aluminium-Verlag, Düsseldorf, Germany, 1988, 87, 112, 145.

Pechini A process for making mixed oxide ceramics from organic precursors. It is based on the ability of certain α-hydroxy-carboxylic acids, such as citric acid, to form polybasic acid chelates with metal ions. The chelates undergo polyesterification when heated with a polyol such as ethylene glycol. Further heating produces a resin, which is then calcined. This method yields very homogeneous products. It was originally developed for making alkaline earth and lead titanates and has since been used for making niobates, ferrites, nickelates, and even ceramic superconductors. Invented in 1963 by M. P. Pechini at the Sprague Electric Company, MA.

> U.S. Patent 3,330,697.
> Falter, L.M., Payne, D.A., Friedmann, T.A., Wright, W.H., and Ginsburg, D.M., *Br. Ceram. Proc.*, 1989, (41), 261.

Pecor *See* Woodall–Duckham.

Pedersen A process for extracting aluminum from bauxite, which also yields metallic iron. The ore is first smelted in an electric furnace with limestone, iron ore, and coke at 1350°C–1400°C to produce a calcium aluminate slag and metallic iron. Aluminum is leached from the slag by sodium carbonate solution, and alumina is then precipitated from the leachate by carbon dioxide. The process requires cheap electricity and a market for the iron. It was invented by H. Pedersen in 1924 and operated at Hoyanger, Norway, from 1928 until the mid-1960s.

> British Patent 232,930.
> Miller, J. and Irgens, A., *Light Met. Met. Ind.*, 1974, **3**, 789.
> O'Connor, D.J., *Alumina Extraction from Non-bauxitic Materials*, Aluminium-Verlag, Düsseldorf, Germany, 1988, 233.

Pekilo [Paecilomyces] A process for making single-cell protein from waste sulfite liquor from the paper industry. The organism is *Paecilomyces variotii*. Developed in Finland from 1973.

> Romantschuk, H. and Lehtomaki, M., *Proc. Biochem.*, 1978, **13**(3), 16.

Penex [**Pen**tane and he**x**ane isomerization] A process for converting *n*-pentane and *n*-hexane and their mixtures into branched-chain pentanes and hexanes of higher octane number by catalytic isomerization. The catalyst is similar to the *Butamer catalyst. The product is used in high-octane gasoline. First commercialized by UOP in 1958. More than 120 units had been commissioned by 2002.

Unzelman, G.H. and Wolf, C.J., in *Petroleum Processing Handbook*, Bland, W.F. and Davidson, R.L., Eds., McGraw-Hill, New York, 1967, 3–48.
Schmidt, R.J., Weiszmann, J.A., and Johnson, J.A., *Oil Gas J.*, 1985, **83**(21), 80.
Hydrocarbon Process. Int. Ed., 1996, **75**(11), 142.
Cusher, N.A., in *Handbook of Petroleum Refining Processes*, 3rd ed., Meyers, R.A., Ed., McGraw-Hill, New York, 2003, 9.15.

Penex-Plus A petroleum refining process that combines the *Penex process with a process for hydrogenating benzene to cyclohexane. Developed by UOP for reducing the benzene content of gasoline; first offered for license in 1991.

Peniakoff A process for extracting aluminum from bauxite or other aluminous ores. The ore is roasted with coke and sodium sulfate in a rotary kiln at 1200°C–1400°C; this converts the aluminum to sodium aluminate, which is leached out with dilute sodium hydroxide solution. The basic reactions are as follows:

$$Na_2SO_4 + 4C = Na_2S + 4CO$$

$$Na_2SO_4 + 2C = Na_2S + 2CO_2$$

$$4Al_2O_3 + Na_2S + 3Na_2SO_4 = NaAlO_2 + 4SO_2$$

The sulfur dioxide is recovered as sulfuric acid and reconverted to sodium sulfate. Alumina hydrate is precipitated from the sodium aluminate by carbon dioxide. The process has not become widely accepted because the product is contaminated by silica, but it was used in Belgium before and after World War I and in Germany in the 1920s and 1940s.

O'Connor, D.J., *Alumina Extraction from Non-bauxitic Materials*, Aluminium-Verlag, Düsseldorf, Germany, 1988, 306.

Penna *See* Woodall–Duckham.

Penniman (1) A process for oxidizing petroleum, or its distillates, to mixtures of acids, phenols, and aldehydes. Powdered coal or coke is added to the petroleum, and air is passed through under high temperature and pressure. Invented by W. B. D. Penniman in 1925.

U.S. Patent 252,327.
Ellis, C., *The Chemistry of Petroleum Derivatives*, The Chemical Catalog Co., New York, 1934, 839.

Penniman (2) Also called **Penniman–Zoph**. A process for making a yellow iron oxide pigment. Hydrated ferric oxide seed is added to a solution of ferrous sulfate and the suspension circulated over scrap iron, with air being passed through. Hydrated ferric oxide deposits on the seed crystals, giving a finely divided, yellow pigment:

$$4FeSO_4 + 6H_2O + O_2 = 4FeO \cdot OH + 4H_2SO_4$$

The pigment can be used in this form, or calcined to a red ferric oxide pigment. Invented by R. S. Penniman and N. M. Zoph in 1917.

U.S. Patents 1,327,061; 1,368,748.
Kirk-Othmer's Encyclopedia of Chemical Technology, 4th ed., Vol. 19, John Wiley & Sons, New York, 1991–1998, 24.
Buxbaum, G. and Pfaff, G., *Industrial Inorganic Pigments*, 3rd ed., VCH, Weinheim, Germany, 2005, 105.

Pennsalt *See* Pennwalt.

Pennwalt Formerly called Pennsalt. A process for making calcium hypochlorite by passing chlorine into an aqueous suspension of calcium hypochlorite and calcium hydroxide. Developed by the Pennsylvania Salt Manufacturing Corporation in 1948.

U.S. Patent 2,441,337.

Pentafining A process for isomerizing pentane in a hydrogen atmosphere, using a platinum catalyst supported on silica–alumina. Developed by the Atlantic Richfield Company.

Unzelman, G.H. and Wolf, C.J., in *Petroleum Processing Handbook*, Bland, W.F. and Davidson, R.L., Eds., McGraw-Hill, New York, 1967, 3–50.

Pentesom *See* C$_5$ Pentesom.

PEP [**P**acol **E**nhancement **P**rocess] A process for selectively removing aromatic hydrocarbons from hydrocarbon mixtures. *See* Pacol.

Pujado, P.R., in *Handbook of Petroleum Refining Processes*, 3rd ed., Meyers, R.A., Ed., McGraw-Hill, New York, 2003, 5.18.

Pep Set A process for making foundry molds developed by Ashland Specialty Chemical, introduced in 1970. In 1990, it was announced that a pilot plant was to be built in cooperation with the USSR and that the process had been licensed in China. *See also* Isocure.

Peracidox A process for removing sulfur dioxide from the tail gases from sulfuric acid plants by absorption in peroxomonosulfuric acid (Caro's acid). The peroxomonosulfuric acid is generated on-site by the electrolytic oxidation of sulfuric acid. Developed by Lurgi and Süd-Chemie and first operated in 1972.

PERC [Possibly named after the **P**ittsburgh **E**nergy **R**esearch **C**enter, now the National Energy Technology Laboratory] A process for making a heavy fuel oil by reacting a slurry of biomass in aqueous sodium carbonate solution with carbon monoxide. Under development in the United States in 1980.

Perchloron A process for making calcium hypochlorite, similar to the *Mathieson (2) process but using more chlorine and yielding a more readily filterable material. Developed by the Pennsylvania Salt Manufacturing Company. The name is also used in Germany as a product name for calcium hypochlorite.

Sheltmire, W.H., in *Chlorine: Its Manufacture, Properties, and Uses*, Sconce, J.S., Ed., Reinhold Publishing, New York, 1962, 523.

Perco An early petroleum sweetening process, using an adsorbent bed containing copper sulfate and sodium chloride.

PERCOS A process for removing sulfur dioxide from waste gases by scrubbing with aqueous hydrogen peroxide. The product is a commercial grade of 30%–60% sulfuric acid. Developed by Adolph Plinke Söhne and Degussa.

Pernert A process for making perchloric acid by reacting sodium perchlorate with hydrochloric acid. Invented by J.C. Pernert in 1946 and operated by the Hooker Electrochemical Company at Niagara Falls.

U.S. Patent 2,392,861.
Schumacher, J.C., *Perchlorates: Their Properties, Manufacture, and Uses*, Reinhold Publishing, New York, 1960, 72.

Perosa A process for extracting beryllium from beryl.

Perox A process for removing hydrogen sulfide from coal gas. The gas is passed through aqueous ammonia containing hydroquinone. Atmospheric oxidation of the resulting solution gives elemental sulfur. Developed in Germany after World War II and still in use, although no longer offered by Krupp.

Pippig, H., *Gas Wasserfach.*, 1953, **94**, 62 (*Chem. Abstr.*, **47**, 5096).
Kohl, A.L. and Riesenfeld, F.C., *Chem. Eng. (N.Y.)*, 1959, **66**, 153.
Kohl, H.L. and Nielsen, R.B., *Gas Purification*, 5th ed., Gulf Publishing Co., Houston, TX, 1997, 762.

perox-pure A process for oxidizing organic contaminants in water by the combined use of hydrogen peroxide and UV radiation. Developed originally for treating contaminated groundwater, it is now also used for purifying recycled water used in semiconductor manufacture and in many chemical manufacturing processes. A variant is known as Rayox. Developed in 1986 by Peroxidation Systems, Tucson, AZ. Calgon Carbon acquired the business from Vulcan Chemicals in December of 1995. By 1996, over 250 installations had been made worldwide.

Masten, S.J. and Davies, S.H.R., in *Environmental Oxidants*, Nriagu, J.O. and Simmons, M.S., Eds., John Wiley & Sons, New York, 1994, 534.
James, S.C., Kovalik, W.W., Jr., and Bassin, J., *Chem. Ind. (London)*, 1995, (13), 492.

Perrin A modification of the *Bessemer process that accomplishes the removal of phosphorus from iron by treating the initial molten metal with a molten mixture of lime, alumina, and fluorspar.

Persson A process for making chlorine dioxide by reducing sodium chlorate with chromium (III) in the presence of sulfuric acid. The chromium (III) becomes oxidized to chromium (VI) and is then reduced back to chromium (III) with sulfur dioxide. This cyclic redox process with chromium avoids complications that would occur if sulfur dioxide itself were used as the reductant. Installed at the Stora Kopparbergs paper mill, Sweden, in 1946.

Sheltmire, W.H., in *Chlorine: its Manufacture, Properties and Uses*, Sconce, J.S., Ed., Reinhold Publishing, New York, 1962, 275, 538.

Pertraction A process for removing organic pollutants from industrial wastewater. The water is contacted with an organic solvent via a hollow-fiber membrane. Developed in 1994 by TNO Institute for Environmental and Energy Technology, in collaboration with Tauw Environmental Consultancy and Hoechst.

Eureka, 1994, **14**(11), 16.

PETC [Pittsburgh Energy Technology Center] A chemical coal-cleaning process based on *oxydesulfurization, in which the oxidant is air and lime is used to fix the sulfur. Developed by the Pittsburgh Energy Technology Center, funded by the U.S. Department of Energy from 1970 to 1981. *See also* Ames (2).

IEA Coal Research, *The Problems of Sulphur*, Butterworths, London, U.K., 1989, 14.

Petit A process for removing hydrogen sulfide and hydrogen cyanide from gas streams by scrubbing with an alkali carbonate solution and regenerating the liquor with carbon dioxide. Invented by T.P.L. Petit.

German Patent 396,353.

Petrifix A process for solidifying aqueous wastes, converting them to a solid form suitable for landfill. Cementitious additives are used, based on the compositions used by the Romans for making Pozzolanic cements. Developed by PEC Engineering, Paris, France. In 1979, it had been used in France and Germany.

Pichat, P., Broadsky, M., and Le Bourg, M., in *Toxic and Hazardous Waste Disposal*, Vol. 1, Pojasek, R.B., Ed., Ann Arbor Science, Ann Arbor, MI, 1979, Chap. 9.

PetroFCC [**Petro**chemicals Fluid Catalytic Cracking] A modified *FCC process that increases the yields of propylene and butene for use in petrochemical manufacture. The modifications are mainly to the engineering and involve a dual reaction zone with a single regenerator. One feature is RxCat technology, which involves mixing some of the spent catalyst with regenerated catalyst in a separate chamber. Developed by UOP in 2000 and licensed for use in the Philippines in 2008. The first Japanese plant to use RxCat technology was built by Fuji Oil in Sogedura.

Eur. Chem. News, 10 Jul 2000, **72**(1917), 25; 7 Mar 2005, **82**(2137), 21.
Hydrocarbon Process. Int. Ed., 2006, **85**(4), 29.
Fahim, M.A., Al-Sahhaf, T.A., and Elkilani, A.S., *Fundamentals of Petroleum Refining*, 2009, 231.
Jpn. Chem. Week, 2008, **49**(2466), 4.

Petrofin [Process enhancement through recovery of olefins] A process for recovering olefins (ethylene and propylene) from polymerization processes by adsorption on zeolite 4A. Developed by BOC and used at Montell's polypropylene plant at Lake Charles, LA. First demonstrated in 1997.

Eur. Chem. News, May 1998 (*Chemscope*), 5.

PetroFlux A refrigeration process for removing liquid hydrocarbons from natural gas. Developed by Costain Engineering. Twenty-eight plants were operating in 2000.

Hydrocarbon Process. Int. Ed., 1996, **75**(4), 132; 2000, **79**(4), 83.

Petrogas A *thermal cracking process for converting heavy petroleum fractions to fuel gas. Developed by Gasco.

Kirk-Othmer's Encyclopedia of Chemical Technology, 3rd ed., Vol. 11, John Wiley & Sons, New York, 1980, 428.

PetroRiser *See* RFCC.

Smith, J.W., in *Handbook of Synfuels Technology*, Meyers, R.A., Ed., McGraw-Hill, New York, 1984, 4–185.

Petrosix [Named after the oil company **Petro**bus and the oil shale company superintendencia de industrializacao do xisto] A pyrolytic process for extracting petroleum from shale, developed in Brazil since 1953. A large demonstration plant was operated in Brazil in the 1970s. The world's largest oil shale pyrolysis reactor is in Sao Mateus do Sul, Brazil. It uses a vertical pyrolysis reactor. By 2008, it had produced 20 million bbl of shale oil. Sulfur is a by-product. Operated by Petrobras.

Oil Gas J., 2008, **106**(41), 16.
Wikipedia, The Free Encyclopedia. Petrosix. Revised 16 Mar 2013. Retrieved 22 Jul 2013.

Petro-Tex A process for oxidizing butenes to maleic anhydride. Developed by the Petro-Tex Chemical Corporation and used at its plant in Houston, TX.

Skinner, W.A. and Tieszin, D., *Ind. Eng. Chem.*, 1961, **53**(7), 557.

Petrox A general term for processes that enhance the yield of organic oxidation processes (e.g., butane to maleic anhydride, propylene to acrylonitrile) by the use of oxygen instead of air. Selectivity is improved at the expense of conversion, and the feedstock hydrocarbon is recycled by the use of a special molecular sieve. Developed by BOC Gases. The Petrox system for acrylonitrile was developed in association with Mitsubishi Chemical.

Chem. Eng. (Rugby, Engl.), 1991, (489), 14.
Eur. Chem. News, 2000, **73**(1937), 25; 2002, **77**(2013), 30.
Chem. Eng. News, 2001, **77**(35), 16; **78**(51), 13.

PFBC [Pressurized Fluidized-Bed Combustion] *See also* AFBC.

Olah, G.A., Goeppart, A., and Prakash, G.K.S., *Beyond Oil and Gas: The Methanol Economy*, Wiley-VCH, Weinheim, Germany, 2006, 16.

PFH [Pressurized Fluidized-bed Hydroretorting] A process for making fuel gas from oil shale. Developed and piloted by the Institute of Gas Technology, Chicago.

Phenosolvan A process for extracting phenols from coke-oven liquor and tar acids from tar by selective solvent extraction with di-isopropyl ether (formerly with *n*-butyl acetate). Developed by Lurgi in 1937.

Wurm, H.-J., *Chem. Ing. Tech.*, 1976, **48**, 840.

Phenoraffin A process for recovering phenols from carbonizer tar and coke-oven tar. The tar is dissolved in aqueous sodium phenolate and extracted with isopropyl ether.

Weissermel, K. and Arpe, H.-J., *Industrial Organic Chemistry*, 4th ed., translated by C.R. Hindley and S. Hawkins, Wiley-VCH, Weinheim, Germany, 2003, 349.

Phillips (1) A process for polymerizing ethylene and other linear olefins and diolefins to make linear polymers. This is a liquid-phase process, operated in a hydrocarbon solvent at an intermediate pressure, using a heterogeneous oxide catalyst, such as chromia on silica–alumina. Developed in the 1950s by the Phillips Petroleum Company, Bartlesville, OK, and first commercialized in 1956 at its plant in Pasadena, TX. In 1991, 77 reaction lines were either operating or under construction worldwide, accounting for 34% of worldwide capacity for linear polyethylene. The polyethylene product was trademarked Marlex.

Belgian Patent 530,617.
U.S. Patent 2,825,721.
Chem. Eng. (N.Y.), 1955, **62**(6), 103.
Clark, A.C., Hogan, J.P., Banks, R.L., and Lanning, W.C., *Ind. Eng. Chem.*, 1956, **48**, 1152.
Clark, A.C. and Hogan, J.P., in *Polyethylene*, 2nd ed., Renfrew, A. and Morgan, P., Eds., Iliffe & Sons, London, U.K.,1960, 29.
Hydrocarbon Process. Int. Ed., 1991, **70**(3), 170.
Schwerdtfeger, E., Buck, R., and McDaniel, M., *Appl. Catal. A, Gen.*, 2012, **423**, 91.
Soares, J.B.P. and McKenna, T.F.L., *Polyolefin Reaction Engineering*, Wiley-VCH, Weinheim, Germany, 2012, 61.

Phillips (2) A fractional crystallization process used to freeze-concentrate beer and fruit juices. Formerly used in the production of *p*-xylene.

McKay, D.L., in *Fractional Solidification*, Zief, M. and Wilcox, W.R., Eds., Marcel Dekker, New York, 1967, Chap. 16.

Phillips (3) A two-stage process for dehydrogenating butane to butadiene.

Phillips Triolefin *See* Triolefin.

Phoredox A modification of the *Activated Sludge sewage treatment process, designed for the separate removal of phosphate. *See also* Phostrip.

Horan, N.J., *Water Waste Treat.*, 1992, **35**(2), 16.

Phorex [**Pho**sphoric acid **ex**traction] A process for purifying phosphoric acid by solvent extraction with *n*-butyl or *n*-amyl alcohol. Developed by Azote et Produits Chimiques, France.

Bergdorf, J. and Fischer, R., *Chem. Eng. Prog.*, 1978, **74**(11), 41.

PHOSAM [**AM**monium **PHOS**phate] Also called Phosam-W and USS Phosam. A method for removing ammonia from coke-oven gas by scrubbing with a solution of ammonium phosphate. The ammonia is recovered by heating the solution:

$$NH_4H_2PO_4 + NH_3 \rightleftharpoons (NH_4)_2HPO_4$$

Developed by P. D. Rice and others at the U.S. Steel Corporation in the 1960s and first used at a coke plant at Clairton, PA. Twenty-four companies had installed the process by 1994.

U.S. Patents 3,024,090; 3,186,795.
Rice, R.D. and Busa, J.V., in *Acid and Sour Gas Treating Processes*, Newman, S.A., Ed., Gulf Publishing Co., Houston, TX, 1985, 786.
Kohl, H.L. and Nielsen, R.B., *Gas Purification*, 5th ed., Gulf Publishing Co., Houston, TX, 1997, 311.

PhosEnergy A process for recovering uranium from phosphate production. Iron is first removed as iron ammonium phosphate. Developed by Uranium Equities (a subsidiary of Urtek, United States) and ANSTO.

U.S. Patent Application 2010-0028226.
TCE (The Chemical Engineer (Rugby)), 2010, (825), 46.

PHOSPAQ [**Phos**phate **Paq**ues] A process for removing phosphate from wastewater by precipitation as struvite (ammonium magnesium phosphate). The struvite can be used as a fertilizer. Developed and offered by Paques, the Netherlands.

Phostrip A modification of the *Activated Sludge sewage treatment process, designed for the separate removal of phosphate. *See also* Phoredox.

Horan, N.J., *Biological Wastewater Treatment Systems*, John Wiley & Sons, Chichester, U.K., 1990, 233.

PHOTHO [**PHO**sphate glass solidification of **THO**rium-bearing reprocessing waste] A process for immobilizing radioactive waste products from the thorium fuel cycle in a phosphate glass for long-term storage. Developed at Kernforschungsanlage Jülich, Germany, from 1968 until it was abandoned in 1977 in favor of *PAMELA.

Lutze, W., in *Radioactive Waste Forms for the Future*, Lutze, W. and Ewing, R.C., Eds., North-Holland, Amsterdam, the Netherlands, 1988, 8.

photographic processes These processes, involving chemical reactions initiated by light, come within the declared scope of this work. However, the history of photography is well documented and does not warrant repetition here. Instead, the named processes that have been, or are being, used commercially are listed in the following text; the reader is referred to the bibliography for references that explain their details. Reprographic processes for line drawings (e.g., Blueprint) are given individual entries. *See* Reprography.

Photographic processes include Agfacolor, albumen, ambrotype, Anscocolor, Autochrome, Autotype, bichromate, bromoil, calotype, carbon, carbro, chromatone, chrysotype, Cibachrome, collodion, cyanotype, daguerreotype, Dufay, Dufaycolor, Duplex, Duxochrome, Dyebro, Ektachrome, Ektacolor, energiatype, Ferraniacolor, ferrotype, Finlay, Flexichrome, Gasparcolor, Gevacolor, Gum, Hicro, hydrotype, Ilfochrome, Ilford, Itek, Joly, Jos-Pe, kallitype, KDB, Kodachrome, Kodacolor, Kotovachrome, Lignose, Lippmann, Lumière, melainotype, Minicolor, Omnicolore, ozobrome, ozotype, Paget, phototype, pinatype, platinotype, Polacolor, Polaroid-Land, polychrome, Printon, Raydex, Sakuracolor, tannin, Thomas, tintype, Utocolor, Uvachrome, Van Dyke, Vivex, Warner–Powrie, wet collodion, woodburytype, and zincotype.

Mees, C.E.K. and James, T.H., *The Theory of the Photographic Process*, Macmillan, New York, 1969.
Coe, B., *Colour Photography: The First Hundred Years, 1840–1940*, Ash & Grant, London, U.K., 1978.
Newhall, B., *The History of Photography from 1839 to the Present Day*, Museum of Modern Art, New York, 1988.
Coote, J.H., *An Illustrated History of Colour Photography*, Fountain Press, Surbiton, U.K., 1993.
Webb, R. and Reed, M., *Spirits of Salts: A Working Guide to Old Photographic Processes*, Aurum Press, London, U.K., 1999.
Hunt, R.W.G., *The Reproduction of Colour*, 6th ed., John Wiley & Sons, Chichester, U.K., 2004.

Physical Vapor Deposition Often abbreviated to PVD. A process for applying a coating of one material to the surface of another, essentially by sublimation. To be distinguished from *chemical vapor deposition, in which chemical changes do occur.

Hocking, M.G., Vasantasree, V., and Sidky, P.S., *Metallic and Ceramic Coatings*, Longman, Harlow, Essex, U.K., 1989.

Pidgeon A process for making magnesium metal by reducing dolomite with ferrosilicon at 1200°C in a vacuum retort:

$$2CaO \cdot MgO + Fe_xSi = 2Mg + Ca_2SiO_4 + xFe$$

Used in World War II.

> Hughes, W.T., Ransley, C.E., and Emley, E.F., in *Advances in Extractive Metallurgy*, Institute of Mining and Metallurgy, London, U.K., 1968, 429.

Pier–Mittasch A high-pressure catalytic process for making methanol from carbon monoxide and hydrogen. Developed by M. Pier and A. Mittasch at BASF in the 1920s.

> Spitz, P.H., *Petrochemicals: the Rise of an Industry*, John Wiley & Sons, New York, 1988, 36.

Pieters *See* Staatsmijnen–Otto.

Pietzsch and Adolph An electrolytic process for making hydrogen peroxide, operated in Germany in 1910 by Elektrochemische Werke München. Its plants in Munich, Bad Lauterberg, and Rhumspringe made the hydrogen peroxide used by the Germans for launching their V1 weapons and for guiding their V2 weapons during World War II. The Bad Lauterberg plant was dismantled after World War II and all its documentation deposited in the Imperial War Museum, London.

> Schumb, W.C., Satterfield, C.N., and Wentworth, R.L., *Hydrogen Peroxide*, Reinhold Publishing, New York, 1955, 136.

pig and ore A colloquial name for the *Siemens Open Hearth process.

pig and scrap A colloquial name for the *Siemens–Martin Open Hearth process.

Pintsch The first commercial oil-gasification process; developed in Germany in the nineteenth century.

Pintsch–Hillebrand An early, two-stage coal gasification process.

> Stief, F., *Gas Wasserfach*, 1932, **75**, 581 (*Chem. Abstr.*, **26**, 5402).
> Müller, H., *Gas Wasserfach*, 1935, **78**, 431 (*Chem. Abstr.*, **29**, 6397).
> van der Hoeven, B.J.C., in *Chemistry of Coal Utilization*, Vol. 2, Lowry, H.H., Ed., John Wiley & Sons, New York, 1945, 1668.

Pirox [**Pi**ttsburgh **Ir**on **Ox**ides] A process for making high-purity magnetite by the controlled reduction of hematite with natural gas. Developed by Pittsburgh Mineral & Environmental Technology and operated in New Brighton, PA, in 2010.

> *Chem. Eng. (N.Y.)*, 2010, **117**(8), 12.
> U.S. Patent 7,744,848.

PIVER [**PI**lote **VEr**re] An early French process for vitrifying nuclear waste. It was a batch process, used between 1969 and 1973 and then from 1979, until it was decommissioned in 1990. Superseded by the continuous *AVH and *AVM processes.

> Ojovan, M.I. and Lee, W.E., *An Introduction to Nuclear Waste Immobilisation*, Elsevier, Amsterdam, the Netherlands, 2005, 233.

Plantrose A process for hydrolyzing low-value biomass to C_5 and C_6 sugars. Supercritical water is used to depolymerize the cellulose, after the hemicellulose has been hydrolyzed and removed. Lignin is separated and removed separately. Developed by Renmatix and announced in 2011. BASF announced a major investment in this company in 2012.

> *Chem. Eng. (N.Y.)*, 2011, **118**(12), 14.
> *ICIS Chem. Bus.*, 2012, **281**(2), 6.

Plasmacat An odor-control process in which the gas is first ionized electrically and then passed into a catalytic converter, in which the organic odorants are oxidized. First produced by Umwelttechnik AG, Switzerland, and further developed in 2002 by Odour Control Systems Ltd., United Kingdom.

Plasmared [**Plasma red**uction] A direct-reduction ironmaking process, using natural gas as the reductant and heated by an electric plasma. Operated in Sweden by SKF from 1981 to 1984. *See* DR.

Platfining The original name for the *LT Unibon process.

Platforming [**Plat**inum re**forming**] A process for converting aliphatic hydrocarbons (acyclic and cyclic) into aromatic hydrocarbons and hydrogen. Methyl cyclopentane can thus be converted to benzene. The catalyst typically incorporates platinum and another metal on an alumina support. Originally, the reactors were side by side and the catalyst had to be regenerated annually. In 1970, continuous catalyst regeneration (*CCR) was introduced, and the three or four reactors were stacked vertically in series (hence another derivation of the name). Invented by V. Haensel at Universal Oil Products, now UOP. First commercialized by that company at Muskegon, MI, in 1949, and now widely licensed. In 2005, 173 units were operating and 30 more were in design or construction. UOP describes Platforming as the most successful reforming process ever offered by any licensor. *See also* RZ Platforming.

Unzelman, G.H. and Wolf, C.J., in *Petroleum Processing Handbook*, Bland, W.F. and Davidson, R.L., Eds., McGraw-Hill, New York, 1967, 3–29.
Little, D.M., *Catalytic Reforming*, PennWell Publishing, Tulsa, OK, 1985, 160.
Peer, R.L., Bennett, R.W., Felch, D.F., and von Schmidt, E., *Catal. Today*, 1993, **18**(4), 473.
Hydrocarbon Process. Int. Ed., 1994, **73**(11), 94.
Lapinski, M., Baird, L., and James, R., in *Handbook of Petroleum Refining Processes*, 3rd ed., Meyers, R.A., Ed., McGraw-Hill, New York, 2003, 4.3.

Platinum Reforming *See* Sovaforming.

Platreating [**Plat**inum hydro**treating**] A *hydrotreating process used in oil refining, using a platinum catalyst.
Unzelman, G.H. and Wolf, C.J., in *Petroleum Processing Handbook*, Bland, W.F. and Davidson, R.L., Eds., McGraw-Hill, New York, 1967, 3–44.

Plattner An early process for extracting gold from auriferous pyrites by chlorination. The resulting gold chloride is extracted by water and reduced with ferrous sulfate:

$$AuCl_3 + 3FeSO_4 = Au + Fe_2(SO_4)_3 + FeCl_3$$

Developed by C. F. Plattner in Germany in 1853.

Mellor, J.W., *Comprehensive Treatise on Inorganic and Theoretical Chemistry*, Vol. 3, Longmans, Green & Co., London, U.K., 1923.

PNC [**P**hoto**N**itrosation of **C**yclohexane] A photochemical process for making caprolactam (a precursor for nylon) from cyclohexane, nitrosyl chloride, and hydrogen chloride. The first photochemical product is cyclohexanone oxime:

$$C_6H_{12} + NOCl + HCl \xrightarrow{UV} C_6H_{10}ONOH + 2HCl$$

This, upon treatment with sulfuric acid, undergoes the Beckmann rearrangement to caprolactam. The nitrosyl chloride is made by reacting nitrosyl sulfuric acid (made from oxides of nitrogen and sulfuric acid) with hydrogen chloride gas:

$$NOHSO_4 + HCl = NOCl + H_2SO_4$$

The process was developed by Toyo Rayon Company (now Toray Industries), Japan, in the 1960s, and is now operated by that company in Nagoya and Tokai.

Aikawa, K., *Hydrocarbon Process.*, 1964, **43**(11), 157.
Horspool, B., in *Light, Chemical Change and Life*, Coyle, J.D., Hill, R.R., and Roberts, D.R., Eds., Open University Press, Milton Keynes, U.K., 1982, 276.
Hydrocarbon Process. Int. Ed., 1989, **68**(11), 97.
Dahlhoff, G., Niederer, J.P.M., and Hölderich, W.P., *Catal. Rev.*, 2001, **43**(4), 389.

Pneumatic *See* Bessemer.

Polyad [**Poly**mer **ad**sorbent] Also written POLYAD. A family of processes for removing volatile organic compounds from air streams by continuous adsorption on an adsorbent and desorption with hot air. Several adsorbents are used, depending on the sorbate, including a macroporous polymer ("Bonopore"). Used for control of emissions and for solvent recovery. Polyad FB uses a fluidized bed. COMBI-AD is a variant for simultaneously removing several solvents, using two different adsorbents. Developed and offered by Chematur Engineering, Sweden. Twelve units had been installed, in several countries, by 1995.

Kohl, H.L. and Nielsen, R.B., *Gas Purification*, 5th ed., Gulf Publishing Co., Houston, TX, 1997, 1113.

Polar An electrolytic process for making titanium metal powder from titanium dioxide. Broadly similar to the *FFC process. Developed by BHP Billiton and acquired by QinetiQ in 2006.

POLF [**P**resaturated **O**ne **L**iquid **F**low] A deep desulfurization process for hydrocarbon fuels in which the oil is presaturated with hydrogen before contacting the Co/Mo catalyst.

Datsevich, L.B. and Mukhortov, D.A., *Catal. Today*, 2007, **120**, 71.
Latz, J., Peters, R., Pasel, J., Datsevich, L., and Jess, A., *Chem. Eng. Sci.*, 2008, **64**, 288.
Stanislaus, A., Marafi, A., and Rana, M.S., *Catal. Today*, 2010, **153**(1–2), 40.

Polybed A version of the *PSA process for separating and purifying gases by selective adsorption, using a number of adsorbent beds and a complex valving system in order to produce a gas of high purity. Used particularly for purifying hydrogen. Developed by the Union Carbide Corporation in the mid-1970s. *See also* Hydrogen Polybed PSA.

Corr, F., Dropp, F., and Rudelstorfer, E., *Hydrocarbon Process. Int. Ed.*, 1979, **58**(3), 119.
Yang, R.T., *Gas Separation by Adsorption Processes*, Butterworths, Guildford, U.K., 1987, 244.
Hydrocarbon Process. Int. Ed. 1996, **75**(4), 120.
Kohl, H.L. and Nielsen, R.B., *Gas Purification*, 5th ed., Gulf Publishing Co., Houston, TX, 1997, 1083.

Polyco A process for converting propylene and butene to liquid fuels, using copper pyrophosphate as the catalyst. The name has also been used as a trade name for a type of polyvinyl acetate made by the Borden Chemical Company.

Polyforming [**Poly**merization re**forming**] An early process for converting "gas oil" to gasoline. It combined *thermal cracking with *polymerization.

Asinger, F., *Mono-olefins: Chemistry and Technology*, translated by B.J., Hazzard, Pergamon Press, Oxford, U.K., 1968, 362.

Polymer Cracking Process *See* SPORT.

Polymer Engineering A catalytic process for making diesel fuel from waste plastics. Developed by C. Koch at Alphakat GMBH, Germany, and now offered by Polymer Engineering, Germany. A zeolite catalyst is used, and the product is called NanoFuel Diesel. Plants have been built in Germany, Mexico, Japan, and South Korea.

Scheirs, J., in *Feedstock Recycling and Pyrolysis of Waste Plastics*, Scheirs, J. and Kaminsky, W., Eds., John Wiley & Sons, Chichester, U.K., 2006, 418.

Polymerization Those polymerization processes having special names that are described in this dictionary are Alfin, Alfol, Alphabutol, Borstar, CP, CX, Dimersol, Exxpol, Gorham, GRS, Hexall, Innovene, Insite, LIPP-SHAC, Mobil–Witco–Shell, MOGD, MPC, MSP3, Naphtachimie, Natta, NORSOLOR, Novolen, Octol, Paralene, Phillips (1), Polyco, SDS, Sclair, Sclairtech, Selectopol, SHOP, SPGK, Spheripol, Standard Oil, UNIPOL, Ziegler (1), and Ziegler–Natta.

Polynaphtha Essence A process for making a linear olefin fraction for making methyl *t*-butyl ether to use as a fuel additive. Developed by IFP in 1996, replacing UOP's *Catpoly process.

Inform. Chim. Hebdo, 1997, (1294), 13.

Polzeniusz–Krauss A process for making calcium cyanamide from calcium carbide by heating it in a nitrogen atmosphere in a channel kiln:

$$CaC_2 + N_2 = CaCN_2 + C$$

This was an early process for fixing nitrogen for use as a fertilizer.

Porter–Clark The original name for the cold lime-soda process. A water-softening process using sodium carbonate and calcium hydroxide. It removes both the noncarbonate and the bicarbonate hardness. Developed by J.H. Porter. *See* Clark.

Porteus A process for conditioning sewage sludge by heating under pressure to 180°C–220°C for approximately 1 h. The solid product is easy to dewater and is sterile. Developed and used in Germany in the 1980s.

POSM [Propylene Oxide Styrene Monomer] A process for making propylene oxide from ethylbenzene. The ethylbenzene is reacted with oxygen and propylene in the presence of a proprietary catalyst. Developed in Russia by JSC Nizhnekamskneftekhim and licensed exclusively by Dow Chemical Company. In 2006, 36% of the world production of propylene oxide was made by this process. *See also* SMPO.

Chem. Eng. (N.Y.), 1998, **105**(7), 21.
Chem. Eng. News, 2006, **84**(41), 22.

POSTech A process for making copolymers of styrene with polyols. A special stabilizer is used, as well as an organic peroxide initiator.

Eur. Chem. News, 1996, **65**(1715), 35.

Pott–Broche A coal liquefaction process in which coal is dissolved in a mixture of tetrahydronaphthalene and cresols and then hydrogenated. Invented by A. Pott and H. Broche at IG Farbenindustrie, Germany in 1927; used by the Ruhrol Company in Germany between 1938 and 1944. *See also* Exxon Donor Solvent.

British Patent 293,808.
French Patent 841,201.
Pott, A. and Broche, H., *Fuel*, 1943, **13**, 91,125,154.

Powerclaus A *flue-gas desulfurization system that applies the *Aquaclaus process to power station effluent gases.

Powerforming A *catalytic reforming process, based on a platinum catalyst. Developed by Esso Research and Engineering Company. First commercialized in Baltimore, MD, in 1955; now widely licensed.

Unzelman, G.H. and Wolf, C.J., in *Petroleum Processing Handbook*, Bland, W.F. and Davidson, R.L., Eds., McGraw-Hill, New York, 1967, 3–32.
Asinger, F., *Mono-olefins: Chemistry and Technology*, translated by B.J. Hazzard, Pergamon Press, Oxford, U.K., 1968, 391.
Little, D.M., *Catalytic Reforming*, PennWell Publishing, Tulsa, OK, 1985, 162.
Hydrocarbon Process. Int. Ed., 1988, **67**(9), 80.

POX [**P**artial **OX**idation] A general term for processes that convert natural gas to *syngas or methanol by partial oxidation. Shell International Petroleum and Lurgi developed one such methanol process in the 1980s, which was used at the Mider refinery in Leuna, Germany. The University of Orléans, France, developed noncatalyzed, plasma-assisted POX processes for making syngas in the 1980s and 1990s. *See also* ***CPOX**.

Leseur, H., Czernichowski, J., and Chapelle, J., *Int. J. Hydrogen Energy*, 1994, **19**(2), 139.
Oil Gas J., 1997, **95**(11), 49.
Fleshman, J.D., in *Handbook of Petroleum Refining Processes*, 3rd ed., Meyers, R.A., Ed., McGraw-Hill, New York, 2003, 6.11.

POZONE A process for making ozone developed by S. G. Chang at the Lawrence Berkeley Laboratory, Berkeley, CA. Elemental phosphorus is emulsified in water at above 45°C, and air is passed through. This generates ozone and orthophosphoric acid:

$$P_4 + 5O_2 = P_4O_{10}$$

$$P_4O_{10} + 6H_2O = 4H_3PO_4$$

Proposed as a source of ozone for removing NO_x and SO_2 from flue gas and for pulp bleaching.

Chem. Eng. (N.Y.), 1994, **101**(11), 25.

Chang, S.-G., Keyuan, H., and Yizhong, W., *J. Environ. Sci.*, 1994, **6**(1), 1.
Wang, H., Shi, Y., Le, L., Wang, S.-M., Wei, J., and Chang, S.-G., *Ind. Eng. Chem. Res.*, 1997, **36**(9), 3656.

Poz-O-Tec A *flue-gas desulfurization process that produces a fibrous form of gypsum, convenient for disposal. In a demonstration project, an artificial reef was built from the product in the Atlantic Ocean near Fire Island, New York. Developed by IU Conversion Systems, Philadelphia, PA.

IEA Coal Research, *The Problems of Sulphur*, Butterworths, London, U.K., 1989, 127.

PPG [**P**ittsburgh **P**late **G**lass Company] A process for making calcium hypochlorite. Hypochlorous acid and chlorine monoxide, generated by reacting chlorine and carbon dioxide with sodium carbonate monohydrate, are passed into lime slurry. Invented in 1938 by I.E. Muskatt and G.H. Cady at the Pittsburgh Plate Glass Company.

U.S. Patent 2,240,344.

PR *See* Sovaforming.

Prayon A *wet process for making phosphoric acid by reacting phosphate rock with sulfuric acid. The by-product is gypsum, calcium sulfate dihydrate. Prayon uses a compartmentalized, multisection, lined, concrete reactor, with finishing tanks in which the gypsum crystals mature. In 1990, one-third of the wet-process phosphoric acid made in the Western world was made in this way. The process was developed in 1977 by the Société de Prayon, Belgium. Variations are known as PH2, PH11, and PH12. One variation uses solvent extraction with *iso*propyl ether and tri-*n*-butyl phosphate.

U.S. Patent 4,188,366.
Forster, J.J., in *Phosphoric Acid*, Vol. 1, Part 2, Slack, A.V., Ed., Marcel Dekker, New York, 1968, 585.
Becker, P., *Phosphates and Phosphoric Acid*, 2nd ed., Marcel Dekker, New York, 1989, 347.
Gard, D.R., in *Encyclopedia of Chemical Processing and Design*, McKetta, J.J. and Cunningham, W.A., Eds., Marcel Dekker, New York, 1990, 35, 453.

Precht *See* Engel–Precht.

PRENFLO [**PR**essurized **EN**trained **FLO**w] A coal gasification process using an entrained-flow gasifier. A mixture of coal dust and oxygen is fed horizontally into a gasifier operating at >2000°C and 25 atm. It is similar to the *Koppers–Totzek process but differs from it mainly in the use of elevated pressure. The solid waste is mostly molten and is collected as a granular slag in a water bath beneath the gasifier; it can be used as a filler in the construction industry. The process can be used to produce *syngas, but the main application foreseen is as a part of a combined-cycle electric power plant. Developed by Krupp Koppers with funding from the Ministry for Research and Technology, FRG, and the Commission of the EEC; the first demonstration plant began operation in Hamburg in 1979 and a second one started up in Fürstenhausen, Germany, in 1986. The first commercial plant was built at Puertollano, Spain, in 1996.

Hydrocarbon Process. Int. Ed., 1986, **65**(4), 100.
Chem. Eng. (N.Y.), 1991, **98**(9), 31.
Rezaiyan, J. and Cheremisinoff, N.P., *Gasification Technologies*, Taylor & Francis, Boca Raton, FL, 2005, 53.

Prenox A pulp-bleaching process using nitrogen dioxide and oxygen instead of chlorine. Developed by AGA, Sweden.

pressure swing adsorption *See* PSA.

PRIAM An electrochemical process for recovering heavy metals. Announced in 1992 by Électricité de France.

Prime-D A process for making ultralow-sulfur diesel fuel. Developed by Axens.

Prime-G, Prime-G+ A deep *hydrodesulfurizing process for removing sulfur compounds prior to *fluid catalytic cracking. It uses a fixed catalyst bed and conventional distillation. Developed by IFP (now Axens) from 1999. The "+" version is an improvement on the original process. In 2001, over 60 units had been licensed and 11 were operating commercially. First commercialized at Gelsenkirchen, Germany, in 2001. By 2012, 140 units had been successfully started up worldwide.

Eur. Chem. News, 29 Nov 1999, **71**(1887), 30.
Pétrole et Gaz Informations, Jan–Feb 2000 (1744), 17, Mar 2000 (1745), 43.
Chem. Eng. (N.Y.), 2001, **108**(12), 23.
Hydrocarbon Process. Int. Ed., Aug 2002, **82**(8), 38; Aug 2005, **84**(8), 31.
Chem. Eng. (N.Y.), 2003, **110**(9), 27.

Primex [**Pr**essureless **i**nfiltration by **me**tal] *See* Lanxide.

PRIMOX A process for injecting oxygen into sewers ("rising mains") in order to prevent the formation of hydrogen sulfide. Developed by BOC.

PRI-SC [**P**eroxide-**R**egenerated **I**ron-**S**ulfide **C**ontrol] A process for removing hydrogen sulfide from sewage. Ferrous sulfide is added in the upper reaches of the system, precipitating iron sulfides and sulfur. Addition of hydrogen peroxide at two sites downstream recovers the iron as ferric hydroxide, which can be used in other water treatment. Developed by U.S. Peroxide in Southern California in the early 2000s.

Walton, J., *Water Wastewater Int.*, 2005, **20**(1), 45.

PROABD A crystallization process developed by BEFS PROKEM, France. In 1994, more than 80 plants were using PROABD distillation and crystallization processes. *See* Ab der-Halden.

Hassene, M., *Asia Pacific Chem.*, 1993, Oct, 30.

PROABD MSC [**M**elt **S**tatic **C**rystallization] A process for purifying *p*-xylene by crystallization. Used in conjunction with *MSTDP. Piloted in France from 1994 to 1996 and proposed for installation in India in 1997 and in Bulgaria in 1998.

> *Chem. Eng. (N.Y.)*, 1996, **103**(9), 23.

Proban A process for rendering textiles flame retardant by treatment with tetrakis hydroxymethyl phosphonium chloride (THPC). Invented at the U.S. Regional Research Laboratory in New Orleans, LA, in the early 1950s. Commercialized in 1955 under the trade name Proban by Proban Ltd., a joint venture between the Bradford Dyers Association and Albright & Wilson. Now offered by Rhodia.

> Podger, H., *Albright & Wilson: The Last 50 Years*, Berwin Books, Studley, Warks., U.K., 2002, 30.

Probion A process for making protein from methanol. Developed by Hoechst–Uhde in the 1980s.

> Weissermel, K. and Arpe, H.-J., *Industrial Organic Chemistry*, 4th ed., translated by C.R. Hindley and S. Hawkins, Wiley-VCH, Weinheim, Germany, 2003, 35.

Prodflex An olefin isomerization process developed by LyondellBasell. It converts mixed olefins to ethylene, propylene, and butenes.

> *Chimie Pharma Hebdo*, 2008, (447), 10.

Prodigy A process for making high-density polyethylene with a bimodal molecular weight distribution that exhibits favorable processing and physical properties. A novel catalyst is used. Developed by Univation in 2002; three licenses had been granted by 2006 and others were being negotiated.

producer A generic name for processes that completely convert solid fuels to gaseous fuels and for the products. Coal or coke is the usual solid fuel. The oxidant was originally air but is now more commonly oxygen. The principal combustible component of the product is carbon monoxide. First developed in the early nineteenth century, these processes became very important. With the development of the natural gas and petroleum industries in the early twentieth century, their importance declined, although there was a revival of interest after the oil crisis of 1973. A common feature of producer gas processes is partial combustion of the solid fuel to provide heat for the reactions. These processes evolved into the many *coal gasification processes.

> van der Hoeven, B.J.C., in *Chemistry of Coal Utilization*, Vol. 2, Lowry, H.H., Ed., John Wiley & Sons, New York, 1945, Chap. 36.

PROESA A process for making sugars, and thus ethanol, from lignocellulose. It combines a steam pretreatment, enzymic hydrolysis, and fermentation. Developed by Chemtex and announced by M&G Group in 2011. Now owned by Beta Renewables. Won the ACHEMA 2012 Innovation Award. Planned to be used in Crescentino, Italy, in late 2012, using reed and straw, and then in N Carolina and Brazil.

> *Spec. Chems.*, 2012, **32**(5), 26.

Progil A thermal process for making phosphoric acid. The phosphorus pentoxide, produced by burning elemental phosphorus, is absorbed in a solution of sodium phosphate; the heat of combustion is partially used in concentrating this solution. Invented by, and named after, Progil S.A.

Propylane An extractive distillation process for removing aromatic hydrocarbons from hydrogenated crude benzene, using propylene carbonate. Developed by Koppers.

Propylox [**Propyl**ene **ox**idation] A process for epoxidizing propylene to propylene oxide (1,2-epoxypropane) using a peroxycarboxylic acid in an organic solvent. The peroxy acid is generated in water and immediately extracted into an organic solvent using an "extractor reactor." Invented in 1975 by A.M. Hilden and P.F. Greenhalgh at Laporte Industries, United Kingdom, and developed by Interox Chemicals, a joint company of Laporte Industries and Solvay. Piloted in Widnes, England, in the 1970s but not commercialized. Similar processes, without special names, have been developed by Bayer and by Olin Corporation. These processes would be economic only on a large scale, in conjunction with dedicated hydrogen peroxide plants.

> Belgian Patent 838,068.
> U.S. Patents 4,071,541; 4,177,196.

Propylur A process for converting C_4 and C_5 hydrocarbons to propylene, using ZSM-5 catalyst in the presence of steam in a fixed bed. Developed by Lurgi in 2000 and licensed exclusively to Linde. Tested on an industrial scale in BP's plant in Worringen, Germany, in 2002.

> *Eur. Chem. News*, 2000, **72**(1902), 47; 2002, **76**(1998), 20.
> *Hydrocarbon Process. Int. Ed.*, 2002, **81**(6), 31.
> *Chem. Eng. (N.Y.)*, 2002, **109**(6), 23; 2004, **111**(3), 20.

Protal *See* metal surface treatment.

Provesteen A microbiological process for making single-cell protein from methanol, ethanol, or whey, developed in the 1980s by Provesta Corporation, a subsidiary of Phillips Petroleum Company. The basis of the process is a special "high-cell-density fermenter," which simplifies the isolation of the product from the water. The organism is the torula yeast; the intended products are specialty flavor enhancers, a high-fiber food bar, a food supplement for horses, and various aquaculture products for shrimp and fish. A plant with an annual capacity of 1500 tons began operation in Bartlesville, OK, in 1989.

> McNaughton, K.T., *Chem. Bus.*, 1989, **11**(5), 18.

PROX [**PR**eferential **OX**idation of CO] A general term used in fuel cell technology for processes that remove CO from fuel gases by preferential oxidation. The usual catalyst is a supported precious metal.

> Piccolo, L., Nassreddine, S., and Morfin, F., *Catal. Today*, 2012, **189**(1), 42.

Proximol A process for making hydrogen by *reforming methanol. Offered by Lurgi.

Pruteen A microbiological process for making single-cell protein from methane or methanol, developed by ICI. The organism is *Methylophilus methylotrophus*, found in the sewers of Naples. A large pilot plant was built in Billingham, England, in the 1970s. The process was never commercialized, but some of the technology was used later in the *Deep Shaft process.

> Weissermel, K. and Arpe, H.-J., *Industrial Organic Chemistry*, 4th ed., translated by C.R. Hindley and S. Hawkins, Wiley-VCH, Weinheim, Germany, 2003, 37.
> Olah, G.A., Goeppart, A., and Prakash, G.K.S., *Beyond Oil and Gas: The Methanol Economy*, Wiley-VCH, Weinheim, Germany, 2006, 253.

PSA [**P**ressure **S**wing **A**dsorption] A general method for separating gases by cyclic adsorption and desorption from a selective adsorbent, at alternating pressures. Invented by C.W. Skarstrom at the Esso Research and Engineering Company in 1958 and subsequently engineered by the Union Carbide Corporation. The Société de L'Air Liquide, France, made a similar development at that time. For separating nitrogen from air, carbon molecular sieve is the preferred adsorbent. For separating oxygen from air, a zeolite is used. Other zeolites have been used for other separations. Many variations on this basic process have been developed for specific gas mixtures and are known by special names; those described in this dictionary are AUTO-PUREX G, Bergbau-Forschung, COPISA, COPSA, DWN, DWO, Generon, HYSEC, KURASEP, LO-FIN, MOLPSA-nitrogen, Moltox, MRH, NitroGEN, OxyGEN, Oxy-Rich, OXYWELL, Polybed, PSPP, Remet, RPSA, Sumitomo-BF. *See also* TSA.

U.S. Patents 2,944,627; 3,155,468.
Skarstrom, C.W., *Ann. N.Y. Acad. Sci.*, 1959, **72**, 751.
Yang, R.T., *Gas Separation by Adsorption Processes*, Butterworths, Guildford, U.K., 1987, 237.
White, D.H., Jr. and Barkley, P.G., *Chem. Eng. Prog.*, 1989, **85**(1), 25.
Ruthven, D.M., Farook, S., and Knaebel, K.S., *Pressure Swing Adsorption*, VCH Publishers, Weinheim, Germany, 1993.
Kohl, H.L. and Nielsen, R.B., *Gas Purification*, 5th ed., Gulf Publishing Co., Houston, TX, 1997, 1081.
LaCava, A.I., Shirley, A.I., and Ramachandran, R., *Chem. Eng. (N.Y.)*, 1998, **105**(6), 110.
Suzuki, M., in *Adsorption and Ion Exchange: Fundamentals and Applications*, LeVan, M.D., Ed., American Institute of Chemical Engineers, New York, 1998, 119.
Sircar, S., *Ind. Eng. Chem. Res.*, 2002, **41**(6), 1389.

PS Claus A process for recovering sulfur from waste gases by a combination of the pressure swing process and the *Claus process.

Eur. Chem. News, 1994, **61**(1611), 28.

PSPP [**P**ressure **S**wing **P**arametric **P**umping] A version of the *PSA process for separating gases by selective adsorption. It operates by rapidly reversing the gas flows through the absorber bed; the pressures are different for each direction of flow. The main use is for generating oxygen-enriched air for medical use. Invented by the Union Carbide Corporation in 1978.

U.S. Patents 4,194,891; 4,194,892.
Keller, G.E., II, in *Industrial Gas Separations*, Whyte, T.E., Yon, C.M., and Wagener, E.H., Eds., American Chemical Society, Washington, DC, 1983.

Puddling A process for making wrought iron from pig iron, based on the partial decarburization of pig iron in a special furnace. Invented by H. Cort in Titchfield, Southampton, England, in 1784 and widely used in the United Kingdom and Europe until the end of the nineteenth century. In 1873, there were 8000 puddling furnaces in the United Kingdom alone.

British Patent 1,420 (1784).
Derry, T.K. and Williams, T.I., *A Short History of Technology*, Oxford University Press, Oxford, U.K., 4th impression, 1979, 477.
Barraclough, K.C., *Steelmaking before Bessemer, Vol. 2, Crucible Steel*, The Metals Society, London, U.K., 1984, 91, 303.

Pulsar A *PSA process for separating oxygen from air. Developed in Canada by Questor Industries and BOC Gases.

Pumpherston [Named after the town near Edinburgh, Scotland, where the process was operated] A process for extracting fuel oil from oil shale. The heart of the process was the Pumpherston retort (also called the Bryson retort), down which the shale fell by gravity and up which air and steam were passed. Ammonia was collected as a by-product. Invented by J. Young in 1850 and operated in Scotland between 1883 and 1962.

British Patent 13,292 (1850).
Smith, J.W., in *Handbook of Synfuels Technology*, Meyers, R.A., Ed., McGraw-Hill, New York, 1984, 4–149.
Russell, P.L., *Oil Shales of the World*, Pergamon Press, Oxford, U.K., 1990, 712.

PuraSiv Hg An adsorptive process for removing mercury vapor from gaseous effluents from the *Castner–Kellner process by *TSA. The adsorbent is a zeolite molecular sieve containing silver. Developed by UOP.

U.S. Patent 4,874,525.

PuraSiv HR A process for removing solvent vapors from air by adsorption on beaded activated carbon contained in a combined, fluidized, moving bed. For water-soluble solvents, the gas used for desorption is nitrogen and the process is known as PuraSiv HR, Type N (not to be confused with Pura Siv-N); for chlorinated hydrocarbons, steam stripping is used and the process is known as PuraSiv HR, Type S. Developed by Kureha Chemical Company and then marketed in the United States by Union Carbide. The process was originally known as GASTAK because it was developed by the **Ta**iyo **K**aken Company, subsequently acquired by Kureha Chemical Company. It is also marketed by Daikin Industries under the name Soldacs.

Chem. Eng. (N.Y.), 1977, **84**(18), 39.

Keller, G.E., II, in *Industrial Gas Separations*, Whyte, T.E., Jr., Yon, C.M., and Wagener, E.H., Eds., American Chemical Society, Washington, DC, 1983.

Yang, R.T., *Gas Separation by Adsorption Processes*, Butterworths, Guildford, U.K., 1987, 217.

Kohl, H.L. and Nielsen, R.B., *Gas Purification*, 5th ed., Gulf Publishing Co., Houston, TX, 1997, 1112.

Pura Siv-N A process for removing nitrogen oxides from the tail gases from nitric acid plants, using an acid-resistant zeolite molecular sieve. Developed by the Union Carbide Corporation in 1971. Not to be confused with PuraSiv HR, Type N (*see* previous entry).

Kohl, H.L. and Nielsen, R.B., *Gas Purification*, 5th ed., Gulf Publishing Co., Houston, TX, 1997, 1073.

PuraSiv S A process for removing sulfur dioxide from the tail gases from sulfuric acid manufacture by adsorption on a special zeolite. Not to be confused with PuraSiv HR, Type S.

Anderson, R.A., in *Molecular Sieves II*, Katzer, J.R., Ed., American Chemical Society, Washington, DC, 1977, 637.

PURASPEC A process for purifying gaseous and liquid hydrocarbons by the use of fixed beds of catalysts and adsorbents, which remove impurities by chemical reaction. The name is used also to denote the special adsorbents and catalysts. Developed in 1990 by ICI Katalco to enable natural gas and natural gas liquids to meet pipeline specifications. Installed in approximately 60 plants worldwide in 1996. Now offered by Johnson Matthey. **PURASPEC 1156** and **PURASPEC 1157** were developed for removing organomercury compounds from gases. About 100 units were operating worldwide in 2000.

Hydrocarbon Process. Int. Ed., 1996, **75**(4), 133; 2000, **79**(4), 84; 2002, **81**(6), 31.

Puratreat A process for removing sulfur compounds from coal-derived *syngas. Developed by BASF and Toyo Engineering.

Chem. Eng. News, 15 Jan 2007, **85**(3), 33.

Purex [**P**lutonium and **u**ranium **r**ecovery by **ex**traction] A process for the solvent extraction of plutonium from solutions of uranium and fission products, obtained by dissolving spent nuclear fuel elements in nitric acid. The solvent is tri-*n*-butyl phosphate (TBP) in kerosene. First operated by the U.S. Atomic Energy Commission at its Savannah River plant, SC, in 1954 and at Hanford, WA, in 1956. Now in operation, with modifications, in several countries. Sites include Savannah River (SC), Cap de la Hague (France), Marcoule (France), Sellafield (England), Karlsruhe (Germany), and Trombay (India). *See also* Recuplex.

Siddall, T.H., III, in *Chemical Processing of Reactor Fuels*, Flagg, J.F., Ed., Academic Press, New York, 1961, 199.

Nuclear Fuel Reprocessing Technology, British Nuclear Fuels PLC, Risley, U.K., 1985.

Büchner, W., Schliebs, R., Winter, G., and Büchel, K.H., *Industrial Inorganic Chemistry*, VCH Publishers, Weinheim, Germany, 1989, 586.

Nash, K.L., in *Separations for the Nuclear Fuel Cycle in the 21st Century*, Lumetta, G.J., Nash, K.L., Clark, S.B., and Friese, J.I., American Chemical Society, Washington, DC, 2006, 24.

Purifier An ammonia synthesis process, developed by C. F. Braun and now licensed by Kellogg Brown & Root. An important feature is the cryogenic removal of inert gases from the system. Seventeen plants were operating in 2005.

PurinNOx A process for reducing emissions of NO_x and particulates from the exhausts of diesel engines, developed by Lubrizol in 2001.

Chem. Week, 2001, **163**(46), 20.

Purisol A process for removing hydrogen sulfide from gases by selective absorption in N-methyl-2-pyrrolidone (NMP). Developed and licensed by Lurgi, particularly for desulfurizing waste gases from *IGCC plants. Eight units were operating in 2000.

Hochgesand, G., *Chem. Ing. Tech.*, 1968, **40**(9/10), 432.
Hydrocarbon Process. Int. Ed., 1975, **54**(4), 92; 1996, **75**(4), 133.
Kohl, H.L. and Nielsen, R.B., *Gas Purification*, 5th ed., Gulf Publishing Co., Houston, TX, 1997, 1210.
Hydrocarbon Process. Int. Ed., 2000, **79**(4), 84.

Purlex An improved version of the *Bufflex process for extracting uranium from its ores. Operated in South Africa.

Eccles, H. and Naylor, A., *Chem. Ind. (London)*, 1987, (6), 174.

Purofer A direct-reduction ironmaking process, using gas as the reductant. Developed by Thyssen Group and first operated in Germany in 1970, but now used only in Iran. *See* DR.

Purox A process for partially combusting organic wastes in a shaft furnace with oxygen, thereby producing a fuel gas and a molten slag. The gas, not diluted with nitrogen, is suitable for use as a chemical feedstock. Developed by Union Carbide Corporation in 1974 and piloted in Charleston, WV.

Masuda, T. and Fisher, T.F., in *Thermal Conversion of Solid Wastes and Biomass*, Jones, J.L. and Radding, S.B., Eds., American Chemical Society, Washington, DC, 1980, 573.
Probstein, R.F. and Hicks, R.E., *Synthetic Fuels*, McGraw-Hill, New York, 1982, 408.

PVD *See* Physical Vapor Deposition.

PxMax A process for selectively disproportionating toluene to *p*-xylene and benzene. Formerly known as Mobil's *MTPX process, commercialized in 1996 by Mobil Oil at Chalmette Refining's Louisiana refinery, and subsequently applied in 1997 at Mobil's grassroots plant in Beaumont, TX. The process uses a highly shape-selective, proprietary zeolite catalyst to produce xylenes rich in *p*-xylene, with a purity exceeding that of currently available technologies, plus coproduct benzene. Licensed to Nippon Petroleum Refining in 2002. Four units were operating in 2005. Retrofit applications are now licensed by ExxonMobil Chemical; grassroots applications by Axens.

Ind. Eng. Chem., 1995, **73**(35), 12.
Eur. Chem. News, 2000, **73**(1927), 36; 2002, **76**(2009), 31.
Chem. Bus. (India), 2000, **14**(11/12), 56.
Chem. Eng. News, 2001, **79**(35), 28–30; 2002, **80**(23), 16.
Hydrocarbon Process. Int. Ed., 2002, **81**(7), 75.
Chem. Week, 27 Jun 2007.

PX-Plus A process for disproportionating toluene to *p*-xylene and benzene. Developed by UOP in the 1990s. Two units were operating in 2005, and another was in design and construction.

Eur. Chem. News, 1997, **67**(1753), 23.
Eur. Chem. News, Proc. Rev., 1997, May, 26.

Pylumin *See* metal surface treatment.

Pyral A process for destroying toxic waste organochlorine compounds. The wastes are mixed with carbon and sodium carbonate and injected into a graphite-lined arc furnace. Metallic sodium, formed by reduction of the sodium carbonate by the carbon, attacks the chlorinated organic compounds, forming sodium chloride. Developed by Hydro-Quebec in the late 1980s but not yet commercialized.

PYRO-A A pyrochemical process proposed for use in nuclear reprocessing for separating transuranic elements from fission products, once the uranium has been removed by the *UREX process. The spent fuel is dissolved in a molten salt bath and electrolyzed. The transuranic elements deposit on the cathode, and the fission products remain in the melt. Developed by the Argonne National Laboratory. *See also* PYRO-B.

PYRO-B A variant of *PYRO-A for use with the products from a "transmuter" nuclear reactor, that is, one that transmutes transuranic elements and does not involve uranium. Developed by the Argonne National Laboratory.

PYROCAT A steam cracking process for converting petroleum into light olefins in which a catalyst is deposited on the walls of the heat-exchanger coils in the cracking furnace. The catalyst is a proprietary promoter on an alumina–calcia base. Based on the *THERMOCAT process, PYROCAT was developed jointly by Veba Oel and Linde from 1996 but has not yet been commercialized.

> *Chem. Eng. (Rugby, Engl.)*, 1997, (638), 24.

Pyrohydrolysis Also called spray roasting. A process for evaporating and calcining metal salt solutions in one step. Originally developed for processing pickle liquors from the steel industry, containing ferrous chloride in hydrochloric acid, but now used for making ceramic raw materials. Recent development of this process has been by the Ruthner Division of Maschinenfabrik Andritz, Austria.

> Kladnig, W.F. and Karner, W., *Am. Ceram. Bull.*, 1990, **69**, 814.

Pyron A process for making iron powder by reducing mill scale, obtained from steelworks, with hydrogen. In operation in the United States, in places where cheap hydrogen is available near steelworks.

Pyroplasma A high-temperature process for destroying toxic liquid wastes such as polychlorinated biphenyls. The liquid is passed through a DC electric arc heater, and the exit gases pass into a refractory-lined chamber where further reactions occur. Developed in the late 1980s by Westinghouse Pyrolysis Systems, United States, and the Kingston Royal Military College, Ontario, Canada, but not commercialized.

> Kolak, N.P., Barton, T.G., Lee, C.C., and Peduto, E.F., *Nucl. Chem. Waste Manag.*, 1987, **7**, 37.

Pyrotol A process for making benzene from pyrolysis gasoline by *hydrocracking. Developed by Houdry Process and Chemical Company. In 1987, 13 units were operating worldwide.

> Lorz, W., Craig, R.G., and Cross, W.J., *Erdoel Kohle Erdgas Petrochem.* 1968, **21**, 610.
> *Hydrocarbon Process. Int. Ed.*, 1970, **49**(9), 223.

PyStR [**Py**rolysis **St**eam **R**eformer, or pyrolytic steam reforming] (pronounced as pie star) A process for generating high-purity hydrogen from various raw materials including coal and biomass. Invented by W. Oullette at Syngas Energy Corp, a subsidiary of Fairchild (now called Energy Quest).

> *Chem. Eng. (N.Y.)*, 2007, **114**(7), 12.

Q

QC *See* TRC.

Q-Max A process for making cumene from benzene and propylene by catalytic alkylation using a proprietary, regenerable zeolite catalyst. Developed by UOP and first installed in 1996 by JLM Chemicals in Illinois. Seven units were operating in 2005, and two more were in design or construction.

Eur. Chem. News, 1996, **66**(1737), 41; 1997, **67**(1755), 16.
Degnan, T.F., Jr., Smith, C.M., and Venkat, C.R., *Appl. Catal. A: Gen.*, 2001, **221**(1–2), 291.
Peterson, G.A. and Schmidt, R.J., in *Handbook of Petroleum Refining Processes*, 3rd ed., Meyers, R.A., Ed., McGraw-Hill, New York, 2003, 1.69.
Schmidt, R.J., *Appl. Catal. A: Gen.*, 2005, **280**(1), 90.

QQ-BOP *See* Steelmaking.

Q-S *See* QSL.

QSL [Queneau–Schumann–Lurgi] A submerged smelting process for extracting lead from its ores and secondary sources. Pellets of sulfide ore concentrate are fed into a bath of molten slag held in a rotating, cylindrical furnace. Oxygen is fed into the bath below the surface of the slag, forming sulfur dioxide and generating heat. Powdered coal is added farther along the pool, reducing the lead oxide to metal. Invented by P.E. Queneau and R. Schumann, Jr.; now offered by Lurgi. Operated in China, Canada, Germany, and Korea.

Morgan, S.W.K., *Zinc and Its Alloys and Compounds*, Ellis Horwood, Chichester, U.K., 1985, 96.
Chem. Eng. (N.Y.), 1990, **97**(4), 55.

QTEROS *See* *AFEX.

Quantum Catalytic Extraction *See* CEP.

Quentin A process for regenerating the ion-exchange resin used in sugar refining using magnesium chloride solution.

Quentin, G., *Zucker*, 1957, **10**, 408 (*Chem. Abstr.*, **52**, 766).
Landi, S. and Mantovani, G., *Sugar Technol. Rev.*, 1975, **3**(1), 67.

Quick Contact *See* TRC.

R

R-2 [**R**apson] A process for making chlorine dioxide by reacting sodium chlorate with sodium chloride and sulfuric acid:

$$2NaClO_3 + 2NaCl + 2H_2SO_4 = 2ClO_2 + Cl_2 + 2Na_2SO_4 + 2H_2O$$

The product gas, mixed with the by-product chlorine, is stripped from the solution by a current of air and passed into water. The chlorine dioxide dissolves, and most of the chlorine does not; the latter is absorbed in a second column containing alkali. Developed by W.H. Rapson at Hooker Chemical Corporation and operated at Springhill, LA, from 1961.

U.S. Patent 2,863,722.
Rapson, W.H., *Tappi*, 1958, **41**(4), 181.
Rapson, W.H. and Partridge, H. de V., *Tappi*, 1961, **44**(10), 698.
Partridge, H. de V., in *Chlorine: Its Manufacture, Properties, and Uses*, Sconce, J.S., Ed., Reinhold Publishing, New York, 1962, 306.

Radenthein *See* Hansgirg.

Radiance A process for removing organic contaminants from the surfaces of semiconductors by irradiation with deep ultraviolet light while simultaneously passing an inert gas over the surface in laminar flow. Invented by A. Englesberg in 1987 and developed by Radiance Services Company, Bethesda, MD.

Kaplan, H., *Photon. Spectra*, 1996, **30**(9), 48.

Raecke *See* Henkel.

RAM [**R**emove **A**rsenic and **M**ercury] A set of processes for removing arsenic, mercury, and lead from hydrocarbon streams prior to their conversion to ethylene. RAM I removes arsenic; RAM II removes arsenic, mercury, and lead; RAM III removes arsenic, mercury, and sulfur. Licensed by Axens. Fifteen units had been licensed by 2005.

Raney Not a process, but a nickel catalyst widely used for hydrogenating organic compounds. It is made from a 50–50 nickel–aluminum alloy by leaching out the aluminum with concentrated aqueous sodium hydroxide. The product has a spongy texture and is highly active. Invented by M. Raney in 1926. The business was acquired by W.R. Grace in 1963.

U.S. Patent 1,628,190.
Bond, G.C., *Catalysis by Metals*, Academic Press, London, U.K., 1982, 34.

Rapid Thermal Processing *See* RTP.

RAPRENO$_x$ [**RAP**id **RE**duction of **NO**$_x$] A process for removing NO$_x$ from flue gases by reaction with cyanuric acid $(HOCN)_3$. The acid decomposes to HOCN, which generates NH$_2$ radicals, which in turn reduce NO to elemental nitrogen.

Miller, J.A., Branch, M.C., and Kee, R.J., *Combust. Flame*, 1981, **43**, 81.
Perry, R.A. and Siebers, D.L., *Nature*, 1986, **324**, 657.
Gmelin Handbook of Inorganic Chemistry & Organometallic Chemistry, 8th ed., *Nitrogen*, Suppl. B1, Springer-Verlag, Berlin, Germany, 1993, 233.

RAR [**R**ecycle **A**bsorption **R**egeneration] A process for extracting traces of sulfur compounds from the effluent gases from the *Claus process by use of a selective amine absorbent. Developed by KTI.

Raschen A process for making sodium cyanide by reacting ammonia with carbon disulfide. Invented by J.B. Raschen at the United Alkali Company in Widnes, England, in the early 1900s.

Raschig (1) A process for making hydrazine by oxidizing ammonia with sodium hypochlorite in the presence of gelatin:

$$NH_3 + NaOCl = NaOH + H_2NCl$$

$$H_2NCl + NH_3 = N_2H_4 + HCl$$

Invented by F. Raschig at Ludwigshafen, Germany, in 1906 and commercialized by Raschig in 1907. The Olin Raschig version has a complex flow chart and does not use gelatin.

> German Patents 192,783; 198,307.
> Raschig, F., *Ber. Dtsch. Chem. Ges.*, 1907, **40**, 4587.
> Reed, R.A., *Hydrazine and Its Derivatives*, Royal Institute of Chemistry, London, U.K., 1957, 2.

Raschig (2) Also called Raschig–Hooker. A two-stage regenerative process for making phenol from benzene. The benzene is first chlorinated with hydrochloric acid in the presence of air, at 200°C–260°C, over a copper catalyst on an alumina base:

$$C_6H_6 + HCl + \tfrac{1}{2}O_2 = C_6H_5Cl + H_2O$$

The resulting chlorobenzene is then hydrolyzed with steam, over an apatite catalyst at about 480°C:

$$C_6H_5Cl + H_2O = C_6H_5OH + HCl$$

The hydrochloric acid is recycled. The process was developed by Raschig at Ludwigshafen, Germany, in the 1930s, based on the work of L. Dusart and Ch. Bardy in 1872. A variation, known as the Hooker–Raschig process, which uses a different catalyst, makes use of the by-product dichlorobenzenes and thus increases the overall yield. There are several commercial routes from benzene to phenol; the Raschig (2) route is now economic only for very large plants in special locations.

> French Patent 698,341.
> U.S. Patents 1,963,761; 2,009,023; 2,035,917.
> Mathes, W., *Angew. Chem.*, 1939, **52**, 591 (*Chem. Abstr.*, **34**, 394).
> Crawford, R.M., *Chem. Eng. (N.Y.)*, 1950, **46**, 483.
> Prahl, W.H., Williams, W.H., and Widiger, A.H., in *Chlorine: Its Manufacture, Properties, and Uses*, Sconce, J.S., Ed., Reinhold Publishing, New York, 1962, 438.
> Weissermel, K. and Arpe, H.-J., *Industrial Organic Chemistry*, 4th ed., translated by C.R. Hindley and S. Hawkins, Wiley-VCH, Weinheim, Germany, 2003, 352.

Raschig (3) A process for making hydroxylamine. Invented by Raschig AG.

> Rayon Not a process but the generic name for regenerated cellulose fibers made by the *Viscose and related processes. *See* Cross–Bevan–Beadle, Cuprammonium, Viscose, Lyocell.

RCA [**R**adio **C**orporation of **A**merica] Also called RCA-2 and HPM. A process for cleaning silicon wafers used in electronics. The wafers are washed successively by three solutions. The first is an alkaline solution of hydrogen peroxide, which oxidizes organic matter. The second is an acid fluoride solution, which removes silica. The third is an acidic solution of hydrogen peroxide, which removes transition metals. The process was developed by RCA Corporation in 1970 and widely used thereafter by the electronics industry.

Kern, W. and Poutinen, D.A., *RCA Rev.*, 1970, **31**, 187.
Christenson, K.K., Smith, S.M., and Werho, D., *Microcontamination*, 1994, **12**(6), 47.

RCC [**R**educed **C**rude oil **C**onversion] A process for converting reduced crude oil (a petroleum fraction) and petroleum residues into high-octane gasoline and other lighter fuels. Based on the *FCC process, but adapted to accommodate higher levels of metal contaminants that can harm the catalyst. Developed by Ashland Oil Company and UOP and commercialized in 1983.

RCD Isomax [**R**educed **C**rude **D**esulfurization] An obsolete process for desulfurizing high-sulfur residual oils. Developed by UOP, later replaced by *RCD Unibon.

RCD Unibon *See* RCD Unionfining.

RCD Unionfining [**R**educed **C**rude **D**esulfurization] The latest version of UOP's process for removing organic sulfur, nitrogen, and metal compounds from heavy petroleum fractions. Formerly called RCD Unibon, which succeeded the Black Oil Conversion process (BOC). Different catalysts are used for different oils. Developed and licensed by UOP. The first commercial unit started operating in Japan in 1967; since then, 27 more units have been licensed.

Marcos, F. and Rosa-Brussin, D., *Catal. Rev. Sci. Eng.*, 1995, **37**(1), 3.
Gillis, D.B., in *Handbook of Petroleum Refining Processes*, 3rd ed., Meyers, R.A., Ed., McGraw-Hill, New York, 2003, 8.43.

RCE A *flue-gas desulfurization process in which the sulfur dioxide is absorbed in aqueous magnesium hydroxide. The product is reacted with calcium chloride to produce gypsum, and the magnesium hydroxide is regenerated by treatment with dolomite. Developed by Refractories Consulting & Engineering, Germany, and piloted in Austria.

RCH [**R**uhr**CH**emie] A process for increasing the octane rating of gasolines by catalytic isomerization of the olefin fraction, the double bonds migrating from the terminal positions. Developed by Ruhr Chemie in the 1940s.

Asinger, F., *Mono-Olefins: Chemistry and Technology*, translated by B.J. Hazzard, Pergamon Press, Oxford, U.K., 1968, 1096.

RCH/RP [**R**uhr**CH**emie/**R**hône **P**oulenc] A variation of the *OXO process in which the triphenyl phosphine (part of the *Wilkinson catalyst) is sulfonated, in order to render the catalyst soluble in water for easier recovery. Invented by E.G. Kuntz and first commercialized in 1984 for the manufacture of butyraldehyde.

Bach, H., Gick, W., Konkol, W., and Wiebus, E., in *Proc. 9th Internat. Conf. Catal.*, Phillips, M.J. and Ternan, M., Eds., Chemical Institute of Canada, Ottawa, Ontario, Canada, 1988.
Chem. Eng. News, 1994, **72**(41), 28.
Beller, M., Cornils, B., Frohning, C.D., and Kohlpaintner, C.W., *J. Mol. Catal.*, 1995, **A104**(1), 32, 48.
Cornils, B. and Kuntz, E.G., *J. Organomet. Chem.*, 1995, **502**, 177.
Wachsen, O., Himmler, K., and Cornils, B., *Catal. Today*, 1998, **42**, 373.
Cornils, B. and Wiebus, E., in *Aqueous-phase Organometallic Catalysis*, 2nd ed., Cornils, B. and Herrman, W.A., Eds., Wiley-VCH, Weinheim, Germany, 2004, Chap. 5.2.

RCP [**R**ecycled **C**lean **P**roducts] A high-temperature process for recycling solid wastes. It combines pyrolysis, smelting, and slag refining with postcombustion in a circulating, fluidized bed. Developed by Von Roll, Switzerland, and operated commercially in Germany since 1997.

Waste Manage. Environ., 1998, **8**(2), 11.

RDS [**R**esiduum **D**e**S**ulfurization] A general name for *hydrotreating processes designed to remove sulfur from the residues from petroleum distillation. *See* RDS Isomax.

Brossard, D.N., in *Handbook of Petroleum Refining Processes*, 3rd ed., Meyers, R.A., Ed., McGraw-Hill, New York, 2003, 8.3.

RDS Isomax [**R**esiduum **DeS**ulfurization] A *hydrodesulfurization process for removing sulfur compounds from petroleum residues while converting the residues to fuel oil. Developed by Chevron Research Company in the early 1970s. Ten units were operating in 1988. *See also* VGO Isomax, VRDS Isomax.

> Scott, J.W., Bridge, A.G., Christensen, R.I., and Gould, G.D., *Oil Gas J.*, 1970, **68**(22), 72.
> Speight, J.G., *The Desulfurization of Heavy Oils and Residua*, Marcel Dekker, New York, 1981, 194.
> *Hydrocarbon Process. Int. Ed.*, 1996, **75**(11), 132.

Readman *See* Furnace.

Recatro A process for making gas from liquid fuels and other gaseous hydrocarbons by catalytic conversion into "rich gas," followed by catalytic *steam reforming. Developed by BASF and Lurgi.

Recoflo An ion-exchange process based on short beds and small beads. Developed by the University of Toronto in the 1960s and commercialized by Eco-Tec, Canada. Used for wastewater recovery and removal of metals from various metallurgical waste streams. In 1988, 500 units had been installed in 27 countries.

> Brown, C.J. and Fletcher, C.J., in *Ion Exchange for Industry*, Streat, M., Ed., Ellis Horwood, Chichester, U.K., 1988, 392.

Recrystallizer A process for recrystallizing sodium chloride from brine. Rock salt is dissolved in brine heated with direct steam. The solution is then partially evaporated under reduced pressure. Invented in 1945 by C.M. Hopper and R.B. Richards at the International Salt Company, Scranton, PA. *See also* Alberger.

> U.S. Patents 2,555,340; 2,876,182.

Rectiflow A multistage, liquid–liquid extraction process for removing nonparaffinic components from lubricating oils. Furfural has been used as the solvent. Developed and used by the Shell Petroleum Company in the 1940s, subsequently abandoned.

> *The Petroleum Handbook*, 3rd ed., Shell Petroleum Co., London, U.K., 1948, 188.

Rectisol A process for removing sulfur compounds from gas mixtures resulting from the partial oxidation of hydrocarbons, based on physical absorption in methanol at low temperatures. Originally developed in 1951 by Lurgi Gesellschaft für Warmetechnik for the SASOL coal gasification plant in South Africa, but now used also for removing sulfur compounds, CO_2, H_2S, HCN, and C_6H_6, and gum-forming hydrocarbons from *syngas and fuel gas. Further developed and now offered by Linde. In 2000, over 100 units were in operation or under construction.

> Herbert, W., *Erdoel Kohle*, 1956, **9**(2), 77.
> Weiss, H., *Gas Sep. Purif.*, 1988, **2**, 171.
> *Hydrocarbon Process. Int. Ed.*, 1992, **71**(4), 125; 1996, **75**(4), 134.
> Kohl, H.L. and Nielsen, R.B., *Gas Purification*, 5th ed., Gulf Publishing Co., Houston, TX, 1997, 1215.
> *Hydrocarbon Process. Int. Ed.*, 2000, **79**(4), 85.

Recuplex A variant of the *Purex process for extracting plutonium, in which the tributyl phosphate is dissolved in carbon tetrachloride in order to make the organic phase denser than the aqueous phase.

REDAR [**RED**uced **AR**omatics] A two-stage hydrogenation process for reducing the nitrogen, sulfur, and polyaromatic content of *FCC hydrocarbon mixtures so that they may be used as diesel fuel. The first stage removes the nitrogen and sulfur compounds. The second stage hydrogenates the aromatic hydrocarbons. Developed by Engelhard and Washington Group International from 2001.

> *Chem. Eng. (N.Y.)*, 2001, **108**(2), 39.

Redex [**Re**cycle **ex**tract **d**ual **ex**traction] A process for improving the cetane rating of diesel fuel by removing heavy aromatic hydrocarbons by solvent extraction.

French Patents 792,281; 1,424,225; 1,424,226.
Benham, A.L., Plummer, M.A., and Robinson, K.W., *Hydrocarbon Process. Int. Ed.*, 1967, **46**(9), 134.

Redox [**Red**uction **ox**idation] A process for separating the components of used nuclear fuel by solvent extraction. It was the first such process to be used and was brought into operation at Hanford, WA, in 1951, but was superseded in 1954 by the *Purex process. The key to the process was the alternate reduction and oxidation of the plutonium, hence the name. The solvent was hexone (4-methyl-2-pentanone, methyl *iso*butyl ketone), so the process was also known as the Hexone process. The aqueous phase contained a high concentration of aluminum nitrate to salt out the uranium and plutonium nitrates into the organic phase. The presence of this aluminum nitrate in the wastes from the process, which made them bulky, was the main reason for the abandonment of the process. *See also* Butex.

Taube, M., *Plutonium*, Macmillan, New York, 1964, 130.
Nuclear Fuel Reprocessing Technology, British Nuclear Fuels, Risley, U.K., 1985.

Reentech A process for converting waste plastics into liquid fuels. A complicated process involving dehydrogenation over a nickel catalyst, followed by *FCC. Developed by Reentech Ltd. in South Korea and now operated on a commercial scale.

U.S. Patent 6,866,830.
Scheirs, J., in *Feedstock Recycling and Pyrolysis of Waste Plastics*, Scheirs, J. and Kaminsky, W., Eds., John Wiley & Sons, Chichester, U.K., 2006, 423.

reforming A general name for the reaction of a hydrocarbon, such as methane, with water and/or carbon dioxide, to produce a mixture of carbon monoxide and hydrogen. If water is used, it is called steam reforming or steam cracking. The reactions are endothermic and require a catalyst:

$$CH_4 + H_2O = CO + 3H_2$$

The usual catalyst is nickel on an oxide support. A second reaction also takes place, the shift reaction, which is also known as the water gas shift reaction:

$$CO + H_2O = CO_2 + H_2$$

The shift reaction can be conducted in a second reactor, catalyzed by a mixture of iron and chromium oxides. The product of reforming is known as synthesis gas, or *syngas, and is mostly used in the manufacture of ammonia and methanol. One of the earliest steam reforming processes was developed in Germany by I.G. Farbenindustrie in 1926. *See also* Catalytic Reforming.

U.S. Patent 1,934,237.
Ridler, D.E. and Twigg, M.V., in *The Catalyst Handbook*, 2nd ed., Twigg, M.V., Ed., Wolfe Publishing, London, U.K., 1989, 225.

REGEN A process for removing mercaptans from hydrocarbon fractions by catalytic oxidation and extraction with aqueous alkali using a bundle of hollow fibers. Developed by the Merichem Company, Houston, TX, and used in 34 plants as of 1991.

Hydrocarbon Process. Int. Ed., 1992, **71**(4), 120; 1996, **75**(4), 126.

RegenMax A single-stage process for regenerating *FCC catalysts. The catalyst flows downward through a reactor; carbon monoxide flows up. The partial combustion of the CO provides the heat. A proprietary baffle separates the upper and lower halves of the reactor, reducing back-mixing. Developed from 1998 by Kellogg Brown & Root and Mobil Technology.

Chem. Eng. (N.Y.), 1999, **106**(4), 19.

REGENOX A catalytic process for oxidizing organic compounds in gaseous effluents. A modified version oxidizes chlorinated and brominated hydrocarbons at 350°C–450°C without forming dioxins. Developed by Haldor Topsoe and first operated by Broomchemie in the Netherlands in 1995. *See* CATOX.

Chem. Eng. (N.Y.), 1995, **102**(9), 17.

RegenOx The trade name for a system for remediating contaminated soils. Two liquids are injected: an alkaline solution of sodium percarbonate and a solution of a catalyst. Claimed to be capable of destroying hydrocarbons, gasoline oxygenates, and chlorinated hydrocarbons. Offered by Regenesis, San Clemente, CA, from 2005.

Reich (1) A process for purifying carbon dioxide obtained by fermentation. The carbon dioxide is first scrubbed with aqueous ethanol, then with aqueous potassium dichromate to oxidize organic compounds, and finally with concentrated sulfuric acid to dry it. Developed in the 1920s by G.T. Reich.

U.S. Patents 1,519,932; 2,225,131.

Reich (2) A complex process for recovering potassium from sugar-processing residues.

Thorpe's Dictionary of Applied Chemistry, 4th ed., Longmans, Green & Co., London, U.K., 1950, 10, 139.

Reichstein–Grüssner A process for making ascorbic acid from D-glucose. It involves catalytic hydrogenation, microbial oxidation, acetal formation, permanganate oxidation, and ring closure. Developed by T. Reichstein and A. Grüssner at ETH, Zurich, in 1933. Still the major manufacturing process for ascorbic acid.

Sheldon, R.A., Arends, I., and Hanefeld, U., *Green Chemistry and Catalysis*, Wiley-VCH, Weinheim, Germany, 2007, 384.
Wikipedia, The Free Encyclopedia. Reichstein. Revised 10 Mar 2013. Retrieved 22 Jul 2013.

Reinluft A *flue-gas desulfurization process using coke. The carbon acts as a catalyst for the oxidation of the sulfur dioxide to sulfur trioxide in the presence of water, and the sulfur trioxide is retained on the coke. The coke is regenerated in another vessel by heating with a hot gas stream, which reduces the sulfur trioxide back to sulfur dioxide and expels it for use in sulfuric acid manufacture. The key to this process is the inexpensive adsorbent. Developed by Reinluft GmbH and Chemiebau Dr. A. Zieren GmbH and marketed as the Reinluft (Clean Air) process. Four plants had been built by 1985.

Bienstock, D., Field, J.H., Katell, S., and Plants, K.D., *J. Air Pollut. Control Assoc.*, 1965, **15**, 459.
Chem. Eng. (N.Y.), 1967, **74**(22), 94.
Kohl, H.L. and Nielsen, R.B., *Gas Purification*, 5th ed., Gulf Publishing Co., Houston, TX, 1997, 638.

Relube A process for removing sulfur and chlorine compounds from waste oils, particularly those contaminated by polychlorinated biphenyls. Developed by Kinetics Technology International, the Netherlands, and operated first in Greece.

Remet [**Re**forming **met**hanol] A process for making high-purity methanol by a combination of *steam reforming and *PSA. Licensed by Tokyo Gas Company and Tokyo Cryogenic Industries Company. Two units were operating in 1990.

Hydrocarbon Process. Int. Ed., 1990, **69**(4), 82.

Renewable Jet *See* Ecofining.

Rentech A process for converting *syngas to hydrocarbons using an iron-based catalyst. Developed in the 2000s by Rentech Inc., which has been working with UOP from 2008. A demonstration plant in Commerce City, CO, has operated since 2008.

Wikipedia, The Free Encyclopedia. Rentech. Revised 2 Jul 2013. Retrieved 22 Jul 2013.

RENUGAS A thermal gasification process for biomass. Developed by the Institute of Gas Technology, Chicago, (now GTI) in the 1980s but not yet commercialized.

Hydrocarbon Process. Int. Ed., 1986, **65**(4), 100.
Chem. Eng. (N.Y.), 1996, **102**(3), 39.

Reppe A family of processes for making a range of aliphatic compounds from acetylene, developed by W. Reppe in IG Farbenindustrie, Germany, before and during World War II. In one of the processes, acetylene is reacted with carbon monoxide to yield acrylic acid:

$$CH\equiv CH + CO + H_2O \rightarrow CH_2=CH-COOH$$

Acrylic esters are formed if alcohols are used instead of water:

$$CH\equiv CH + CO + ROH \rightarrow CH_2=CH-COOR$$

Nickel carbonyl is the catalyst for these reactions. In another Reppe process, acetylene is reacted with formaldehyde to yield butyndiol, which can be converted to butadiene for the manufacture of the synthetic rubber *Buna; the catalyst is nickel cyanide:

$$CH\equiv CH + 2CH_2O \rightarrow HOCH_2-C\equiv C-CII_2OH$$

When, in the 1950s, ethylene became the preferred feedstock for making petrochemicals, most of these acetylene-based processes became obsolete.

German Patents 725,326; 728,466.
U.S. Patents 2,806,040; 2,809,976; 2,925,436; 3,023,237.
Reppe, J.W., *Acetylene Chemistry*, translated, Charles A. Meyer & Co., New York, 1949.
Reppe, J.W., *Experentia*, 1949, **5**, 93.
Miller, S.A., *Acetylene: Its Properties, Manufacture and Uses*, Vol. 1, Ernest Benn, London, U.K., 1965.
Morris, P.J.T., *Chem. Ind. (London)*, 1983, (18), 713.
Eur. Chem. News, Process Rev. Suppl., 1988, Oct, 10.
Tang, C.-M., Zeng, Y., Yang, X.-G., Lei, Y.-C., and Wang, G.-Y., *J. Mol. Catal. A: Chem.*, 2009, **314**, (1–2), 15.

reprography The processes listed in the following text, described elsewhere in this dictionary, are mostly for reproducing line drawings, rather than pictures. There is, however, some overlap with *photography. These processes are Bluecprint, Diazo, Dual-Spectrum, Dyeline, Eichner, Extafax, Kalvar, Ozalid, Thermofax, and Van Dyke.

Kirk-Othmer's Encyclopedia of Chemical Technology, 3rd ed., Vol. 20, John Wiley & Sons, New York, 1980, 128.

Reptile-96 [**Rep**lacement **rutile**] A version of the *Wendell Dunn process for removing iron from ilmenite by selective chlorination. The product contains 96% rutile, hence the name. A large demonstration plant was completed by Heubach in Ankleshwar, India, in 2005. The product can be used as a feedstock for the *Chloride process or as an inexpensive off-white pigment.

Republic Steel *See* DR.

RESID-fining [**RESID**uum re**fining**] A *hydrodesulfurization process adapted for petroleum residues. Developed by Esso Research & Engineering Company and licensed by them and Union Oil Company of California. A proprietary catalyst is used in a fixed bed. As of 1988, eight plants had been designed.

Speight, J.G., *The Desulfurization of Heavy Oils and Residua*, Marcel Dekker, New York, 1981, 190.
Hydrocarbon Process. Int. Ed., 1988, **67**(9), 70; 1994, **73**(11), 135.

RESOX A process for converting sulfur dioxide in dilute gas streams to elemental sulfur. The use of coal to remove sulfur dioxide from gas streams was described as early as 1879. Bergbau-Forschung developed a process for reversibly adsorbing sulfur dioxide on activated coke in the 1950s, and Foster Wheeler Corporation modified it in the late 1960s to produce elemental sulfur. In the first stage, the sulfur dioxide is converted to sulfuric acid in the pores of the coke, and this is reduced back to sulfur dioxide in a second stage. The modified version uses coal instead of coke, and the reduction product is sulfur vapor, which is condensed as a liquid. The first demonstration plant was built by Foster Wheeler in Florida in 1974. *See also* Trail.

British Patent 189 (1879).
Steiner, P., Jüntgen, H., and Knoblauch, K., in *Sulfur Removal and Recovery from Industrial Processes*, Pfeiffer, J.B., Ed., American Chemical Society, Washington, DC, 1975, 185.

RESS [**R**apid **E**xpansion of **S**upercritical **S**olutions] A process for depositing a film of solid material on a surface. The substance is dissolved in supercritical carbon dioxide. When the pressure is suddenly reduced, the fluid reverts to the gaseous state and the solute is deposited on the walls of the vessel. Used for size reduction, coating, and microencapsulation. First described in 1879. Developed in 1983 by R.D. Smith at the Battelle Pacific Northwest National Laboratory, Richland, WA.

U.S. Patent 4,582,731.
Matson, D.W., Peterson, R.C., and Smith, R.D., *Mater. Lett.*, 1986, **4**, 429.
Petersen, R.C., Matson, D.W., and Smith, D.R., *J. Am. Chem. Soc.*, 1986, **108**, 2102.
Matson, D.W. and Smith, R.D., *J. Am. Ceram. Soc.*, 1989, **72**, 877.
Weber, M. and Thies, M.C., in *Superfluid Technology in Materials Science and Technology*, Sun, Y.-P., Ed., Marcel Dekker, New York, 2002, 387–437.

Resulf A process for removing residual sulfur compounds from refinery tail gases. The residues are hydrogenated to hydrogen sulfide, which is absorbed in an aqueous solution of an amine, such as methyl diethanolamine. Licensed by TPA. Thirty-seven units had been licensed and designed by 2000.

Hydrocarbon Process. Int. Ed., 1996, **75**(4), 134; 2000, **79**(4), 85.

ReVAP [**Re**duced **V**olatility **A**lkylation **P**rocess] A process for improving the safety of *alkylation processes catalyzed by hydrofluoric acid. A proprietary additive curtails the emission of the acid aerosol, which forms in the event of a leak. The additive has a high boiling point and is miscible with acid but is sparingly soluble in hydrocarbons. Developed by Phillips Petroleum Corporation and ExxonMobil and first installed at Woods Cross, UT, in 1996. *See also* Alkad.

Chem. Mark. Rep., 1996, **250**(3), 7.
Hydrocarbon Process. Int. Ed., 2000, **79**(12), 33.

REVERSE-methane A process for producing high-temperature heat from exhaust gases containing small concentrations of combustible components (e.g., methane in coal mine exhaust gases). Combustion takes place in a fixed catalytic bed in which the flow is periodically reversed.

CATTECH, 2002, **6**(3), 97.

Rexene A process for making polypropylene. Developed by Appryl, a joint venture of BP and Atochem.

Chem. Br., 1996, **32**(8), 7.

Rexforming A petroleum refining process that combines *Platforming with an aromatics extraction process using ethylene glycol. Developed in the 1950s by Universal Oil Products.

Unzelman, G.H. and Wolf, C.J., in *Petroleum Processing Handbook*, Bland, W.F. and Davidson, R.L., Eds., McGraw-Hill, New York, 1967, 3–37.
Asinger, F., *Mono-olefins: Chemistry and Technology*, translated by B.J. Hazzard, Pergamon Press, Oxford, U.K., 1968, 391.

Reynolds Metal A process for extracting aluminum from clay by leaching with nitric acid. An essential feature is the pelletizing of the clay by calcination with kaolin to provide particles that will not disintegrate during leaching. Aluminum nitrate nonahydrate is crystallized from the leachate and thermally decomposed in several stages designed to conserve the nitric acid and nitrogen oxides. Developed by the Reynolds Metal Company, United States, between 1973 and 1988, but not yet commercialized.

U.S. Patents 3,804,598; 4,251,265; 4,256,714.
O'Connor, D.J., *Alumina Extraction from Non-bauxitic Materials*, Aluminium-Verlag, Düsseldorf, Germany, 1988, 163.

RFCC [**R**esidual **F**luid **C**atalytic **C**racking] A process for *cracking residues from petroleum distillation. It uses an ultrastable zeolite catalyst with two-stage regeneration. Originally developed in the early 1980s by Total Petroleum in Kansas and Oklahoma under the name *R2R. Further developed by IFP with Stone and Webster and now more commonly known as RFCC. Twenty-six units had been licensed by 2003. A new version, incorporating PetroRiser technology, was announced by Axens in 2008. It offers higher yields of propylene.

Chauvel, A., Delmon, B., and Hölderich, W.F., *Appl. Catal. A: Gen.*, 1994, **115**, 173.
Letzsch, W.S., in *Handbook of Petroleum Refining Processes*, 3rd ed., Meyers, R.A., Ed., McGraw-Hill, New York, 2003, 3.71.
Chem. Eng. (N.Y.), 2008, **115**(7), 14.

RH *See* Steelmaking.

Rheniforming [**Rhen**ium re**forming**] A *catalytic reforming process developed by Chevron Research Company. The catalyst formulation includes rhenium. Announced in 1967 and first commercialized in 1970; by 1988, 73 units had been licensed.

McCoy, C.S. and Munk, P., *Chem. Eng. Prog.*, 1971, **67**(10), 78.
Hughes, T.R., Jacobson, R.L., Gibson, K.R., Schornack, L.G., and McCabe, J.R., *Hydrocarbon Process. Int. Ed.*, 1976, **55**(5), 75.
Little, D.M., *Catalytic Reforming*, PennWell Publishing, Tulsa, OK, 1985, 166.
Hydrocarbon Process. Int. Ed., 1988, **67**(9), 79.

Rhenipal A sewage sludge treatment process. One of three proprietary additives is used before dewatering. The resulting filter cake is smaller in volume and has less odor, and its heavy metals are insoluble. Offered by Rhenipal, United Kingdom, a joint venture between National Power and Dirk European.

RH-FR *See* Steelmaking.

RH-OB *See* Steelmaking.

Rhodaks A process for removing hydrogen cyanide from coke-oven gas, developed by Rhodia. *See also* Fumaks–Rhodaks.

Rhodia A commercial process for making vanillin from guaiacol, developed by Rhodia in the 1970s.

Sheldon, R.A., Arends, I., and Hanefeld, U., *Green Chemistry and Catalysis*, Wiley-VCH, Weinheim, Germany, 2007, 40.
Wikipedia, The Free Encyclopedia. Vanillin. Revised 10 Jul 2013. Retrieved 22 Jul 2013.

Rhône-Poulenc Also called the RP process. This large French chemical manufacturer is perhaps best known for its process for making oxalic acid by oxidizing propylene with nitric acid. Nitratolactic acid is an intermediate. The process, invented in 1966, is operated on a large scale at Chalampé, France.

U.S. Patent 3,549,696.
French Patent 1,501,725.
British Patents 1,154,061; 1,159,066.

Rhône-Poulenc/Melle Bezons A process for making acetic acid by oxidizing acetaldehyde with oxygen in air. Removal of the nitrogen would incur a cost penalty.

Weissermel, K. and Arpe, H.-J., *Industrial Organic Chemistry*, 4th ed., translated by C.R. Hindley and S. Hawkins, Wiley-VCH, Weinheim, Germany, 2003, 174.

R-HYC A *hydrocracking process.

Marcos, F. and Rosa-Brussin, D., *Catal. Rev. Sci. Eng.*, 1995, **37**(1), 3.

Richards *See* RSRP.

Riedel A process for making vanillin from guaiacol derived from catechol. Invented by J.D. Riedel in 1932.

British Patent 401,562.

Riedel–Pfleiderer *See* AO.

Riley–Morgan A coal gasification process, based on a cylindrical gas producer developed by C.H. Morgan in Worcester, MA, in 1880. By 1964, the Morgan Construction Company had installed more than 9000 such units in a number of industries worldwide. In 1971, the Riley Stoker Corporation acquired the manufacturing rights to the technology and made a number of modifications to the design. Air (or oxygen) and steam are passed through a fixed bed of coal, supported on a rotating ash pan. The temperature is kept below the melting point of the ash.

Hebden, D. and Stroud, H.J.F., in *Chemistry of Coal Utilization*, 2nd Suppl. Vol., Elliott, M.A., Ed., John Wiley & Sons, New York, 1981, 1619.

RIMNAT A process for making a fertilizer from domestic waste by treatment with the zeolite phillipsite.

Ciambelli, P., Corbo, P., Liberti, L., and Lopez, A., in *Occurrence, Properties, and Utilization of Natural Zeolites: Proceedings of the 2nd International Conference on Natural Zeolites, Budapest*, Kallo, D. and Sherry, H.S., Eds., 1985, Akad. Kiado, Budapest, 1988, 501.

Rincker–Wolter A process for making hydrogen by the thermal decomposition of oils and tars over hot coke. Invented by F.G.C. Rincker and L. Wolter in Germany in 1904.

German Patent 174,253.
Ellis, C., *The Chemistry of Petroleum Derivatives*, The Chemical Catalog Co., New York, 1934, 208.

RIP [**R**esin **I**n **P**ulp] A general term for hydrometallurgical processes in which an ion-exchange resin is mixed with a suspension of a ground ore in water. The desired metal is selectively extracted into the resin. *See* CIP (1).

Streat, M. and Naden, D., in *Ion Exchange and Sorption Processes in Hydrometallurgy*, Streat, M. and Naden, D., Eds., John Wiley & Sons, London, U.K., 1987, 35.

Rittman An early process for making aromatic hydrocarbons by thermally cracking petroleum naphtha. *See also* Hall.

British Patents 9,162; 9,163 (1915).
Ellis, C., *The Chemistry of Petroleum Derivatives*, The Chemical Catalog Co., New York, 1934, 165.
Asinger, F., *Mono-olefins: Chemistry and Technology*, translated by B.J. Hazzard, Pergamon Press, Oxford, U.K., 1968, 137.

RKN A process for making hydrogen from hydrocarbon gases (from natural gas to naphtha) by *steam reforming. Developed by Haldor Topsoe in the 1960s; as of 1975, 24 plants were operating.

Hydrocarbon Process. Int. Ed., 1975, **54**(4), 132.

RLE [**R**oasting, **L**eaching, **E**lectrowinning] A process for extracting copper from sulfide ores using the three named processes. Developed by Hecla Mining Company, AZ, in 1969.

Griffith, W.A., Day, H.E., Jordan, T.S., and Nyman, V.C., *J. Met.*, 1975, **27**(2), 17.

RM [**R**alph **M**. Parsons] A process for methanating synthesis gas, that is, converting a mixture of carbon monoxide and hydrogen to mainly methane and carbon dioxide. Six adiabatic reactors are used in series, and steam is injected at the inlet. Under development by the R.M. Parsons Company in 1975.

Benson, H.E., in *Chemistry of Coal Utilization*, 2nd Suppl. Vol., Elliott, M.A., Ed., John Wiley & Sons, New York, 1981, 1795.

R-N [**R**epublic Steel Corporation and **N**ational Lead Company] An ironmaking process developed by these two American companies in the 1960s. Granulated ore is reduced with coal in a rotating kiln, heated by burning the coal at the lower end. Three plants were operating in 1970.

Robinson–Bindley *See* Synthetic Oils.

Rodgers An early process for making potassium cyanide by fusing together potassium ferrocyanide and potassium carbonate.

Roelen *See* OXO.

Röhm A process for making sodium cyanide, engineered by Uhde. A plant was commissioned in Kwinana, Western Australia, in 1988.

Roka A process for making acetone by passing a mixture of ethanol and steam over a catalyst containing iron and calcium:

$$2C_2H_5-OH + H_2O = CH_3-CO-CH_3 + 4H_2 + CO_2$$

Invented by K. Roka at Holzverkohlungs-Industrie in 1924 and operated by British Industrial Solvents at Hull, United Kingdom, in the 1930s.

German Patent 475,428.
U.S. Patent 1,663,350.
Morgan, G.T. and Pratt, D.D., *British Chemical Industry*, Edward Arnold & Co., London, U.K., 1938, 315.

ROSE (1) [**R**esiduum **O**il **S**upercritical **E**xtraction] A process for extracting asphaltenes and resins from petroleum residues using supercritical propane or isobutane as the extractant. Developed by Kerr-McGee Corporation in the 1970s, first licensed in 1979, and sold to the M.W. Kellogg Company in 1995. Thirty-three units had been licensed or designed by 2003.

Gearhart, J.A. and Garwin, L., *Hydrocarbon Process. Int. Ed.*, 1976, **55**(5), 125.
Hydrocarbon Process. Int. Ed., 1978, **57**(9), 200.
Chem. Eng. (N.Y.), 1989, **96**(7), 35.
Hydrocarbon Process. Int. Ed., 1996, **75**(11), 106.
Teixeira, M.A.G. and Goncalvez, M.L.A. in *CO$_2$ Conversion and Utilization,* Song, C., Gaffrey, A.M., and Fujimoto, K., Eds., American Chemical Society, Washington, DC, 2002, 389.
Abdel-Halim, T. and Floyd, R., in *Handbook of Petroleum Refining Processes*, 3rd ed., Meyers, R.A., Ed., McGraw-Hill, New York, 2003, 10.15.

Rose (2) A process for extracting gold from the residues from zinc production. The residues are fused with a mixture of borax and silica, and air blown through. The base metals oxidize and pass into the slag.

Rothamsted A process for extracting protein from leaves. Green plant juice is coagulated at 70°C, yielding a lipid–protein concentrate. Developed at the eponymous research institute in the United Kingdom.

Davys, M.N.G. and Pirie, N.W., *J. Agric. Eng. Res.*, 1963, **8**, 70.
Pirie, N.W., *Science*, 1966, **152**, 1701.
Kromus, S., Kamm, B., Kamm, M., Fowler, P., and Narodoslawsky, M., in *Biorefineries: Industrial Processes and Products*, Kamm, B., Gruber, P.R., and Kamm, M., Eds., Wiley-VCH, Weinheim, Germany, 2005, **I**, 256.

Rotor An oxygen steelmaking process, similar to the *Kaldo process but using a furnace rotating about a horizontal axis. In this method of operation, the refractory lining is cooled by the molten metal and slag and therefore lasts longer. Developed in Oberhausen, Germany, in the 1950s. *See also* DR.

British Patent 726,368.
Osborne, A.K., *An Encyclopedia of the Iron and Steel Industry*, 2nd ed., The Technical Press, London, U.K., 1967, 472.

Royco *See* EZ-Oil Generator.

Rozan A variation of the *Pattinson process for extracting silver from lead, in which steam is blown through the molten metal. This oxidizes the zinc and antimony, which come to the surface and are removed.

RPSA [**R**apid-**P**ressure **S**wing **A**dsorption] A version of the *PSA process that uses fast pressure cycles known as parametric pumping. The molecular sieve adsorbent for this duty has to be of a smaller grain size than that for PSA. Developed by the Union Carbide Corporation.

Pritchard, C.L. and Simpson, G.K., *Trans. Inst. Chem. Eng.*, 1986, **64**(6), 467.

R2R [**1 R**eactor, **2 R**egenerators] A fluid-bed process for *cracking residues from petroleum distillation that uses two stages of regeneration. Originally developed by Total Petroleum in Kansas and Oklahoma. Further developed by IFP with Stone and Webster; now more commonly known as *RFCC.

R-SO$_x$ A process for removing sulfur dioxide from flue gases by injecting hydrated lime with a proprietary additive. Developed in the 1980s by Fossil Energy Services International but not commercialized.

Kohl, H.L. and Nielsen, R.B., *Gas Purification*, 5th ed., Gulf Publishing Co., Houston, TX, 1997, 620.

RSRP [**R**ichards **S**ulphur **R**ecovery **P**rocess] A proposed modification of the *Claus process, in which liquid sulfur is used to cool the catalyst bed. Developed jointly by the Alberta Energy Company and the Hudson's Bay Oil & Gas Company, but not reported to have been commercialized.

Kerr, R.K., Sit, S.-P., Jagodzinski, R.F., and Dillon, J., *Oil Gas J.*, 1982, **80**(30), 230.
Kohl, H.L. and Nielsen, R.B., *Gas Purification*, 5th ed., Gulf Publishing Co., Houston, TX, 1997, 697.

RTP (1) [**R**apid **T**hermal **P**rocessing] A process for converting heavy petroleum oil and bitumen to less viscous liquids, which are easier to pump through long-distance pipelines. Developed by Ensyn Petroleum Canada and proposed to be piloted in Alberta in 2002.

Oil Gas J., 2002, **100**(13), 26.

RTP (2) [**R**apid **T**hermal **P**rocessing] A process for converting biomass into liquid biofuels. The biomass, usually forest residues or agricultural by-products, is rapidly heated to about 500°C in the absence of oxygen in a fluidized bed similar to the beds used in the *FCC process. Hot sand vaporizes the biomass that is then rapidly quenched. The resulting bio-oil can be used directly or upgraded to transportation fuels. First developed in Canada by Ensyn Petroleum Canada and further developed by Envergent Technologies, a joint venture of Ensyn with UOP. In 2012, it was in use in seven commercial plants in Canada and the United States. The construction of a demonstration plant in Hawaii, with finance from the U.S. Department of Energy, was started in 2011.

Ruhr Chemie Ruhr Chemie was an important German chemical company in the 1930s and 1940s. It was perhaps best known for its process for making acetylene by pyrolyzing hydrocarbons. *See also* Wulff.

Ruhr Chemie–Lurgi A variant of the *Fischer–Tropsch process, developed at Ruhr Chemie and Lurgi Gesellschaft für Warmetechnik in Germany during World War II.

Asinger, F., *Paraffins: Chemistry and Technology*, translated by B.J. Hazzard, Pergamon Press, Oxford, U.K., 1968, 168.

Ruhr Chemie/Rhone-Poulenc *See* RCH/RP.

Ruhrgas (1) A coal gasification process developed and used in Germany. Pulverized coal is gasified by a blast of preheated air in a vertical shaft, the temperature reaching approximately 1500°C.

Ruhrgas (2) A process for thermally decomposing oil shale, developed by Lurgi.

Ruhrkohle/VEBA Oel-hydrogenation A coal liquefaction process based on the *IG-Hydrogenation process.

Runcol [**Run**corn thiodigly**col**] Not a process but the code name for a mixture of chemical warfare agents made in Runcorn, England, and Rhydymwyn, North Wales, during World War II. One of them was sulfur mustard, made from thiodiglycol. The other was nitrogen mustard.

Rüping Also known as the Empty Cell process. A method for treating timber with a creosote preservative. The wood is first exposed to compressed air and then impregnated at a higher pressure. *See also* Bethell.

Russell A process for extracting silver from argentite, Ag_2S, using a solution of sodium thiosulfate and cupric sulfate. Invented by E.H. Russell in 1884, following his use of this solution to remove sodium sulfide from soda ash.

U.S. Patent 295,815.

Ruthner A process for recovering hydrochloric acid and iron oxide from steel pickling liquors. Invented in 1968 by A. Hake and P. Borsody at Ruthner Industrieplanungs, Vienna. *See also* Dravo–Ruthner.

U.S. Patent 3,495,945.
Austrian Patent 284,062.

RVG Chlorsorb *See* Chlorsorb.

RWD [**R**eaction **W**ith **D**istillation] A general chemical process in which a chemical reaction takes place within a distillation column, of which the packing is also the catalyst. First used in the 1920s. *See* Ethermax.

DeGarmo, J.L., Parulekar, V.N., and Pinjala, V., *Chem. Eng. Prog.*, 1992, **88**(3), 43.

RxCat *See* PetroFCC.

Ryan–Holmes A cryogenic, extractive distillation process using liquid carbon dioxide, in which a light hydrocarbon is added to suppress the freezing of the carbon dioxide. Licensed by Process Systems International: nine licenses had been granted by 1992.

Chiu, C.-W., *Hydrocarbon Process. Int. Ed.*, 1990, **69**(1), 69.
Hydrocarbon Process. Int. Ed., 1992, **71**(4), 126.
Kohl, H.L. and Nielsen, R.B., *Gas Purification*, 5th ed., Gulf Publishing Co., Houston, TX, 1997, 1342.

RZ Platforming A *Platforming process using a novel catalyst. The catalyst is a zeolite containing platinum, which maximizes the production of benzene and toluene. First commercialized in 1998.

Lapinski, M., Baird, L., and James, R., in *Handbook of Petroleum RefiningProcesses*, 3rd ed., Meyers, R.A., Ed., McGraw-Hill, New York, 2003, 4.25.

S

Saarburg–Holter A *flue-gas desulfurization process.

Sulphur, 1979, (141), 34.

Saarburg–OTTO A coal gasification process. Powdered coal, together with steam and oxygen, is injected tangentially into a gasifier containing molten slag. Gasification in a bath of molten slag was invented by R. Rummel in the 1950s and developed by Dr. C. Otto & Company in Germany in the early 1960s. In 1976, Saarbergwerke and Otto agreed to a joint development program, which culminated in the building of a large demonstration unit at Völklingen/Fürstenhausen, Germany, which was operated from 1979 to 1982.

Rummel, R., *Coke Gas*, 1959, **21**(247) 493 (*Chem. Abstr.*, **54**, 11438).
Eur. Chem. News, Petrochem. Suppl., 1981, Dec, 14.
Mueller, R. and Pitz, H., in *Handbook of Synfuels Technology*, Meyers, R.A., Ed., McGraw-Hill, New York, 1984, 3–195.

SAB *See* Steelmaking.

SABA [**S**pherical **A**gglomeration–**B**acterial **A**dsorption] A microbiological process for leaching iron pyrite from coal. The bacterium *Thiobacillus ferrooxidans* adsorbs on the surface of the pyrite crystals, oxidizing them with the formation of soluble ferrous sulfate. Developed by the Canadian Center for Mineral and Energy Technology, Ottawa. In 1990, the process had been developed only on the laboratory scale, using coal from Eastern Canada.

McCready, G.G.L., in *Bioprocessing and Biotreatment of Coal*, Wise, D.L., Ed., Marcel Dekker, New York, 1990, 685.

SABAR [**S**trong **A**cid **B**y **A**zeotropic **R**ectification] A process for making nitric acid by the atmospheric oxidation of ammonia. The nitrous gases from the oxidation are absorbed in azeotropic nitric acid in the presence of oxygen under pressure:

$$2NO_2 + \tfrac{1}{2}O_2 + H_2O = 2HNO_3$$

Developed by Davy McKee, which built plants from 1974 to 1986. *See also* CONIA.

Hellmer, L., *Chem. Eng. Prog.*, 1972, **68**(4), 67.
Hydrocarbon Process. Int. Ed., 1989, **68**(11), 106.
Büchner, W., Schliebs, R., Winter, G., and Büchel, K.H., *Industrial Inorganic Chemistry*, VCH Publishers, Weinheim, Germany, 1989, 63.

Sabatier–Normann *See* Normann.

Sablin *See* Alpha-Sablin.

saccharification A general name for processes that convert wood to useful organic chemicals by hydrolysis of the polysaccharides in the wood to monomeric sugars. Exemplified by *Bergius–Rheinau, *Madison, and *Scholler–Tornesch. First operated in Germany in 1901 and in the United States in 1909.

Riegel's Handbook of Industrial Chemistry, 9th ed., Kent, J.A., Ed., Van Nostrand Reinhold, New York, 1992, 254.
Kamm, B., Kamm, M., Gruber, P.R., and Kromus, S., in *Biorefineries: Industrial Processes and Products*, Kamm, B., Gruber, P.R., and Kamm, M., Eds., Wiley-VCH, Weinheim, Germany, 2005, **I**, 5.

Sachsse Also called the flame cracking process and the Sachsse–Bartholomé process. A process for making acetylene by the partial combustion of methane. The product gases are quenched rapidly, and the acetylene is extracted with methyl pyrrolidone. First operated by IG Farbenindustrie at Oppau, Germany, in 1942. Worldwide, 13 plants used the process, of which 7 were still in operation in 1991.

Sachsse, H., *Chem. Ing. Tech.*, 1949, **21**, 129; 1954, **26**, 245.
Miller, S.A., *Acetylene: Its Properties, Manufacture and Uses*, Vol. 1, Ernest Benn, London, U.K., 1965, 419.
Tedeschi, R.J., *Acetylene-based Chemicals from Coal and Other Natural Resources*, Marcel Dekker, New York, 1982, 20.
Weissermel, K. and Arpe, H.-J., *Industrial Organic Chemistry*, 4th ed., translated by C.R. Hindley and S. Hawkins, Wiley-VCH, Weinheim, Germany, 2003, 96.

SAFe A *BAF process offered by PWT Projects. The biological medium is supported on a bed of expanded shale.

Stephenson, T., Mann, A., and Upton, J., *Chem. Ind. (London)*, 1993, (14), 533.

Salex A process for purifying sodium chloride. Continuous countercurrent extraction with brine removes the impurities without dissolving the salt. There are three variants: SALEX-B, SALEX-C, and SALEX-M. Developed by Krebs Swiss and first operated in 1982.

Sedivy, V.M., *Ind. Miner. (London)*, 1996, (343), 73.

Salsigne A *Cyanide process for extracting gold from ores containing arsenic or antimony. Pretreatment with a lime slurry reduces cyanide losses.

Yannopoulos, J.C., *The Extractive Metallurgy of Gold*, Van Nostrand Reinhold, New York, 1991, 156.

Salt An obsolete, two-stage process for obtaining chlorine and sodium nitrate from sodium chloride by the intermediary of nitrosyl chloride. In the first stage, the sodium chloride was reacted with nitric acid, producing nitrosyl chloride and chlorine:

$$3NaCl + 4HNO_3 = NOCl + Cl_2 + 2H_2O + 3NaNO_3$$

In the second, the nitrosyl chloride was either reacted with sodium carbonate

$$3NOCl + 2Na_2CO_3 = NaNO_3 + 3NaCl + 2CO_2 + 2NO$$

or oxidized with oxygen

$$2NOCl + O_2 = N_2O_4 + Cl_2$$

The sodium nitrate was used as a fertilizer. The evolution of the process was complex; the book referenced in the following lists 63 patents relating to it. The process was installed by the Solvay Process Company at Hopewell, VA, in the 1930s. It was in operation there, subsequently under the management of the Allied Chemical Corporation, until the 1950s.

Fogler, M.F., *Chlorine: Its Manufacture, Properties and Uses*, Sconce, J.S., Ed., Reinhold Publishing, New York, 1962, 235.

Salt Cake One of the two processes comprising the *Leblanc process for making sodium carbonate. Salt cake was the colloquial name for sodium sulfate.

SAMEX A process for removing traces of mercury from the waste brine from the *Chlor-Alkali process.

Nogueira, E.D., Regife, J.M., and Melendo, J.F.J., in *Modern Chlor-Alkali Technology*, Vol. 1, Coulter, M.O., Ed., Ellis Horwood, Chichester, U.K., 1980, 85.

Samica *See* Bardet.

sandwich desulfurization A *hydrotreating process for removing sulfur compounds from petroleum streams. The sulfur compounds are first hydrogenated and then absorbed in a train of three catalyst beds: the "sandwich." In the first bed, zinc oxide absorbs hydrogen sulfide and reactive sulfur compounds; in the second, cobalt molybdate on alumina hydrogenates nonreactive thiophenes, forming hydrogen sulfide; in the third, zinc oxide absorbs the hydrogen sulfide from the second bed. Developed and offered by ICI, particularly for use in the *ICI Steam Naphtha Reforming process.

SANEX [Selective ActiNide EXtraction] A process for removing lanthanide elements from actinides in the *Purex process. Not yet fully developed. The name is used also for a range of toiletries.

Sanitech *See* Sorbtech.

Saniter A modification of the *Basic Open Hearth process for reducing the sulfur content of the steel product by adding relatively large quantities of limestone and calcium chloride. Invented by E.H. Saniter at the Wigan Coal & Iron Company in England in 1892 and used there and in Germany for about 20 years, until superseded by the use of calcium fluoride as a flux.

> British Patent 8,612 (1892).
> Barraclough, K.C., *Steelmaking 1850–1900*, The Institute of Metals, London, U.K., 1990, 271.

SAPHYR A process for controlling odors from sewage sludge. Developed by Anjou Recherche (VIVENDI), France, in 1999.

SAPIC [Société d'Applications de Procédés Industriels et Chimiques, Asnières, France] A process used in metal foundries for curing resin–sand mixtures used in making molds. The resin is usually an unsaturated polyester resin. In the SAPIC process, the resin is hardened by means of an organic peroxide or hydrogen peroxide, which is activated by sulfur dioxide gas when required.

saponification [Latin, *sapo*, soap] A process for making soap by the alkaline hydrolysis of animal or vegetable fats, using aqueous sodium or potassium hydroxide. Glycerol is a coproduct. The term is now more generally used for any alkaline ester hydrolysis:

$$R-COOR' + NaOH = R-COONa + R'OH$$

where R is a long-chain alkyl group and ROH is an alcohol or polyol such as glycerol. *See also* Kettle, Twitchell.

> Russell, C.A., in *Chemistry, Society and Environment*, Russell, C.A., Ed., Royal Society of Chemistry, London, U.K., 2000, 209.

Sapoxal A pulp-bleaching process used in the paper industry. The bleaching agent is oxygen, and the process is operated under alkaline conditions.

Sapozhnikov *See* carbonization.

Sapper An obsolete process for making phthalic anhydride by oxidizing *o*-xylene, using a mercury sulfate catalyst. Invented by E. Sapper in 1891 in the course of searching for a commercial route to indigo and used until the catalytic gas-phase oxidation of naphthalene was introduced in 1925.

> Franck, H.-G. and Stadelhofer, J.W., *Industrial Aromatic Chemistry*, Springer-Verlag, Berlin, Germany, 1988, 266.

SAR [Sulfuric Acid Recovery] A process for purifying and concentrating used sulfuric acid for reuse. The acid is heated with oxygen at 1040°C to convert the acid to sulfur dioxide. This is then oxidized over a vanadium-containing catalyst to sulfur trioxide, which is dissolved in fresh sulfuric acid to give 98% acid. Developed by Air Liquide and ICI. First demonstrated in 1991 at a methyl methacrylate plant in Taiwan.

Eur. Chem. News, 1991, **57**(1501), 34.
Chem. Br., 1992, **28**(3), 216.

Sarex (1) [**Sa**ccharide **ex**traction] A version of the *Sorbex process, for separating fructose from mixtures of fructose and glucose. The usual feed is corn syrup. The adsorbent is either a proprietary zeolite or an ion-exchange resin. Unlike all the other Sorbex processes, the solvent is water. The process depends on the tendency of calcium and magnesium ions to complex with fructose. The patents describe several methods for minimizing the dissolution of silica from the zeolite. The process is intended for use with a glucose isomerization unit so that the sole product from corn syrup is fructose. Invented by UOP in 1976; by 2003, five plants had been licensed.

British Patent 1,574,915.
U.S. Patent 4,248,737.
Ching, C.B. and Ruthven, D.M., *Zeolites*, 1988, **8**, 68.

SAREX (2) A process for recovering oil from wash liquors. Developed by Separation & Recovery Systems.

Hydrocarbon Process. Int. Ed., 1993, **72**(8), 98.

SAROX A process for utilizing the waste ammonium bisulfate from the manufacture of methyl methacrylate. High-temperature pyrolysis yields ammonia and sulfur trioxide. Developed by ICI and Air Liquide.

Wittcoff, H.A., Reuben, B.G., and Plotkin, J.S., *Industrial Organic Chemicals*, 2nd ed., Wiley-Interscience, Hoboken, NJ, 2004, 188.

SARP [**S**ulfuric **A**cid **R**ecovery **P**rocess] A method for recovering sulfuric acid that has been used for *alkylation, for reuse. The acid is reacted with propylene, yielding dipropyl sulfate, which is extracted from the acid tar with isobutane. It is not necessary to hydrolyze the sulfate to sulfuric acid because the sulfate itself is an active alkylation catalyst.

Oil Gas J., 1967, **65**(1), 48.

SAS [**S**asol **A**dvanced **S**ynthol] A process for converting synthesis gas to petroleum and light olefins. Developed and operated by Sasol in South Africa.

Oil Gas J., 1997, **95**(25), 16; 2000, **98**(2), 42.

SASOL [**S**uid-**A**frikaanse **S**intetiese **OL**ie] Not a process but a large coal gasification complex in South Africa, operated by the South African Oil and Gas Corporation. It first operated in 1955 but took several years to be fully commissioned. A Lurgi fixed-bed gasification unit is used for the primary process. Downstream processes include the following ones, described under their respective names: Arge, Fischer–Tropsch, Rectisol, Sulfolin, Synthol.

Report of the Committee on Coal Derivatives, HMSO, CMND 1120, 1960.
Mako, P.F. and Samuel, W.A., in *Handbook of Synfuels Technology*, Meyers, R.A., Ed., McGraw-Hill, New York, 1984, 2–7.
Chem. Eng. (N.Y.), 1995, **102**(12), 70.

Satco A process for recovering contaminated sulfuric acid. The acid is cracked at 1000°C, and the resulting sulfur dioxide is cooled, purified, dried, and reconverted to sulfuric acid. Developed by Rhône-Poulenc in the 1990s. Nine plants were operating in 1997.

Chem. Eng. News, 1997, **75**(27), 16.

Satin A *CVD process for depositing silicon nitride films on semiconductor devices at the relatively low temperature of 500°C. The identity of the gaseous reactant is proprietary. Developed by Aviza Technology and Air Liquide in 2004.

Chem. Eng. News, 2005, **83**(32), 20.

SAX [**S**olution **A**tomization and Crystalli**x**ation by Sonication] A process for controlling the crystallization and surface geometries of micron-sized particles. The suspension in a nonsolvent is treated with ultrasonic radiation in a *Prosonitron* flow reactor. Used for making drugs in forms that can be atomized and inhaled. Developed by the University of Bath and Prosonix.

O'Driscoll, C., *Chem. Ind. (London)*, 2007, (17), 18.

SBA [**S**ociété **B**elge de l'**A**zote] A process for making acetylene by the partial combustion of methane. It differs from similar processes in using liquid ammonia as a selective solvent for the product. Invented by F.F.A. Braconier and J.J.L.E. Riga at the Société Belge de l'Azote et des Produits Chimiques, Liège, and first operated at Marly, Belgium, in 1958.

U.S. Patent 3,019,271.
Miller, S.A., *Acetylene: Its Properties, Manufacture and Uses*, Vol. 1, Ernest Benn, London, U.K., 1965, 465.
Tedeschi, R.J., *Acetylene-based Chemicals from Coal and Other Natural Resources*, Marcel Dekker, New York, 1982, 22.

SBA-HT [**S**ociété **B**elge de l'**A**zote–**H**aldor **T**opsoe] A process for converting LPG to *syngas rich in hydrogen. Two *cracking processes are conducted in two zones of one reactor. In the first zone, the LPG is autothermally cracked with steam and oxygen. In the second, the products from the first are catalytically cracked. It was operated in France and Belgium in the 1960s.

SBA-Kellogg A pyrolytic process for making ethylene and acetylene from saturated hydrocarbon gases, similar to the *Hoechst HTP process but with the addition of steam, which increases the yield of C_2 gases.

Barry, M.J., Fox, J.M., Grover, S.S., Braconier, F., and Leroux, P., *Chem. Eng. Prog.*, 1960, **56**(1), 39.
Asinger, F., *Mono-olefins: Chemistry and Technology*, translated by B.J. Hazzard, Pergamon Press, Oxford, U.K., 1968, 180.

SBK [**S**inclair–**B**aker–**K**ellogg] A petroleum reforming process that uses a regenerable platinum catalyst.

SCA-Billerud A variation of the *Sulfite process for making paper from wood, in which the waste hydrogen sulfide is burned to sulfur dioxide, which is used to make sulfuric acid.

Rydholm, S.A., *Pulping Processes*, Interscience, New York, 1965, 809.

SCANfining [**S**elective **Ca**talytic **N**aphtha Hydro**fin**ing] A selective catalytic *hydrotreating process for reducing the sulfur content of naphtha. Developed by ExxonMobil and Albemarle. The catalyst, developed by ExxonMobil and Akzo Nobel, contains cobalt and molybdenum. The key feature is its prevention of the mercaptans reversion reaction—the formation of mercaptans from olefins and hydrogen sulfide. The process also minimizes olefin saturation and hydrogen loss. Operated at a number of refineries since 1990.

U.S. Patents 5,985,136; 6,013,598; 6,126,814.
Chem. Eng. (N.Y.), 2000, **107**(10), 17.
Hydrocarbon Process. Int. Ed., 2001, **80**(2), 39; **80**(10), 102-P; 2002, **81**(4), 90.
Sulphur, 12 Oct 2001, (276), 12.

SCANfining II An improved version of *SCANfining. In 2005, over 30 units were either announced or operating.

Kerby, M.C., Degnan, Jr., T.F., Marler, D.O., and Beck, J.S., *Catal. Today*, 2005, **104**, 59.

SCG *See* Shell Coal Gasification.

Schaffner Also called Schaffner–Helbig. A process for recovering sulfur from the residues from the *Leblanc process. Operated in Aussig, Bohemia, in the 1860s. *See also* Mond.

Chem. Ind. (London), 3 Jun 1985, 367.

Scheibler A process for extracting sucrose from molasses residues, based on the precipitation of strontium saccharate. Operated in Germany in the late nineteenth century.

Scheidemandel [German, meaning "separating almond"] A process for converting bone glue to a bead form. The hot, concentrated glue solution is dropped down a tower filled with an immiscible liquid such as a petroleum fraction or tetrachloroethylene.

Schenk–Wenzel *See* carbonization.

Schiller A predecessor of the *ICI Low-Pressure Methanol process, developed by IG Farben.

> Weissermel, K. and Arpe, H.-J., *Industrial Organic Chemistry*, 4th ed., translated by C.R. Hindley and S. Hawkins, Wiley-VCH, Weinheim, Germany, 2003, 18.

Schlempe [German, meaning "residue from fermentation processes"] An obsolete process for making sodium cyanide by calcining sugar beet residues. These residues contain betaine, $(CH_3)_3NCH_2COO$, which yields HCN when heated above 1000°C.

> *Kirk-Othmer's Encyclopedia of Chemical Technology*, 3rd ed., Vol. 7, John Wiley & Sons, New York, 1979, 324.

Schloesing–Rolland An early variant of the *ammonia soda process for making sodium carbonate. Operated near Paris in 1857 and then in Middlesbrough, England, for approximately 20 years until supplanted by the *Solvay process.

> Watts, J.I., *The First Fifty Years of Brunner, Mond & Co.*, Brunner, Mond, Winnington, U.K., 1923, 18.
> Hardie, D.W.F. and Pratt, J.D., *A History of the Modern British Chemical Industry*, Pergamon Press, Oxford, U.K., 1966, 85.

Schmidt Also called Meissner and Schmidt–Meissner. The first continuous process for nitrating glycerol to nitroglycerol. Invented by A. Schmidt in Austria in 1927. *See also* Biazzi.

Schmidt–Meissner *See* Schmidt.

Schoch A process for making acetylene by subjecting aliphatic hydrocarbons to a silent electric discharge. Developed by E.P. Schoch at the University of Texas.

> Daniels, L.S., *Pet. Refin.*, 1950, **29**(9), 221.

Scholler A process for making glucose from wood by acid hydrolysis of the cellulose component under pressure. Invented by H. Scholler in Germany in 1929. Operated in Germany in the 1920s and in Russia in the 1970s. *See also* Madison–Scholler, Scholler–Tornesch.

> French Patents 706,678; 777,824.
> Faith, W.L., *Ind. Eng. Chem.*, 1945, **37**, 9.
> Worthy, W., *Chem. Eng. News*, 1981, **59**(14), 35.
> Katzen, R. and Schell, D.J., in *Biorefineries: Industrial Processes and Products*, Kamm, B., Gruber, P.R., and Kamm, M., Eds., Wiley-VCH, Weinheim, Germany, 2005, **I**, 131.

Scholler–Tornesch A process for making single-cell protein from wood. The wood is *saccharified by heating with dilute sulfuric acid. The resulting sugars are then fermented to ethanol, using the common yeast *Saccharomyces cerevisiae*. The process was developed and used in Germany in the 1920s and 1930s. *See also* Heiskenskjold.

> Litchfield, J.H., *CHEMTECH*, 1978, **8**, 218.
> *Riegel's Handbook of Industrial Chemistry*, 9th ed., Kent, J.A., Ed., Van Nostrand Reinhold, New York, 1992, 255.

Schönherr Also called Schönherr–Hessberger. A variation of the Birkeland–Eyde arc process for making nitric acid from air, developed by O. Schönherr at BASF. Schönherr stabilized the arc by confining it in a spirally moving airstream. Superseded by the *Haber–Bosch process.

Finch, G.I., *Proc. Phys. Soc.*, 1926, **39**(1), 464.
Chilton, T.H., *Strong Water: Nitric Acid, Its Sources, Methods of Manufacture, and Uses*, MIT Press, Cambridge, MA, 1968, 114.
Leigh, G.J., *The World's Greatest Fix: A History of Nitrogen and Agriculture*, Oxford University Press, Oxford, U.K., 2004, 133, 153.

Schröder-Grillo Also known as Grillo-Schröder. An early version of the *Contact process for making sulfuric acid. The catalyst was magnesium sulfate impregnated with platinum. The process was invented in 1899 by A. Hecksher at the New Jersey Zinc Company and first used at its plant in Mineral Point, WI, in 1901; this was the first use of the Contact process in the United States. In the United Kingdom, it was first used in Widnes in 1917. *See also* Mannheim (2).

Miles, F.D., *The Manufacture of Sulfuric Acid (Contact Process)*, Gurney & Jackson, London, U.K., 1925, Chap. 8.

Schroeder An electrolytic process for recovering chlorine from waste hydrochloric acid.

Schulte A process for removing ammonia from the gases produced in the coking of coal. The ammonia is removed by scrubbing with water, then distilled from the water and incinerated. Invented by E.V. Schulte and commercialized by the Koppers Company.

Schulte, E.V., *Blast Furn. Coke Oven Raw Mater. Proc.*, 1958, **17**, 237 (*Chem. Abstr.*, **53**, 9944).

Schwarting Also called Uhde/Schwarting. An anaerobic fermentation process for treating aqueous wastes containing high concentrations of organic materials. Two fermenters are used, operated at different temperatures and acidities. In the first, insoluble materials are brought into solution, and most of the organic matter is converted to acids and alcohols. In the second, methane and carbon dioxide are produced. Developed in Germany by Geratebau Schwarting and the Fraunhofer Institute for Boundary Layer Research and engineered and offered by Uhde. Three plants were operating in Germany in 1994 for treating sewage sludge and wastewater.

Schwenzfeier–Pomelée A process for purifying beryllium and producing glassy beryllium fluoride. Beryllium hydroxide is dissolved in aqueous ammonium hydrogen fluoride, various metal impurities are removed by successive precipitations, and ammonium fluoroberyllate is crystallized under vacuum. When this is heated, ammonium fluoride vaporizes and molten beryllium fluoride remains.

Scientific Design A chemical engineering company, founded in New York in 1946 by R. Landau and H. Rehnberg. It developed many processes, of which the first and perhaps the best known was that for oxidizing ethylene to ethylene oxide, using a silver catalyst. Later, it merged with Halcon Corporation and become the Halcon SD Group. *See* Halcon, Oxirane.

Spitz, P.H., *Petrochemicals: The Rise of an Industry*, John Wiley & Sons, New York, 1988, 319.

Sclair A process for polymerizing ethylene. Depending on the comonomer used, the product can be linear low-density polyethylene (LLDPE) or high-density polyethylene (HDPE). Developed by DuPont in 1960 and widely licensed. Engineered by Uhde under the name *Sclairtech. Nova Chemicals (Alberta) acquired the technology in 1994.

Sclairtech An advanced version of the *Sclair ethylene polymerization process, using a *Ziegler–Natta catalyst and multiple reactors. Announced in 1996. By 2005, more than 12 plants had been built or were in design or under construction. Licensed by Nova Chemicals (International).

Eur. Chem. News, 1996, **66**(1744), 4.

Sconox A catalytic process for oxidizing oxides of nitrogen and carbon monoxide. The catalyst is in the form of a ceramic honeycomb coated with platinum and containing potassium carbonate. The platinum oxidizes the carbon monoxide to carbon dioxide, and the potassium carbonate absorbs the NO_x. Developed in 1995 by Sunlaw Energy Corporation, CA, and Advanced Catalytic Systems, TN. Now offered by Goal Line Environmental Technologies, Knoxville, TN.

> *Chem. Eng. (N.Y.)*, 1995, **102**(12), 25.

Score [**S**elective **c**racking **o**ptimum **re**covery] A process for making ethylene by *cracking ethane or naphtha. It combines Exxon Chemical's low-residence time technology with Brown & Root's cracking technology. Developed by Kellogg Brown & Root in 1999 and used at Dow's refinery in Freeport, TX, in 2003. BP planned to use it when it expanded its Chocolate Bayou ethylene plant for completion in 2005.

> *Chem. Eng. (N.Y.)*, 1999, **106**(4), 27.
> *Hydrocarbon Process. Int. Ed.*, 1999, **78**(4), 33; 2005, **84**(11), 29.
> *Eur. Chem. News*, 2002, **77**(2032), 29.

Scorodite It is the mineral ferric arsenate, $FeAsO_4 \cdot 2H_2O$. Because it is very sparingly soluble in water, the synthetic material has been used from 1997 in several laboratory processes for removing arsenic from hydrometallurgical solutions.

> Filippou, D. and Demopoulos, G.P., *J. Met.* 1997, 52.
> Swash, P.M. and Monhemius, A.J., in *Effluent Treatment in the Mining Industry*, Castro, S.H., Varga, F., and Sanchez, M.A., Eds., University of Conceptiòn, Chile, 1998, p. 119.

Scot [**S**hell **C**laus **o**ff-gas **t**reatment] A variation on the *Claus process for removing hydrogen sulfide from gas streams, in which residual sulfur dioxide in the off-gases is reduced with methane or hydrogen and the resulting hydrogen sulfide is returned to the start of the process. Other features are the catalytic conversion of organic sulfur compounds to hydrogen sulfide and the use of an alkanolamine for selectively absorbing the hydrogen sulfide. Developed by Shell International Petroleum Maatschappij. In 1996, 130 units were operating. *See also* Beavon.

> Naber, J.E., Wesselingh, J.A., and Groendaal, W., *Chem. Eng. Prog.*, 1973, **69**(12), 29.
> Swaim, C.D., Jr., in *Sulfur Removal and Recovery from Industrial Processes*, Pfeiffer, J.B., Ed., American Chemical Society, Washington, DC, 1975, 111.
> *Hydrocarbon Process. Int. Ed.*, 1996, **75**(4), 136.
> Kohl, H.L. and Nielsen, R.B., *Gas Purification*, 5th ed., Gulf Publishing Co., Houston, TX, 1997, 719.
> *Hydrocarbon Process. Int. Ed.*, 2000, **79**(4), 86.

SCR [**S**elective **C**atalytic **R**eduction] A general term for processes that destroy nitrogen oxides in gaseous effluents by reacting them with ammonia in the presence of a catalyst:

$$4NO + 4NH_3 + O_2 = 4N_2 + 6H_2O$$

$$6NO_2 + 8NH_3 = 7N_2 + 12H_2O$$

For clean gaseous effluents, such as those from nitric acid plants, the preferred catalyst is mordenite. For flue gases containing fly ash, the preferred catalyst is titania–vanadia. The process was developed in Japan in the mid-1970s by a consortium of Hitachi, Babcock–Hitachi, and the Mitsubishi Chemical Company and by the Sakai Chemical Industry Company. It is widely used in power stations in Japan and Germany. A modification of the process, using aqueous urea instead of ammonia gas, is now used to destroy nitrogen oxides in the exhaust gases from trucks; the trade name for the urea solution is AdBlue. *See also* DENOX, SINOX, SNCR.

> Matsuda, S., Takeuchi, M., Hishinuma, T., Nakajima, F., Narita, T., Watanabe, Y., and Imanari, M., *J. Air Pollut. Control Assoc.*, 1978, **28**, 350.
> Offen, G.R., Eskinazi, D., McElroy, M.W., and Maulbetsch, J.S., *J. Air Pollut. Control Assoc.*, 1987, **37**, 864.

Ritzert, G., *Tech. Mitt.*, 1987, **80**, 602.

Bosch, H. and Janssen, F., *Catal. Today*, 1988, **2**, 392.

Mukherjee, A.B., in *Environmental Oxidants*, Nriagu, J.O. and Simmons, M.S., Eds., John Wiley & Sons, New York, 1994, 585.

Hydrocarbon Process. Int. Ed., 1994, **73**(8), 67.

Kohl, A.L. and Nielsen, R.B., *Gas Purification*, 5th ed., Gulf Publishing Co., Houston, TX, 1997, 904.

Radojevic, M., *Chem. Br.*, 1998, **34**(3), 30.

Kirk-Othmer's Encyclopedia of Chemical Technology, 5th ed., John Wiley & Sons, New York, 2001, **10**, 97.

SCWO [**S**uper**C**ritical **W**ater **O**xidation] A generic name for processes that destroy organic wastes in water by oxidation under supercritical conditions. The first such process was *MODAR, invented in 1980. Since then, several other companies have introduced competing processes. The first commercial plant for treating wastewater by SCWO was built by Eco Waste Technologies in 1994 in Austin, TX. The first commercial SCWO plant for treating sewage sludge was commissioned in Kobe, Japan, in 2000. There has been a pilot unit in Karlskoga, Sweden, since 1998. *See* AquaCritox.

Oil Gas J., 1994, **92**(44), 44.

Luck, F., *Catal. Today*, 1996, **27**(1–2), 195.

Chem. Eng. (N.Y.), 1996, **103**(3), 21.

Ding, Z.Y., Frisch, M.A., Li, L., and Gloyna, E.F., *Ind. Eng. Chem. Res.*, 1996, **35**(10), 3257.

Chem. Eng. (Rugby, Engl.), 1996, (604), 9.

Anitescu, G. and Tavlarides, L.L., *Ind. Eng. Chem. Res.*, 2000, **39**(3), 583.

Chem. Eng. (N.Y.), 2001, **108**(7), 15.

Water & Waste Treatment, 2002, **45**(7), 26.

Dinjus, E. and Krupp, A., in *High Pressure Chemistry*, van Eldik, R. and Klärner, F.G., Eds., 2002, Wiley-VCH, Weinheim, Germany, 422–445.

SDA [**S**pray **D**ryer **A**bsorber] A *flue-gas desulfurization process in which an aqueous suspension of lime is injected into a spray dryer. Basically similar to *DRYPAC. Developed by Niro Atomizer, Denmark. In 1986, it was in use in 16 plants in Austria, Denmark, Germany, Italy, Sweden, China, and United States.

SDR *See* DR.

SDS (1) [**S**ulzer **D**ainippon **S**umitomo] A continuous process for polymerizing styrene. The reactants are mixed in a static mixer, which gives a very uniform time and temperature history for the polymer. Developed jointly by Sulzer, Dainippon Ink & Chemicals, and Sumitomo Heavy Industries. Offered for license in 1990.

Eur. Chem. News, 23 Jul 1990, 20.

SDS (2) [**S**hell **D**enox **S**ystem] An alternative name for *DENOX.

SDW [**S**olvent **D**e**W**axing] A general term for processes that remove linear paraffinic hydrocarbons from petroleum fractions by solvent extraction.

Seaboard A process for removing hydrogen sulfide from coke-oven and oil refinery gases by scrubbing with aqueous sodium carbonate solution. The solution is regenerated by blowing air through it. In its original version, the hydrogen sulfide was simply vented to the atmosphere. In later developments, the air containing the hydrogen sulfide was used as the combustion air for boilers or was passed into the coke oven. Developed by Koppers Company in 1920; two plants were operating in the United States in 1992, but the process is now obsolete.

British Patent 391,833.

Sperr, F.W., Jr., *Proc. Am. Gas. Assoc.*, 1921, Sept. (*Chem. Abstr.*, **16**, 482).

Kohl, H.L. and Nielsen, R.B., *Gas Purification*, 5th ed., Gulf Publishing Co., Houston, TX, 1997, 381.

Seacoke A process for making tar and coke by carbonizing mixtures of coal and petroleum residuum. The tar would be used in an oil refinery, and the coke would be used for generating electricity. The process was sponsored by the U.S. Office of Coal Research from 1964 to 1969; the work was carried out by FMC Corporation,

Atlantic Richfield Company, and Blaw-Knox Company. Results from the pilot plant were encouraging, but the project was abandoned because the benefits were judged insufficient to justify the complexity.

> Aristoff, E., Rieve, R.W., and Shalit, H., in *Chemistry of Coal Utilization*, 2nd Suppl. Vol., Elliott, M.A., Ed., John Wiley & Sons, New York, 1981, 984.

Séailles–Dyckerhoff A process for extracting aluminum from clays and other aluminous minerals rich in silica. The ore is calcined with limestone in a rotary kiln. The product is leached with aqueous sodium carbonate, yielding sodium aluminate solution, from which alumina is precipitated by carbon dioxide. Invented by J.C. Séailles and W.R.G. Dyckerhoff in 1938 and piloted in Tennessee in 1942, South Carolina in 1945, and Germany during the World War II. *See also* Pedersen.

> U.S. Patent 2,248,826.
> O'Connor, D.J., *Alumina Extraction from Non-bauxitic Materials*, Aluminium-Verlag, Düsseldorf, Germany, 1988, 233.

Sealosafe A family of processes for encapsulating inorganic and organic wastes in a cementitious material suitable for landfill. The product, known as Stablex, is made from a cement and an aluminosilicate and may incorporate pulverized fly ash. Developed by C. Chappell in the United Kingdom in the 1970s and now operated in a number of countries. Offered by the Stablex Corporation, Radnor, PA. The environmental acceptability of the product has since been criticized.

> Schofield, J.T., in *Toxic and Hazardous Waste Disposal*, Vol. 1, Pojasek, R.J., Ed., Ann Arbor Science, Ann Arbor, MI, 1979, Chap. 15.
> *Environmental Data Services Report*, 1989, (173), 8; 1995, (240), 15.

Sec-Feed A process for converting used lubricating oils into feeds for *catalytic cracking. Essentially, the process removes water, light hydrocarbons, phosphorus and silicon compounds, and heavy metals. Developed by Chemical Engineering Partners in 1997 but not known to have been commercialized. Chemical Engineering Partners is the engineering associate of Evergreen Oil, Newark, CA, which has operated a used lubricating oil recycling plant there since 1986 and licensed similar facilities in Indonesia and Australia.

> *Oil Gas J.*, 1997, **95**(49), 61.

Sedifloc A water-treatment process based on flocculation. Offered by Lurgi.

Sedimat A water-treatment process based on sedimentation. Offered by Lurgi.

SEGAS [SouthEastern GAS] Also written as Seagas. An oil and gasification process in which oil is sprayed into a stream of superheated steam and cracked over a calcia–magnesia catalyst. Developed by the Southeastern Gas Board, United Kingdom. SEGAS is also the abbreviated name of the Spanish Egyptian Gas Corp.

> Stanier, H. and McKean, J.B., *Institution of Gas Engineers*, Publn. No. 457, 1954 (*Chem. Abstr.*, **49**, 3508).
> Ward, E.R., *Institution of Gas Engineers*, Publn. No. 515, 1957 (*Chem. Abstr.*, **52**, 3308).
> Claxton, G., *Benzoles: Production and Uses*, National Benzole & Allied Products Association, London, U.K., 1961, 94.
> British Petroleum Co., *Gas Making and Natural Gas*, British Petroleum Co., London, U.K., 1972, 83.

Selectoforming A process for increasing the octane rating of a petroleum fraction by selectively *cracking the *n*-pentane and *n*-hexane in it. The catalyst is a metal-loaded synthetic zeolite. Developed by Mobil Corporation and first commercialized in the mid-1960s.

> Chen, N.Y. and Degnan, T.F., *Chem. Eng. Prog.*, 1988, **84**(2), 32.

Selectopol A process for converting isobutene into "polymer gasoline," that is, a mixture of branched-chain C_6–C_{12} hydrocarbons, using an acid catalyst. Offered for license by the Institut Français du Pétrole.

> *Hydrocarbon Process. Int. Ed.*, 1980, **59**(9), 219.

SELECTOR *See* OXO.

Selectox Also called BSR/Selectox. A process for converting hydrogen sulfide in refinery gases to liquid elemental sulfur, without the need for a reaction furnace. The gases are passed over a fixed bed of a proprietary catalyst (Selectox 33) at 160°C–370°C. Claimed to be better than the *Claus process in several respects. Often used in conjunction with the *Beavon process. Developed by the Union Oil Company of California and the Ralph M. Parsons Company and first operated in 1978. Twenty-one units had been installed by 2000.

Beavon, D.K., Hass, R.H., and Muke, B., *Oil Gas J.*, 1979, **77**(11), 76.
Hass, R.H., Ingalls, M.N., Trinker, T.A., Goar, B.G., and Purgason, R.S., *Hydrocarbon Process. Int. Ed.*, 1981, **60**(5), 104.
Wieckowska, J., *Catal. Today*, 1995, **24**(4), 444.
Hydrocarbon Process. Int. Ed., 1996, **75**(4), 136; 2000, **79**(4), 87.
Kohl, H.L. and Nielsen, R.B., *Gas Purification*, 5th ed., Gulf Publishing Co., Houston, TX, 1997, 711.
Jones, S.G. and Bertram, R.V., *Oil Gas J.*, 2001, **99**(35), 44, 48, 50, 52.

SELEX-Asp [**SEL**ective **EX**traction of **Asp**haltene] A process for removing asphaltenes from vacuum residual oils using supercritical solvent extraction with *n*-pentane. Developed by ISTP Canada and the China University of Petroleum.

Oil & Gas J., 2010, **108**(12), 52.
U.S. Patent 7,597,794.

Selexol Also called Selexol solvent. A process for removing acid gases from hydrocarbon gas streams by selective absorption in polyethylene glycol dimethyl ether (DMPEG). The ether absorbs carbon dioxide, hydrogen sulfide, carbonyl sulfide, and mercaptans. Absorption takes place in a countercurrent extraction column under pressure. The solvent is regenerated by "flashing" (evaporating) or by "stripping" (passing an inert gas through it). Selexol has been used for removing carbon dioxide from *syngas, natural gas, and coal gas. Developed by Allied Chemical Corporation in the 1960s, further developed by Union Carbide Corporation, and now offered by Dow Chemical. Over 55 units had been installed in 2000. *See also* SOLINOX.

Sweny, J.W. and Valentine, J.P., *Chem. Eng. (N.Y.)*, 1970, **77**, 54.
Valentine, J.P., *Oil Gas J.*, 1974, **72**(46), 60.
Judd, D.K., *Hydrocarbon Process. Int. Ed.*, 1978, **57**(4), 122; 1996, **75**(4), 137; 2000, **79**(4), 86.
Kohl, H.L. and Nielsen, R.B., *Gas Purification*, 5th ed., Gulf Publishing Co., Houston, TX, 1997, 1202.

Selexsorb A five-stage process for purifying ethylene before converting it to polyethylene. Developed by the Industrial Chemicals Division of the Aluminum Company of America. More than 80 units had been installed by 2000. The name is now used for a family of selective adsorbents based on alumina produced by Alcoa. The range includes Selexsorb CD, CDO, CDX, COS, SPCl, HCl.

Hydrocarbon Process. Int. Ed., 1996, **75**(4), 137; 2000, **79**(4), 87.

SELOP A range of processes for selectively hydrogenating olefins. Different catalysts, containing palladium on alumina, are used for different feedstock compositions. Developed by BASF and used in its Antwerp, Belgium, plant since 1994.

Büchele, W., Roos, H., Wanjek, H., and Müller, H.J., *Catal. Today*, 1996, **30**, 33.
Chem. Week Supplement, 12 Sep 2001, S8, S9.

SELOX (1) A generic name for selective oxidation processes for removing CO from hydrogen for use in fuel cells.

Ribeiro, N.F.P., Mendes, F.M.T., Perez, C.A.C., Souza, M.M.V.M., and Schmal, M., *Appl. Catal. A: Gen.*, 2008, **347**(1), 62.

Selox (2) [**Sel**ective **ox**idation] A process for selectively oxidizing methane to *syngas using a proprietary heterogeneous catalyst at temperatures up to 1000°C. Developed on a laboratory scale by TRW, CA, partly financed by the U.S. Department of Energy in 1983.

Chem. Eng. News, 1984, **62**(2), 5.
Chem. Eng. (N.Y.), 1984, **91**(13), 157.

Semet-Solvay The Semet-Solvay Engineering Corporation, Syracuse, NY, was best known for its coke-oven technology, developed from the end of the nineteenth century. The eponymous process was a cyclic, noncatalytic process for making fuel gas from oil.

British Petroleum, *Gasmaking* (revised ed.), British Petroleum, London, U.K., 1965, 75.

Sendzimir *See* metal surface treatment.

SEPACLAUS [**SEPA**ration **CLAUS**] An integrated process for removing hydrogen sulfide from coke-oven gases and converting it to elemental sulfur by the *Claus process. It also catalytically decomposes the ammonia present. Offered by Krupp Koppers.

Sepaflot A process for removing solids and oil from wastewater by a combination of *Activated Sludge treatment and dissolved air flotation. Offered by Lurgi.

Separex A membrane process for separating methane from landfill gas.

Oil Gas J., 1998, **96**(43), 45.

Sepasolv MPE [**M**ethyl iso**P**ropyl **E**ster] A variation on the *Selexol process, using the methyl isopropyl ethers of polyethylene glycol as the solvent. Developed by BASF. Four commercial plants were operating in 1985, removing hydrogen sulfide from natural gas.

Wölfer, W., *Hydrocarbon Process. Int. Ed.*, 1982, **61**(11), 193.
Kohl, H.L. and Nielsen, R.B., *Gas Purification*, 5th ed., Gulf Publishing Co., Houston, TX, 1997, 1210.

Sequential Continuous Catalytic Reforming *See* CCR.

SER [**S**orption-**E**nhanced **R**eaction] A modification of the *steam-reforming process for making hydrogen, in which a selective adsorbent removes the carbon dioxide, thereby reducing the temperature required from 800°C–1000°C to 500°C–800°C. Developed by Air Products and Chemicals Ltd. in 1998.

Chem. Eng. (N.Y.), 1998, **105**(5), 116.

Serpek A nitrogen fixation process using aluminum nitride. A mixture of bauxite and coke is heated in nitrogen at 1800°C to produce aluminum nitride; this yields ammonia on hydrolysis by boiling with aqueous potassium aluminate:

$$Al_2O_3 + 3C + N_2 = 2AlN + 3CO$$

$$2AlN + 6H_2O = 2NH_3 + 2Al(OH)_3$$

Invented by O. Serpek in Germany in 1909.

British Patents 15,996; 15,997 (1909).
Chilton, T.H., *Strong Water: Nitric Acid, Its Sources, Methods of Manufacture, and Uses*, MIT Press, Cambridge, MA, 1968, 62.

SFE [**S**upercritical **F**luid **E**xtraction] *See* ROSE (1).

SFGD [**S**hell **F**lue-**G**as **D**esulfurization] A *flue-gas desulfurization process using a fixed bed of copper on alumina. The sulfur dioxide is desorbed with a reducing gas.

Dautzenberg, F.M., Nader, J.E., and van Ginneken, A.J.J., *Chem. Eng. Prog.*, 1991, **67**(8), 86.
Speight, J.G., *Gas Processing*, Butterworth Heinemann, Oxford, U.K., 1993, 316.

SGP [Shell Gasification Process] A process for converting fuel oils into *syngas; the sulfur is isolated as the element. Developed by Shell, Amsterdam, from the 1950s and now widely used. Licensed by Lurgi Öl Gas Chemie.

SHAC Not a process but a range of catalysts developed by Dow Chemical Co. for making polypropylene.

shale oil extraction Those named processes described in this dictionary are Galoter, HY-TORT, Kiviter, Paraho, Petrosix, PFH, Pumpherston, TOSCO II, TOSCOAL.

Shanks An obsolete process for extracting sodium nitrate from caliche, a Chilean mineral. The ore was leached with sodium chloride solution at 70°C, and the sodium chloride and nitrate were separated by fractional crystallization. *See also* Guggenheim.

SHARON [Single Reactor system for Ammonia Removal Over Nitrate] A process for removing ammonia from wastewater that is rich in nitrogen. Developed in the Netherlands by the Technical University in Delft, Zuiveringsschap Hollandse Eilanden en Waarden, and Grontmij Consulting Engineers. Marketed in the United States from 2002 by Lotepro Environmental Systems and Services.

Water Wastewater Int., 2002, **16**(4), 9.

Sharple A process for purifying paraffin wax by crystallization from a petroleum fraction at low temperature.

Sharples A continuous soapmaking process using centrifuges, invented in 1939 by A.T. Scott at the Sharples Corporation, Philadelphia.

U.S. Patents 2,300,749; 2,300,751; 2,336,893.

Shawinigan A process for making acetic acid by oxidizing acetaldehyde by atmospheric oxygen in the presence of manganese acetate. Operated by the Shawinigan Chemical Company, at Shawinigan Falls, Québec, using acetaldehyde made from acetylene.

Cadenhead, A.F.G., *Chem. Metall. Eng.*, 1933, **40**(4), 184.
Dumas, T. and Bulani, W., *Oxidation of Petrochemicals: Chemistry and Technology*, Applied Science Publishers, London, U.K., 1974, 26.

Shell–Adip *See* Adip.

Shell Coal Gasification Also called **SCG** and **SCGP**. A coal gasification process in which powdered coal is reacted with oxygen and steam at high temperature and pressure. The high temperature causes most of the ash to melt and flow down the reactor wall into a water-filled compartment. The product gas is mainly a mixture of carbon monoxide and hydrogen, suitable for the synthesis of ammonia, methanol, and hydrocarbons. Piloted at the Shell laboratory, Amsterdam, and at the Deutsche Shell refinery at Hamburg, from 1978. The first commercial plant was built in 1993 at Buggenum, the Netherlands. At least 13 plants have since been built in China. *See also* Shell–Koppers.

Vogt, E.V., Weller, P.J., and Vanderburgt, M.J., in *Handbook of Synfuels Technology*, Meyers, R.A., Ed., McGraw-Hill, New York, 1984, 3–27.
Cornils, B., in *Chemicals from Coal: New Processes*, Payne, K.R., Ed., John Wiley & Sons, Chichester, U.K., 1987, 19.
Rezaiyan, J. and Cheremisinoff, N.P., *Gasification Technologies*, Taylor & Francis, Boca Raton, FL, 2005, 43.
Chem. Eng. (N.Y.), 2006, **113**(2), 28.

Shell Deacon An improved version of the *Deacon process for oxidizing hydrogen chloride to chlorine, using a catalyst containing the mixed chlorides of copper, potassium, and lanthanides. Formerly operated in the Netherlands and still operated in India.

Tozuka, Y., in *Science and Technology in Catalysis*, Izumi, Y., Aral, Y., and Iwamoto, M., Eds., Elsevier, Amsterdam, the Netherlands, 1994, 45.

Weissermel, K. and Arpe, H.-J., *Industrial Organic Chemistry*, 4th ed., translated by C.R. Hindley and S. Hawkins, Wiley-VCH, Weinheim, Germany, 2003, 221.

Wittcoff, H.A., Reuben, B.G., and Plotkin, J.S., *Industrial Organic Chemicals*, 2nd ed., Wiley-Interscience, Hoboken, NJ, 2004, 119.

Shell Glycerol The Shell Development Company has developed three routes for making glycerol from propylene. The first begins by chlorinating propylene to make allyl chloride, which is converted to glycerol via epichlorhydrin. The second and third both involve acrolein as an intermediate, the second reacting it with 2-propanol and the third with 2-butanol. The second of these processes became known as the Shell Glycerol process. The successive reactions are as follows:

$$\text{propylene} + \text{oxygen} \rightarrow \text{acrolein} + \text{water}$$

$$\text{acrolein} + \text{2-propanol} \rightarrow \text{allyl alcohol} + \text{acetone}$$

$$\text{allyl alcohol} + \text{hydrogen peroxide} \rightarrow \text{glycerol}$$

The third route was used at the Shell plant at Norco, LA, until it was closed in 1980.

Shell–Koppers A coal gasification process, using steam and air or oxygen. Operated at the Shell oil refinery in Hamburg. It evolved into the *Shell Coal Gasification process.

Eur. Chem. News, Petrochem. Suppl., 14 Dec, 1981.

Shell Paques A version of the *THIOPAQ process for removing H_2S from gas streams by use of natural bacteria (*Thiobacilli*). It was awarded a prize by IChemE in 2002.

Hydrocarbon Process. Int. Ed., 2000, **79**(4), 82.

Shellperm A process for rendering sandy soils impermeable to water by pumping in a bitumen emulsion with a coagulant that is effective after a predetermined period. Used in the construction of dams.

The Petroleum Handbook, 3rd ed., Shell Petroleum Co., London, U.K., 1948, 586.

Sheppard *See* metal surface treatment.

Sherardizing [After the inventor, **Sherard** Cowper-Cowles] A process for coating iron articles with zinc. The articles are placed in a sealed drum with zinc dust and sand. The drum is rotated and maintained at a temperature below the melting point of zinc. The mechanism is not understood. In 1990, the world consumption of zinc for this process was several thousand tons. *See* metal surface treatment.

British Standard BS 4921.

Sherpol A process for making polypropylene, developed and licensed by Himont.

Sherritt–Cominco A process for extracting copper from chalcopyrite, $CuFeS_2$. The ore is reduced with hydrogen, the iron leached out with sulfuric acid, the residual Cu_5FeS_4 dissolved in concentrated sulfuric acid, and the copper isolated by electrowinning or hydrogen reduction. Pilot testing was complete in 1976.

Maschmeyer, D.E.G., Kawulka, P., Milner, E.F.G., and Swinkels, G.M., *J. Met.*, 1978, **27**(7), 27.

Sherritt-Gordon The Canadian company, Sherritt-Gordon Mines, has developed a number of hydrometallurgical leaching processes known by its company name. The essential feature of these processes is based on the observation of F.A. Forward that sulfide ores will dissolve in aqueous ammonia in the presence of oxygen. One such process is for leaching nickel from sulfide ores, using aqueous ammonia and oxygen under pressure. The hexammine nickel (II) ion is formed:

$$NiS + 2O_2 + 6NH_3 = Ni(NH_3)_6^{2+} + SO_4^{2-}$$

Another Sherritt-Gordon process is for leaching zinc from sulfide ores, again using oxygen under pressure.

Morgan, S.W.K., *Zinc and Its Alloys and Compounds*, Ellis Horwood, Chichester, U.K., 1985, 124.
Gupta, C.K. and Mukherjee, T.K., *Hydrometallurgy in Extraction Processes*, Vol. 1, CRC Press, Boca Raton, FL, 1990, 26, 87.

shift *See* Reforming.

Shimer *See* metal surface treatment.

SHOP [**Sh**ell **H**igher **O**lefins **P**rocess] A process for producing α-olefins by oligomeiizing ethylene, using a proprietary rhodium–phosphine catalyst. The α-olefins can then be isomerized to internal olefins as required. Invented by W. Keim in the Institut für Technische Chemie und Petrochemie, Aachen, West Germany, in the 1970s. The first plant was built in Geismar, LA, in 1979; the second in Stanlow, Cheshire, UK, in 1982. Licensed worldwide by a consortium of Union Carbide, Davy-McKee, and Johnson Matthey.

Freitas, E.R. and Gum, C.R., *Chem. Eng. Prog.*, 1979, **75**(1), 73.
Sherwood, M., *Chem. Ind. (London)*, 1982, (24), 994.
Keim, W., *Angew. Chem. Int. Ed.*, 1990, **29**, 235.
Keim, W., *Green Chem.*, 2003, **5**, 105.
Weissermel, K. and Arpe, H.-J., *Industrial Organic Chemistry*, 4th ed., translated by C.R. Hindley and S. Hawkins, Wiley-VCH, Weinheim, Germany, 2003, 87.
Imamoglu, Y. and Bencze, L., Eds., *Novel Metathesis Chemistry*, Kluwer Academic Publishers, Dordrecht, the Netherlands, 2003.
Vogt, D., in *Aqueous-phase Organometallic Catalysis*, 2nd ed., Cornils, B. and Herrman, W.A., Eds., Wiley-VCH, Weinheim, Germany, 2004, Chap. 7.1.
Wittcoff, H.A., Reuben, B.G., and Plotkin, J.S., *Industrial Organic Chemicals*, 2nd ed., Wiley-Interscience, Hoboken, NJ, 2004, 116.
Steinborn, D., *Fundamentals of Organometallic Catalysis*, 2012, Wiley-VCH, Weinheim, Germany, 156–158.

Shoppler A process for extracting tungsten from scheelite. The ore is fused with sodium carbonate, forming sodium tungstate. This is leached with water, the solution acidified with hydrochloric acid, and hydrated tungsten oxide precipitated by boiling. The metal is produced by reducing the oxide with carbon at a high temperature.

SHP [**S**elective **H**ydrogenation **P**rocess] A process for purifying 1-butene by selective hydrogenation of C_4 streams in petroleum refineries. A heterogeneous palladium catalyst is used. Developed in Hüls and used in 1989 in Germany, the United States, and Japan. The licensing rights were acquired by UOP in 1991.

Derrien, M., Bronner, C., Cosyns, J., and Leger, G., *Hydrocarbon Process. Int. Ed.*, 1979, **58**(5), 175.
McCulloch, B., Luebke, C., and Meister, J., in *Handbook of Petroleum Refining Processes*, 3rd ed., Meyers, R.A., Ed., McGraw-Hill, New York, 2003, 8.26.

SHS [**S**elf-propagating **H**igh-temperature **S**ynthesis] A process for manufacturing ceramics and intermetallic compounds by exothermic reactions in which the heat of reaction is large enough to sustain the propagation of a combustion wave through the reactants. The reactants can be mixed powders or a powder with a gas. It has been used to synthesize TiB_2, ZrB_2, TiC, and AlN from the elements and to make a number of refractory composites of oxides with metals. Invented in 1967 at the Institute of Chemical Physics, Moscow, by I.P. Borovinskaya, A.G. Merzhanov, and V.M. Shkiro. The Institute of Scientific and Industrial Research at Osaka University has studied the process conducted in various gases under pressure.

U.S. Patent 3,726,643.
British Patent 1,321,084.
Merzhanov, A.G. and Borovinskaya, I.P., *Doklady Akad. Nauk SSSR (Engl. Transl.)*, 1972, **204**(2), 429.
Crider, J.F., *Ceram. Eng. Sci. Proc.*, 1982, **3**, 519.
Merzhanov, A.G., in *Combustion and Plasma Synthesis of High-temperature Materials*, Munir, Z.A. and Holt, J.B.,
    Eds., VCH Publishers, Weinheim, Germany, 1990, 1.
International Journal of Self-Propagating High-temperature Synthesis, Allerton Press, New York, 1992.
Avakayan, P.B., Nereseyan, M.D., and Merzhanov, A.G., *Am. Ceram. Soc. Bull.*, 1996, **75**(2), 50.
Parkin, I.P., *Chem. Ind. (London)*, 1997, (18), 725.
He, C. and Stangle, G.C., *J. Mater. Res.*, 1998, **13**(1), 135.

SHU [**S**aarberg–**H**olter–L**U**rgi] A *flue-gas desulfurization process using wet limestone as the scrubbing medium, assisted by the addition of dilute formic acid. Developed by the companies named and used in 11 power stations in Germany and Turkey in 1987.

Shukhov The first thermal cracking process for petroleum. Invented in Russia in 1891 by V. Shukhov and S. Gavrulov and still in use in Russia in the 1930s. Largely replaced by *catalytic cracking, but a version survives as the *Eureka process.

Russian Patent 12926, 1891.
Wikipedia, *Shukhov cracking process; Vladimir Shukhov*.

SIAPE [**S**ociété **I**ndustrielle d'**A**cide **P**hosphorique et d'**E**ngrais] A *wet process for making phosphoric acid, based on a pair of coupled, stirred reactors. Operated by the eponymous company in Tunisia, formed in 1952, now the Groupe Chimique Tunisien.

French Patent 1,592,005.
Becker, P., *Phosphates and Phosphoric Acid*, 2nd ed., Marcel Dekker, New York, 1989, 349.

Siemens A method for making ultrapure silicon for semiconductors by thermally decomposing trichlorosilane. Invented in 1954 by F. Bischof at Siemens-Halska. It was the major process used worldwide in 1993.

German Patent 1,102,117.
Yaws, C.L. and Hopper, J.R., in *Chemical Processing Handbook*, Marcel Dekker, New York, 1993, 939.

Siemens–Martin A predecessor of the *Siemens Open Hearth process. *See* Open Hearth.

Siemens Open Hearth *See* Open Hearth.

Sieurin An early process for making sponge iron. Layers of iron ore concentrate, lime, and coal were heated in covered crucibles.

SIIL A direct reduction ironmaking process, using coal as the reductant. In 1997, five plants were operating in India and one in Peru. *See* DR.

Silamit P3 A cyclic *Catalytic Reforming process for making town gas from oil, similar to the*UGI process. Developed by Gaz de France and built by Silamit Indugas, Düsseldorf.

Gas Making and Natural Gas, British Petroleum Co., London, U.K., 1972, 94.

Silex [**S**eparation of **i**sotopes by **l**aser **ex**citation] A process for separating isotopes by exposing vapors to laser light, which ionizes one isotope but not others. The treated vapors are passed through an electric field, which separates the ionized species from the others. Developed in Australia from 1990 by M. Goldsworthy and H. Struve, originally for separating uranium isotopes using uranium hexafluoride vapor but expected to find application with other elements, including zirconium and silicon. Now operated by Silex Systems, which first had an agreement with the United States Enrichment Corporation and later with General Electric.

Boureston, J. and Ferguson, C.D., *Bull. Atomic Sci.*, 2005, **61**(2), 14.

SilvaGas A thermal process for gasifying biomass. Wood chips and hot sand are mixed in a circulating fluidized bed at 1000°C under atmospheric pressure. Steam is injected to induce mixing. The products are H_2, CO, CH_4, CO_2, C_2H_4, and char. The char and sand are separated, and the char is burned in another reactor. Intended for operation with a gas turbine. Developed by Future Energy Resources Corporation and demonstrated in Burlington, VT, in 2000.

Chem. Eng. (N.Y.), 2000, **107**(10), 23.

Silver II An electrolytic oxidation process for destroying traces of organic substances in water. The oxidizing agent is the silver ion in a nitric acid environment. Developed by AEA Technology, Oxford, United Kingdom, and intended for destroying war gases. A disadvantage of the process is that organochlorine compounds, although destroyed, produce some silver chloride, which must be recovered. A ruthenium-based electrochemical system, also developed by AEA Technology, does not suffer from this; it is particularly suitable for treating dilute, salt-containing organic wastes in water.

Davidson, L., Quinn, Y., and Steele, D.F., *Platinum Metals Rev.*, 1998, **42**(3), 90.
Chem. Eng. (N.Y.), 1996, **103**(5), 41.
Eur. Chem. News, 1996, **66**(1726), 23.
Eur. Chem. News, CHEMSCOPE, 1997, Jul, 18.
Chem. Eng. (Rugby, Engl.), 1998, (659), 10.
Environment Bus. Mag., 2001, (71), 12.

Silverguard An ion-exchange process for removing traces of iodide from the acetic acid made by the *AO Plus process. Developed by Celanese in the 1980s.

Plotkin, J., *Eur. Chem. News*, 2003, **78**(2047), 22.

Simons An electrochemical method for fluorinating organic compounds. First developed by J.H. Simons at Pennsylvania State College in 1941, but not announced until 1948 for reasons of national security. A direct current is passed through a solution of an organic compound in anhydrous hydrofluoric acid, hydrogen is evolved at the cathode, and the organic material is fluorinated at the anode.

U.S. Patents 2,519,983; 2,594,272; 2,616,927.
Simons, J.H., *J. Electrochem. Soc.*, 1949, **95**, 47.
Simons, J.H., in *Fluorine Chemistry*, Vol. 1, Simons, J.H., Ed., Academic Press, New York, 1950, 414.
Burdon, J. and Tatlow, J.C., *Adv. Fluorine Chem.*, 1960, **1**, 129.
Pletcher, D. and Walsh, F.C., *Industrial Electrochemistry*, 2nd ed., Chapman & Hall, London, U.K., 1960, 319.

SIMOX [Separation by IMplanted OXygen] In the manufacture of semiconductor devices, a thin layer of insulating silicon dioxide is formed beneath the surface of a monocrystalline silicon substrate to form a buried dielectric layer. The layer is formed by implanting oxygen ions at 500°C–600°C and then annealing at >1000°C.

U.S. Patent 6,423,975.

Simplex A process for reducing the carbon content of ferrochrome, an alloy of iron and chromium. Some of the alloy is oxidized by heating in air, and the oxidized alloy is mixed in appropriate proportions with the remainder; on heating the mixture in a vacuum furnace, the carbon volatilizes as carbon monoxide.

Sinclair–Baker A naphtha reforming process, catalyzed by platinum on alumina. *See also* Reforming.

SINI [SItola and NIzovkin] Also known as the double steeping process and Double Mercerization. A variation of the *Viscose process for making regenerated cellulose fibers, in which the treatment with sodium hydroxide is done in two stages, at different concentrations. Invented by G. Sitola and V. Nizovkin around 1976.

Sihtola, H., *Paper Pulp*, 1976, **58**(9), 534 (*Chem. Abstr.*, **86**, 56986).

SINOx [**SI**emens **NO**x system] A process for removing nitrogen oxides and dioxins from the exhausts of stationary diesel engines and truck engines, based on the *SCR process. The catalyst is based on titania and is in the form of a honeycomb. The reducing agent is ammonia, generated from an aqueous solution of urea. Acquired by Johnson Matthey in 2007. *See* SCR.

> Hums, E., Joisten, M., Müller, R., Sigling, R., and Spielmann, H., *Catal. Today*, 1996, **27**(1–2), 29.

Sinterna A process for stabilizing the calcium sulfate/sulfite waste produced by *FGD so that it may be used for landfill. The waste is mixed with ash from a coal-fired power station, pelletized, and sintered. Developed in the 1970s by Battelle Columbus Laboratories, OH, under contract with Industrial Resources. *See also* Fersona.

> U.S. Patent 3,962,080.
> Dulin, J.M., in *Toxic and Hazardous Waste Disposal*, Vol. 1, Pojasek, R.J., Ed., Ann Arbor Science, Ann Arbor, MI, 1979, Chap. 18.

Siroc [**Si**licate **roc**k] Also known as the one-shot system. A chemical grouting system for hardening ground formations. Aqueous solutions of sodium silicate and formamide are mixed and injected into the ground; the formamide slowly reacts with the silicate, precipitating hydrated silica, which binds the soil particles together. Invented in 1961 by the Diamond Alkali Company. *See also* Joosten.

> U.S. Patent 2,968,572.

Sirofloc [**CSIRO floc**culation] A process for purifying municipal water supplies by flocculation, developed by the Commonwealth Scientific and Industrial Research Organization (CSIRO), Australia. Powdered magnetite, which has been given a special surface treatment, is added to the water. Particulate impurities, such as clays, adhere to these particles by colloidal forces. The magnetite particles are removed by a magnetic separator, chemically cleaned, and reused. The effluent from this process occupies a much smaller volume than the traditional alum sludge. The first installation outside Australia was at the Redmires Water Works, Yorkshire, United Kingdom, in 1988. The process should also be applicable to the effluents from sewage works, and pilot trials of such a process were being held in Melbourne and Sydney in 1992. In 1995, it was in use by South West Water, United Kingdom. Now offered by Aker Kvaerner.

> *New Sci.*, 1990, **127**(1725), 44.

Sirola A variation of the *Sulfite process for making paper from wood, in which two "cooking" processes are used; the first is alkaline and the second acid. *See also* Kramfors, Stora.

Sirosmelt [**CSIRO smelt**ing] A copper smelting process developed in the 1970s by J. Floyd at the Commonwealth Scientific and Industrial Research Organization (CSIRO), Australia. It formed the basis for the *Isasmelt process.

Sirotherm An ion-exchange process for desalinating brackish waters, in which regeneration is accomplished by heating the resin to approximately 80°C rather than by reverse ion exchange. Developed in the 1960s by D.E. Weiss and others at the Chemical Technology Division of the Commonwealth Scientific and Industrial Research Organization, Melbourne, Australia, and ICI Australia. The process was piloted by ICI Australia but eventually abandoned because it was unreliable.

> Calomon, C., in *Ion Exchange for Pollution Control*, Calomon, C. and Gold, H., Eds., CRC Press, Boca Raton, FL, 1979, 96.

Siurin *See* Hoganas.

SiVARA [**Si**lica No**VARA**] A process for making objects of high-purity silica glass. Fumed silica, made by the *AEROSIL process, is mixed with silicon tetraethoxide, water, and a hydrolysis catalyst. This forms a sol, which is poured into a mold where it spontaneously gels. The water is replaced by an organic solvent and the object dried and calcined. Developed by Degussa Novara Technology in 2005 and licensed to Silicaglas Ilmenau in Germany.

U.S. Patent 6,799,442.
Costa, L. and Kerner, D., *J. Sol-Gel Sci. Tech.*, 2003, **26**, 63.
Costa, F., Costa, L., Gini. L., Sparpaglione, M., and Abbondanza, L., *J. Non-Cryst. Solids*, 2004, (345–346), 787.
Chem. Eng. (N.Y.), 2005, **112**(12), 18.

Skarstrom *See* PSA.

SKF A *DR process for making iron. Powdered iron ore and coal are injected through a plasma arc heater into a vertical-shaft furnace. *See* DR.

SKIP [**SK**eletal **I**somerization **P**rocess] A process for converting linear butenes into *iso*butene. Developed by Texas Olefins in the 1990s and operated by that company in Houston, TX.

Morgan, M.L., *Chem. Ind. (London)*, 1998, (3), 90.

Skygas A process for gasifying carbon-containing wastes. The wastes are passed through a shaft furnace, heated by an arc burning between carbon electrodes. The wastes are gasified with the addition of water, and the off-gases are further processed in a secondary, coke-filled, electrically heated reactor. The product is a medium-BTU gas that can be used for raising steam, generating electricity, or making ammonia or methanol. Developed jointly by Xytel-Bechtel and Montana Precision Mining and now offered by MPM Technologies.

Processing (Sutton, England), 1989, Aug/Sept, 25.

slow-cooled matte A process for extracting platinum metals from copper–nickel matte. The molten matte is cooled slowly, over several days. This causes the platinum metals to enter a nickel–iron phase, which can be separated magnetically from the other components. Operated by Rustenburg Platinum Mines in South Africa and INCO in Canada.

SL/RN A *DR ironmaking process, using coal as the reductant in a rotary kiln. First operated in New Zealand in 1970 and now in operation in several other countries. Six plants were in operation as of 1997. The *Codir process is similar to this process. *See* DR.

SLRP [Sequential Liquid-Lignin **R**ecovery and **P**urification] A continuous, three-stage process for recovering lignin from the black liquor from the *Kraft pulping process. The first stage reduces the pH from 14 to 9 by contact with CO$_2$. The second stage removes the lignin as a dense lower liquid phase and acidifies it. The third stage separates the lignin by extraction with water. Developed and piloted by the Liquid Lignin Company in South Carolina in 2012 and proposed for installation in a commercial mill in 2013.

Chem. Eng. (N.Y.), 2012, **119**(5), 11.

Slurrisweet A process for removing sulfur dioxide from gas streams by absorption in an aqueous suspension of iron oxide particles. Developed by Gas Sweetener Associates from the 1980s but later abandoned.

Kohl, H.L. and Nielsen, R.B., *Gas Purification*, 5th ed., Gulf Publishing Co., Houston, TX, 1997, 1309.

SM *See* SMART SM.

S-Magyp A *flue-gas desulfurization process. The sulfur dioxide is absorbed in a slurry of magnesium hydroxide, forming magnesium sulfate. This is reacted with calcium hydroxide, generating magnesium hydroxide, which is recycled, and gypsum, which is sold. Developed by Fuji Machinery Company and Toyo Engineering Corporation and first operated in a cement factory in Japan in 1999.

SMART *See* SMART SM.

SMART H$_2$ [**S**team **M**ethane **A**dvanced **R**eformer **T**echnology] A process for making hydrogen by the steam reforming of methane. It differs from similar systems in housing the catalyst within a proprietary heat exchanger. Developed by Mannesmann KTI in 1996; it was planned for installation in Salisbury, MD, in late 1997.

SMART SM [**S**tyrene **M**onomer **A**dvanced **R**eheat **T**echnology] A process for making styrene by dehydrogenating ethylbenzene. It uses "oxidative reheating"—in situ reheating of process gas between endothermic stages of dehydrogenation, which uses less energy than previous processes. Developed and licensed jointly by UOP and ABB Lummus Crest. Seven units were operating in 2005. *See also* Styro-Plus.

> Ward, D.J., Black, S.M., Imai, T., Sato, Y., Nakayama, N., Tokano, H., and Egawa, K., *Hydrocarbon Process. Int. Ed.*, 1987, **66**(3), 47; **66**(11), 87.
> *Eur. Chem. News*, 1990, 1/8 Jan, 23.
> Cavani, F. and Trifiro, F., *Appl. Catal. A: Gen.*, 1995, **133**, 237.

SMDS [**S**hell **M**iddle **D**istillate **S**ynthesis] A three-stage process for converting natural gas to liquid fuels. The first stage uses *reforming to convert the natural gas to *syngas. The syngas is converted to heavy paraffins by the *Fischer–Tropsch process; this mixture is converted to hydrocarbons of the required boiling range by hydroconversion. The process was commercialized in Sarawak, Malaysia, in the early 1990s by a consortium of Shell Gas, Mitsubishi Corporation, Petronas, and the Sarawak state government. The plant suffered an explosion in 1998 but was rebuilt.

> van der Burgt, M., van Klinken, J., and Sie, T., *Pet. Rev.*, 1990, **44**(516), 204.
> *Oil Gas J.*, 1990, **88**(40), Suppl., 15.
> Sie, S.T., Senden, M.M.G., and van Wechem, H.M.W., *Catal. Today*, 1991, **8**, 371.
> Chauvel, A., Delmon, B., and Hölderich, W.F., *Appl. Catal. A: Gen.*, 1994, **115**, 186.
> Schrauwen, F.J.M., in *Handbook of Petroleum Refining Processes*, 3rd ed., Meyers, R.A., Ed., McGraw-Hill, New York, 2003, 15.25.
> Wittcoff, H.A., Reuben, B.G., and Plotkin, J.S., *Industrial Organic Chemicals*, 2nd ed., Wiley-Interscience, Hoboken, NJ, 2004, 383.

Smopex A process for removing metals from aqueous solution by adsorption on a fibrous scavenger. Developed by Johnson Matthey in 2005.

SMPO [**S**tyrene **M**onomer **P**ropylene **O**xide] A process for making propylene oxide by the catalytic epoxidation of propylene. The catalyst contains a compound of vanadium, tungsten, molybdenum, or titanium on a silica support. Developed by Shell and operated in the Netherlands since 1978. *See also* POSM.

> Weissermel, K. and Arpe, H.-J., *Industrial Organic Chemistry*, 4th ed., translated by C.R. Hindley and S. Hawkins, Wiley-VCH, Weinheim, Germany, 2003, 272.

Smuda A process for pyrolyzing waste plastics (preferably polyolefins) with the production of diesel fuel and gasoline. A disposable catalyst is used, preferably nickel silicate. Developed by H.W. Smuda (also spelled Zmuda). A large plant has operated in Zabrze, Poland, since 1997.

> U.S. Patents 6,255,547; 6,777,581.
> Scheirs, J., in *Feedstock Recycling and Pyrolysis of Waste Plastics*, Scheirs, J. and Kaminsky, W., Eds., John Wiley & Sons, Chichester, U.K., 2006, 414.

SNAM (1) [**S**ocietà **Na**zionale **M**etanodotti] A urea synthesis process, developed in the late 1960s and now widely used worldwide.

> Slack, A.V. and Blouin, B.M., *CHEMTECH*, 1971, **1**, 32.
> *Hydrocarbon Process. Int. Ed.*, 1979, **58**(11), 248.

SNAM (2) An *Ammoxidation process for converting propylene to acrylonitrile. The catalyst is based on molybdenum/vanadium or bismuth, operated in a fluidized bed. Operated in Europe in 1968.

SNCR [**S**elective **N**on**C**atalytic **R**eduction] A generic term for processes that remove oxides of nitrogen from flue gases by noncatalytic chemical reactions. These include the reaction with ammonia at high temperature (1300°C–1900°C) and the reaction with urea. *See* NOxOut, SCR.

> Radojevic, M., *Chem. Br.*, 1998, **34**(3), 30.

SNOW [**SN**amprogetti–d**OW**] A process for making styrene by the dehydrogenation of ethylbenzene. The catalyst is a zeolite containing gallium and a trace of platinum. An alternative feed is a mixture of ethane and benzene. The reactor is a "fast riser," followed by a stripper and a catalyst regenerator. Developed by Snamprogetti and Russian workers and demonstrated by Dow Chemicals in a large plant in Freeport, TX.

Sanfilippo, D., *Catal. Today*, 2011, **178**(1), 149.

SNOX A combined *flue-gas desulfurization and denitrification process. The NO_x is first removed by the *SCR process, and then the SO_2 is catalytically oxidized to SO_3 and converted to sulfuric acid by the *WSA process. Developed by Haldor Topsoe and first operated at a power station in Denmark in the 1990s.

Chem. Eng. (Rugby, Engl.), 1989, (462), 29.
Rostrup-Nielsen, J.R., *Catal. Today*, 1993, **18**, 141.
Hydrocarbon Process. Int. Ed., 1993, **72**(8), 80.
Appl. Catal. B: Environmental, 1994, **3**(4), N28.
Environ. Sci. Technol., 1994, **28**(2), 88.

SNPA *See* Sulfreen.

SNPA-DEA A process for removing hydrogen sulfide and carbon dioxide from gas streams by absorption in aqueous diethanolamine. Developed by the Société Nationale des Pétroles d'Aquitaine at the gas field in Lacq, France, and still used there and in Canada.

Canadian Patent 651,379.
Wendt, C.J., Jr. and Dailey, L.W., *Hydrocarbon Process. Int. Ed.*, 1967, **46**(10), 155.
Hydrocarbon Process. Int. Ed., 1975, **54**(4), 95.
Kohl, H.L. and Nielsen, R.B., *Gas Purification*, 5th ed., Gulf Publishing Co., Houston, TX, 1997, 51.

Sobolevsky A process for converting native platinum to malleable platinum by pressing and heating. Developed by P.G. Sobolevsky in Russia in the 1820s.

McDonald, D., *A History of Platinum*, Johnson Matthey, London, U.K., 1960, 159.

Socony-Vacuum This U.S. oil company, now part of ExxonMobil, invented many processes, but the one bearing the company name was that for making thiophene from butane and elemental sulfur at 560°C. It was operated by the Pennwalt Company in the 1950s and 1960s but then abandoned.

Soda Iron An obsolete process for removing organic sulfur compounds from *syngas by adsorption on a composition made by heating a natural iron oxide ore with sodium carbonate. Developed and operated in Germany in the 1930s.

Kohl, H.L. and Nielsen, R.B., *Gas Purification*, 5th ed., Gulf Publishing Co., Houston, TX, 1997, 1164.

Sofnolime A process for removing acid gases from gas streams by adsorption on a proprietary granular solid called Sofnolime RG. Developed by Molecular Products and operated in 15 plants worldwide in 1994.

Kohl, H.L. and Nielsen, R.B., *Gas Purification*, 5th ed., Gulf Publishing Co., Houston, TX, 1997, 1309.

SOFT [**S**plit-**O**lefin **F**eed **T**echnology] An improved version of Phillips' HF *alkylation process.

Rhodes, A.K., *Oil Gas J.*, 1994, **92**(34), 56.

SOHIO [**S**tandard **OHIO**] The Standard Oil Company of Ohio (later BP Chemicals America) has developed many processes, but its *Ammoxidation process, for converting propylene to acrylonitrile, is the one mostly associated with its name. The original catalyst was bismuth phosphomolybdate, but the composition has evolved over the years and now may contain one or more of the following: Fe, Ni, Co. First operated in the

United States in 1960, it is the predominant process for making acrylonitrile used in the world today. Honored by a National Historic Chemical Landmark of the American Chemical Society at the BP (now INEOS) Headquarters in Warrenville Heights, OH.

U.S. Patent 2,941,007.
Jacobs, M., *Ind. Eng. Chem.*, 1996, **74**(41), 40.
Weissermel, K. and Arpe, H.-J., *Industrial Organic Chemistry*, 4th ed., translated by C.R. Hindley and S. Hawkins, Wiley-VCH, Weinheim, Germany, 2003, 307.

Solatech [Solvay Revatech] A process for treating the solid residue from the *NEUTREC process. Developed by Solvay and Revatech. *See also* NEUTREC.

Eur. Chem. News, 1999, **71**(1878), 34.

Soldacs *See* PuraSiv HR.

Solexol A process for refining fats and oils by solvent extraction into propane. Developed by the M.W. Kellogg Company. Not to be confused with *Selexol.

Passino, H.J., *Ind. Eng. Chem.*, 1949, **41**, 280.

Sol-Gel A family of processes for making oxide ceramics and glasses from colloidal hydrated oxides or hydroxides. The colloids are converted to gels by several methods, and the gels are then dehydrated and calcined. First used in the late 1950s in the nuclear energy industry at Oak Ridge National Laboratory, TN. It was then used for making uranium dioxide in the form of ceramic spheres for fuel elements. Later, many other metal oxides were made in this way. The process has also been used for making metal oxides in the form of thin films, fibers, and special shapes for electronics. *See also* Stoeber.

Klein, L.C., Ed., *Sol-Gel Technology*, Noyes Publications, Park Ridge, NJ, 1988.
Segal, D., *Chemical Synthesis of Advanced Ceramic Materials*, Cambridge University Press, Cambridge, U.K., 1989, Chaps. 4, 5.

Solid Lime *See* Imatra.

SOLINOX [SO$_x$ LINde NO$_x$] A process for removing both NO$_x$ and SO$_x$ from flue gases. The SO$_x$ is removed by scrubbing with tetraethylene glycol dimethyl ether, circulated in a packed tower (the *Selexol process). The NO$_x$ is destroyed by selective catalytic reduction (*SCR). The sorbent is regenerated with steam; the SO$_x$ is recovered for conversion to sulfuric acid. Developed by Linde in 1985 and used in a lead smelter in Austria and several power stations in Germany. In 1990, it was announced that the process would be used at the titanium pigment plant in the Netherlands operated by Sachtleben.

Merrick, D. and Vernon, J., *Chem. Ind. (London)*, 1989, (3), 56.
Hydrocarbon Process. Int. Ed., 1990, **69**(4), 92.

Soliroc A process for solidifying aqueous wastes, converting them to a solid form suitable for landfill. The waste is initially acidic. Sodium silicate, cement, and lime are added, converting the liquid to a gel that hardens in several days. Developed by the Dutch company Esdex and used on a large scale in several countries. Soliroc Metal is the name of a Canadian manufacturer of vacuum cleaners.

Conner, J.R. and Hoeffner, S.L., *Crit. Rev. Environ. Sci Technol.*, 1998, **28**(4), 352.

SOLNOX A process for manufacturing nitric acid developed by Pechiney–Ugine–Kuhlmann.

U.S. Patent 8,263,036.
Chem. Eng. (N.Y.), 1977, **84**(9), 70.

Solutizer Also spelled Solutiser. An early process for removing sulfur compounds from fuel oil by solvent extraction with alkaline sodium isobutyrate. Developed by Shell Development Company.

Hydrocarbon Process., 1964, **43**(9), 213.
Unzelman, G.H. and Wolf, C.J., in *Petroleum Processing Handbook*, Bland, W.F. and Davidson, R.L., Eds., McGraw-Hill, New York, 1967, 3–116.

SOLVAHL A petroleum deasphalting process developed by IFP and now licensed by Axens. *See* ASVAHL.

Chauvel, A., Delmon, B., and Hölderich, W.F., *Appl. Catal. A: Gen.*, 1994, **115**, 186.

Solvay (1) An important process for making sodium carbonate, invented by E. Solvay in Belgium in 1861 and still used worldwide. *See* ammonia soda.

British Patent 3,131 (1863).

Solvay (2) A pulp-bleaching process using chlorine dioxide, generated by reducing sodium chlorate with methanol. In 1962, 20 paper mills in the United States were using this process. *See also* Mathieson (1).

Sheltmire, W.H., in *Chlorine: Its Manufacture, Properties and Uses*, Sconce, J.S., Ed., Reinhold Publishing, New York, 1962, 304,538.

Solvex [**Solv**ent **ex**traction] A process for extracting platinum metals from their ores by solvent extraction. Invented by Johnson Matthey in 1980 and piloted by that company, together with Rustenburg Platinum Holdings, since 1983. Commercialized in South Africa in the late 1980s.

Bruce, J.T., *Platinum Met. Rev.*, 1996, **40**(1), 4.

SOLVOX Not a process but a piece of equipment for dissolving oxygen in water. SOLVOX-I uses a venturi jet mixer; SOLVOX-B uses a perforated hose. It is mostly used for treating wastewater, but it is used also for potable water. Developed and provided by Linde, Germany.

Water Wastewater Int., Feb/Mar 2005, 37.

Solv-X A pulp-bleaching process that prevents the transition metal catalyzed side reactions of hydrogen peroxide that would otherwise occur.

Pulp Pap. Eur., 1997, **5**(2), 32.

Sonocracking Also called **SulphCo**. A group of processes for pretreating petroleum by ultrasound. The basic process is oxidative desulfurization, but conversion of petroleum residua to usable lighter oils is also claimed. Invented by R.W. Gunnerman in Reno, NV, and developed from 2000 by SulphCo, Houston, TX. The first experimental plant was commissioned in 2006 in Fujairah, UAE. Indonesian distributor PT Isis Megah has bought the technology but did not install it. A demonstration plant was planned to be built by Enterprise Products in Mont Belvieu, TX, in 2010, but this plant was damaged by an explosion in February 2011. A successful installation at the IPLOM refinery near Genoa, Italy, has been described. SulphCo filed for bankruptcy in September 2011.

A high-powered ultrasonic generator and transducer are included in the technology.
Gunnerman has won a number of international awards for several innovations, including the Albert Einstein Award in Physics in 2002.

Chem. Eng. (N.Y.), Apr 2001, 23; 2005, **112**(12), 15.
Chem. Eng. (N.Y.), 2008, **115**(3), 19.
Stanislaus, A., Marafi, A., and Rana, M.S., *Catal. Today*, 2010, **153**(1–2), 44.
U.S. Patents 6,402,939; 6,500,219; 7,300,566; 7,408,290.

SONOX A process for simultaneously removing SO_2 and NO_x from flue gas. A slurry of lime or limestone, with a proprietary nitrogen-based additive, is injected into the furnace. Developed by Ontario Hydro (the largest electricity supplier in North America) and marketed by Research-Cottrell.

> *Eur. Chem. News*, 1990, **55**(1444), 27; 1991, **56**(1471), 33.

Soot A process for making a very pure "preform" of silica glass for drawing into optical fiber. Invented in 1970 by D.B. Keck and P.C. Schultz at Corning Glass Works, New York. The silica made in this way was so much purer than the silicate glasses used previously that it revolutionized fiber-optic communications technology, permitting the progressive replacement of copper wire by optical fiber for telephone links.

> U.S. Patents 3,711,262; 3,737,292.
> Schultz, P.C., *Am. Ceram. Soc. Bull.*, 1973, **52**, 383.
> *Kirk-Othmer's Encyclopedia of Chemical Technology*, 4th ed., Vol. 12, John Wiley & Sons, New York, 1991–1998, 615.
> Hecht, J., *City of Light, the Story of Fiber Optics*, Oxford University Press, Oxford, New York, 1999, Chap. 11.

SOR [**S**elective **O**lefin **R**ecovery] A process for recovering or purifying olefins by selective absorption in silver nitrate solution. The silver solution is passed through hollow-fiber microporous membranes, and the olefin-containing gases are passed along the outside of the fibers. Developed by BP Chemicals and engineered by Stone & Webster and piloted in Grangemouth, Scotland, and Toledo, OH.

> *Eur. Chem. News*, 1996, **65**(1709), 59.
> *Hydrocarbon Process. Int. Ed.*, 1996, **75**(5), 29.

SORB33 A process for removing arsenic from public water supplies by adsorption on granular ferric oxide. The adsorbent is Bayoxide 33, a proprietary product made by Bayer. Developed by Bayer and Severn Trent PLC and first demonstrated at the Burton Joyce waterworks in Nottingham, United Kingdom. In 2004, the process was in use at 15 plants in the United Kingdom and planned for use in 6 demonstration plants in the United States.

> *Water Wastewater Int.*, 2002, **17**(2), 11–13.
> *Eur. Chem. News*, 2003, **79**(2081), 26.

Sorbex A family of separation processes, based on continuous liquid chromatography. The adsorbent, chosen for its selectivity and usually a zeolite, is contained in a chromatographic column with a number of ports spaced along its length. These ports, operated by a proprietary rotary valve, function alternately as inlets and outlets, thus permitting simulated countercurrent liquid–solid adsorption to be operated in a continuous mode. Variations of this process for specific mixtures, each with its preferred adsorbent and desorbent, are described under their own names, namely, Cresex, Cymex, Ebex, Molex, MX Sorbex, Olex, Parex, Sarex, and Sorbutene. Invented in 1957 by C.G. Gerhold, D.B. Broughton, and their colleagues at Universal Oil Products Company (now UOP, owned by Honeywell).

> U.S. Patent 2,985,589.
> Broughton, D.B., *Chem. Eng. Prog.*, 1968, **64**(8), 60.
> Spitz, P.H., *Petrochemicals: The Rise of an Industry*, John Wiley & Sons, New York, 1988, 191.
> Johnson, J.A. and Oroskar, A.R., in *Zeolites as Catalysts, Sorbents and Detergent Builders*, Karge, H.G. and Weitkamp, J., Eds., Elsevier, Amsterdam, the Netherlands, 1989.
> Johnson, J.A. and Kabza, R.G., in *Advances in Separation Processes*, American Institute of Chemical Engineers, New York, 1990, 35.
> Johnson, J.A., in *Handbook of Petroleum Refining Processes*, 3rd ed., Meyers, R.A., Ed., McGraw-Hill, New York, 2003, 10.29.

Sorbtech Formerly called Sanitech. A process for removing both sulfur dioxide and oxides of nitrogen from flue gases by adsorption. The sorbent is made from magnesia and expanded vermiculite. Developed by Solvent Technologies Corp. from 1985 and piloted in the 1990s but not commercialized. Today, the company is best known for its processes for removing mercury vapor from flue gases by adsorption on brominated activated carbon.

> Kohl, H.L. and Nielsen, R.B., *Gas Purification*, 5th ed., Gulf Publishing Co., Houston, TX, 1997, 630, 932.

Sorbutene A version of the *Sorbex process, for extracting 1-butene from mixed C_4 hydrocarbons. Offered for license by UOP.

Friedlander, R.H., in *Handbook of Petroleum Refining Processes*, 1st ed., Meyers, R.A., Ed., McGraw-Hill, New York, 1986, 8–101.

Sovafining [**So**cony-**Va**cuum re**fining**] A *hydrofining process developed by the Socony-Vacuum Oil Company, now ExxonMobil. *See* Hydrofining.

Sovaforming [**So**cony-**Va**cuum re**forming**] A *Catalytic Reforming process, using a platinum catalyst in a fixed bed. Developed by the Socony-Vacuum Oil Company in 1954. Subsequently renamed Platinum Reforming or PR.

Pet. Process., 1955, **10**, 1174.
Unzelman, G.H. and Wolf, C.J., in *Petroleum Processing Handbook*, Bland, W.F. and Davidson, R.L., Eds., McGraw-Hill, New York, 1967, 3–30.

SOXAL A *flue-gas desulfurization process, basically similar to the *Wellman–Lord process but regenerating the bisulfite solution in a membrane electrochemical cell. Developed by Allied Signal Group and piloted at the Florida Power & Light Company, Miami, FL, in 1991. A larger demonstration plant was planned for Dunkirk, New York.

Chem. Week, 1986, **139**(22), 106.
Chem. Mark Rep., 1991, **239**(5), 5; **240**(22), 9.

SPARG A process for the CO_2 reforming of methane, producing carbon monoxide and hydrogen, using a partially presulfided nickel catalyst. Developed by Haldor Topsoe in the 1990s for the direct reduction of iron ore, but not commercialized.

Rostrup-Nielsen, J., *J. Catal.*, 1984, **85**, 31.
Udengaard, N.R., Hansen, H.H.-B., and Hansen, D.C., *Oil Gas J.*, 1992, **90**, 62.
Armor, J.N., *Appl. Catal. A Gen.*, 1999, **176**(2), 170.
Rostrup-Nielsen, J., *Catal. Today*, 2006, **111**(1–2), 6.

SPD [**S**lurry-**P**hase **D**istillate] A process for making diesel fuel, kerosene, and naphtha from natural gas. Developed by Sasol from the *Fischer–Tropsch process and first commercialized in South Africa in 1993. A joint venture with Haldor Topsoe for the further development and commercialization of the process was announced in 1996. Commercialization in Nigeria was announced in 1998. An agreement with Engelhard to provide the catalyst was announced in 2000.

Hydrocarbon Process. Int. Ed., 1996, **75**(4), 46.

SPGK [**S**hell **P**oly**G**asoline and **K**erosene] A process for oligomerizing C_2–C_5 olefins to liquid transport fuels. The reaction takes place at 200°C–280°C and is catalyzed by a zeolite. Developed by the Shell Petroleum Company, Amsterdam, in 1990.

Eur. Chem. News, 1991, **56**(1455), 23.
Chem. Eng. (Rugby, Engl.), 1991, (489), 12.

SPHER [**S**hell **P**ellet **H**eat **E**xchange **R**etorting] A process for extracting oil from shale. The process is conducted in a fluidized bed in which heat is transferred by inert pellets of two sizes.

Chem. Eng. News, 1980, **58**(37), 42.

Spherilene [**Spheri**cal polyethy**lene**] A process for making polyethylene in the form of spherical granules. The catalyst combines a *Ziegler–Natta catalyst with a silane. Developed by Montecatini, Italy. Spherilene C uses two reactors in series. Spherilene S uses a single reactor. The first plant was built by Himont in Lake Charles, LA, in 1993. Fifteen plants had been licensed by 2007. The process is now licensed by Basell Polyolefins.

Eur. Chem. News, 1991, **56**(1476), 27; 1992, **58**(1530), 18.

Spheripol A process for making polypropylene and propylene copolymers. Homopolymerization is conducted in the liquid phase in a loop tubular reactor; copolymerization is conducted in the gas phase in a fluidized-bed reactor. The catalyst is treated with a special silane. The product is in the form of beads of up to 5 mm in diameter. Developed by Montecatini, Italy, and first licensed by Himont, United States, and Mitsui Petrochemical Industries, Japan. Ninety-four plants were operating in 2005. Now offered for license by Basell Polyolefins. *See also* Addipol.

Chem. Week, 1987, **141**(21), 99.
Hydrocarbon Process. Int. Ed., 1989, **68**(11), 109; 1991, **70**(3), 173.
Bertucco, A. and Vetter, G., Eds., *High Pressure Process Technology*, Elsevier, Amsterdam, the Netherlands, 2001, 348.
Soares, J.B.P. and McKenna, T.F.L., *Polyolefin Reaction Engineering*, Wiley-VCH, Weinheim, Germany, 2012, 125.

Spherizone A development of the *Spheripol process for making polypropylene. A special zone is established in the continuous circulating reactor in which additional monomer or a comonomer can be introduced in order to make polymers with different properties. The copolymers can be monomodal, bimodal, random, or twin-random. Nine licenses had been issued in 2006.

Chem. Ind. (London), 2002, (21), 10.
Chem. Week, 23 Jun 2004, **166**(21).
Chem. Ind. (London), 21 Aug 2006, (16), 11.

SPIREX [**Sp**outed **I**ron **R**eduction MIDR**EX**] A *DR process for making iron powder or hot briquetted iron from iron ore fines. Three stages are used. The first is a circulating fluidized-bed preheater whose turbulent conditions reduce the particle size of the ore. The second and third stages achieve the reduction in fluidized beds, fed by reducing gases from a *MIDREX reformer. Developed by Midrex Direct Reduction Corporation and Kobe Steel. The project was abandoned in the late 1990s for commercial reasons.

SP-Isoether [**Snam**Progetti] A process for making an isooctane alkylate on a plant designed to make methyl *t*-butyl ether. It is based on the selective dimerization of isobutylene, followed by hydrogenation to isooctane. Developed by Snamprogetti in 2000. *See also* Nexoctane.

Eur. Chem. News, 2000, **72**(1894), 39.

SPM *See* DR.

SPOC [**S**elective **P**artial **O**xidation **C**laus] Developed by ConocoPhillips (or MsTGC) and licensed by GTC Technology. An improved version of the *Claus process that avoids the use of a furnace. Licensed to GTC Technology by ConocoPhillips in 2012.

U.S. Patent 7,074,375.

SPOR [**S**ulfur removal by **P**artial **O**xidation and **R**eduction] A process for converting H_2S and SO_2 to elemental sulfur by the use of stoichiometric amounts of oxygen and sulfur.

Chung, J.S., Paik, S.C., Kim, H.S., Lee, D.S., and Nam, I.S., *Catal. Today*, 1997, **35**(1–2), 37.

SPORT [**S**ustainable **P**olymers-to-**O**lefins **R**ecycling **T**echnology] Also called the BP Polymer Cracking Process. A catalytic process for converting waste polymers to hydrocarbon mixtures, which can be used to make new polymers. The mixed waste plastics are introduced into a fluidized bed of sand in an inert atmosphere, heated externally to about 500°C. The plastics melt and coat the sand particles and then undergo pyrolysis to lighter hydrocarbons, which leave the reactor with the inert gas. The resulting gas mixture, after separation from the inert gas, is fed to a conventional catalytic cracker. Piloted by BP at Grangemouth, United Kingdom, from 2002.

Arena, U. and Mastellone, M.L., in *Feedstock Recycling and Pyrolysis of Waste Plastics*, Scheirs, J. and Kaminsky, W., Eds., John Wiley & Sons, Chichester, U.K., 2006, 467.

Spühlgas [German, meaning *rinsing gas*] Also known as Lurgi Spühlgas. A low-temperature process for carbonizing lump or briquetted materials such as coal, lignite, peat, wood, and oil shale. The heat is provided by burning the gas, which is also generated. Developed by Lurgi in Germany in 1925, originally to provide liquid hydrocarbons and later to provide industrial coke too, and now offered for license. As of 1981, 270 plants had been built in a number of countries.

Rammler, R.W. and Weiss, H.-J., in *Handbook of Synfuels Technology*, Meyers, R.A., Ed., McGraw-Hill, New York, 1984, 4–12.

Squibb A process for making acetone by passing acetic acid vapors through an iron tube at 500°C–600°C packed with pumice and precipitated barium carbonate. Invented by E.R. Squibb in 1895 and commercialized in the early twentieth century until supplanted by the *Weizmann process. Other ketones have since been made by a similar process.

Squibb, E.R., *J. Am. Chem. Soc.*, 1895, **17**(3), 187.
Kochakian, C.D., *J. Chem. Ed.*, 1932, **9**(9), 1649.
Jackson, M.A. and Cermak, S.C., *Appl. Catal. A: Gen.*, 2012, **431**, 157.
Wikipedia, The Free Encyclopedia. Ketonic Decarboxylation. Revised 12 Dec 2013. Retrieved 22 Jul 2013.

Squire and Messel An early process for making oleum from sulfuric acid produced by the *Chamber process. The acid was decomposed at red heat to sulfur dioxide, oxygen, and steam; the steam was condensed out; and the remaining gases passed over platinized pumice to form sulfur trioxide, which was absorbed in more chamber acid. Invented by W.S. Squire and R. Messel in 1875 in London and first commercialized there. Messel was one of the founders of the Society of Chemical Industry, United Kingdom, and is still commemorated in that society by the biennial award of a medal.

British Patent 3,278 (1875).
Morgan, G.T. and Pratt, D.D., *British Chemical Industry*, Edward Arnold & Co., London, U.K., 1938, 39.

SRB [Sulfate-Reducing Bacteria] A process for removing heavy metals and sulfates from contaminated water by the use of special bacteria. Ethanol is provided as a cheap source of energy for the bacteria. The sulfur is precipitated as heavy metal sulfides. Developed by Shell Research, Sittingbourne, United Kingdom, and first operated in the Netherlands in 1991.

Barnes, L.J., Janssen, F.J., Sherrin, J., Versteegh, J.H., Koch, R.O., and Scheerin, P.J.H., *Trans. Inst. Chem. Eng., Chem. Eng. Res. Dev.*, 1991, **69A**, 184.
Chem. Br., 1991, **27**, 884.

SRC [Solvent Refined Coal] Also known as P&M and PAMCO. A coal liquefaction process developed by the Spencer Chemical Company and the Pittsburgh & Midway Coal Mining Company, which later became part of the Gulf Oil Corporation. It was based on the *Pott–Broche process. Two versions of the process became known as Gulf SRC I and Gulf SRC II. The purpose is to produce a fuel of low sulfur content. Crushed coal is suspended in a mixture of cresol and tetrahydronaphthalene and hydrogenated under pressure with no added catalyst. The process is actually catalyzed by the finely divided iron pyrites naturally present in the coal. The undissolved residue is filtered off under pressure. Piloted in the 1970s at Tacoma and Fort Lewis, WA, and at Wilsonville, AL. SRC II is a variant, using more severe cracking conditions, which produces heavy heating oil.

Brant, V.L. and Schmid, B.K., *Chem. Eng. Prog.*, 1969, **65**(12), 55.
Pastor, G.R., Keetley, D.J., and Naylor, J.D., *Chem. Eng. Prog.*, 1976, **72**(8), 67.
Alpert, S.B. and Wolk, R.H., in *Chemistry of Coal Utilization*, 2nd Suppl. Vol., Elliott, M.A., Ed., John Wiley & Sons, New York, 1981, 1926.

Phillips, M.T., Bronfenbrenner, J.C., Kuhns, A.R., O'Leary, J.R., Snyder, G.D., and Znaimer, S., in *Handbook of Synfuels Technology*, Meyers, R.A., Ed., McGraw-Hill, New York, 1984, 1–3.

Davies, G.O., in *Chemicals from Coal: New Developments*, Payne, K.R., Ed., Blackwell Scientific Publications, Oxford, U.K., 1985, 101.

Schulze, J. and Gaensslen, H., in *Chemicals from Coal: New Developments*, Payne, K.R., Ed., Blackwell Scientific Publications, Oxford, U.K., 1985, 54.

SREP [**S**ynthetic **R**utile **E**nhancement **P**rocess] An improved version of the *Becher process for upgrading titanium ores. It is designed to reduce the level of radioactive contamination in the product. Developed in the early 1990s by Renison Goldfields, Australia, in collaboration with the CSIRO, Perth, and first operated in Narngulu in 1996.

SRET A two-stage process for making synthetic rutile from ilmenite, developed in Australia by Iluka Resources. The first stage is pyrometallurgical in which the ilmenite is heated in a rotary kiln in the presence of a reductant. The second stage is hydrometallurgical in which the iron is removed by oxidation and leaching. The final product contains between 88% and 95% of TiO_2.

SREX [**St**ontium **EX**traction] A process for removing strontium-90 from aqueous wastes from nuclear fuel processing, by solvent extraction into a solution of 18-crown-6 in octanol. Developed in 1990 by E.P. Horwitz at the Argonne National Laboratory, Chicago, IL.

SRP [**S**ulfonation-**R**eduction-**P**eroxide] A pulp-bleaching process that combines reduction with sodium borohydride and oxidation with hydrogen peroxide. Developed by Atochem in 1991 and licensed to Morton International, Chicago, IL.

Eur. Chem. News, 1991, **57**(1480), 38.

SRU A variation of the *Claus process, for use in the presence of ammonia. It uses a special reactor, designed to avoid plugging by compounds of ammonia with sulfur trioxide. Designed by JGC Corporation. Thirty-one units were in operation as of 1992.

Hydrocarbon Process. Int. Ed., 1996, **75**(4), 138.

SSANfining A gasoline-desulfurizing process developed by ExxonMobil Research and Engineering Company. Licensed to 11 companies as of June, 2001.

Sulphur, 17 May 2001, (274), 32.

SSC [**S**tazione **S**perimentale per i **C**ombustibili] A cyclic *Catalytic Reforming process for making gas from oil. Developed by SSC, Milan, Italy.

Gas Making and Natural Gas, British Petroleum Co., London, U.K., 1972, 96.

SSF [**S**imultaneous **S**accharification and **F**ermentation] *See* Emert.

SSP [**S**olid **S**tate **P**olymerization] Also called **UOP Sinco SSP**. A process for making articles from PET (polyethylene terephthalate) and related polymers. The key step is controlled crystallization of the polymer under optimum conditions. Developed jointly by UOP (United States) and Sinco (Italy) and widely adopted since 1986.

SSPD [**S**asol **S**lurry-**P**hase **D**istillate] A process for converting natural gas to diesel fuel, kerosene, and naphtha. Operated by Sasol in South Africa since 1993. Three stages are involved. In the first, natural gas is converted to synthesis gas by *reforming. In the second, the synthesis gas is converted to waxy hydrocarbons in a slurry-phase reactor. In the third, the waxes are upgraded to middle distillates. *See also* Arge.

Oil Gas J., 1997, **95**(25), 16.

SSX [**S**ynergistic **S**olvent **E**xtraction] A general term for metallurgical processes that use two extractants, one which does the extraction and another which improves its selectivity. Developed by CSIRO for separating cobalt from nickel.

tce (The Chemical Engineer), 2008–2009, (810), 34.

S-T *See* VAW-Sulfite.

Staatsmijnen–Otto Also known as the Pieters process, after the inventor, H.A.J. Pieters. A process for removing hydrogen sulfide from coal gas by scrubbing with an aqueous solution containing a suspension of iron cyanide complexes known as iron blue. The product is elemental sulfur, which separates as a froth and is purified by heating with water in an autoclave. Staatsmijnen is a Dutch producer of smokeless fuels, which has also given its name to a briquetting process.

U.S. Patent 2,169,282.
Kohl, H.L. and Nielsen, R.B., *Gas Purification*, 5th ed., Gulf Publishing Co., Houston, TX, 1997, 746.

Stack *See* Dutch.

Stackpol 150 Also known as **IFP Stackpol 150**. A *flue-gas desulfurization process. The sulfur dioxide is removed by scrubbing with aqueous ammonia, and the sulfur is then recovered by a four-stage process. Developed by the Institut Français du Pétrole.

Sulphur, 1976, (125), 43.
Environ. Sci. Technol., 1977, **11**(1), 22.
Hydrocarbon Process. Int. Ed., 1986, **65**(4), 97.

Stamicarbon [**Sta**atsmijnen **carbon**] Stamicarbon BV is the licensing subsidiary of the Dutch chemical company DSM. It offers a number of processes, including *HPO and *HSO. Historically, the process for which the company was best known was a coal carbonization process. Today, a process for making urea is probably its most important one.

Stamylex A process for making LLDPE using an octene solution. Developed by DSM.

Standard Oil A process for polymerizing ethylene and other linear olefins and diolefins to make linear polymers. This is a liquid-phase process, operated in a hydrocarbon solvent at an intermediate pressure, using a heterogeneous catalyst such as nickel oxide on carbon, or vanadia or molybdena on alumina. Licensed to Furukawa Chemical Industry Company at Kawasaki, Japan.

Raff, R.A.V. and Allison, J.B., *Polyethylene*, Interscience, New York, 1956, 68.
Peters, E.F., Zletz, A., and Evering, B.L., *Ind. Eng. Chem.*, 1957, **49**, 1879.

STAR [**ST**eam **A**ctive **R**eforming] A *Catalytic Reforming process for converting aliphatic hydrocarbons to olefins or aromatic hydrocarbons. Hydrocarbons containing five or fewer carbon atoms are converted to olefins. Those containing six or more are dehydrocyclized to aromatic hydrocarbons. The reactions take place in the vapor phase, in a fixed catalyst bed, in the presence of steam. The catalyst is platinum/tin alloy on zinc calcium aluminate. Developed by Phillips Petroleum Company. The first commercial plant was built for Coastal Chemicals in Cheyenne, WY, in 1992; another was built for Polibutenos Argentinos in 1996. The process is now offered by Uhde.

Brinkmeyer, F.M., Rohr, D.F., Olbrich, M.E., and Drehman, L.E., *Oil Gas J.*, 1983, c**81**(13), 75.
Little, D.M., *Catalytic Reforming*, PennWell Publishing, Tulsa, OK, 1985, 171.
Hutson, T., Jr. and McCarthy, W.C., in *Handbook of Petroleum Refining Processes*, 1st ed., Meyers, R.A., Ed., McGraw-Hill, New York, 1986, 4–29.
Hu, Y.C., in *Chemical Processing Handbook*, Marcel Dekker, New York, 1993, 807.
Hydrocarbon Process. Int. Ed., 2000, **79**(2), 84-F–84-H.
Chem. Eng. (N.Y.), 2004, **111**(3), 24.
ICIS Chem. Bus., 2012, **281**(14), 37.

STDP [Selective Toluene Disproportionation Process] A process for converting toluene to mixed xylenes, predominately *p*-xylene. It takes place in the presence of hydrogen over a ZSM-5-type catalyst. Developed by Mobil in the 1980s and first operated by Enichem.

> *Chem. Eng. News*, 27 Aug 2001, **79**(35), 28–30.
> Weissermel, K. and Arpe, H.-J., *Industrial Organic Chemistry*, 4th ed., translated by C.R. Hindley and S. Hawkins, Wiley-VCH, Weinheim, Germany, 2003, 335.

STEAG [STeinen Elektrizitat AG] A process for producing both electric power and gas from coal, developed by the German company named. Installed at Leuna, Germany.

> Dainton, A.D., in *Coal and Modern Coal Processing*, Pitt, G.J. and Millward, G.R., Eds., Academic Press, London, U.K., 1979, 135.

steam cracking *See* Reforming.

steam reforming *See* Reforming. Those steam-reforming processes having special names described in this dictionary are Fauser–Montecatini, ICI Low-Pressure Methanol, ICI Steam Naphtha Reforming, Kureha/Union Carbide, MRG, MS, Onia-Gegi, Pyrotol, and Remet.

steelmaking Because of its size and importance, the iron and steel industry has attracted many inventors and developers, and many of their processes are known by special names. The more important of these, listed as follows, are mostly described in individual entries: Acid Bessemer, Acid Open Hearth, AOD, Basic Bessemer, Basic Open Hearth, Bertrand Thiel, Bessemer, Chenot, Corex, Crucible, DH, D-LM, FIOR, HIB, Hoesch, Huntsman, LBE, L-D, Open Hearth, Perrin, pig and ore, pig and scrap, Puddling, Saniter, Siemens–Martin, SL/RN, Talbot, Thomas, Uchatius, and Wiberg-Sodefors.

The steelmaking processes listed as follows are more specialized or of lesser importance and are not described elsewhere in this work: ACAR [Allis-Chalmers Agglomeration–Reduction], ASEA–SKF [named after two Swedish companies], CAB [Capped Oxygen Bubbling], CAS [Composition Adjustment by Sealed oxygen bubbling], Hornsey, KIP [Kimetsu Injection Process], Leckie, LF [Ladle-Furnace], Moffat, Monell, Morse, Nesbitt, Nippon Steel, OBM, QQ-BOP, RH [Ruhrstahl-Heraeus], RH-FR [Ruhrstahl-Heraeus], RH-OB [Ruhrstahl-Heraeus, Oxygen Blowing], SAB [Sealed Argon Bubbling], TN [Thiessen Niederrhein], Tysland–Hole, VAD [Vacuum Arc Degassing], VAR [Vacuum Arc Remelting], VOD [Vacuum Oxygen Decarburization].

> Osborne, A.K., *An Encyclopedia of the Iron & Steel Industry*, 2nd ed., Technical Press, London, U.K., 1967.
> *Kirk-Othmer's Encyclopedia of Chemical Technology*, 4th ed., Vol. 22, John Wiley & Sons, New York, 1991–1998, 765.

Steffens A process for separating sugar from beet sugar molasses by adding calcium hydroxide to precipitate calcium saccharate. Treatment of the liquor with carbon dioxide precipitates calcium carbonate and regenerates the sucrose. Invented in Vienna in 1883 by C. Steffens. *See also* Boivin–Louiseau, Scheibler.

> British Patents 967 (1883); 2,416 (1883).

Stelling *See* DR.

Stengel A process for making ammonium nitrate by reacting ammonia vapor with 60% nitric acid in a packed tower reactor. Air blown through the reactor reduces the moisture content to the desired value, and the product flows to the bottom of the reactor, where it is discharged on to a moving, water-cooled belt. Invented by A. Stengel and first operated by Commercial Solvents Corporation in Sterlington, LA, in the 1950s.

> U.S. Patent 2,568,901.
> Dorsey, J.J., Jr., *Ind. Eng. Chem.*, 1955, **47**, 11.
> Hester, A.S., Dorsey, J.J., Jr., and Kaufman, J.T., in *Modern Chemical Processes*, Vol. 4, Reinhold Publishing, New York, 1956, 36.

STEREAU A *BAF process.

> Stephenson, T., Mann, A., and Upton, J., *Chem. Ind. (London)*, 1993, (14), 533.

Stevens *See* carbonization.

STEX [**ST**yrene **EX**traction] A process for extracting styrene from pyrolysis gasoline. Developed by Toray.

Chem. Eng. News, 27 Aug 2001, **79**(35), 28–30.
Weissermel, K. and Arpe, H.-J., *Industrial Organic Chemistry*, 4th ed., translated by C.R. Hindley and S. Hawkins, Wiley-VCH, Weinheim, Germany, 2003, 343.

Still A method for increasing the yield of light oil formed in the carbonization of coal. Some of the gas produced is passed through the partially carbonized coal in a cooler part of the bed. Developed by C. Still and used in Recklinghausen, Germany, in the 1930s for producing motor fuel. *See also* Carl Still.

U.S. Patents 1,810,629; 1,937,853.
Dean, H., *Fuel*, 1934, **13**, 112.

Still Otto A process for removing hydrogen sulfide from coke-oven gas by scrubbing with aqueous ammonia. Developed from the *Carl Still (2) process.

Kohl, H.L. and Nielsen, R.B., *Gas Purification*, 5th ed., Gulf Publishing Co., Houston, TX, 1997, 322.

St. Joseph Also known as St. Joe. A process for extracting zinc from zinc sulfide ore by electrothermic reduction, practiced by the St. Joseph Lead Company at its Josephtown, PA, refinery in the 1930s. A mixture of zinc blended with coke was heated by passing electricity through the mixture. The zinc vapor thus produced was condensed in a bath of molten zinc. The name has also been applied to a similar lead extraction process.

Cocks, E.J. and Walters, B., *A History of the Zinc Smelting Industry in Britain*, George G. Harrap, London, U.K., 1968, 164.
Morgan, S.W.K., *Zinc and Its Alloys and Compounds*, Ellis Horwood, Chichester, U.K., 1985, 69.

Stöber A process for making metal oxides in the form of small spheres of uniform diameter by the controlled hydrolysis of metal alkoxides. First used in 1968 to make silica spheres from alkyl silicates. The products can be used to make high-quality oxide ceramics. *See also* Sol-Gel.

Stöber, W., Fink, A., and Bohn, E., *J. Colloid Interface Sci.*, 1968, **26**, 62.

Stockbarger *See* Bridgman.

Stoic Also called Foster Wheeler–Stoic and FW–Stoic. A two-stage, nonslagging coal gasification process, operated under atmospheric pressure and using air as the oxidant. Initially developed by Stoic Combustion Limited Pty, South Africa; licensed and further developed by Foster Wheeler Corporation, United States. First used in South Africa in 1950; now widely used in Europe, the United States, and South Africa.

Brand, R.G. and Bress, D.F., in *Handbook of Synthetic Fuels Technology*, Meyers, R.A., Ed., McGraw-Hill, New York, 1984, 3–8.

Stone & Webster/Ionics A *flue-gas desulfurization process in which the sulfur dioxide is absorbed in aqueous sodium hydroxide, forming sodium sulfite and bisulfite, the sulfur dioxide is liberated by the addition of sulfuric acid, and the reagents are regenerated electrolytically. Designed by Stone & Webster Engineering Corporation and Ionics Inc. Operated in a demonstration plant in Milwaukee, WI, in 1974, but not commercialized.

Humphries, J.J., Jr. and McRae, W.A., *Proc. Am. Power Conf.*, 1970, **32**, 663 (*Chem. Abstr.*, **74**, 34348).
Kohl, H.L. and Nielsen, R.B., *Gas Purification*, 5th ed., Gulf Publishing Co., Houston, TX, 1997, 561.

Stora A two-stage variation on the *Sulfite papermaking process, in which the acidity of the second stage is increased by adding sulfur dioxide. *See also* Kramfors, Sirola.

Higham, R.R.A., *A Handbook of Papermaking*, 2nd ed., Business Books, London, U.K., 1968, 261.

Stratco A process for making a high-octane gasoline component by alkylation of C_3–C_5 hydrocarbons with isobutane, catalyzed by refrigerated sulfuric acid. The product is known as alkylate. Widely used worldwide.

Weitkamp, J. and Maixner, S., *Erdoel Kohle*, 1983, **36**, 523.
Graves, D.C., in *Handbook of Petroleum Refining Processes*, 3rd ed., Meyers, R.A., Ed., McGraw-Hill, New York, 2003, 1.11.

Strategic-Udy *See* DR.

Strehlenert A process for treating sulfite liquors from papermaking. Invented by R.W. Strehlenert in Sweden ca. 1912.

Svensk Chem. Tidskr., 1913.
Swedish Patent 34,941 (1911).
U.S. Patent 1,149,420.

Stretford A process for removing hydrogen sulfide and organic sulfur compounds from coal gas and general refinery streams by air oxidation to elementary sulfur, using a cyclic process involving an aqueous solution of a vanadium catalyst and anthraquinone disulfonic acid. Developed by T. Nicklin in the late 1950s by the North Western Gas Board (later British Gas) and then at the Clayton Aniline Company, in Stretford, near Manchester, United Kingdom. It is the principal process used today; more than 100 plants were operating, worldwide, in 1992.

U.S. Patent 2,997,439.
Nicklin, T. and Brunner, E., *Hydrocarbon Process. Int. Ed.*, 1961, **40**(12), 141.
Wilson, B.M. and Newell, R.D., in *Acid and Sour Gas Treating Processes*, Newman, S.A., Ed., Gulf Publishing Co., Houston, TX, 1985, 342.
Dalrymple, D.A., Trofe, T.W., and Evans, J.M., *Chem. Eng. Prog.*, 1989, **85**(3), 43.
Hydrocarbon Process. Int. Ed., 1996, **75**(4), 138; 2000, **79**(4), 89.
Kohl, H.L. and Nielsen, R.B., *Gas Purification*, 5th ed., Gulf Publishing Co., Houston, TX, 1997, 769.

Sturzelberg *See* DR.

Styromax Not a process but a proprietary name for a range of catalysts for making styrene, developed and sold by the Nissan Girdler Catalyst Company.

Jpn. Chem. Week, 9 Jul 1998, **39**(1983), 1.

Styro-Plus An early version of the *SMART SM process.

Sucro-Blanc A process for decolorizing sugar solution by the addition of calcium hypochlorite.

Spencer, G.L. and Meade, G.P., *Cane Sugar Handbook*, 8th ed., John Wiley & Sons, New York, 1945, 331.

Suida An extractive distillation process for concentrating the dilute acetic acid obtained from the manufacture of cellulose acetate. It was originally used for separating the products of wood pyrolysis. Invented in 1926 by H. Suida in Vienna and operated in the 1930s.

U.S. Patents 1,624,812; 1,697,738; 1,703,020.
Faith, W.L., Keyes, D.B., and Clark, R.L., *Industrial Chemicals*, 1st ed., John Wiley & Sons, New York, 1950, 11.

Sulfa-Check A process for removing acid gases from hydrocarbon gas streams by absorption in aqueous sodium nitrite. Developed by NL Industries. More than 100 units were operating in the United States and Canada in 1986. *See also* Hondo HS-100.

Chem. Eng. (N.Y.), 1987, **94**(2), 159.
Kohl, H.L. and Nielsen, R.B., *Gas Purification*, 5th ed., Gulf Publishing Co., Houston, TX, 1997, 1312.

Sulfacid A process for removing sulfur dioxide from effluent gases using active carbon. The gas is first contacted with dilute sulfuric acid from the adsorption step. It then passes through a fixed bed of active carbon on which water is being sprayed. The resulting dilute sulfuric acid is recovered. Suitable for effluents from the *Contact process and flue gases. Developed by Lurgi, Bergbau-Forschung, and Babcock & Wilcox.

Oil Week, 5 July, 1971, 9.jmm.

Kohl, H.L. and Nielsen, R.B., *Gas Purification*, 5th ed., Gulf Publishing Co., Houston, TX, 1997, 634.

Sulfa Guard Similar to *Sulfa-scrub but reportedly more cost effective.

Sulfa-scrub A process for removing low concentrations of hydrogen sulfide from gas streams by reaction with hexahydrotriazine. The product is water soluble, noncorrosive, and nonhazardous. Developed by the Quaker Petroleum Chemical Company in 1991.

U.S. Patent 4,978,512.

Chem. Eng. (N.Y.), 1991, **98**(9), 43.

Dillon, E.T., *Hydrocarbon Process. Int. Ed.*, 1991, **70**(12), 65.

Sulfate (1) A process for making titanium dioxide pigment from ilmenite. The ilmenite is digested with sulfuric acid, yielding a solution of titanyl and ferrous sulfates:

$$FeTiO_3 + 2H_2SO_4 = TiOSO_4 + FeSO_4 + 2H_2O$$

Much of the ferrous sulfate is crystallized out and discarded. The titanium is hydrolyzed by boiling, yielding hydrated titanium dioxide, which is then calcined:

$$TiOSO_4 + xH_2O = TiO_2xH_2O + H_2SO_4$$

$$TiO_2xH_2O = TiO_2 + xH_2O$$

The process can be adjusted to yield the product in either the anatase or rutile crystal modification, by use of proprietary nuclei at the precipitation stage and by adding small quantities of other materials at the calcination stage. The whole process is much more complex than has been indicated, involving at least 20 processing steps, every one of which is critical for the development of optimum pigment properties. Disposal of the waste sulfuric acid and ferrous sulfate has been a major problem for the industry for many years. In the 1980s, international legislation compelled the manufacturers to recycle or neutralize the acid.

Barksdale, J., *Titanium: Its Occurrence, Chemistry, and Technology*, 2nd ed., Ronald Press, New York, 1966, Chaps. 13–16.

Egerton, T.A. and Tetlow, A., in *Industrial Inorganic Chemicals: Production and Use*, Thompson, R., Ed., Royal Society of Chemistry, Cambridge, U.K., 1995, 360.

Buxbaum, G. and Pfaff, G., *Industrial Inorganic Pigments*, 3rd ed., VCH, Weinheim, Germany, 2005, 59.

Sulfate (2) An acid papermaking process, also known as the *Kraft process.

SulfaTreat A process for removing hydrogen sulfide and mercaptans from natural gas or carbon dioxide streams, using a proprietary solid absorbent, which is subsequently dumped. First developed by Gas Sweetener Associates, now licensed by Sulfa Treat, a business unit of MI-SWACO, MO. More than 1000 plants were operating in 2000.

Samuels, A., *Oil Gas J.*, 1990, **88**(6), 44.

Hydrocarbon Process. Int. Ed., 1996, **75**(4), 140; 2000, **79**(4), 90.

SulfCo See *Sonocracking.

SulFerox A process for removing hydrogen sulfide and organic sulfur compounds from hydrocarbons, similar to the *Stretford process but using an aqueous solution containing chelated iron and proprietary additives. The product is elemental sulfur. The basic reactions are

$$2Fe^{3+} + H_2S = 2Fe^{2+} + 2H^+ + S$$

$$2Fe^{2+} + 2H^+ + \tfrac{1}{2}O_2 = 2Fe^{3+} + H_2O$$

Developed and jointly licensed by Shell Oil Company and Dow Chemical Company. Introduced in 1987; 29 units were operating in 2000. The first application in coke making, in a plant near Pittsburgh, PA, was announced in 1996.

Fong, H.L., Kushner, D.S., and Scott, R.T., *Oil Gas J.*, 1987, **85**(21), 54.
Dalrymple, D.A., Trofe, T.W., and Evans, J.M., *Chem. Eng. Prog.*, 1989, **85**(3), 43.
Chem. Eng. (N.Y.), 1991, **98**(9), 44.
Hydrocarbon Process. Int. Ed., 1996, **75**(4), 140; 2000, **79**(4), 89.

Sulfex [**Sulf**ide **ex**traction] A process for removing heavy metals from waste streams by adding ferrous sulfide to precipitate them as their sulfides. Developed by the Permutit Company and now owned by U.S. Filter. Not to be confused with *SULF-X or *Sulph-X.

Chem. Eng. (N.Y.), 1983, **90**(10), 23.

Sulfiban A process for removing hydrogen sulfide from coke-oven gases by scrubbing with monoethanolamine. Developed by the Bethlehem Steel Corporation and BS&B Process Systems and tested in a demonstration plant of the former company in the mid-1970s. Not commercialized.

Massey, M.J. and Dunlap, R.W., *J. Air Pollut. Control Assoc.*, 1975, **25**(10), 1019.
Kohl, A.L. and Riesenfeld, F.C., *Gas Purification*, 4th ed., Gulf Publishing Co., Houston, TX, 1985, 104.

SULFICAT A method for presufiding *HDS catalysts. Developed by Eurocat. Piloted in 1982 and commercialized in France in 1986.

Wilson, J.H. and Berrebi, G., in *Catalysis 1987*, Ward, J.W., Ed., Elsevier, Amsterdam, the Netherlands, 1988, 393.
Chauvel, A., Delmon, B., and Hölderich, W.F., *Appl. Catal. A: Gen.*, 1994, **115**, 184.

Sulfidine A process for removing sulfur dioxide from smelter gases by reaction with a suspension of xylidine in water. Developed by the Gesellschaft für Chemische Industrie, Basel, and Metallgesellschaft, Frankfurt, and used in Germany in the 1930s; now probably obsolete.

Katz, M. and Cole, R.J., *Ind. Eng. Chem.*, 1950, **42**, 2263.
Kohl, H.L. and Nielsen, R.B., *Gas Purification*, 5th ed., Gulf Publishing Co., Houston, TX, 1997, 591.

Sulfining A process for removing sulfur compounds from petroleum distillates by treatment with sulfuric acid, followed by electrostatic phase separation.

Unzelman, G.H. and Wolf, C.J., in *Petroleum Processing Handbook*, Bland, W.F. and Davidson, R.L., Eds., McGraw-Hill, New York, 1967, 3–137.

Sulfinol A process for removing hydrogen sulfide, carbon dioxide, carbonyl sulfide, and organic sulfur compounds from natural gas by scrubbing with di-isopropanolamine dissolved in a mixture of sulfolane and water. Developed in the 1960s by Shell International Research Mij NV, Netherlands, and Shell Development Company, Houston. In 2000, over 200 units had been licensed worldwide.

Deal, G.H., Jr., Evans, H.D., Oliver, E.D., and Papadopoulos, M.N., *Pet. Refin.*, 1959, **38**(9), 185.

Dunn, C.L., Freitas, E.R., Goodenbour, J.W., Henderson, H.T., and Papadopoulos, M.N., *Hydrocarbon Process. Int. Ed.*, 1964, **43**(3), 150; 1996, **75**(4), 142; 2000, **79**(4), 90.

Kohl, H.L. and Nielsen, R.B., *Gas Purification*, 5th ed., Gulf Publishing Co., Houston, TX, 1997, 1225.

Sulfint A process for removing hydrogen sulfide from industrial gases by scrubbing with an aqueous solution of the iron–EDTA complex. The solution is regenerated by air oxidation, liberating sulfur dioxide, and reverse osmosis, which separates the dissolved salts from the iron complex. Developed by Integral Engineering, Vienna, and licensed by Le Gaz Integral Enterprise. Ten plants had been built by 1993.

Mackinger, H., Rossati, F., and Schmidt, G., *Hydrocarbon Process. Int. Ed.*, 1982, **61**(3), 169.

Hydrocarbon Process. Int. Ed., 1992, **71**(4), 136.

Kohl, H.L. and Nielsen, R.B., *Gas Purification*, 5th ed., Gulf Publishing Co., Houston, TX, 1997, 823.

SULFINT HP A variant of the *Sulfint process, which operated at high pressure. It incorporates a high-pressure filtration step to remove particles of sulfur from the redox solution before it is depressurized. Developed by IFP and Le Gaz Integral and piloted at Soings-en-Sologne, France, in 2001.

Oil Gas J., 2001, **99**(48), 46.

Sulphur, 2001 (273), 46–55.

Sulfite An acid papermaking process. Wood chips are digested in a hot sulfite solution, made by dissolving sulfur dioxide in a suspension of calcium or magnesium carbonate or hydroxide. The lignin in the wood is thereby converted to soluble calcium or magnesium lignosulfonate. Invented by B.C. Tilghman in Philadelphia in the 1860s and commercialized by C.D. Ekman in Sweden in the 1870s and widely used thereafter.

U.S. Patent 70,485.

Grant, J., *Cellulose Pulp and Allied Products*, Leonard Hill, London, U.K., 1958, Chaps. 2, 7.

Wenzel, H.F.J., *Sulfite Pulping Technology*, Lockwood Trade Journal Co., New York, 1965.

sulfite/bisulfite *See* Burkheiser.

Sulfolane A process for removing aromatic hydrocarbons from petroleum fractions by liquid–liquid extraction using sulfolane (tetramethylene sulfone; tetrahydrothiophene-1,1-dioxide) at approximately 190°C. Developed by Shell Development Company in 1959 and first commercialized in 1962; now licensed through UOP. By 2005, 134 units had been licensed worldwide. The process replaced the *Udex process. Sulfolane is used for another purpose in the *Sulfinol process.

Deal, C.H., Evans, H.D., Oliver, E.D., and Papadopulos, M.N., *Pet. Refin.*, 1959, **38**(9), 185.

Beardmore, F.S. and Kosters, W.C.G., *J. Inst. Pet.*, 1963, **49**(469), 1.

Unzelman, G.H. and Wolf, C.J., in *Petroleum Processing Handbook*, Bland, W.F. and Davidson, R.L., Eds., McGraw-Hill, New York, 1967, 3–108.

Stoodt, T.J. and Negiz, A., in *Handbook of Petroleum Refining Processes*, 3rd ed., Meyers, R.A., Ed., McGraw-Hill, New York, 2003, 2.13.

Sulfolin A process for removing sulfur compounds from hydrocarbons, similar to the *Stretford process, but including vanadium and an organic nitrogen compound in the catalytic solution. Developed jointly by SASOL and Linde. First commercialized in 1978; six plants were operating in 1994, including one for treating the *Rectisol off-gas in the *SASOL plant in South Africa and one in the Dakota Gasification Plant.

Dalrymple, D.A., Trofe, T.W., and Evans, J.M., *Chem. Eng. Prog.*, 1989, **85**(3), 43.

Hydrocarbon Process. Int. Ed., 1992, **71**(4), 137.

Kohl, H.L. and Nielsen, R.B., *Gas Purification*, 5th ed., Gulf Publishing Co., Houston, TX, 1997, 797.

Sulfosorbon A process for removing hydrogen sulfide and carbon disulfide from the gaseous effluent from the *Viscose process. Offered by Lurgi.

SULFOX A process for the catalytic oxidation of aqueous sulfides, developed by UOP but not commercialized.

Sulphur, 1974, (117), 40.
Hydrocarbon Process. Int. Ed., 1975, **54**(4), 106.

Sulfreen Also known as the SNPA process. A variation of the *Claus process for removing hydrogen sulfide from gas streams by reaction with sulfur dioxide to produce elemental sulfur. It differs from the Claus process in using a lower temperature, causing the sulfur to be retained on the catalyst. The catalyst was originally carbon, but alumina is used now. Developed by Lurgi Gesellschaft für Warme- und Chemotechnik and the Société Nationale des Pétroles d'Aquitaine. More than 53 units were operating or being designed in 2000. *See also* HYDROSULFREEN, Oxysulfreen.

Krill, H. and Storp, K., *Chem. Eng. (N.Y.)*, 1973, **80**(17), 84.
Davis, G.W., *Oil Gas J.*, 1985, **83**(8), 108.
Hydrocarbon Process. Int. Ed., 1996, **75**(4), 142; 2000, **79**(4), 90.
Kohl, H.L. and Nielsen, R.B., *Gas Purification*, 5th ed., Gulf Publishing Co., Houston, TX, 1997, 699.

SULFREX A catalytic process for removing sulfur compounds from LPG, gasoline, and kerosene. Developed by Total and IFP in the 1980s and operated in Tenguiz, CIS, since 1991.

Chauvel, A., Delmon, B., and Hölderich, W.F., *Appl. Catal. A: Gen.*, 1994, **115**, 1783.

Sulften A process for removing hydrogen sulfide from the tail gases from the *Claus process. The solvent (Ucarsol HS 103) was developed by the Union Carbide Corporation; the process was developed by Ford Bacon and Davis, Dallas, TX.

Chem. Eng. (N.Y.), 1984, **91**(13), 150.

Sulfur-X A process for removing sulfur compounds (principally thiophene) from naphtha by solvent extraction with sulfolane. Developed by UOP and announced in 2002. This process does not require hydrogen, which gives it an advantage over competing desulfurization processes.

Chem. Eng. (N.Y.), 2003, **110**(9), 29.
Nafis, D.A. and Houde, E.J., in *Handbook of Petroleum Refining Processes*, 3rd ed., Meyers, R.A., Ed., McGraw-Hill, New York, 2003, 11.75.

SULF-X [**SULF**ur e**X**traction] A regenerable *flue-gas desulfurization process in which the sulfur dioxide is absorbed by aqueous sodium sulfide in a bed packed with pyrite. Ferrous sulfate is produced; this is removed by centrifugation and calcined with coke and fresh pyrite. Sulfur vapor is evolved and condensed, and the residue is reused in the scrubber. Developed by Pittsburgh Environmental and Energy Systems. Piloted in the mid-1980s but not commercialized. Not to be confused with *Sulfex or *Sulph-X.

Kohl, H.L. and Nielsen, R.B., *Gas Purification*, 5th ed., Gulf Publishing Co., Houston, TX, 1997, 585.

Sulpel [**Sul**fur **pel**letization] A process for making sulfur pellets. Molten sulfur is injected through nozzles into water containing a trace of a proprietary additive, which gives the resulting pellets a smooth, waterproof surface. Developed and offered by Humphreys and Glasgow, United Kingdom; nine plants had been engineered as of 1992.

Sulph-X A process for trapping sulfur in coal combustion. A proprietary mixture of inorganic salts, including sodium chloride, is mixed with the coal and combines with the sulfur dioxide so that it remains fixed in the ash instead of evolving with the combustion gases. Invented in China and developed in Australia in the 1990s by Coal Corporation Pty. Not to be confused with *Sulfex or *SULF-X.

Sultrol A *flue-gas desulfurization process, similar to the *Flakt–Boliden process, but using potassium citrate instead of sodium citrate. Developed by Pfizer and announced in 1985, but not known to have been commercialized.

Sulzer A family of processes for separating or purifying organic chemicals by melt crystallization without using solvents. Two systems are in use: static crystallization and falling-film crystallization. The latter is proprietary to Sulzer Chemtec, Switzerland.

Wynn, N.P., *Chem. Eng. Prog.*, 1992, **88**(3), 52.

Sulzer-MWB [MetalWerk AG Buchs] An obsolete process for recovering naphthalene from a coal-tar fraction by multistage fractional crystallization from the melt. Formerly operated by Rutgerswerke at Castrop-Rauxel, Germany.

U.S. Patent 3,621,664.
Franck, H.G. and Stadelhofer, J.W., *Industrial Aromatic Chemistry*, Springer-Verlag, Berlin, Germany, 1988, 303.

Sumitomo-BF A *PSA hydrogen purification process using a carbon molecular sieve as the selective adsorbent. Developed by Sumitomo, Japan.

Suzuki, M., in *Adsorption and Ion Exchange: Fundamentals and Applications*, LeVan, M.D., Ed., American Institute of Chemical Engineers, New York, 1998, 122.

Sunliquid An enzymic process for making ethanol from straw. Developed by Süd-Chemie in Munich from 2009. A larger demonstration plant was being built by Clariant, a subsidiary of Süd-Chemie, in Straubing, Germany, in 2012.

Superclaus [**Super**ior **Claus**] A superior version of the *Claus process. Hydrogen sulfide is catalytically oxidized to elemental sulfur, using air and water. The first generation of catalysts used iron and chromium oxides on α-alumina. The second generation used iron oxide on silica. The latest version, Superclaus-99, uses a different catalyst, which produces less sulfur dioxide in the oxidation stage. Developed in the Netherlands by Comprimo, V.E.G. Gasinstitut, and the University of Utrecht and operated in Germany since 1988. Over 80 units were operating or under construction in 2000. See also *Ecoclaus.

Lagas, J.A., Borsboom, J., and Heijkoop, G., *Hydrocarbon Process. Int. Ed.*, 1989, **68**(4), 40.
Goar, B.G., Lagas, J.A., Borsboom, J., and Heijkoop, G., *Sulphur*, 1992, (220), 44.
Wieckowska, J., *Catal. Today*, 1995, **24**(4), 442.
Hydrocarbon Process. Int. Ed., 1996, **75**(4), 144; 2000, **79**(4), 92.

Supercondensed Mode A method of operating gas-phase olefin polymerization plants. *See* High Productivity.

Super Drizo *See* Drizo.

SUPERFLEX (1) A modified *FCC process designed to maximize the production of propylene from C_4 to C_8 refinery streams. Developed by Arco Chemical and commercialized by Sasol in South Africa in 2005. KBR was the exclusive licensor in 2005. *See also* Maxofin. SuperFlex is also the trade name for a range of polyurethane suspension components. There are numerous miscellaneous other uses of this name.

Chem. Br., 1993, 110.
Eur. Chem. News, 2002, **76**(1998), 20; 2005, **82**(2137), 22.
Chem. Eng. (N.Y.), 2004, **111**(4), 8.

Superflex (2) A LyondellBasell process for isomerizing olefin mixtures to ethylene, propylene, and butenes.

Super Hy-Pro *See* Hy-Pro.

Super Morphylane An improved version of *Morphylane being developed by Krupp Uhde in 1999.

Eur. Chem. News, 1999, **70**(1867), 27.

SUPER-SCOT An improved version of the *SCOT process that emits smaller quantities of sulfur compounds. Engineered by Stork Engineers and Contractors.

Supersorbon The trade name for grades of activated carbon made in Germany for use in gas masks during World War II. The carbons are now provided by F.B. Silbermann, Germany. Also loosely used as the name of a process developed by Lurgi, which uses this adsorbent for removing organic vapors from gas streams. These organic materials are subsequently recovered by steam treatment.

Super Sour A process for recovering hydrogen sulfide from aqueous wastes produced in oil refineries. It is evaporated in a hot-feed flash drum at about 60° and absorbed in an amine. The amine is not identified. Developed at the Indian Oil Company's refinery in Gujarat and used there since 2009. The name is also used for a type of marijuana.

> *Oil & Gas J.*, 2009, **107**(18), 44.

Supramics A family of processes for making ceramic materials by reacting cement (or other metal oxide hydroxide mixtures) with supercritical carbon dioxide.

> Jones, R., in *Green Engineering*, Anastas, P.T., Heine, L.G., and Williamson, T.C., Eds., American Chemical Society, Washington, DC, 2000.

SURE [**SU**lfur **RE**covery] A version of the *Claus process in which the capacity of the plant is increased by using air enriched in oxygen in the production of the sulfur dioxide. There are two versions: SURE SSB [Side Stream Burner] and SURE DC [Double Combustion]. In the first, a small portion of the feed stream containing hydrogen sulfide is burned substoichiometrically in a second burner; in the second, the hydrogen sulfide is oxidized in two stages, with cooling and sulfur separation between them. Both of these modifications are to provide the hydrogen sulfide–sulfur dioxide mixture at the optimum ratio and temperature for the Claus reaction. Developed by the Ralph M. Parsons Company and BOC Group and now offered by Worley Parsons (Australia). One plant was operating in Japan in 1992. A large development plant was installed at a carbon disulfide plant at Stretford, United Kingdom, in 1996. Twelve plants were operating, and 12 more were under design in 2000.

> *Chem. Eng. (Rugby, Engl.)*, 1991, (494), 13.
> *Eur. Chem. News*, 1997, **67**(1746), 19.
> *Hydrocarbon Process. Int. Ed.*, 1996, **75**(4), 144; 2000, **79**(4), 92.

SURECAT A method for prereducing and passivating nickel catalysts. Developed in 1990 by Eurocat for ATOCHEM.

> Rommelaere, F., Rondi, J.C., Dufresne, P., Rabehasaina, H., Boitiaux, J.P., and Sarrazin, P., *Bull. Soc. Chim. Belg.*, 1991, **100**, 897.

SUREMAX An improved version of the *SURE process, offering increased plant capacity and recovery. Developed by BOC and announced in 2001 and now offered by Worley Parsons (Australia).

> *Chem. Eng. (Rugby, Engl.)*, 2001, (718), 28.
> *Hydrocarbon Process. Int. Ed.*, 2003, **82**(1), 73.

Suspensoid An early *catalytic cracking process in which the silica–alumina catalyst was suspended in the petroleum. First operated in Ontario in 1940.

> Asinger, F., *Mono-olefins: Chemistry and Technology*, translated by B.J. Hazzard, Pergamon Press, Oxford, U.K., 1968, 22.
> Unzelman, G.H. and Wolf, C.J., in *Petroleum Processing Handbook*, Bland, W.F. and Davidson, R.L., Eds., McGraw-Hill, New York, 1967, 3–11.

SVP [**S**ingle-**V**essel **P**rocess] A process for making chlorine dioxide by reacting sodium chlorate with hydrochloric acid. Invented in 1971 by the Hooker Chemical Corporation, Niagara Falls. A modern variation, SVP-LITE, is now offered by Akzo Nobel.

> U.S. Patent 3,816,077.

Sweetfrac A process for removing sulfur compounds from naphtha. The process is essentially the UOP *Merox process, positioned upstream of the naphtha splitter in conventional petroleum refining.

> Nafis, D.A. and Houde, E.J., in *Handbook of Petroleum Refining Processes*, 3rd ed., Meyers, R.A., Ed., McGraw-Hill, New York, 2003, 11.67.

Swift A process for making monoammonium phosphate from liquid ammonia and phosphoric acid. The reactants are mixed with a special nozzle. The slurry product from the neutralization is injected into the top of a heated tower, water flashes off as the product falls, and the powdered product collects at the base of the column. *See also* Gardinier.

SWITGTHERM A catalytic process for oxidizing volatile organic compounds (VOCs). It involves regenerative heat exchange, which permits autothermal operation at VOC concentrations in the range 250–650 ppm. Developed in Poland and now used in over 100 installations there.

> Haber, J. and Borowiak, M., *Appl. Catal. A: Gen.*, 1997, **155**(2), 293.

SX A common abbreviation for Solvent eXtraction, as used in hydrometallurgy.

SYDEC [Selective Yield DElayed Coking] A *Thermal Cracking process that converts petroleum residues to petroleum coke and lighter hydrocarbons. Developed by Foster Wheeler North America Corporation.

> Elliott, J.D., *Oil Gas J.*, 1991, **89**(5), 41.
> *Hydrocarbon Process. Int. Ed.*, 1994, **73**(11), 96.

Sydox [Sydney oxidation] A process for destroying polychlorinated biphenyls by oxidation. A catalyst containing ruthenium is used, and the temperature is kept below 100°C to prevent the formation of dioxins. Developed by J. Beattie at the University of Sydney in the 1980s; by 1991, it had not been piloted.

> Beder, S., *New Sci.*, 1991, **130**(1772), 36.

Symrise A process for making enantiomerically pure (−) menthol from a mixture of menthol isomers. Formerly called the Haarmann and Reimer process. Developed by the German company Symrise, which acquired Haarmann and Reimer. The proprietary heterogeneous catalyst (not described) contains nickel.

> Symrise has also patented an enzymic process for making (−) menthol that uses a lipase.
> U.S. Patent 6,306,500.
> Etzold, B. and Jess, A., *Chim. Oggi*, 2008, **26**(3), 21.

Symyx-Dow A process for making polyolefins using a hafnium-based metallocene catalyst that had been identified by the Symyx combinatorial method. Not commercialized in 2003.

> *J. Am. Chem. Soc.*, 2003, **125**, 4306.

SynCrude-Syncoal Also written as **SC2**. A medium-temperature coal pyrolysis process for use with coals containing >30% water and >32% other volatile matter. The product is a synthetic crude oil plus a char. Developed by Convert Coal, CA, in 2006, but the entire process had not been piloted then.

> *Chem. Eng. (N.Y.)*, 2006, **113**(13), 10.

Synfining A process for upgrading *Fischer–Tropsch liquid hydrocarbons into middle distillate products such as synthetic diesel and jet fuels. Developed by Syntroleum Corp. since 1984. See also *Bio-Synfining.

syngas Also called synthesis gas. Not a process but a general name for mixtures of carbon monoxide and hydrogen made from petroleum fractions or coal. Widely used as a feed for synthesizing organic chemicals by catalytic processes.

Cornils, B., in *Chemicals from Coal: New Processes*, Payne, K.R., Ed., John Wiley & Sons, Chichester, U.K., 1987, 1, 45, 93.

Rostrup-Nielsen, J.R., *Catal. Today*, 1993, **18**(4), 305.

Eur. Chem. News CHEMSCOPE, 1997, Sept, 24.

SynGen A cold plasma process for obtaining a hydrogen-rich gas, low in sulfur, from a crude hydrocarbon source, rich in sulfur. Developed in 2002 by Synergy Technologies Corporation, United States, and proposed for generating hydrogen-rich gases for use in fuel cells. There are several unrelated uses for the name.

Synol A version of the *Fischer–Tropsch process developed in Germany during World War II. It used a different catalyst and produced a larger fraction of alcohols and olefins.

Storch, H.H., Golumbic, N., and Anderson, R.B., *The Fischer-Tropsch and Related Syntheses*, John Wiley & Sons, New York, 1951, 559.

Synroc [**Syn**thetic **rock**] A process for immobilizing radioactive wastes by incorporating them in a synthetic rock. Invented in 1978 by A.E. Ringwood in the Australian National University, Canberra, and subsequently developed further in many other laboratories in several countries. The "rock" has four main components, all containing titanium: perovskite, zirconolite, hollandite, and rutile. A nonradioactive pilot plant was designed and built at the Australian Nuclear Science and Technology Organization (ANSTO), Sydney, and operated from 1988 to 1991. The project is continuing at ANSTO with emphasis on Synroc processing science and active plant design. In 1990, the Australian company Nuclear Waste Management Pty., which has nonexclusive rights, proposed to build a pilot plant in the USSR, but this plan was probably abandoned. A test with highly active waste was conducted in 1997.

Ringwood, A.E., Kesson, S.E., Reeve, K.D., Levins, D.M., and Ramm, E.J., in *Radioactive Waste Forms for the Future*, Lutze, W. and Ewing, R.C., Eds., North-Holland, Amsterdam, the Netherlands, 1988, 233.

Ojovan, M.I. and Lee, W.E., *An Introduction to Nuclear Waste Immobilisation*, Elsevier, Amsterdam, the Netherlands, 2005, 256.

SynSat [**Syn**ergetic **Sat**uration] A process for removing aromatic hydrocarbons and sulfur compounds from diesel fuel. Developed by ABB Lummus Crest and Criterion Catalyst Company. Six units were operating in 1996.

Suchanek, A.J., *Oil Gas J.*, 1990, 7 May, 109.

Stanislaus, A. and Cooper, B.H., *Catal. Rev. Sci. Eng.*, 1994, **36**(1), 113.

Hydrocarbon Process. Int. Ed., 1996, **75**(11), 130; 1997, xxx, 56.

Synthane A coal gasification process using steam and oxygen in a fluidized bed. An unusual feature is the large volume of hot gas recycled. Developed by the U.S. Bureau of Mines from 1961. A pilot plant, designed by the C.E. Lummus Company, was built at Bruceton, PA, in 1976.

Hebden, D. and Stroud, H.J.F., in *Chemistry of Coal Utilization*, 2nd Suppl. Vol., Elliott, M.A., Ed., John Wiley & Sons, New York, 1981, 1692.

synthesis gas *See* Syngas.

Synthetic Oils Also known as the Robinson–Bindley process. A variation on the *Fischer–Tropsch process that uses a different catalyst and a different H_2/CO ratio and yields a higher proportion of olefins in the product. Piloted by Synthetic Oils, United Kingdom, in the 1930s, but not commercialized.

Mydleton, W.W., *J. Inst. Fuel*, 1938, **11**, 477.

Synthine [**Synth**etic benz**in**] An early version of the *Fischer–Tropsch process in which a mixture of carbon monoxide and hydrogen was passed over an iron catalyst and thereby converted to a complex mixture of oxygenates.

Fischer, F. and Tropsch, H., *Ber. Dtsch. Chem. Ges.*, 1923, **56**, 2428.

Lane, J.C., *Pet. Refin.*, 1946, **25**(8), 87; **25**(9), 423; **25**(10), 493; **25**(11), 587 (*Chem. Abstr.*, **42**, 9118).

Synthoil A coal liquefaction process in which coal, suspended in oil from the process, is hydrogenated over a cobalt–molybdenum catalyst on alumina. The process was piloted by the Pittsburgh Energy Research Center at Bruceton, PA, in the 1970s using several types of coal, but it was abandoned in 1978. *See also* CSF, H-Coal.

Yavorsky, P.M., Akhtar, S., and Freidman, S., *Chem. Eng. Prog.*, 1973, **69**(3), 51.
Yavorsky, P.M., Akhtar, S., Lacey, J.J., Weintraub, M., and Reznik, A.A., *Chem. Eng. Prog.*, 1975, **71**(4), 79.

Synthol A version of the *Fischer–Tropsch process, for making liquid fuels and organic chemicals from *syngas. Developed by Pullman Kellogg between 1940 and 1960. First operated at the SASOL plant in South Africa in 1955. The name was also used for the product from the original Fischer–Tropsch process, developed in the 1920s. *See also* Synol.

Garrett, L.W., Jr., *Chem. Eng. Prog.*, 1960, **56**(4), 39.
Hydrocarbon Process. Int. Ed., 1963, **42**(11), 225.
Asinger, F., *Paraffins: Chemistry and Technology*, translated by B.J. Hazzard, Pergamon Press, Oxford, U.K., 1968, 96.
Mako, P.F. and Samuel, W.A., in *Handbook of Synfuels Technology*, Meyers, R.A., Ed., McGraw-Hill, New York, 1984, 2–13.

Synthracite *See* carbonization.

Syntroleum A process for converting natural gas into liquid fuels. Developed since 1984 by Syntroleum Corp. Supplanted by *Synfining. Piloted on a large scale since 1990. It is especially suitable for small plants in remote locations. Unlike the *FT process, it uses air as the oxidant. A further development has been *Bio-Synfining, a process for making diesel from biomass by a combination of pyrolysis and syngas technologies. Developed by Pacific Renewable Fuels and Chemicals and Red Lion Bio-Energy.

Agee, M., *Appl. Catal. A: Gen.*, 1997, **155**(1), N5.
Hydrocarbon Process. Int. Ed., 1998, **77**(1), 41.
Weick, L. and Nimmo, M., in *Handbook of Petroleum Refining Processes*, 3rd ed., Meyers, R.A., Ed., McGraw-Hill, New York, 2003, 15.15.
Biorefining Magazine, Sep 2011.
Wikipedia, The Free Encyclopedia. Syntroleum. Revised 15 Mar 2013. Retrieved 22 Jul 2013.

S Zorb A process for removing sulfur compounds from diesel and gasoline by adsorption on a proprietary adsorbent. The adsorbent, believed to contain a reduced metal that is converted to its sulfide, is contained in a fluidized-bed reactor with continuous regeneration. Regeneration is by oxidation to sulfur dioxide, which can then be used in various ways. Developed by Phillips Petroleum Company in the 1980s, originally for removing sulfur from furnace gases. A demonstration unit was built in Borger, TX, in 2001. Now licensed by ConocoPhillips, which had granted more than 40 licenses in 2003. The technology was sold to SINOPEC TECH in 2007. By 2010, there were 13 units in commercial operation.

Hydrocarbon Process. Int. Ed., 1999, **78**(10), 37.
Sulphur, Jul-Aug 2001 (275), 16.
Oil Gas J., 2002, **99**, 74.
Chem. Eng. (N.Y.), 2003, **110**(9), 29.
Song, C., *Catal. Today*, 2003, **86**(1–4), 227.
Sughrue, E. and Parsons, J.S., in *Handbook of Petroleum Refining Processes*, 3rd ed., Meyers, R.A., Ed., McGraw-Hill, New York, 2003, 11.51.
Ito, E. and van Veen, J.A.R., *Catal. Today*, 2006, **116**(4), 447.

T

TAC-9 [**T**oray **A**dvanced **C**atalyst] A catalytic process for making mixed xylenes from mixed aromatic hydrocarbons. Developed by Toray Industries and now licensed exclusively to UOP. Three commercial units were operating in 2005.

Oil Gas J., 1998, **96**(48), 35.
Eur. Chem. News, 1998, **69**(1837), 35.

Taciuk *See* carbonization.

Tainton A metallurgical process in which sulfides are converted to sulfates by heating in a controlled quantity of air, and the sulfates so produced are dissolved in water. Used for removing zinc from silver and lead ores.

Takahax A variation of the *Stretford process for removing hydrogen sulfide from gas streams, in which naphthaquinone sulfonic acid is used in place of anthraquinone disulfonic acid. Four variants have been devised: types A and B use ammonia as the alkali; types C and D use sodium hydroxide or carbonate. Developed by the Tokyo Gas Company and licensed in the United States by the General Electric Environmental Services. Many plants are operating in Japan.

Hydrocarbon Process. Int. Ed., 1975, **54**(4), 105.
Kohl, H.L. and Nielsen, R.B., *Gas Purification*, 5th ed., Gulf Publishing Co., Houston, TX, 1997, 765.

Takasago A catalytic process for the enantioselective isomerization of allylic amines. The catalyst is a chiral rhodium complex. Used in the manufacture of (−) menthol. Named after Takasago International Corporation, the Japanese company that commercialized the process in 1983. Nobel prizewinner R. Noyori was employed by that company.

Ojima, I., ed., *Catalytic Asymmetric Synthesis*, VCH Publishers, New York, 1993, 42.
Sheldon, R.A., Arends, I., and Hanefeld, U., *Green Chemistry and Catalysis*, Wiley-VCH, Weinheim, Germany, 2007, 36.
Wikipedia, The Free Encyclopedia. Takasago International Corporation. Revised 17 Jul 2013. Retrieved 22 Jul 2013.

Talafloc A process for decolorizing cane sugar syrup by precipitating the coloring impurities with a long-chain quaternary ammonium salt. Invented in 1967 by M.C. Bennett at Tate and Lyle, United Kingdom.

British Patent 1,224,990.
U.S. Patent 3,698,951.

Talalay A process for making foam rubber. The foaming gas is either carbon dioxide or oxygen generated from hydrogen peroxide and an enzyme. After foaming, the latex particles are coagulated by freezing and introducing carbon dioxide. Invented by L. Talalay at the B.F. Goodrich Company in 1959. *See also* Dunlop.

U.S. Patent 2,984,631.
Madge, E.W., *Latex Foam Rubber*, John Wiley & Sons, New York, 1962.

Talbot A semicontinuous steelmaking process that combines the *Bessemer and *Open Hearth processes. Molten pig iron from a Bessemer converter is poured into an Open Hearth furnace containing fresh ore and lime. Impurities in the pig iron oxidize and enter the slag. The process improves the yield of steel and the throughput of the plant. Introduced by B. Talbot at Pencoyd, PA, in 1900 and subsequently adopted in Europe.

Barraclough, K.C., *Steelmaking 1850–1900*, The Institute of Metals, London, U.K., 1990, 283.

Talodura A process for clarifying cane sugar syrup. Calcium phosphate is precipitated in the syrup and flocculated with a polyacrylamide. Invented in 1973 by J.T. Rundell and P.R. Pottage at Tate & Lyle, United Kingdom.

British Patent 1,428,790.

TALSPEAK A process for separating lanthanides from trivalent actinides.

Weaver, B. and Kappelmann, F.A., *J. Inorg. Nucl. Chem.*, 1968, **30**, 263.
Nash, K.L., in *Separations for the Nuclear Fuel Cycle in the 21st Century*, Lumetta, G.J., Nash, K.L., Clark, S.B., and Friese, J.I., Eds., American Chemical Society, Washington, DC, 2006, 31.

TAREX A process for destroying hydrocarbons, organic wastes, and hydrogen chloride by controlled combustion in a special combustion chamber that can withstand sudden pressure surges. Engineered by KEU Energie- und Umwelttechnik, Germany.

TAS [AutoThermal Aerobic Stabilization] An autothermal, aerobic process for stabilizing activated sewage sludge. Offered by Linde, Munich.

Tatoray [Transalkylation aromatics **Toray**] A process for transalkylating toluene, trimethylbenzenes, or both into a mixture of benzene and xylenes. Operated in the vapor phase, with hydrogen, in a fixed bed containing a zeolite catalyst. Developed jointly by Toray Industries and UOP and now licensed by UOP. First operated commercially in Japan in 1969; as of 2005, 44 units had been licensed and 40 were operating.

Otani, S., *Chem. Eng. (N.Y.)*, 1970, **77**(16), 118.
Iwamura, T., Otani, S., and Sato, M., *Bull. Jpn. Pet. Inst.*, 1971, **13**, 116 (*Chem. Abstr.*, **75**, 131292).
Hydrocarbon Process. Int. Ed., 1975, **54**(11), 115.
Johnson, J.A., in *Handbook of Petroleum Refining Processes*, 3rd ed., Meyers, R.A., Ed., McGraw-Hill, New York, 2003, 2.5.

T2BX [**T**oluene **to B**enzene and **X**ylenes] A process for converting toluene to a mixture of benzene and xylenes. Toluene vapor, mixed with hydrogen, is passed over a zeolite catalyst at 370°C–430°C; the hydrogen is separated and the products fractionated. The production of benzene is its main purpose. Developed by Cosden Technology, United States.

Chem. Eng. (N.Y.), 1987, **94**(2), 159.

TCC [**T**hermofor **C**atalytic **C**racking] *See* Thermofor.

TCF [**T**otally **C**hlorine-**F**ree] A generic term for pulp-bleaching processes that do not use chlorine in any form. Oxidants and enzymes are used. *See also* ECF.

Chem. Eng. (N.Y.), 1997, **104**(4), 33.
Nelson, P.J., in *Environmentally Friendly Technologies for the Pulp and Paper Industries*, Young, R.A. and Akhar, M., Eds., John Wiley & Sons, New York, 1998, 215.

TCX A catalytic process for making ethanol from a variety of carbon sources via acetic acid. Developed by Celanese International Corp, piloted in Texas, and announced in 2010. Not legal in the United States because the ethanol is not made from a renewable source.

U.S. Patent 7,863,489.
Chem. Eng. News, 2011, **89**(43), 20.
ICIS Chem. Bus., 2012, **281**(13), 13.
Wikipedia, The Free Encyclopedia. TCX Technology. Revised 3 Dec 2012. Retrieved 22 Jul 2013.

TDP [**T**oluene **D**is**P**roportionation] A general name for catalytic processes for converting toluene to a mixture of xylene isomers and benzene. One proprietary version is *MTDP.

TEES [**T**hermochemical **E**nvironmental **E**nergy **S**ystem] A catalytic process for destroying organic wastes in aqueous systems by thermochemical gasification. High temperatures and pressures are used. The catalyst is nickel metal supported on sodium carbonate; the products are mostly methane, carbon dioxide, and hydrogen. Developed by Battelle Pacific Northwest Laboratory, Richland, WA, in the late 1980s and now licensed by Onsite and Offsite, Inc.

> U.S. Patent 5,019,135.
> *Process Eng. (London)*, 1989, Oct, 25.
> Elliott, D.C., Neuenschwander, G.G., Phelps, M.R., Hart, T.R., Zacher, A.H., and Silva, L.J., *Ind. Eng. Chem. Res.*, 1999, **38**(3), 879.

Teijin A process for oxidizing *p*-xylene to terephthalic acid. The catalyst used is a soluble cobalt compound, used at a high concentration. Developed by Teijin, Tokyo.

> Yoshimura, T., *Chem. Eng. (N.Y.)*, 1969, **76**(10), 78.
> Raghavendrachar, P. and Ramachandran, S., *Ind. Eng. Chem. Res.*, 1992, **31**, 453.

Tekkosha An electrolytic process for obtaining sodium from the sodium amalgam formed in the *Chlor-Alkali process. The electrolyte is a fused mixture of sodium hydroxide, sodium iodide, and sodium cyanide. The sodium deposits at the iron cathode. Developed by Tekkosha Company, Japan, in the 1960s and commercialized in 1971.

> Yamaguchi, T., *Chem. Econ. Eng. Rev.*, 1972, **4**(1), 24.
> Nakamura, T. and Fukuchi, Y., *J. Met.*, 1972, **24**(8), 25.

Tencel *See* Lyocel.

Tenex-Plus A process for increasing the octane number of gasoline by a combination of hydrogenation and isomerization. *See also* BenSat.

> Weissermel, K. and Arpe, H.-J., *Industrial Organic Chemistry*, 4th ed., translated by C.R. Hindley and S. Hawkins, Wiley-VCH, Weinheim, Germany, 2003, 349.

Tennessee Eastman *See* Eastman.

Tenteleff Also spelled Tentelew. An early version of the *Contact process for making sulfuric acid. The catalyst was platinum supported on asbestos. Invented in 1907 and operated by the Gesellschaft der Tentelewschen Chemischen Fabrik, St. Petersburg.

> British Patents 12,213 (1907); 14,670 (1911).
> Miles, F.D., *The Manufacture of Sulfuric Acid (Contact Process)*, Gurney & Jackson, London, U.K., 1925, Chap. 9.

Terra-Crete A process for stabilizing the calcium sulfate/sulfite waste from *flue-gas desulfurization, so that it may be used for landfill. Calcination converts the calcium sulfite to cementitious material, to which proprietary additions are made. Developed by SFT Corporation, York, PA. *See also* Terra-Tite.

> Valiga, R., in *Toxic and Hazardous Waste Disposal*, Vol. 1, Pojasek, R.J., Ed., Ann Arbor Science, Ann Arbor, MI, 1979, Chap. 10.

Terra-Tite A process for stabilizing the calcium sulfate/sulfite waste from *flue-gas desulfurization, so that it may be used for landfill. Proprietary cementitious additives are used. Developed by the Stabatrol Corporation, Norristown, PA. *See also* Terra-Crete.

> Smith, R.H., in *Toxic and Hazardous Waste Disposal*, Vol. 1, Pojasek, R.J., Ed., Ann Arbor Science, Ann Arbor, MI, 1979, Chap. 8.

TERRox A biological process for decontaminating soil that has been polluted by hydrocarbons and other organic chemicals. Developed by Degussa and operated at its plant in Knapsack, Germany, from 1996.

TERVAHL A *visbreaking process, developed by *ASVAHL. One plant was operating in France in 1988. *See also* HYVAHL.

> *Hydrocarbon Process. Int. Ed.*, 1994, **73**(11), 148.
> Chauvel, A., Delmon, B., and Hölderich, W.F., *Appl. Catal. A: Gen.*, 1994, **115**, 186.

Tesima A plasma process for making nanoparticles of oxides, carbides, and nitrides. Two or more materials are fed simultaneously to the plasma. Developed by QinetiQ.

> *Chem. World*, 2005, **2**(6), 47.

Testrup *See* Normann.

Tetra [**Tetra**ethylene glycol] Also called Tetraextraction. A process for removing aromatic hydrocarbons from petroleum fractions by liquid–liquid extraction using tetraethylene glycol. Developed by Union Carbide Corporation as an improvement on the *Udex process. Seventeen units had been licensed by 1986, when the process was supplanted by the *Carom process. These processes are now offered or serviced by UOP.

> *Hydrocarbon Process. Int. Ed.*, 1980, **59**(9), 204.
> Symoniac, M.F., Ganju, Y.N., and Vidueira, J.A., *Hydrocarbon Process. Int. Ed.*, 1981, **60**(9), 139.
> Vidueira, J.A., in *Handbook of Solvent Extraction*, Lo, C.C., Baird, M.H.I., and Hanson, C., Eds., John Wiley & Sons, Chichester, U.K., 1983, 18.2.2.
> Stoodt, T.J. and Negiz, A., in *Handbook of Petroleum Refining Processes*, 3rd ed., Meyers, R.A., Ed., McGraw-Hill, New York, 2003, 2.22.

TETRA HDS [**H**igh-**D**ensity **S**olids] A process for aiding the removal of heavy metals from wastewaters. It is a physical process, which controls the characteristics of heavy metal hydroxide precipitates so that they settle quicker. The precipitates have a hydrophobic surface, so they are easy to dewater. Developed and licensed by Tetra Technologies, Houston, TX. Widely used by the iron and steel industry in the United States. Not to be confused with *hydrodesulfurization, which is often abbreviated to HDS.

Tetronics A process for treating dusts from electrical arc furnaces for making steel and nonferrous metals. Volatile metals (zinc, lead, cadmium) are recovered, and the residual slag is nontoxic and suitable for landfill. The dusts, mixed with coal dust and a flux, are fed to a furnace heated by a plasma gun. The metal oxides present are selectively reduced, and the vapors of zinc, lead, and cadmium are condensed in a modified *Imperial Smelting furnace. Developed by Tetronics Research and Development Company, United Kingdom, and first commercialized for steel dusts at Florida Steel, Jackson, TN, in 1989. Seven plants were operating in several countries in 1992.

> Chapman, C.D. and Cowx, P.M., *Steel Times*, 1991, Jun, 301.

Texaco The Texaco Oil Company has developed many processes, but the one most associated with its name is a coal gasification system. Powdered coal, in the form of a water slurry, together with oxygen, is fed to the gasifier. The water moderates the temperature of the reaction. Development began at Texaco's laboratory in Los Angeles in 1948 and has continued ever since. The first demonstration unit was built in Morgantown, WV, in 1957. The process is now operated in the United States, Germany, and Japan and is licensed in China. In some plants, the gas produced is used for producing chemicals.

> Schlinger, W.D., in *Handbook of Synfuels Technology*, Meyers, R.A., Ed., McGraw-Hill, New York, 1984, 3–5.
> Rezaiyan, J. and Cheremisinoff, N.P., *Gasification Technologies*, Taylor & Francis, Boca Raton, FL, 2005, 47.

Texaco Selective Finishing *See* TSF.

TEXTREAT A process for removing hydrogen sulfide and carbon dioxide from gas streams by scrubbing with a formulated alkanolamine solution. Developed by Texaco Chemical Company.

> Kohl, H.L. and Nielsen, R.B., *Gas Purification*, 5th ed., Gulf Publishing Co., Houston, TX, 1997, 42.

Thann A process for making crystalline calcium hypochlorite by passing chlorine into an aqueous slurry of calcium hydroxide. There are several such processes; in this one, some of the filtrate is recycled in order to produce larger crystals. Invented by J. Ourisson in France in 1936.

French Patent 825,903.
British Patent 487,009.

THD [Toluene HydroDealkylation] A process for converting toluene to benzene, developed by the Gulf Oil Corporation.

THDA [Thermal HydroDeAlkylation] A process for dealkylating alkyl benzenes to produce benzene. The by-product is mainly methane. Developed by UOP and licensed by that company.

Stoodt, T.J. and Negiz, A., in *Handbook of Petroleum Refining Processes*, 3rd ed., Meyers, R.A., Ed., McGraw-Hill, New York, 2003, 2.25.

Thénard A process for making white lead pigment (basic lead carbonate) by boiling litharge (lead monoxide) with lead acetate solution and passing carbon dioxide gas into the suspension. L.J. Thénard (1777–1857) was an eminent French chemist, more famous for his invention of Thénard's blue (cobalt aluminate).

Thermal Black One of the processes used to make carbon black. The feedstock is usually natural gas. The gas is pyrolyzed in one of a pair of refractory reactors, which has been preheated by burning part of the feed and hydrogen from the process. When the temperature has fallen, the functions of the reactors are interchanged.

Kühner, G. and Voll, M., in *Carbon Black Science and Technology*, Donnet, J.-B., Bansai, R.C., and Wang, M.-J., Eds., Marcel Dekker, New York, 1993, 59.
Buxbaum, G. and Pfaff, G., *Industrial Inorganic Pigments*, 3rd ed., VCH, Weinheim, Germany, 2005, 175.

Thermal Cracking The pyrolysis of petroleum fractions to produce lower molecular weight materials. Developed by J.A. Dubbs and C.P. Dubbs in 1909 and demonstrated on a larger scale in Kansas in 1919. Such processes with special names that are described in this dictionary are ACR, Burton, Carburol, Cross, Dubbs, Dubrovai, Fleming, FLEXICOKING, Flexicracking, FLUID COKING, Gyro, Hall, Hoechst HTP, HSC, Jenkins, Knox, MHDV, TERVAHL, TPC, Tube and Tank, TVP, and Visbreaking. *See also* Catalytic Cracking.

Thermal DeNOx A noncatalytic version of *Denox developed by Exxon Research and Engineering Company and first operated in Japan in 1974. By 1999, it had been operated in 60 units in three countries.

U.S. Patent 3,900,554.

Thermatomic An early process for making carbon black by the incomplete combustion of natural gas. Operated in the 1920s and 1930s by the Thermatomic Carbon Company, Pittsburgh.

Ellis, C., *The Chemistry of Petroleum Derivatives*, The Chemical Catalog Co., New York, 1934, 210.

Thermatrix A process for destroying organic vapors in waste streams by oxidation in a packed-bed reactor containing a porous, inert ceramic matrix. The operating temperature is 870°C–980°C, at which temperature hydrocarbons are oxidized to carbon dioxide and water and chlorinated, fluorinated, or sulfonated hydrocarbons are converted to HCl, HF, or SO_2, respectively. Quantities of carbon monoxide and NO_x produced are minimal. Developed by Thermatrix, CA, in the 1990s. By 1996, it had been installed in over 50 plants in a variety of industries. Although the capital cost is greater than that of a similar catalytic oxidation unit, the running cost is claimed to be less.

Hohl, H.M., *Oil Gas J.*, 1996, **94**(45), 77.

Thermical A process for making aluminum metal, developed by Calsmelt Pty Ltd, Melbourne. Operates at 500°C lower than the *Hall–Heroult process. Alumina is mixed with molten aluminum in the presence of carbon to form aluminum carbide. This is heated with more alumina to produce aluminum and carbon dioxide. Not yet piloted.

> *Chem. Eng. (N.Y.)*, 2010, **117**(8), 9.

THERMIDOR *See* HYVAHL.

Thermit Also spelled Thermite and also called the Goldschmidt process. The reaction of metallic aluminum with a metal oxide is very exothermic and can be used to liberate other metals from their oxides or simply as a source of heat. In the latter case, iron oxide is used:

$$2Al + Fe_2O_3 = Al_2O_3 + 2Fe$$

Refractory metals such as chromium, manganese, and cobalt are made in this way. The process was invented by H. Goldschmidt at the German company Th. Goldschmidt, Essen, in 1898. *See also* Goldschmidt.

THERMOCAT A petroleum *cracking process that combines fixed-bed *catalytic cracking with steam cracking. Developed by Veba Oel and Linde from 1994. *See* PYROCAT.

Thermofax An early thermographic copying process using paper impregnated with a ferric salt, a heavy metal sulfide, and a phenolic compound.

> U.S. Patents 2,663,654; 2,663,655; 2,663,656; 2,663,657.
> *Chem. Eng. News*, 1964, 13 Jul, 115; 20 Jul, 85.

Thermofor This name was first used in the 1930s for the equipment and process for burning off the carbon that deposits on the clays used for purifying mineral oils. The lumps of clay were regenerated by passing through a hot reactor. The Socony Vacuum Oil Company subsequently used this technology as the basis for its range of processes for regenerating cracking catalysts. These included Thermofor Catalytic Cracking (*TCC), Thermofor Catalytic Reforming (*TCR), Thermofor Pyrolytic Cracking (*TPC), and *Airlift Thermofor Catalytic Cracking. The first Thermofor cracking process was commercialized in 1943. Socony operated a Thermofor catalytic reformer from 1955; the catalyst was chromia on alumina.

> Enos, J.L., *Petroleum Progress and Profits*, MIT Press, Cambridge, MA, 1962, 165.
> Unzelman, G.H. and Wolf, C.J., in *Petroleum Processing Handbook*, Bland, W.F. and Davidson, R.L., Eds., McGraw-Hill, New York, 1967, 3–7, 3–34.
> *The Petroleum Handbook*, 6th ed., Elsevier, Amsterdam, the Netherlands, 1983, 286.

ThermoFuel (1) A process for making diesel fuel from waste plastic. Preferred plastics are polyethylene and polystyrene. The plastic is first melted in an extruder and then pyrolyzed continuously in a cylindrical chamber at 370°C–420°C, giving a C_8 to C_{26} hydrocarbon mixture having a peak at C_{16}. An important feature is the incorporation of a catalytic reaction tower after the main pyrolysis reactor, which incorporates metal plates made from a proprietary catalytic metal alloy. Distillation yields an average of 930 L of diesel per ton of waste plastic. Developed by Ozmotech, Australia, and now offered by EnviroSmart Technologies of Roosendaal, the Netherlands. In 2006, there were plans for 31 installations in Europe to be made over the next 4 years.

> *Chem. Eng. (N.Y.)*, 2006, **113**(5), 18.
> Scheirs, J., in *Feedstock Recycling and Pyrolysis of Waste Plastics*, Scheirs, J. and Kaminsky, W., Eds., John Wiley & Sons, Chichester, U.K., 2006, 407.

ThermoFuel (2) A process for making fuel oil by the hydrothermal treatment of sewage sludge. Developed in 2005 by ThermoEnergy Corporation.

THERMOSELECT A high-temperature integrated process for making a syngas, glassy materials, and certain metals from various wastes. The main aim is to minimize landfill. Developed in Switzerland between 1985 and 1992 and demonstrated in Fondotoce, Italy, between 1992 and 1999. Energie Baden-Württemburg AG built a larger plant in Karlsruhe in 1999 but abandoned it in 2004. Several plants were built in Japan between 1999 and 2006 but their present status is unclear. The process may be economic only in Japan because landfill costs are high there.

> U.S. Patents 5,282,431; 8,198,339.
> Themelis, N.J., *Waste Manage. World*, Jul 2007.

Thermosoft A water-softening process for treating waters that are high in dissolved solids and alkalinity, intended for oilfield steam injection. Naturally occurring bicarbonate is added and the temperature is raised to 180°C–200°C, causing the hydroxides of calcium and magnesium to precipitate. Invented by T. Bertness and licensed to USFilter in 1995.

> *Chem. Eng. (N.Y.)*, 1995, **102**(11), 23.

Thermosorption A process for recovering hydrocarbons from wet natural gas by adsorption on activated carbon. Offered by Lurgi. *See also* Supersorbon.

THGP [Texaco Hydrogen Generation Process] A process for making pure, high-pressure hydrogen from various gaseous and light hydrocarbons. Partial combustion of the hydrocarbons yields *syngas. The carbon monoxide in this mixture is converted to carbon dioxide by the *shift reaction and removed by *PSA. Seen as an alternative to *steam reforming. Offered by Texaco Development Corporation.

> *Hydrocarbon Process. Int. Ed.*, 1996, **75**(4), 149.

Thiofex An early process for refining benzole by treatment with sulfuric acid at a rising temperature, followed by sodium carbonate. Invented in 1947 by T. Scott at Refiners Limited and used in the United Kingdom.

> British Patent 642,772.
> Claxton, G., *Benzoles: Production and Uses*, National Benzole & Allied Products Association, London, U.K., 1961, 434.

Thiolex A process for removing hydrogen sulfide from a light hydrocarbon liquid by extraction with aqueous sodium hydroxide passed through a bundle of hollow fibers immersed in it. Developed by the Merichem Company, Houston, TX. In 1991, 52 units were operating. Variations are known as Thiolex/Regen and Thiolex/Regen/Mericat. *See also* Mericat.

> *Hydrocarbon Process. Int. Ed.*, 1984, **63**(4), 87; 1996, **75**(4), 126.

Thionate *See* Feld.

THIOPAQ A biological process for removing hydrogen sulfide from gas streams. The gas is scrubbed with aqueous sodium hydroxide, forming sulfide ions, and the solution is regenerated in a bioreactor in which bacteria convert the sulfide to elemental sulfur and regenerate the alkali. Sixty units were installed, worldwide, between 1993 and 2006 for a variety of industries. Invented and licensed by Paques, the Netherlands. UOP engineered the process for the oil industry and installed its first such unit in Egypt in 1999. *See also* Shell Paques.

> *Oil Gas J.*, 1997, **95**(29), 54.
> *Eur. Chem. News*, 8 Nov 1999, **7**(1884), 47.
> *Sulphur*, Nov–Dec 1999, (265), 12–14.
> *Hydrocarbon Process. Int. Ed.*, 2000, **79**(4), 82.
> *Chem. Week*, 2001, **163**(40), 32.

Thoma A process for alkylating aniline with methanol or ethanol to produce mixtures of mono- and di-alkylanilines. Operated in hot, concentrated phosphoric acid in a vertical tubular reactor. The proportions of secondary and tertiary amines can be partly controlled by controlling the ratios of the reactants; the products are separated by fractional distillation. Invented in 1954 by M. Thoma in Germany.

U.S. Patent 2,991,311.

Thomas Also called the Basic Bessemer process. A variation of the *Bessemer process for steelmaking, for use with ores rich in phosphorus, in which the converter is lined with calcined dolomite (magnesium and calcium oxides) and limestone is added to the charge; the phosphorus remains with the slag. Invented by the cousins Sidney Gilchrist Thomas and Percy Carlisle Gilchrist (hence the alternative name for the process: Thomas and Gilchrist) in 1877. Developed at Blaenavon steelworks in South Wales (now a world heritage site) and first commercialized at the works of Bolckow, Vaughan & Company in Middlesbrough, England. Widely used for the treatment of iron ores rich in phosphorus. *See also* Bessemer.

British Patent 4,422 (1877).
Barraclough, K.C., *Steelmaking 1850–1900*, The Institute of Metals, London, U.K., 1990, 207, 222.

Thomas and Gilchrist *See* Thomas.

Thompson–Stewart A process for making basic lead carbonate ("white lead," 2PbCO$_3$Pb(OH)$_2$) by reacting lead monoxide ("litharge") with acetic acid and then with carbon dioxide. Basic lead acetate is an intermediate. *See also* Dutch, Carter.

Thorex [**Thor**ium **ex**traction] A process for separating the products from the nuclear breeder reaction, in which uranium-233 is produced by the neutron bombardment of thorium-232. It uses solvent extraction into tri-*n*-butyl phosphate. Developed at the Oak Ridge National Laboratory, TN, in the early 1960s. *See also* Butex, Purex, Redox.

Nash, K.L., in *Separations for the Nuclear Fuel Cycle in the 21st Century*, Lumetta, G.J., Nash, K.L., Clark, S.B., and Friese, J.I., Eds., American Chemical Society, Washington, DC, 2006, 25.

Thoroughbred A family of *flue-gas desulfurization processes. Commercialized versions have been called Thoroughbred-101, Thoroughbred-102, and Thoroughbred-121. Another name used for the last is CT-121. They are integrated *flue-gas desulfurization processes that achieve limestone neutralization, oxidation, and conversion to gypsum in a complex jet bubbling reactor. Developed by the Chiyoda Chemical Engineering and Construction Company, Japan, and widely used in Japan and the United States. As of 1986, CT-121 had been installed in eight plants in the United States and Japan.

U.S. Patents 4,099,925; 4,156,712 (reactor); 4,178,348; 4,203,954 (process).
Idemura, H., *Chem. Econ. Eng. Rev.*, 1974, **6**(8), 22.
Tamaki, A., *Chem. Eng. Prog.*, 1975, **71**(5), 55.
Kohl, H.L. and Nielsen, R.B., *Gas Purification*, 5th ed., Gulf Publishing Co., Houston, TX, 1997, 585.

Thum A variation of the *Balbech process for separating silver from gold, in which the electrodes are held vertically. The anodes are contained in cloth bags to retain the slimes; silver deposits at the cathodes and is periodically scraped off.

Thylox A process for removing hydrogen sulfide from refinery and coke-oven gases by absorption in a solution of sodium ammonium thioarsenate. The solution is regenerated by blowing air through it, precipitating elemental sulfur, which is filtered off:

$$(NH_4)_3AsO_2S_2 + H_2S = (NH_4)_3AsOS_3 + H_2O$$

$$2(NH_4)_3AsOS_3 + O_2 = 2(NH_4)_3AsO_2S_2 + 2S$$

Invented in 1926 by H.A. Gollmar and D.L. Jacobson at the Koppers Company. Although the process had shortcomings—it removed only 90%–96% of the hydrogen sulfide and the sulfur produced was quite impure—it had been used in 10 installations in the United States by 1950.

U.S. Patents 1,719,177; 1,719,180; 1,719,762.
Kohl, A.L. and Riesenfeld, F.C., *Chem. Eng. (N.Y.)*, 1959, **66**, 152.
Kohl, H.L. and Nielsen, R.B., *Gas Purification*, 5th ed., Gulf Publishing Co., Houston, TX, 1997, 748.

Thyssen–Galoczy A slagging coal gasification process.

TIGAS [**T**opsoe **I**ntegrated **GA**soline **S**ynthesis] A multistage process for converting natural gas to gasoline that combined a *syngas process with an *MTG process. Developed by Haldor Topsoe and piloted in Houston from 1984 to 1987. Not commercialized, but used in 1995 as the basis for a process for making dimethyl ether for use as a diesel fuel. A demonstration plant aimed at making gasoline from wood, via dimethyl ether as an intermediate, has been running since 2009 at the Gas Technology Institute, Des Plaines, IL, in conjunction with UPM-Kymmene and ConocoPhillips.

Topp-Joergensen, J., *Stud. Surf. Sci. Catal.*, 1987, **107**, 99.
Rouhi, A.M., *Chem. Eng. News*, 1995, **73**(22), 37.
Rostrup-Nielsen, J., *Catal. Today*, 2006, **111**(1–2), 8.
Bjørgen, M., Joensen, F., Holm, M.S., Olsbye, U., Lillerud, K.P., and Svelle, S., *Appl. Catal. A: Gen.*, 2008, **345**(1), 44.
Chem. Eng. (N.Y.), 2010, **117**(1), 16.

Tin Sol A process for making a stannic oxide sol by electrodialysis. Invented in 1973 by H.P. Wilson of the Vulcan Materials Company, Birmingham, AL.

U.S. Patent 3,723,273.

TIP [**T**otal **I**somerization **P**rocess] *See* Total Isomerization.

TiRO [**Ti**tanium **CSIRO**] A continuous process for making titanium metal. Titanium tetrachloride is reduced with magnesium metal at a temperature between the melting points of magnesium (650°C) and magnesium chloride (712°C). The product is a composite of magnesium chloride with titanium particles, which are separated in a second process. Developed by C. Doblin and G.A. Wellwood at CSIRO, Melbourne, in 2007 but not yet piloted.

U.S. Patent 7,837,759.
Chem. Eng. (N.Y.), 2007, **114**(12), 18.

Tisco A direct reduction ironmaking process using coal as the reductant. Operated on a small scale in India since 1986.

Titanizing [From the Greek, Titan, meaning a person of superhuman strength; often incorrectly assumed to be derived from the name of the element titanium] A process for hardening the surfaces of glass vessels by coating them with a layer of titanium dioxide or tin dioxide. The oxides are deposited from the vapors of the respective tetrachlorides or tetraalkoxides by chemical vapor deposition (*see* CVD). Invented by S.M. Budd at United Glass, United Kingdom, and widely used for strengthening glass bottles.

British Patents 1,115,342; 1,187,784.

Titanox FR [**F**lame **R**esistant] A process for making cellulose textiles flame resistant by treating them with titanyl acetate chloride and antimony oxychloride. Invented in 1951 by W.F. Sullivan and I.M. Panik at the National Lead Company, New York. *See also* Erifon.

U.S. Patent 2,658,000.

TN *See* Steelmaking.

Tokuyama A process for hydrating propene to propenediol, catalyzed by an aqueous solution of 12-tungstosilicic acid. Developed by Tokuyama (a Japanese company) and commercialized in Japan on a 3000 tpa scale since 1972.

Nisono, M. and Nojivi, N., *Catal. Today*, 1997, 33, 371.

Topnir [**Top**ological **n**ear **infra**red] Not a chemical process but an instrumental process for online monitoring of hydrocarbon process streams by near-infrared spectroscopy. Developed by BP and now offered by ABB.

Eur. Chem. News, 1997, **67**(1753), 23.

Toray (1) A large Japanese chemical manufacturer, perhaps best known for its process for synthesizing L-lysine for use as a dietary supplement. The starting material is cyclohexene, which is converted in five steps to racemic lysine. An enzymic process isolates the desired optical isomer; the other is recycled.

Toray (2) A process for making terephthalic acid by oxidizing *p*-xylene using a cobalt catalyst promoted by paraldehyde.

Raghavendrachar, P. and Ramachandran, S., *Ind. Eng. Chem. Res.*, 1992, **31**, 453.

Toray Aromax *See* Aromax.

torrefaction Also spelled torrification. Named after Evangelista Torricelli (1608–1647), who was Galileo's secretary and is better known for his invention of the barometer. The Torricellian vacuum is the space above the mercury in a mercury barometer, which contains only mercury vapor. Torrefaction has come to mean drying by the application of heat, not necessarily under reduced pressure. Current usage is for a process for upgrading biomass by heating at atmospheric pressure in the absence of oxygen to 200°C–300°C. Developed from 2005 at the Energy Research Centre of the Netherlands (ENC). Not yet commercialized.

TOSCO II A process for extracting oil from shale by contacting it with hot ceramic or steel balls in a rotating drum. Based on an invention made in 1922 by F. Puening in Pittsburgh, who used hot iron balls to provide heat for the destructive distillation of lignite, shale, peat, and bituminous coal. Further developed by the Oil Shale Corporation in the 1950s. The Tosco Corporation, in association with Exxon Corporation, continued the work in Colorado from the 1960s, until the U.S. oil shale projects were discontinued in 1982.

U.S. Patents 1,698,345–1,698,349, inclusive.
Klass, D.L., *ChemTech*, 1975, **5**, 499.
Waitman, C.S., Braddock, R.L., and Siebert, T.E., in *Handbook of Synfuels Technology*, Meyers, R.A., Ed., McGraw-Hill, New York, 1984, 4–45.

TOSCOAL [**T**he **O**il **S**hale **Cor**poration] A low-temperature carbonization process for producing liquid fuels from oil shales. Developed by the Oil Shale Corporation in the 1960s. *See* TOSCO II.

Carlson, F.B., Yardumian, L.H., and Atwood, M.T., *Chem. Eng. Prog.*, 1973, **69**(3), 50.
Probstein, R.F. and Hicks, R.E., *Synthetic Fuels*, McGraw-Hill, New York, 1982, 259.

Total Isomerization Also called TIP. An integrated process that combines light paraffin isomerization using a zeolite catalyst, with the *IsoSiv process, which separates the unconverted normal paraffins so that they can be returned to the reactor. Developed by Union Carbide Corporation and now licensed by UOP. The first plant was operated in Japan in 1975; by 2002, more than 30 units were operating.

Hydrocarbon Process. Int. Ed., 1980, **59**(5), 110; 1988, **67**(9), 82.
Cusher, N.A., in *Handbook of Petroleum Refining Processes*, 3rd ed., Meyers, R.A., Ed., McGraw-Hill, New York, 2003, 9.36.

Toth Also called TAC [**T**oth **A**luminum **C**orporation] A method proposed for making aluminum metal from clay. The dried clay, mixed with coke, is chlorinated to yield aluminum trichloride and silicon tetrachloride. The volatile chlorides are separated by distillation, and the aluminum chloride is then reduced with manganese metal:

$$Al_xSi_yO_z + (3x + 4y)/2\ Cl_2 + z/2\ C = xAlCl_3 + ySiCl_4 + z/2\ CO_2$$

$$2AlCl_3 + 3Mn = 2Al + 3MnCl_2$$

The manganese chloride is then reduced with hydrogen:

$$MnCl_2 + H_2 = Mn + 2HCl$$

and the manganese metal reused. In another version, the manganese chloride is oxidized in oxygen to manganese oxide, which is then reduced to the metal using carbon. Invented in 1969 by C. Toth and piloted by the Applied Aluminum Research Corporation in Baton Rouge, LA, in the mid-1970s, but not commercialized. The practicality and economics of this complex cyclic process remain controversial.

U.S. Patents 3,615,359; 3,918,960; 4,035,169; 4,035,180.
Chem. Eng. News, 1973, **51**(9), 11.
Grjotheim, K., Krohn, C., Malinovsky, M., Matiaskovsky, K., and Thonstad, J., *Aluminum Electrolysis: Fundamentals of the Hall-Héroult Process*, CRC Press, Boca Raton, FL, 1982, 13.
Eur. Chem. News, 1984, **42**(1120), 10.
Chem. Mark. Rep., 1997, **251**(13), 5.

Tower Biology A biological waste treatment process, developed from the *Activated Sludge process. The sludge is contained in a tall tower, at the base of which oxygen is injected as small bubbles. The bubbles are almost completely absorbed by the time they reach the surface of the liquid. The system uses less energy than does surface aeration. Developed by Bayer in Germany. The first plant was built in 1977 at a wastewater treatment plant at its Brunsbüttel plant. The second installation was in 1980 in its Leverkusen plant. The process was subsequently adopted in India and then elsewhere.

Chem. Eng. (N.Y.), 1992, **99**(12), 101.

Townsend A process for removing hydrogen sulfide from natural gas by absorption in triethylene glycol containing sulfur dioxide. Part of the sulfur produced is burned to sulfur dioxide in order to provide this solution. The hydrogen sulfide and sulfur dioxide react in the presence of water to generate elemental sulfur. Invented in 1958 by F.M. Townsend; piloted in Canada but never commercialized.

U.S. Patent 3,170,766.
Townsend, F.M. and Reid, L.S., *Pet. Refin.*, 1958, **37**, 263.
Kohl, H.L. and Nielsen, R.B., *Gas Purification*, 5th ed., Gulf Publishing Co., Houston, TX, 1997, 841.

TPC [**T**hermofor **P**yrolytic **C**racking] A continuous process for thermally cracking petroleum fractions on a moving bed of hot pebbles. Developed by the Socony Vacuum Oil Company (now part of ExxonMobil). *See also* Thermofor.

Eastwood, S.C. and Potas, A.E., *Pet. Eng.*, 1948, **19**(12), 43.
Asinger, F., *Mono-Olefins: Chemistry and Technology*, translated by B.J. Hazzard, Pergamon Press, Oxford, U.K., 1968, 163.

Trail A process for recovering elemental sulfur from sulfur dioxide by reduction with carbon:

$$SO_2 + C = CO_2 + S$$

$$CO_2 + C = 2CO$$

Carbonyl sulfide is an intermediate in this reaction. A mixture of sulfur dioxide and oxygen was blown into the bottom of a coke-fired reduction furnace, and sulfur vapor condensed from the off-gases. Trail is the location of a large mine and smelter in British Columbia. The process was originally used in the 1930s for abating air pollution from the smelter, but when the demand for sulfuric acid for fertilizer production increased in 1943, it became obsolete. *See also* Boliden (1), RESOX.

Katz, M. and Cole, R.J., *Ind. Eng. Chem.*, 1950, **42**, 2264.

Tramex [**Tra**nsuranic **m**etal (or **am**ine) **ex**traction] A process for separating transuranic elements from fission products by solvent extraction from chloride solutions into a tertiary amine solution. Developed at Oak Ridge National Laboratory, TN, for processing irradiated plutonium.

Leuze, R.E. and Lloyd, M.H., *Prog. Nucl. Energy*, 1970, (Ser. III), **4**, 596.
Nash, K.L., in *Separations for the Nuclear Fuel Cycle in the 21st Century*, Lumetta, G.J., Nash, K.L., Clark, S.B., and Friese, J.I., Eds., American Chemical Society, Washington, DC, 2006, 30.

Trans4m S A catalytic skeletal isomerization process that transforms *n*-butane to *iso*butene and *n*-pentene to *iso*amylene. Developed by LyondellBasell and used in its refinery in Texas since 1995.

transalkylation In organic chemistry, any reaction in which an alkyl group is transferred from one molecule to another. In process chemistry, the word has a more limited meaning, generally restricted to aromatic rearrangements such as the conversion of toluene to benzene and C_8 aromatic hydrocarbons. Examples of such processes are *LTD, *Tatoray.

Transcat An *oxychlorination process for making vinyl chloride from ethane and chlorine:

$$2C_2H_6 + Cl_2 + \tfrac{3}{2}O_2 = 2CH_2CHCl + 3H_2O$$

The catalyst, and the source of the oxygen, is cupric oxide dissolved in a molten mixture of cupric chloride and potassium chloride. Developed by Lummus Corporation.

Weissermel, K. and Arpe, H.-J., *Industrial Organic Chemistry*, 4th ed., translated by C.R. Hindley and S. Hawkins, Wiley-VCH, Weinheim, Germany, 2003, 223.

TransPlus A transalkylation process for making mixed xylenes from heavy aromatics and toluene. Developed by Mobil Technology and Chinese Petroleum Corporation and first operated in Taiwan in 1997. Three units were operating in 2005. It is now offered by Axens.

Eur. Chem. News, 1997, **68**(1783), 33.

TRC [**T**hermal **R**egenerative **C**racking] Also known as Quick Contact and QC. A process for making olefins from petroleum fractions by rapid *thermal cracking of petroleum residues. The feed is cracked by passing through a hot, fluidized bed of micron-sized refractory particles, with a contact time of a quarter of a second. Developed by Gulf Chemical, Gulf Canada, and Stone & Webster and piloted at Cedar Bayou, TX, from 1981.

Hu, Y.C., in *Chemical Processing Handbook*, Marcel Dekker, New York, 1993, 782.
Eur. Chem. News, 1996, **65**(1710), 24.
Picciotti, M., *Oil Gas J.*, 1997, **95**(25), 55.

Treadwell A process for extracting copper from chalcopyrite by leaching with the stoichiometric quantity of sulfuric acid:

$$CuFeS_2 + 4H_2SO_4 = CuSO_4 + FeSO_4 + 2SO_2 + 2S + 4H_2O$$

Developed by the Anaconda Company in 1968 but not commercialized.

Prater, J.D., Queneau, P.B., and Hudson, T.J., *J. Met.*, 1970, **22**(12), 23.
Gupta, C.K. and Mukherjee, T.K., *Hydrometallurgy in Extraction Processes*, Vol. 1, CRC Press, Boca Raton, FL, 1990, 78.

treating A general name for processes that remove S-, N-, and O-compounds from petroleum streams.

TREMP [Topsøe Recycle Energy-efficient Methanation Process] A process for making substitute natural gas from a variety of carbon feedstocks. In addition, it produces superheated steam for use in a steam turbine. Developed by Haldor Topsøe from the *EVA-ADAM process. Licensed to CPI Xinjiang Energy Co. in 2012 for completion in 2014.

Harms, H., Hohlein, B., and Skov, A., *Chem. Ing. Tech.*, 1980, **52**(6), 504.
Chem. Eng. (N.Y.), 2009, **116**(13), 12.
Chem. Eng. (N.Y.), 2010, **117**(8), 14.

Trencor Also called Trentham Trencor. A wet-scrubbing process for removing residual sulfur dioxide and hydrogen sulfide from the tail gas from the *Claus process.

Speight, J.G., *Gas Processing*, Butterworth Heinemann, Oxford, U.K., 1993, 327.

Trentham Trencor *See* Trencor.

Trickle Hydrodesulfurization A process for removing sulfur, nitrogen, and heavy metal compounds from petroleum distillates before *catalytic cracking. The preheated feed is hydrogenated, without a catalyst, in an adiabatic reactor at 315°C–430°C. Developed by Shell Development Company. As of 1978, 91 units had been installed.

Hoog, H., Klinkert, H.G., and Schaafsma, A., *Pet. Refin.*, 1953, **32**(5), 137.
Hydrocarbon Process., 1964, **43**(9), 194.

TRI-NOx A process for removing NO_x and nitric acid mists from the waste gases from the manufacture of electronic devices. It uses multistage scrubbing with a proprietary solvent. Developed by Wacker Siltronic Corporation, based on a scrubber engineered and manufactured by Tri-Mer Corporation.

Chem. Eng. (N.Y.), 1996, **103**(12), 123.
Kohl, H.L. and Nielsen, R.B., *Gas Purification*, 5th ed., Gulf Publishing Co., Houston, TX, 1997, 935.

Triolefin Also called **Phillips Triolefin**. A process for disproportionating propylene into a mixture of ethylene and 2-butene. The reaction takes place at 160°C over a cobalt–molybdenum catalyst on an alumina base. Developed by the Phillips Petroleum Company from 1963. A commercial plant was built by Gulf Oil Canada in 1966 and operated by Shawinigan between 1966 and 1972 before closing for economic reasons. *See also* Olefin Conversion Technology, Meta-4.

U.S. Patent 3,236,912.
Logan, R.S. and Banks, R.L., *Hydrocarbon Process. Int. Ed.*, 1968, **47**(6), 135.
Hydrocarbon Process. Int. Ed., 1971, **50**(11), 140; 1991, **70**(3), 144.
Imamoglu, Y. and Bencze, L., Eds., *Novel Metathesis Chemistry*, Kluwer Academic Publishers, Dordrecht, the Netherlands, 2003.
Weissermel, K. and Arpe, H.-J., *Industrial Organic Chemistry*, 4th ed., translated by C.R. Hindley and S. Hawkins, Wiley-VCH, Weinheim, Germany, 2003, 67.

TRS [Thermal Reactor System] A process for converting coke-oven gas to a gas mixture suitable for use in the *Midrex process. It involves partial oxidation with a high-velocity oxygen jet that breaks down the heavy hydrocarbons and tars. Developed by Midrex and Praxair and licensed by them. Expected to be ready for commercial application in 2013. It is expected to be applicable to the *MXCOL process too.

TRUEX [TRansUranium EXtraction] A process for removing transuranic elements and lanthanide fission products during the processing of nuclear fuel by solvent extraction. The solvent is a complex phosphine oxide mixed with tributyl phosphate and diluted with *n*-dodecane. By removing the transuranic elements, the alpha activity of the waste is greatly reduced and the residue is easier to dispose of. Developed by E.P. Horwitz at the Argonne National Laboratory, Chicago, IL. *See also* SREX, UREX[+].

> Nash, K.L., in *Separations for the Nuclear Fuel Cycle in the 21st Century*, Lumetta, G.J., Nash, K.L., Clark, S.B., and Friese, J.I., Eds., American Chemical Society, Washington, DC, 2006, 26.

TRW Gravichem A modification of the *TRW Meyers process in which the coal fraction that contains less pyrite is first removed by sedimentation in the ferric sulfate solution.

> IEA Coal Research, *The Problems of Sulphur*, Butterworths, London, U.K., 1989, 30.

TRW Meyers [Named after the three CalTech professors who founded the company: **T**hompson, **R**amo, and **W**ooldridge] A chemical method for desulfurizing coal. The iron pyrite is leached out with a hot, aqueous solution of ferric sulfate, liberating elemental sulfur. The resulting ferrous sulfate solution is reoxidized with air or oxygen:

$$Fe_2(SO_4)_3 + FeS_2 = 3FeSO_4 + 2S$$

$$7Fe_2(SO_4)_3 + 8H_2O + FeS_2 = 15FeSO_4 + 8H_2SO_4$$

$$4FeSO_4 + 2H_2SO_4 + O_2 = 2Fe_2(SO_4)_3 + 2H_2O$$

> Meyers, R.A., *Hydrocarbon Process. Int. Ed.*, 1975, **54**(6), 93.
> IEA Coal Research, *The Problems of Sulphur*, Butterworths, London, U.K., 1989, 28.

TSA [**T**hermal (or **T**emperature) **S**wing **A**dsorption] A method for separating gases by cyclic adsorption and desorption from a selective adsorbent at alternating temperatures. Less commonly used than *PSA.

> Yang, R.T., *Gas Separation by Adsorption Processes*, Butterworths, Guildford, U.K., 1987, 204.
> Sherman, J.D. and Yon, C.M., in *Kirk-Othmer Encyclopedia of Chemical Technology*, 4th ed., Vol. 1, John Wiley & Sons, New York, 1991–1998, 546.

TSF [**T**exaco **S**elective **F**inishing] A process for separating linear from branched-chain aliphatic hydrocarbons by *PSA using zeolite 5A as the adsorbent. The desorbent is a hydrocarbon having two to four carbon atoms less than the feed. Developed by Texaco in the late 1950s. Believed to be still in operation in its Trinidad refinery in 1990.

> Franz, W.F., Christensen, E.R., May, J.E., and Hess, H.V., *Oil Gas J.*, 1959, **57**(15), 112.
> Cooper, D.E., Griswold, H.E., Lewis, R.M., and Stokeld, R.W., *Eur. Chem. News*, 1966, **10**(254), Suppl., 23.

TSR (1) [**T**race **S**ulfur **R**emoval] A process for removing sulfur compounds from naphtha so that they will not poison catalytic reformers. A proprietary solid absorbent is used. Developed by the Union Oil Company of California and first commercialized in 1983 at the Unocal oil refinery in San Francisco.

TSR (2) [**T**iomin **S**ynthetic **R**utile] A process for removing much of the iron from ilmenite in order to make a feedstock for titanium pigment manufacture. The ore is successively oxidized and reduced in fluidized beds and then leached with hydrochloric acid. The product contains approximately 95% of titanium dioxide. Developed by Tiomin Resources, Canada, in the 1990s. Planned to be tested on a large scale in Queensland, Australia.

> U.S. Patent 5,885,324.
> *Ind. Min.*, 1998, (371), 19.

T-Star (1) A *hydrotreating process introduced by Texaco in 1993.

> *Hydrocarbon Process. Int. Ed.*, 1997, **76**(2), 45.

T-Star (2) A catalytic *hydrocracking process using an ebullated bed reactor containing an extruded Ni/Mo-based catalyst. Developed by Axens North America, based on the *H-Oil process. Planned to be used in a coal-to-liquids plant in Inner Mongolia from 2005.

> *Chem. Eng. (N.Y.)*, 2003, **110**(6), 17.

TTH [TiefTemperaturHydrierung; German, meaning low-temperature hydrogenation] A petroleum refining process for converting tars and middle distillates into lower-boiling fractions. The catalyst is a combination of tungsten and nickel sulfides. Developed in Germany by IG Farbenindustrie. *See also* MTH.

> Weisser, O. and Landa, S., *Sulphide Catalysts: Their Properties and Applications*, Pergamon Press, Oxford, U.K., 1973, 333.

Tube and Tank A continuous process for thermally cracking petroleum, developed by Standard Oil of Indiana in the 1920s.

> Ellis, C., *The Chemistry of Petroleum Derivatives*, The Chemical Catalog Co., New York, 1934, 109.
> Enos, J.L., *Petroleum Progress and Profits*, MIT Press, Cambridge, MA, 1962, Chap. 3.

Turbotak A process for removing residual sulfur from the tail gases from the *Claus process. Developed by TurboSonic Technologies.

> *Sulphur*, Jul–Aug 1999, (263), 53–54, 56.

TVP [True Vapor Phase] A *thermal cracking process in which vaporized petroleum oil is contacted with a hotter gas such that the temperature of the gas mixture is approximately 500°C. Used in the 1930s but supplanted by various *catalytic cracking processes.

> *The Petroleum Handbook*, 3rd ed., Shell Petroleum Co., London, U.K., 1948, 172.

TwinRec A two-stage process for recovering energy and materials from various industrial wastes. The first stage, operated at 500°C–600°C in a reducing atmosphere, separates combustible materials from metallic and inert materials. The second stage, operated at 1350°C–1450°C, produces a molten ash, which solidifies to a granular material suitable for the construction industry. Developed by Ebara Corporation and commercialized in 20 plants in Japan by 2004.

> Arena, U. and Mastellone, M.L., in *Feedstock Recycling and Pyrolysis of Waste Plastics*, Scheirs, J. and Kaminsky, W., Eds., John Wiley & Sons, Chichester, U.K., 2006, 469.

Twitchell An early process for the acid-catalyzed hydrolysis of animal and vegetable fats for making glycerol and soap. The catalyst was a mixture of sulfonated oleic and naphthenic acids and sulfuric acid, known as "Twitchell saponifier." Invented in 1897 by E. Twitchell and commercialized by Joslin-Schmidt Company, Cincinnati, OH. Twitchell was awarded the Perkin Medal for this invention by the ACS in 1917. The British soapmakers at that time, Joseph Crosfield and Sons, did not use the process because the products were considered to be too dark in color.

> U.S. Patents 601,603; 628,503; 1,170,468.
> Twitchell, E., *J. Am. Chem. Soc.*, 1900, **22**, 22; 1906, **28**, 196.
> Mills, V. and McClain, H.K., *Ind. Eng. Chem.*, 1949, **41**, 1982.
> Musson, A.E., *Enterprise in Soap and Chemicals*, Manchester University Press, Manchester, U.K., 1965, 174.

Tyrer A process for making phenol by first sulfonating benzene. Benzene vapor was passed through hot sulfuric acid; the excess of benzene served to remove the water formed in the reaction. The benzene sulfonic acid was then hydrolyzed by fusion with sodium hydroxide. Invented by D. Tyrer in 1916. *See also* Dennis–Bull.

> U.S. Patents 1,210,725; 1,210,726.
> Kenyon, R.L. and Boehmer, N., in *Modern Chemical Processes*, Vol. 2, Reinhold Publishing, New York, 1952, 35.

Tysland–Hole *See* Steelmaking.

U

U2A [**U**rea **to A**mmonia] A process for converting urea to ammonia for use in controlling emissions of NO_x from industrial power units. It generates ammonia on demand, which is safer than transporting ammonia. Developed by the Wahlco Air Systems division of Thermatrix, and Research-Cottrell.

Eur. Chem. News, 2000, **72**(1898), 39.

UBC [**U**pgraded **B**rown **C**oal] A process for upgrading brown coal for use in power plants. The pulverized coal is mixed with light petroleum and asphalt and heated to >140°C, which expels the water and solvent, leaving a porous material containing asphalt in the pores. This material is briquetted and is then suitable for use in a power plant. Developed by Kobe Steel from 1993, piloted in Indonesia in 2004, and expected to be commercialized in 2012.

Chem. Eng. (N.Y.), 2006, **113**(9), 13.

UCAP A process for selectively removing residual sulfur dioxide from the tail gas from the *Claus process. It had not been commercialized by 1983.

Yon, C.M., Atwood, G.R., and Swaim, C.D., Jr., *Hydrocarbon Process. Int. Ed.*, 1979, **58**(7), 197.

UCAR [**U**nion **C**arbide **CAR**bon dioxide] A process for removing carbon dioxide from gas streams by scrubbing with methyl diethanolamine. Use of a proprietary corrosion inhibitor permits higher concentrations of the amine to be used than in similar processes.

Hawkes, E.N. and Mago, B.F., *Hydrocarbon Process. Int. Ed.*, 1971, **50**(8), 109.

UCARSOL [**U**nion **CAR**bide **SOL**vent] *See* HS.

UCB-MC1 [**U**nion **C**himique–**C**hemische **B**edrijven and **M**inistry of **C**hemical **I**ndustry for the USSR] An *EHD process for making adiponitrile, differing from the *Monsanto process in using an emulsion of acrylonitrile and in not using a membrane.

UCBSRP [**U**niversity of **C**alifornia **B**erkeley **S**ulfur **R**ecovery **P**rocess] A process for removing hydrogen sulfide and sulfur dioxide from natural gas. These gases react together in a polyglycol solvent with a catalyst, yielding elemental sulfur. Not piloted as of 1997.

Kohl, H.L. and Nielsen, R.B., *Gas Purification*, 5th ed., Gulf Publishing Co., Houston, TX, 1997, 846.

Uchatius A modification of the *Huntsman process for making steel. Cast iron was first granulated by pouring the molten metal into water. The granules, mixed with fresh iron ore containing manganese and fireclay, were then heated as in the Huntsman process. The molten product was poured into molds. Invented in Austria by F. Uchatius, a captain in the Austrian Army, and operated at the Newburn Steelworks in Tyneside, England, from 1856 to 1876. In Vikmanshyttan, Sweden, it was operated from 1859 to 1929.

British Patent 2,189 (1855).
Barraclough, K.C., *Steelmaking before Bessemer, Vol. 2, Crucible Steel*, The Metals Society, London, U.K., 1984, 71.
Barraclough, K.C., *Steelmaking 1850–1900*, The Institute of Metals, London, U.K., 1990, 36.

U-COAL A coal gasification process developed by UBE Industries, Japan.

Udex [**U**niversal **d**ow **ex**traction] A process for removing aromatic hydrocarbons from petroleum fractions by liquid–liquid extraction with glycols, followed by extractive distillation. The glycol used originally was diethylene glycol, but later this was replaced by triethylene glycol and tetraethylene glycol. Extraction is conducted at 140°C–150°C under a pressure of 10 atm. Developed jointly by Dow Chemical Company and UOP in the 1940s and used on a large scale since 1952. Between 1950 and 1965, UOP licensed 82 Udex units. Largely replaced by the *Sulfolane process in the 1960s. *See also* Tetra.

Grote, H.W., *Chem. Eng. Prog.*, 1958, **54**(8), 43.
Beardmore, F.S. and Kosters, W.C.G., *J. Inst. Pet.*, 1963, **49**(469), 1.
Hoover, T.S., *Hydrocarbon Process. Int. Ed.*, 1969, **48**(12), 131.
Somekh, G.S. and Friedlander, B.I., *Hydrocarbon Process. Int. Ed.*, 1969, **48**(12), 127.
Achilladelis, B., *Chem. Ind. (London)*, 1975, (8), 343.
Franck, H.-G. and Stadelhofer, J.W., *Industrial Aromatic Chemistry*, Springer-Verlag, Berlin, Germany, 1988, 107.

Udic–Rheinau A process for making sugars from wood by hydrolysis with hydrochloric acid. Operated in Germany. *See also* Bergius (2).

Katzen, R. and Schell, D.J., in *Biorefineries: Industrial Processes and Products*, Kamm, B., Gruber, P.R., and Kamm, M., Eds., Wiley-VCH, Weinheim, Germany, 2005, **I**, 131.

Ufer A process for refining the light oil produced in coal carbonization. The oil is washed with sulfuric acid, and a controlled amount of water is then added to the mixture. The complex mixture of reaction products ("resins") enters the oil phase; the dilute sulfuric acid can be used directly for making ammonium sulfate. Invented in Germany in 1924 by A. Ufer; operated in Germany and Canada in the 1920s and 1930s.

German Patent 489,753.
British Patent 251,117.
Claxton, G., *Benzoles: Production and Uses*, National Benzole & Allied Products Association, London, U.K., 1961, 433.

U-GAS [**U**tility **GAS**] A process for gasifying carbonaceous products, such as coal, oil, forest wastes, and municipal solid wastes, by reacting them with steam and oxygen (or air) at 950°C–1100°C, under 3–35 atm. The product gases are hydrogen, carbon monoxide, carbon dioxide, and some methane. The ash is agglomerated into rough spheres, hence the name of the equipment: ash-agglomerating gasifier. Developed by the Institute of Gas Technology, Chicago, from 1974 and tested there on a variety of coals. First licensed in 1989 to Tampella, a Finnish corporation involved in forest products, which built a plant at Messukyla, Finland, in 1990.

Now further developed and offered by System Energy Systems (Houston), under license from the GTI.
Dainton, A.D., in *Coal and Modern Coal Processing*, Pitt, G.J. and Millward, G.R., Eds., Academic Press, London, U.K., 1979.
Patel, J.G., *Int. J. Energy Res.*, 1980, **4**, 149.
Chem. Eng. (N.Y.), 1996, **103**(3), 41.

UGI [**U**nited **G**as **I**mprovement Company] Also called Ugite. A regenerative *catalytic reforming process for making town gas and liquid hydrocarbons from oil. The catalyst was a fixed bed of hot, refractory pebbles. Developed by UGI Company, Philadelphia, PA, in the early 1940s.

Asinger, F., *Mono-olefins: Chemistry and Technology*, translated by B.J. Hazzard, Pergamon Press, Oxford, U.K., 1968, 168.
Gas Making and Natural Gas, British Petroleum Co., London, U.K., 1972, 93.

UGINE *See* DR.

Ugite *See* UGI.

Uhde–Hibernia A process for making a mixed ammonium nitrate—ammonium sulfate fertilizer (ASN)—which is less liable to explode than ammonium nitrate. Sulfuric acid is added to aqueous ammonium nitrate and ammonia gas passed in. The double salt crystallizes out. Additives are used to improve the handling characteristics of the product. Developed by Hibernia and licensed to Friedrich Uhde. *See also* Victor.

> *Nitrogen*, 1968, (53), 27.

Uhde/Schwarting *See* Schwarting.

Ultracat A version of the *FCC process, developed by Standard Oil of Indiana in the 1970s.

Ultrafining Two *hydrodesulfurization processes developed by Standard Oil of Indiana, one for petroleum residua and one for vacuum gas oil.

> Unzelman, G.H. and Wolf, C.J., in *Petroleum Processing Handbook*, Bland, W.F. and Davidson, R.I.., Eds., McGraw-Hill, New York, 1967, 3–40.

Ultraforming A *catalytic reforming process developed by Standard Oil of Indiana and licensed by Amoco Oil Company. The catalyst contains platinum and rhenium, contained in a "swing" reactor—one that can be isolated from the rest of the equipment so that the catalyst can be regenerated while the unit is operating. The first unit was commissioned in 1954.

> Unzelman, G.H. and Wolf, C.J., in *Petroleum Processing Handbook*, Bland, W.F. and Davidson, R.L., Eds., McGraw-Hill, New York, 1967, 3–32.
> Little, D.M., *Catalytic Reforming*, PennWell Publishing, Tulsa, OK, 1985, 169.

Ultra-Orthoflow An *FCC process that converts petroleum distillates and heavier fractions to products of lower molecular weight. Developed by M.W. Kellogg Company. Over 100 units were operating in 1988.

> *Hydrocarbon Process. Int. Ed.*, 1988, **67**(9), 67.

ULTROX A process for removing traces of organic compounds from groundwater or wastewater by oxidizing them with ozone, hydrogen peroxide, or both, under the influence of ultraviolet radiation. Invented and developed by Ultrox International in the 1970s and first demonstrated in 1989 with contaminated groundwater from a former drum recycling plant in San Jose, CA. Further developed with the General Electric Company as its *GEODE process. Ultrox International was bought by Zimpro Environmental (now USFilter) in 1993.

> U.S. Patent 4,792,407.
> *Chem. Eng. (N.Y.)*, 1989, **96**(4), 19.
> Hughes, S., *Water Waste Treat.*, 1992, **35**(7), 26.
> Masten, S.J. and Davies, S.H.R., in *Environmental Oxidants*, Nriagu, J.O. and Simmons, M.S., Eds., John Wiley & Sons, New York, 1994, 533.

UMATAC A process for extracting hydrocarbons from tar sands. The sand is heated in a rotating kiln in which the tar is thermally cracked. Developed in Calgary, Canada, in the 1970s by UMATAC Industrial Processes. It had not been commercialized by 1984.

> Bowman, C.W., Phillips, R.S., and Turner, L.R., in *Handbook of Synfuels Technology*, Meyers, R.A., Ed., McGraw-Hill, New York, 1984, 5–42.

UNEX [**UN**iversal **EX**tractant] A process for removing all the major radioactive elements from nuclear processing liquors in one step, using a mixture of complex extractants. Developed from 1994 by an international team from the Idaho National Engineering and Environmental Laboratory (United States) and the Khlopin Radium Institute (Russia).

Unibon A family of related processes offered by UOP for interconverting hydrocarbons using combinations of *hydrotreating and *hydrocracking. *See* AH Unibon, HB Unibon, HC Unibon, LPG Unibon, LT Unibon, RCD Unibon.

UNICARB [**UNI**on Carbide **CARB**on dioxide] A process for spray painting objects using supercritical carbon dioxide as the solvent. Developed by Union Carbide Company and workers at Johns Hopkins University.

Ind. Health Hazards Update, Apr 1995.
Chem. Br., 1998, **34**(8), 38.
Green Chem., 1999, **1**(5), G130.

Unicoil An early thermal process for cracking petroleum.

Asinger, F., *Mono-olefins: Chemistry and Technology*, translated by B.J. Hazzard, Pergamon Press, Oxford, U.K., 1968, 339.

Unicracking A *hydrocracking process for simultaneously hydrogenating and cracking various liquid petroleum fractions to form hydrocarbon mixtures of lower molecular weight. The catalyst contains both acidic and hydrogenation functions. Jointly developed and marketed by UOP and Union Oil Company of California. In 1990, 84 units were operating. The technology was acquired by UOP in 1995 and is widely used worldwide.

Adams, N.R., Watkins, C.H., and Stine, L.O., *Chem. Eng. Prog.*, 1961, **57**(12), 55.
Speight, J.G., *The Desulfurization of Heavy Oils and Residua*, Marcel Dekker, New York, 1981, 181.
Hydrocarbon Process. Int. Ed., 1994, **73**(11), 128.
Ackelson, D., in *Handbook of Petroleum Refining Processes*, 3rd ed., Meyers, R.A., Ed., McGraw-Hill, New York, 2003, 7.23.

Unicracking/DW [**D**e**W**axing] A version of the *Unicracking process developed for upgrading waxy petroleum fractions.

UNIDAK A process for extracting naphthalene from reformer residues in petroleum refining. It includes a dealkylation stage to convert the naphthalene homologues to naphthalene. The process temperature is approximately 600°C; the catalyst is based on cobalt/molybdenum. Developed by the Union Oil Company of California.

Hydrocarbon Process., 1963, **42**(11), 232.

Unifining A *hydrodesulfurization process developed jointly by UOP and the Union Oil Company of California. It is now incorporated in the UOP *hydrotreating and UOP *Unibon processes.

Claxton, G., *Benzoles: Production and Uses*, National Benzole & Allied Products Assoc., London, U.K., 1961, 452.
Unzelman, G.H. and Wolf, C.J., in *Petroleum Processing Handbook*, Bland, W.F. and Davidson, R.L., Eds., McGraw-Hill, New York, 1967, 3–39.

Unionfining A group of petroleum *hydrodesulfurization and *hydrodenitrogenation processes developed by the Union Oil Company of California, primarily for making premium-quality diesel fuel. In 1991, 90 such units were operating. One variant is for purifying naphthalene by selective hydrogenation. The naphthalene vapor is hydrogenated at 400°C over a cobalt–molybdenum catalyst, thereby converting the sulfur in thionaphthalene to hydrogen sulfide. The technology was acquired by UOP in 1995.

Hydrocarbon Process. Int. Ed., 1988, **67**(9), 79.
Eur. Chem. News, 1995, **63**(1653), 24.
Kokayeff, P., in *Handbook of Petroleum Refining Processes*, 3rd ed., Meyers, R.A., Ed., McGraw-Hill, New York, 2003, 8.31.

UNIPOL [**UNI**on Carbide **POL**ymerization] A process for polymerizing ethylene to polyethylene and propylene to polypropylene. It is a low-pressure, gas-phase, fluidized-bed process, in contrast to the *Ziegler–Natta process, which is conducted in the liquid phase. The catalyst powder is continuously added to the bed, and the granular product is continuously withdrawn. A comonomer, such as 1-butene, is normally used.

The polyethylene process was developed by F.J. Karol and his colleagues at Union Carbide Corporation; the polypropylene process was developed jointly with the Shell Chemical Company. The development of the ethylene process started in the mid-1960s; the propylene process was first commercialized in 1983. Eleven licenses were signed with Chinese companies between 2006 and 2012. Nearly 25% of world production of polyethylene was made by this process in 2012. Nearly 100 reactor lines in use in 25 countries produced more than 18 million tons of polyethylene in 2012.

It is now available through Univation Technologies, the joint licensing subsidiary of Dow Chemical Company and ExxonMobil Chemical Company. A supported metallocene catalyst is used today.

U.S. Patents 4,003,712; 4,011,382.
Chem. Eng. Int. (N.Y.), 1979, **86**(26), 80.
Karol, F.J., *ChemTech*, 1983, **13**, 222.
Karol, F.J. and Jacobsen, F.I., in *Catalytic Polymerization of Olefins*, Keii, T. and Soga, K., Eds., Elsevier, Amsterdam, the Netherlands, 1986, 323.
Hydrocarbon Process. Int. Ed., 1991, **70**(3), 173.
Burdett, I.D., *ChemTech*, 1992, **22**(10), 616.
Soares, J.B.P. and McKenna, T.F.L., *Polyolefin Reaction Engineering*, Wiley-VCH, Weinheim, Germany, 2012.

Unisar [**Uni**on **s**aturation of **ar**omatics] A process for hydrogenating aromatic hydrocarbons in petroleum fractions, using a noble metal heterogeneous catalyst. Developed by the Union Oil Company of California. The first commercial unit opened in Beaumont, TX, in 1969. Twenty units had been licensed by 2003.

Hydrocarbon Process. Int. Ed., 1970, **49**(9), 231.
Hydrocarbon Process., *Refinery Process Handbook*, 1982, 137.
Gowdy, H.W., in *Handbook of Petroleum Refining Processes*, 3rd ed., Meyers, R.A., Ed., McGraw-Hill, New York, 2003, 8.59.

Unisol A process for extracting organic sulfur and nitrogen compounds from petroleum fractions by solvent extraction with aqueous sodium or potassium hydroxide containing methanol. First operated in Montana in 1942.

Unzelman, G.H. and Wolf, C.J., in *Petroleum Processing Handbook*, Bland, W.F. and Davidson, R.L., Eds., McGraw-Hill, New York, 1967, 3–118.

Unisulf [**Un**ocal **sulf**ur removal] A process for removing sulfur compounds from petroleum fractions; similar to the *Stretford process but including in the catalytic solution vanadium, a thiocyanate, a carboxylate (usually citrate), and an aromatic sulfonate complexing agent. Developed by the Union Oil Company of California in 1979, commercialized in 1985, and operated in four commercial plants.

U.S. Patent 4,283,379.
Dalrymple, D.A., Trofe, T.W., and Evans, J.M., *Chem. Eng. Prog.*, 1989, **85**(3), 43.
Kohl, H.L. and Nielsen, R.B., *Gas Purification*, 5th ed., Gulf Publishing Co., Houston, TX, 1997, 802.

UNOX A modification of the *Activated Sludge sewage-treatment process for treating domestic effluents, based on the use of oxygen instead of air, in closed reaction tanks. The preferred source of oxygen depends on the size of the plant: small plants use liquid oxygen, medium-sized plants use the *PSA process, and large plants have cryogenic generators. Developed by the Union Carbide Corporation in the late 1960s and now licensed to a number of other companies through Lotepro Environmental Systems and Services.

Lewandowski, T.P., *Water Pollut. Control*, 1974, **73**(6), 647.
Gray, N.F., *Activated Sludge: Theory and Practice*, Oxford University Press, Oxford, U.K., 1990, 119.

UnPET [**Un**ited Resource Recovery **P**oly**E**thylene **T**erephthalate] A process for recycling waste PET. The PET is first hydrolyzed by hot aqueous alkali, and the product then undergoes five complex processes. The only products leaving the process are terephthalic acid, ethylene glycol, carbon dioxide, and water. Developed from 1996 by the United Resource Recovery Corporation, Spartanburg, SC, and operated on a large scale in Switzerland, Germany, and the United States.

U.S. Patent 5,580,905.

Yoshioka, T. and Grause, G., in *Feedstock Recycling and Pyrolysis of Waste Plastics*, Scheirs, J. and Kaminsky, W., Eds., John Wiley & Sons, Chichester, U.K., 2006, 648.

Urbain A process for activating charcoal by heating it with phosphoric acid and then washing with hydrochloric acid. Invented by E. Urbain in 1923; the product was used in the 1920s and 1930s for recovering benzole vapor.

British Patent 218,242.

French Patent 579,596.

Claxton, G., *Benzoles: Production and Uses*, National Benzole & Allied Products Assoc., London, U.K., 1961, 417.

Urea 2000plus A process for making urea from ammonium carbamate, using a novel pool reactor. Developed by DSM and first offered for license in 1996. Stamicarbon is the licensing arm of DSM and now offers the technology. Over 200 plants had been built by 2005.

Eur. Chem. News, 1996, **65**(1716), 22; 2001, **1970**(75), 27,28.

UREX [**UR**anium **EX**traction] A solvent extraction process for extracting uranium and technetium from used nuclear fuel while rejecting all the transuranic elements. Based on the *Purex process, which uses tributyl phosphate in a hydrocarbon mixture, but incorporating acetohydroxamic acid, which complexes the Pu and Np and thereby prevents them from being extracted. Developed by the Westinghouse Savannah River Company in 2003. Associated processes are *NPEX, *TRUEX, and *Cyanex 301.

Rudisill, T.S., Thompson, M.C., Norato, M.A., Kessinger, G.F., Pierce, R.A., and Johnson, J.D., *Trans. Am. Nuclear Soc.*, 2003, 88, 374.

UREX⁺ A solvent extraction process for separating the components of used nuclear fuel so that the unreacted fraction can be reused in an advanced burner reactor. Based on the *Purex process, which uses tributyl phosphate in *n*-dodecane, but using multistage, centrifugal contactors. Developed by the Argonne National Laboratory, Chicago, IL, from 2003 and proposed for use by 2014.

USC [**U**ltra**S**elective **C**onversion] A front-end process for improving the operation of catalytic crackers for making ethylene. Developed and offered by Stone & Webster Engineering Corporation.

USCO A direct-reduction ironmaking process, using coal gas as the reductant. Operated in South Africa for several years since 1985 but now abandoned. *See* DR.

USS Phosam *See* Phosam.

UTI A process for making urea from ammonia and carbon dioxide, using heat recycle. Invented in 1970 by I. Mavrovic in New York.

U.S. Patents 3,759,992; 3,952,055.

Frayer, J.Y., *Chem. Eng. (Rugby, Engl.)*, 1973, **80**(12), 72.

UVOX [**U**ltra**V**iolet **OX**idation] A process for purifying seawater before clarification, using chlorine and exposure to sunlight. Developed in South Africa.

Hebden, D. and Botha, G.R., *Desalination*, 1980, **32**, 115.

V

VACASULF [**VAC**uum de**SULF**urization] A process for removing hydrogen sulfide from coke-oven gas by scrubbing with aqueous potassium carbonate:

$$K_2CO_3 + H_2S = KHS + KHCO_3$$

Hydrogen sulfide is recovered from the scrubbing solution under vacuum, hence the name. It is then either oxidized with air and the sulfur dioxide used for making sulfuric acid or converted to elemental sulfur by the *Claus process. The process is suitable only for gases not containing ammonia. Developed by Krupp Koppers, Germany. Three units were being built in 1993.

Vacuum Carbonate An improved version of the *Seabord process for removing hydrogen sulfide from refinery gases, in which the hydrogen sulfide is stripped from the sodium carbonate solution by steam instead of by air. Developed by the Koppers Company, Pittsburgh, in 1939; two plants were using this process in the United States in 1950.

U.S. Patents 2,379,076; 2,464,805; 2,242,323.
Reed, R.M. and Updegraff, N.C., *Ind. Eng. Chem.*, 1950, **42**, 2271.
Claxton, G., *Benzoles: Production and Uses*, National Benzole & Allied Products Assoc., London, U.K., 1961, 212.

VAD *See* Steelmaking.

Valorga [**Val**orisation optimale des dechets **organ**iques] A process for treating household waste by anaerobic digestion with the production of methane. Developed in the 1990s at the Languedoc University of Science & Technology and now operated by Valorga International in seven European countries.

Waste Manage. Environ., 1996, **6**(3), 40.

van Arkel and de Boer Also called the Iodide process. A process for producing or purifying a metal by thermal decomposition of its iodide on a hot tungsten filament. Modern high-intensity halide lamps are based on this reaction. Originally used for making small quantities of tungsten, titanium, and zirconium, but today probably used only for the preparation of ultrapure chromium. First used in 1925 by A.E. van Arkel at the Philips Gloeilampenfabrieken in the Netherlands for making tungsten; subsequently used by him and J.H. de Boer for making other metals. An extension of the process has been to make metal nitrides by passing a mixture of the metal halide with nitrogen and hydrogen over a hot tungsten wire.

U.S. Patent 1,671,213.
van Arkel, A.E., *Physica*, 1923, **3**, 76.
van Arkel, A.E. and de Boer, J.H., *Z. Anorg. Allg. Chem.*, 1925, **148**, 345.
Rolsten, R.F., *Iodide Metals and Metal Iodides*, John Wiley & Sons, New York, 1961.

Van Dyke A reprographic process, based on the photoreduction of ferric ammonium oxalate to ferrous ammonium oxalate, which in turn reduces silver nitrate to silver metal.

Kosar, J., *Light Sensitive Systems*, John Wiley & Sons, New York, 1965.

Van Ruymbeke (1) A process for recovering glycerol from the residual liquor from the *kettle soapmaking process. After separating the solid soap, the liquor is heated with basic ferric sulfate, thereby precipitating the residual carboxylic acids as their insoluble ferric salts. After removing these by filtration, the liquor is concentrated by vacuum evaporation and the glycerol distilled out under vacuum. *See also* Garrigue.

> Martin, G. and Cooke, E.I., in *Industrial and Manufacturing Chemistry*, Cooke, E.I., Ed., Technical Press, Kingston Hill, Surrey, U.K., 1952, 122.

Van Ruymbeke (2) A process for dehydrating 95% aqueous ethanol by countercurrent extraction of the vapor with glycerol.

> British Patents 184,036; 184,129.
> U.S. Patent 1,459,699.

VaporSep A family of separation processes, based on membranes that are selectively permeable to organic vapors. Developed by Membrane Technology & Research, CA, in the 1990s and used by DSM in its polypropylene plant to separate propylene from nitrogen. The membrane is a three-layer sandwich, packaged in spiral modules.

> *Eur. Chem. News*, 1996, **65**(1700), 24; 1997, **67**(1771), 16.

VAR *See* Steelmaking.

Varga A complex process for hydrogenating brown coal and high molecular weight asphaltenes. The process uses hydrogen at a high pressure, in the presence of an iron oxide catalyst. Invented by J. Varga in Budapest and operated in Germany.

> Varga, J., *Brennstoff-Chemie*, 1928, **9**, 277.
> Varga, J., Rabo, G., and Zalai, A., *Brennstoff-Chemie*, 1956, **37**, 244.
> Weisser, O. and Landa, S., *Sulphide Catalysts: Their Properties and Applications*, Pergamon Press, Oxford, U.K., 1973, 296.

VARICOL A continuous chromatographic process for separating natural products. It generally resembles the *Sorbex process, which uses a simulated moving bed, but the column is divided into a larger number of independent sectors.

> Ludemann-Hombourger, O., Nicoud, R.M., and Bailly, M., *Sep. Sci. Technol.*, 2000, **35**(12), 1829.
> Toumi, A., Hanisch, F., and Engell, S., *Ind. Eng. Chem. Res.*, 2002, **41**(17), 4328.

VAROX An air separation process, allowing for variable oxygen demand. Developed by Linde from 1984.

VAW Lurgi [Vereinigte AluminiumWerke] An energy-efficient process for extracting aluminum from bauxite. Extraction is done in a pipe reactor, and the aluminum hydroxide is converted to the oxide in a fluidized bed.

VAW Sulfite [Vereinigte AluminiumWerke] Also known as the S-T process. A process for extracting aluminum from clay by sulfurous acid. Basic aluminum sulfite, $Al_2O_3 \cdot 2SO_2 \cdot 5H_2O$, is crystallized from the purified leachate and thermally decomposed. The process was operated on a commercial scale by Vereinigte Aluminiumwerke at Lauterwerk, Germany, during World War II.

> O'Connor, D.J., *Alumina Extraction from Non-bauxitic Materials*, Aluminium-Verlag, Düsseldorf, Germany, 1988, 188.

VCC *See* Veba-Combi Cracking.

Veba A process for *hydrocracking bitumen, developed by Veba Oel, Germany.

> Graeser, U. and Niemann, K., *Oil Gas J.*, 1982, **82**(12), 121.

Veba-Combi Cracking Also called VCC. A *Bergius–Pier high-pressure thermal hydrotreating process. The catalyst is usually a promoted iron oxide, operated in a slurry, but an added catalyst may not be necessary.

Used in Germany during World War II. A version developed by Veba Oel Entwicklungsgesellschaft mbH was operated in Bottrop, Germany, from 1988 to 1994, using vacuum residues from crude oil distillation and *visbreaking operations, with plastics from municipal wastes, and with chlorinated organic compounds.

U.S. Patent 4,851,107.

Bowman, C.W., Phillips, R.S., and Turner, L.B., in *Handbook of Synfuels Technology*, Meyers, R.A., Ed., McGraw-Hill, New York, 1984, 5–74.

Chauvel, A., Delmon, B., and Hölderich, W.F., *Appl. Catal. A: Gen.*, 1994, **115**, 173.

Ventron A process for removing mercury from aqueous wastes containing organic mercury compounds. Chlorine is passed in, converting organic mercury compounds to inorganic compounds, and the mercury is then reduced to the metallic state with sodium borohydride.

Rosenzweig, M.D., *Chem. Eng. (N.Y.)*, 1971, **78**(5), 70.

VERA [**VE**rglasungsanlage für **R**adioaktive **A**bfälle] A continuous process for immobilizing nuclear waste by incorporating it in a borosilicate glass made by spray calcination. Developed at the Kernforschungszentrum Karlsruhe, Germany, from the mid-1960s. The process was abandoned in the mid-1970s, but some of the technology was later used in *PAMELA.

Lutze, W., in *Radioactive Waste Forms for the Future*, Lutze, W. and Ewing, R.C., Eds., North-Holland, Amsterdam, the Netherlands, 1988, 7.

Verneuil A process for growing single crystals of refractory compounds. The powdered material is dropped through an oxyhydrogen flame, and the product, consisting of microscopic molten droplets, is collected on a seed crystal. As the liquid mass reaches a cooler zone, it crystallizes in the form of a single crystal, known as a boule. Invented by A.V.L. Verneuil at the Museum of Natural History, Paris, who made synthetic rubies and sapphires in this way.

Verneuil, A.V.L., *Ann. Chim. Phys.*, 1904, **3**, 20.

Merker, L., *Min. Eng. (N.Y.)*, 1955, **7**, 645.

Elwell, D., *Man-made Gemstones*, Ellis Horwood, Chichester, U.K., 1979, 34.

Vere, A.W., *Crystal Growth: Principles and Progress*, Plenum Press, New York, 1987, 67.

VerTech A sewage-treatment process in which the sludge is digested under aerobic conditions at high temperature and pressure in a vertical shaft approximately one mile deep. Developed in the 1980s by VerTech Treatment Systems, the Netherlands. Piloted in 1985 at Longmont, CO, and installed at Apeldoorn, the Netherlands, in 1994. *See also* Deep Shaft.

Water Bull., 31 Mar 1994, 6.

Downie, N.A., *Industrial Gases*, Blackie Academic & Professional, London, U.K., 1997, 438.

Luck, F., *Catal. Today*, 1999, **53**(1), 83.

VGO Isomax [**V**acuum **G**as **O**il] A *hydrodesulfurization process adapted for treating vacuum gas oil, a petroleum fraction. Developed by Chevron Research Company in the early 1970s. In 1972, five plants were in operation and six were under construction. *See also* RDS Isomax and VRDS Isomax.

Hydrocarbon Process. Int. Ed., 1972, **51**(9), 184.

Viad An early, two-stage coal gasification process.

Victor A process for making a mixed ammonium nitrate–ammonium sulfate fertilizer (ASN), which is less liable to explode than ammonium nitrate. Ammonium sulfate is mixed with nitric acid, and the mixture is ammoniated in an evaporator. Developed by Gewerkschaft Victor–Chemische Werke and used by that company at Castrop Rauxel, Germany. *See also* Uhde–Hibernia.

Nitrogen, 1968, (53), 27.

Vinnolit A process for making polyvinyl chloride by suspension polymerization. Developed by Vinnolit, Germany, and engineered by Uhde. Widely licensed worldwide.

Vinyloop A process for recycling polyvinyl chloride (PVC). Used PVC is ground, dissolved in a proprietary solvent, and reprecipitated in a granular form. Developed by Solvay in 1999 and first commercialized by a joint venture at Ferrara, Italy. The first plant, at Ferrara, was opened in 2004. A joint venture between Solvay and Kobelco Eco-Solutions (a subsidiary of Kobe Steel), VINYLOOP, was formed to commercialize the process in Japan; its plant opened in 2006.

Chem. Eng. (N.Y.), 1999, **106**(10), 17.
Chem. Eng. (Rugby, Engl.), 6 Jul 2000, (705), 11.
Chem. Week, 2001, **163**(34), 16.
Chem. Eng. News, 2002, **80**(9), 15.
Eur. Chem. News, 2002, **77**(2012), 28.

Visbreaking A *thermal cracking process that reduces the viscosity of the residues from petroleum distillation, so that they may be handled at lower temperatures. It is essentially a high-temperature, noncatalytic pyrolytic process conducted in the presence of steam. *See also* HSC.

Ballard, W.P., Cottingham, G.I., and Cooper, T.A., in *Encyclopedia of Chemical Processing and Design*, McKetta, J.J. and Cunningham, W.A., Eds., Marcel Dekker, New York, 1981, **13**, 172.
The Petroleum Handbook, 6th ed., Elsevier, Amsterdam, the Netherlands, 1983, 280.
Negin, K.M. and Van Tine, F.M., in *Handbook of Petroleum Refining Processes*, 3rd ed., Meyers, R.A., Ed., McGraw-Hill, New York, 2003, 12.91.

Viscose Also known as the Cross–Bevan–Beadle process. A process for making regenerated cellulose fibers. The product has been known by the generic name "rayon" since 1924. Cellulose, from cotton or wood, is first reacted with sodium hydroxide (*Mercerization), yielding alkali cellulose. This is dissolved in carbon disulfide, yielding cellulose xanthate, which is dissolved in sodium hydroxide solution. Injection of this solution (known as "viscose" because of its high viscosity) into a bath of acid regenerates the cellulose. The process was invented by C.F. Cross, E.J. Bevan, and C. Beadle in London in 1882, further developed in the United States in the 1890s, and then widely adopted worldwide. *See also* SINI.

Cross, C.F., Bevan, E.J., and Beadle, C., *J. Soc. Chem. Ind.*, 1892, **12**, 516.
Moncrieff, R.W., *Man-made Fibres*, 6th ed., Butterworth Scientific, London, U.K., 1975, 162.
Kotek, R., in *Handbook of Fiber Chemistry*, 3rd ed., Lewin, M., Ed., CRC Press, Boca Raton, FL, 2006, 711–764.

VITOX A process for providing oxygen to a microbiological process such as sewage treatment. The heart of the process is a subsurface mixer–oxygenator, developed by the British Oxygen Company for uprating the oxygenation capacity of overloaded sewage plants. It is now used as an integral design feature in new plants. The oxygen is introduced at the neck of a venturi and forms extremely fine bubbles. The process operates in open tanks, unlike the *UNOX process, which uses closed tanks. The equipment has been used also for dissolving carbon dioxide in water, for hardening the water with lime. In 1991, the process was used in over 100 sewage plants in the United Kingdom and another 200 in the rest of the world.

Gould, F.J. and Stringer, P.R., in *Effluent Treatment and Disposal*, Institution of Chemical Engineers, Rugby, U.K., 1986, 33.
Gray, N.F., *Activated Sludge: Theory and Practice*, Oxford University Press, Oxford, U.K., 1990, 120.

Vitrifix A vitrification process for converting asbestos to a harmless glassy substance, suitable for use as a construction material. Developed by Vitrifix of North America and commercialized in the United Kingdom.

Roberts, D., *IEEE Trans. Ind. Appl.*, 1989, **25**(3), 451.

Vniios A process for catalytically pyrolyzing hydrocarbons to low molecular weight alkenes, similar to catalytic cracking but more efficient. The catalyst is either potassium vanadate on corundum or indium oxide on pumice. Developed by the All-Union Research Institute for Organic Synthesis, Moscow.

Oil Gas J., 1997, **95**(25), 54.

VOD *See* Steelmaking.

Voest *See* DR.

Voloxidation A process for removing volatile fission products and tritium from irradiated nuclear fuel in advance of other processing. Being developed from 2004 at the Argonne National Laboratory.

Volto *See* Elektrion.

von Heyden One of several processes for oxidizing naphthalene to phthalic anhydride. It operates with a fixed bed of vanadium–molybdenum oxide catalyst. Another version of the von Heyden process has been developed by Wacker Chemie for oxidizing *o*-xylene to phthalic anhydride and is licensed by that company. In 1989, 65 plants had been built or were under construction. BASF and Lurgi now offer their versions of the process.

> *Hydrocarbon Process. Int. Ed.*, 1989, **68**(11), 107.

VPSA [**V**acuum **P**ressure **S**wing **A**dsorption] Also known as **VSA**. A version of *PSA in which the adsorbed gas fraction is desorbed by reducing the pressure, rather than by displacement by another gas. Proprietary versions developed by the Linde Division of Union Carbide Corporation are known as *OxyGEN and *NitroGEN. Invented by L'Air Liquide, France, in 1957.

> French Patent 1,223,261.
> U.S. Patent 3,155,468.
> Young, R.T., *Gas Separation by Adsorption Processes*, Butterworths, Guildford, U.K., 1987.
> *Chem. Eng. (N.Y.)*, 1989, **96**(10), 17.

VRC [**V**alorisation des **R**ésidues **C**hlorés] A process for incinerating chlorine-containing organic residues. The hydrochloric acid produced is condensed and sold (hence the name). Developed by Atochem, France, in 1975, operated at Saint-Auban, and planned for installation in Poland in 1992.

> Robin, A., in *Chemical Waste Handling and Treatment*, Muller, K.R., Ed., Springer-Verlag, Berlin, Germany, 1985, 268.
> *Eur. Chem. News*, 1990, **55**(1450), 30.

VRDS [**V**acuum **R**esidua **DeS**ulfurization] A general name for *hydrotreating processes for removing sulfur from the residues from the vacuum distillation of petroleum residues. *See also* VRDS Isomax.

> Brossard, D.N., in *Handbook of Petroleum Refining Processes*, 3rd ed., Meyers, R.A., Ed., McGraw-Hill, New York, 2003, 8.3.

VRDS Isomax [**V**acuum **R**esidua **DeS**ulfurization] A *hydrodesulfurization process adapted for processing the residues from the vacuum distillation of the least volatile fraction of petroleum. An extension of the *RDS Isomax process, developed and piloted by Chevron Research Company in the early 1970s. In 1988, one unit was under construction and another was being engineered.

> Speight, J.G., *The Desulfurization of Heavy Oils and Residua*, Marcel Dekker, New York, 1981, 194.

VSA [**V**acuum **S**wing **A**dsorption] *See* VPSA.

V-tex An efficient scrubbing system for removing acid gases and particulates from effluent gases. Developed in 1999 by AEA Technology, United Kingdom.

Vulcanization The treatment of natural rubber with sulfur to reduce its tackiness and improve its strength and elasticity. Invented independently by C. Goodyear and N. Hayward in the United States in 1839 and by T. Hancock in London in 1842 to 1843. Goodyear was honored for his invention by Napoleon III, but he died in a debtors' prison in Paris. Various chemicals other than elemental sulfur are effective, for example, sulfur monochloride, selenium, and p-quinone dioxime. The chemical mechanism of this process is still not fully understood: some believe that traces of zinc, derived from the zinc oxide used in compounding, are essential.

Duerdon, F., *Thomas Hancock: An Appreciation, Plas. Rubber Int.*, 1986, **11**(3), 22.

Chapman, A.V. and Porter, M., in *Natural Rubber Science and Technology*, Roberts, A.D., Ed., Oxford University Press, Oxford, U.K., 1988, Chap. 12.

Slack, C., *Noble Obsession*, Theia Books, New York, 2002.

W

Wacker (1) A general process for oxidizing aliphatic hydrocarbons to aldehydes or ketones by the use of oxygen, catalyzed by an aqueous solution of mixed palladium and copper chlorides. Ethylene is thus oxidized to acetaldehyde. If the reaction is conducted in acetic acid, the product is vinyl acetate. The process can be operated with the catalyst in solution or with the catalyst deposited on a support, such as activated carbon. There has been a considerable amount of fundamental research on the reaction mechanism, which is believed to proceed by alternate oxidation and reduction of the palladium:

$$CH_2{=}CH_2 + PdCl_2 + H_2O = CH_3CHO + Pd + 2HCl$$

$$Pd + 2HCl + \tfrac{1}{2}O_2 = PdCl_2 + H_2O$$

The naming of this process has been confused because of various corporate relationships. The basic invention was created in 1957 at the Consortium für Elektrochemische Industrie, Munich, a wholly owned subsidiary of Wacker Chemie. It has therefore been called both the Wacker process and the Consortium process. But for many years, Wacker Chemie has had a close relationship with Farbwerke Hoechst, and the latter company has participated in some of the development and licensing activities, so two other names have come to be used: Wacker–Hoechst and Hoechst–Wacker. The five inventors (J. Schmidt, W. Hafner, J. Sedlmeier, R. Jira, and R. Rüttinger) received the DECHEMA prize in 1962 for this invention. The acetaldehyde process was first operated commercially in 1960. In 1997, this process was used in making 85% of the world's production of acetaldehyde. Although Wacker Chemie still makes vinyl acetate, it no longer uses the Wacker process to do so. The process is said to be have been the first industrially applied organometallic reaction.

German Patents 1,049,845; 1,06.
Smidt, J., Hafner, W., Jira, R., Sedlmeier, J., Sieber, R., Rüttinger, R., and Kojer, H., *Angew. Chem.*, 1959, **71**(5), 176.
Chem. Eng. News., 1961, **39**(16), 52.
Jira, R., in *Ethylene and Its Industrial Derivatives*, Miller, S.A., Ed., Ernest Benn, London, U.K., 1969, 639.
Lowry, R.P. and Aquilo, A., *Hydrocarbon Process. Int. Ed.*, 1974, **53**(11), 105.
Weissermel, K. and Arpe, H.-J., *Industrial Organic Chemistry*, 4th ed., translated by C.R. Hindley and S. Hawkins, Wiley-VCH, Weinheim, Germany, 2003, 166.
Hartings, M.R., *Nat. Chem.*, 2012, **4**(9), 764.

Wacker (2) A process for making sodium salicylate by reacting sodium phenate with carbon dioxide.

Lindsey, A.S. and Jeskey, H., *Chem. Rev.*, 1957, **57**, 583.

Wacker–Hoechst *See* Wacker (1).

Waelz A process for extracting zinc and lead from lean ores, using a large rotary kiln. Developed by Metallurgische Gesellschaft and Fried. Krupp Grusonwerk at Magdeburg, Germany, in 1926. The process is used also for extracting zinc and lead from the dusts from electric arc furnaces. The pelletized dusts are mixed with 25% coke and 15% sand and heated in a rotary kiln to 1200°C. Lead and zinc volatilize and are collected as dusts that, after bricketting, can be treated by the *Imperial Smelting process.

Cocks, E.J. and Walters, B., *A History of the Zinc Smelting Industry in Britain*, George G. Harrap, London, U.K., 1968, 61, 150.
Morgan, S.W.K., *Zinc and Its Alloys and Compounds*, Ellis Horwood, Chichester, U.K., 1985, 141.

Wah Chang *See* Benilite.

Walker A process for partially oxidizing natural gas or LPG, forming a mixture of methanol, formaldehyde, and acetaldehyde. Air is the oxidant and aluminum phosphate the catalyst. Invented by J.C. Walker in the 1920s and operated by the Cities Service Corporation, OK, in the 1950s.

> U.S. Patent 2,186,688.
> Walker, J.C. and Malakoff, H.L., *Oil Gas J.*, 1946, **45**(33), 59.
> Meyer, R.E., *Oil Gas J.*, 1955, **54**(7), 82.

Walterization *See* metal surface treatment.

Walthal An obsolete process for obtaining alumina from clay. The clay was roasted and extracted with sulfuric acid, and the aluminum sulfate dried and calcined.

Walther Also called Walther Ammonia and AMASOX. A *flue-gas desulfurization process in which the gas is scrubbed with aqueous ammonia. Two scrubbing stages are used, operating at different pH values. The by-product is suitable for use as a fertilizer. Developed and licensed by Walther & Company. Two plants were operating in Germany in 1987, one of which was experiencing serious operational problems.

> Merrick, D. and Vernon, J., *Chem. Ind. (London)*, 1989, 3, 55.

WAO [Wet Air Oxidation] A generic name for processes for destroying aqueous hazardous wastes by heating in air or oxygen. One example is its use for oxidizing sodium sulfide in alkaline scrubber effluents from olefin plants, developed by Nippon Petrochemicals and offered by Toyo Engineering.

> Mishra, S.V., Mahajani, V.V., and Joshi, J.B., *Ind. Eng. Chem. Res.*, 1995, **34**, 2.
> Luck, F., *Catal. Today*, 1999, **53**(1), 81.
> Dinjus, E. and Krupp, A., in *High Pressure Chemistry*, van Eldik, R. and Klärner, F.G., Eds., Wiley-VCH, Weinheim, Germany, 2002.

Warner A novel process for extracting zinc from sulfide ores. Two linked furnaces are used. In the first, the ore is reacted with metallic copper:

$$ZnS + 2Cu = Zn + Cu_2S$$

The resulting zinc vapor is condensed to liquid and run off. The copper sulfide is oxidized to copper metal in the second furnace:

$$CuS + O_2 = Cu + SO_2$$

The molten copper is circulated back to the first furnace. The sulfur dioxide is converted to sulfuric acid for sale. Overall, the process uses much less energy than other zinc extraction processes. Developed in the 1980s by N.A. Warner at the University of Birmingham, United Kingdom, but not piloted as of 1992.

> British Patent 2,048,309.
> Gray, P.M.J., *Min. Mag.*, 1992, Jan, 14.

Washoe A process for extracting silver from sulfide ores. The ore is heated with aqueous sodium chloride in an iron pot. The chloride dissolves the silver and the iron reduces it; addition of mercury gives silver amalgam:

$$2AgCl + Fe = 2Ag + FeCl_2$$

This is a variation of the *Patio and *Cazo processes. Invented around 1860 at the Comstock mines, Nevada, and named after the district where it was developed. Mark Twain described the operations in his autobiographical novel *Roughing It* (Vol. 1, Chap. 36).

> Dennis, W.H., *A Hundred Years of Metallurgy*, Gerald Duckworth, London, U.K., 1963, 285.

Watercatox Not an established process, but a project of the Fifth Framework Program of the European Union. The purpose was to develop catalytic processes for destroying organic residues in water by wet air oxidation (*WAO). The chosen system used a tubular catalytic membrane reactor for contacting the aqueous solution with air. Several companies and research institutes participated in this project from 2000, and the process was piloted with several real industrial liquid effluents.

Iojoiu, E.E., Landrivon, E., Raeder, H., Torp, E.G., Miachon, S., and Dalmon, J.-A., *Catal. Today*, 2006, **118**(1–2), 246.

water gas A generic name applied to two processes and their products. The original process, dating to the end of the eighteenth century, makes a fuel gas by passing steam over a carbonaceous fuel that has been heated by partial combustion. The product is a mixture of carbon monoxide, carbon dioxide, and hydrogen. It is also known as blue gas, because it burns with a blue flame. Enrichment of blue gas by adding hydrocarbons was invented by T.S.C. Lowe and first commercialized at Phoenixville, PA, in 1874. The product, known as carbureted water gas, was used mainly as an additive to coal gas; in 1931, 13% of the town gas distributed in Great Britain was made by this process. *See also* Blaugas.

Parker, A., *J. Soc. Chem. Ind.*, 1927, **46**, 72.
Morgan, J.J., in *Chemistry of Coal Utilization*, Vol. 2, Lowry, H.H., Ed., John Wiley & Sons, New York, 1945, Chap. 37.
Peebles, M.W.H., *The Evolution of the Gas Industry*, Macmillan Press, London, U.K., 1980, 14.

water gas shift *See* Reforming.

W-D *See* Woodall–Duckham.

WD-IGI [Woodall–Duckham Il Gas Internazionale] A two-stage, nonslagging coal gasification process. In the first stage, tar and volatile matter is removed; in the second, steam and air (or oxygen) gasify the coke, producing a mixture of carbon monoxide, hydrogen, and nitrogen (if air is used). The process is based on a design by Il Gas Integrale, Milan, Italy, developed in the 1950s. In 1984, over 100 plants had been installed in Europe, South Africa, and Australia.

Jones, D.M., in *Handbook of Synfuels Technology*, Meyers, R.A., Ed., McGraw-Hill, New York, 1984, 3–169.

Weber *See* carbonization.

Weissenstein An electrolytic process for making hydrogen peroxide by the electrolysis of sulfuric acid. Peroxodisulfuric acid, $H_2S_2O_8$, is formed first; this is then hydrolyzed via peroxymonosulfuric acid:

$$2H_2SO_4 = H_2S_2O_8 + H_2$$

$$H_2S_2O_8 + H_2O = H_2SO_5 + H_2SO_4$$

$$H_2SO_5 + H_2O = H_2O_2 + H_2SO_4$$

First operated in 1908 at the Österreichische Chemische Werke, Weissenstein, Austria, and then by Degussa, Germany. This process, as well as the other electrolytic processes, was made obsolete by the invention of the *AO process.

Schumb, W.C., Satterfield, C.N., and Wentworth, R.L., *Hydrogen Peroxide*, Reinhold Publishing, New York, 1955, 132.

Weizmann A process for producing acetone and *n*-butanol by the fermentation of carbohydrates by bacteria isolated from soil or cereals. Later work has shown that effective bacteria are *Clostridium acetobutylicum* and *Bacillus Granulobacter pectinorum*. Used in Britain in World War I for the manufacture of acetone, needed for the production of cordite. Subsequently operated by Commercial Solvents Corporation in Terre Haute, IN, and in two plants in Canada. Later abandoned in favor of synthetic processes. Invented by C. Weizmann in the University of Manchester in 1915, based on earlier work at the Pasteur Institute by A. Fernbach and E.H. Strange (hence the alternative name Fernbach–Strange–Weizmann). The money that Weizmann obtained from royalties on this process was used in founding the state of Israel, of which he was the first president. *See also* ABE.

British Patents 21,073 (1912) (Fernbach and Strange); 4,845 (1915) (Weizmann).
U.S. Patent 1,315,585.
Goodman, P., Ed., *Chaim Weizmann: A Tribute on His Seventieth Birthday*, Victor Gollancz, London, U.K., 1945.
Rhodes, R., *The Making of the Atomic Bomb*, Simon & Schuster, London, U.K., 1986, 87–90.
New Ency. Brit., 1988, **12**, 565.
Benfey, T., *Chem. Ind. (London)*, 1992, (21), 827.
Bud, R., *The Uses of Life: A History of Biotechnology*, Cambridge University Press, Cambridge, U.K., 1993, 37–46.
Chem. Brit., 2002, **38**(12), 34.

Weldon An early process for making chlorine by oxidizing hydrochloric acid (from the *Leblanc process) with manganese dioxide. The mixture was heated with steam in stone tanks. Manganese was recovered from the liquor by precipitation with calcium hydroxide and subsequent oxidation by air:

$$4HCl + MnO_2 = MnCl_2 + Cl_2 + 2H_2O$$

$$MnCl_2 + Ca(OH)_2 = Mn(OH)_2 + CaCl_2$$

$$2Mn(OH)_2 + O_2 = 2MnO_2 + 2H_2O$$

The process was complicated by the formation of calcium manganite, $CaMn_2O_6$, known as Weldon mud. Invented by W. Weldon in 1866 and developed at St. Helens, United Kingdom, from 1868 to 1870. Operated in competition with the *Deacon process until both were overtaken by the electrolytic process for making chlorine from brine. Weldon mud has been used as a catalyst for oxidizing the hydrogen sulfide in coal gas to elemental sulfur.

British Patents 1,948 (1866); 133 (1867).
Hardie, D.W.F., *A History of the Chemical Industry in Widnes*, Imperial Chemical Industries, Widnes, U.K., 1950, 66.

Welland A process for making nitroguanidine, an explosive. Cyanamide dimer is reacted with ammonium nitrate to form guanidine nitrate, which forms nitroguanidine when dehydrated by heating with 96% sulfuric acid. *See also* Marqueyrol and Loriette.

Smith, G.B.L., Sabetta, V.J., and Steinbach, O.F., Jr., *Ind. Eng. Chem.*, 1931, **23**, 1124.

Wellman A coal gasification process, widely used since its introduction in the 1950s. Air and steam, at atmospheric pressure, are passed through a fixed bed of coal supported on a rotating ash bed. *See also* Riley–Morgan.

Hebden, D. and Stroud, H.J.F., in *Chemistry of Coal Utilization*, 2nd Suppl. Vol., Elliott, M.A., Ed., John Wiley & Sons, New York, 1981, 1616.

Wellman–Galusha A coal gasification process using a fixed bed; the dry ash is removed through a revolving grate.

van der Hoeven, B.J.C., in *Chemistry of Coal Utilization*, Vol. 2, Lowry, H.H., Ed., John Wiley & Sons, New York, 1945, 1659.

Wellman–Lord The most widely used regenerable *flue-gas desulfurization process. The sulfur dioxide is absorbed in sodium sulfite solution in a wet spray scrubber, forming sodium bisulfite:

$$Na_2SO_3 + H_2O + SO_2 = 2NaHSO_3$$

The solution is regenerated by heat to provide a sulfur-rich gas, which can be used to make elemental sulfur, sulfuric acid, or sulfur dioxide. A small amount of sodium sulfate is produced, which must be crystallized out and disposed of. Initially, the process used the potassium salts. Developed in the late 1960s. Thirty-eight plants had been built by 1992.

British Patent 1,557,295.
Hydrocarbon Process. Int. Ed., 1975, **54**(4), 111.
Ford, P.G., in *The Problem of Acid Emissions*, Institution of Chemical Engineers, Rugby, England, 1988, 151.
Kohl, H.L. and Nielsen, R.B., *Gas Purification*, 5th ed., Gulf Publishing Co., Houston, TX, 1997, 554.

Welsh The Welsh process is the general name given to the complex copper smelting operations carried out in Swansea, South Wales, from around 1800 until the introduction of larger smelters at the end of the nineteenth century. The heart of the Welsh process was a reverberatory furnace in which all the operations of roasting, fusing, and refining were conducted. It was superseded by the development of much larger furnaces, initially in the United States, and by the use of Bessemer-type converters for the final stages.

Dennis, W.H., *A Hundred Years of Metallurgy*, Gerald Duckworth, London, U.K., 1963, 129.

Wendell Dunn A family of chlorine beneficiation processes based on selective chlorination of ores in a fluidized bed. Developed by W.E. Dunn of Chlorine Technology in Australia in the 1970s, primarily for beneficiating ilmenite. The first such commercial ilmenite beneficiation plant, completed in 1991, was that of Benechlor Chemicals Private, Madras, India. The entire process, including recovery of chlorine by oxidizing the ferrous chloride, was piloted by Heubach in Ankleshwar, India, in 2001, the process being called the *Reptile Process and the product being called Reptile 96 [**Rep**lacement ru**tile** containing **96%** rutile].

U.S. Patents 3,699,206; 4,349,516.
German Patents 2,103,478; 2,220,870; 2,221,006.

Western Gas One of the processes for making water gas, in which heavy oil was introduced. This enabled the cost to be optimized, depending on the relative prices of coal and oil. In 1937, 45 installations used the process in the United States. *See also* water gas, Willien–Stein.

Hartzel, F.W. and Lueders, C.J., *Proc. Am. Gas Assoc.*, 1932, **14**, 882 (*Chem. Abstr.*, **27**, 3319).

Westinghouse A proposed thermochemical process for decomposing water to oxygen and hydrogen by electrolysis, coupled with the high-temperature decomposition of sulfuric acid:

$$2H_2O + SO_2 \rightarrow H_2 + H_2SO_4$$

$$H_2SO_4 \overset{870°C}{\rightarrow} H_2O + SO_2 + \tfrac{1}{2}O_2$$

Demonstrated only on the laboratory scale. It was developed in the 1970s as a potentially economic method of obtaining hydrogen fuel from a high-temperature source.

Williams, L.O., *Hydrogen Power: An Introduction to Hydrogen Energy and Its Applications*, Pergamon Press, Oxford, U.K., 1980, 85.

Westvaco (1) A variation of the *Claus process for removing hydrogen sulfide from gas streams, in which the sulfur dioxide is catalytically oxidized to sulfur trioxide over activated carbon at 75°C–150°C. The adsorbed sulfur trioxide is hydrated to sulfuric acid and then converted back to sulfur dioxide by reaction with the hydrogen sulfide at a higher temperature.

Ball, F.G., Brown, G.N., Davis, J.E., Repik, A.J., and Torrence, S.L., *Hydrocarbon Process. Int. Ed.*, 1972, **51**(10), 125.
Sulphur, 1974, (111), 51.
Oil Gas J., 1978, **76**(37), 88.

Westvaco (2) A process proposed for making chlorine by electrolyzing aqueous copper chloride. Invented in 1928 by F. S. Low at Chlorine Products, New York. Piloted by Westvaco in the 1940s, but not commercialized.

U.S. Patent 1,746,542.

Roberts, C.P., *Chem. Eng. Prog.*, 1950, **46**(9), 456.

Berkey, F.M., in *Chlorine: Its Manufacture, Properties and Uses*, Sconce, J.S., Ed., Reinhold Publishing, New York, 1962, 220.

Wetherill *See* American.

wetox [**wet ox**idation] A generic name for processes for oxidizing organic wastes, based on the use of hydrogen peroxide and a catalyst. Examples are *WINWOX and *WOX.

Wet Process A process for making phosphoric acid by treating phosphate rock with an acid. The acid is usually sulfuric acid, but hydrochloric and nitric acids are used commercially in special circumstances: in Israel, where by-product hydrochloric acid is available, and in Norway and Switzerland, where nitric acid is made by cheap hydroelectric power. The basic process with sulfuric acid is

$$3H_2SO_4 + Ca_3(PO_4)_2 = 3CaSO_4 + 2H_3PO_4$$

The calcium sulfate by-product separates as either the dihydrate or the hemihydrate, depending on the conditions. The process originates from the work of J.B. Lawes, who in 1842 patented a method of making a fertilizer by treating bones with sulfuric acid. Many variations are practiced today. *See also* Dorr and Haifa.

Childs, A.F., in *The Modern Inorganic Chemicals Industry*, Thompson, R., Ed., The Chemical Society, London, U.K., 1977, 386.

Becker, P., *Phosphates and Phosphoric Acid*, 2nd ed., Marcel Dekker, New York, 1989.

McCoubry, J.C., in *Industrial Inorganic Chemicals: Production and Use*, Thompson, R., Ed., Royal Society of Chemistry, Cambridge, U.K., 1995, 379.

Wiberg-Soderfors A direct reduction process for extracting iron. *See* DR.

Wiewiorowski A process proposed for removing hydrogen sulfide from industrial gases by reacting it with sulfur dioxide in molten sulfur in the presence of an amine catalyst. Invented by T.K. Wieriorowski at the Freeport Sulfur Company, but never commercialized.

U.S. Patent 3,447,903.

Chem. Eng. News, 1970, **48**(18), 68.

Kohl, H.L. and Nielsen, R.B., *Gas Purification*, 5th ed., Gulf Publishing Co., Houston, TX, 1997, 846.

Wilbuschewitsch *See* Normann.

Wilkinson Not a process but a catalyst, often stated to have been invented by Nobel laureate Geoffrey Wilkinson, but also invented independently in 1965 by two ICI chemists: R.S. Coffee and B.J. Smith. The catalyst is *tris*(triphenylphosphine) rhodium chloride, $RhCl(PPh_3)_3$. It catalyzes several important reactions, including hydrogenation. In 1974, Wilkinson developed a hydroformylation process for making butyraldehyde, which used a related rhodium complex, $RhHCO(PPh_3)_3$. Wilkinson's early career was as a nuclear chemist, and he is believed to have discovered more isotopes than anyone else. His Nobel Prize, awarded jointly with E.O. Fischer, was for establishing the molecular structure of ferrocene. He also coauthored a best-selling textbook of inorganic chemistry.

British Patent 1,121,642 (catalyst).

U.S. Patent 4,108,905 (hydroformylation).

Young, J.F., Osbourne, J.A., Jardine, F.H., and Wilkinson, G., *J. Chem. Soc., Chem. Commun.*, 1965, 131.

Osbourne, J.A., Jardine, F.H., Young, J.F., and Wilkinson, G., *J. Chem. Soc. A.*, 1966, 1711.

Willhoft A proposed process for making aluminum chloride from the solid waste from paper mills. The waste is mainly a mixture of clay with cellulose. It is dried and calcined in an inert atmosphere, giving a mixture of clay and carbon that chlorinates readily. Conceived by E.M.A. Willhoft and briefly examined by the Research Association for the Paper and Board, Printing and Packaging Industries (PIRA) in England in 1977, but not piloted.

British Patent 1,472,683.

Willien–Stein A method for increasing the hydrogen content of water gas by introducing gas oil at one stage in the process. Invented by L.J. Willien and L. Stein in 1929. Piloted at three locations in the United States in the 1930s, but apparently not adopted on a large scale. *See also* Water Gas, Western Gas.

Canadian Patent 305,227.
Morgan, J.J., in *Chemistry of Coal Utilization*, Vol. 2, Lowry, H.H., Ed., John Wiley & Sons, New York, 1945, 1741.

Willson A process for making calcium carbide by heating calcium oxide with tar or carbon in an electric furnace:

$$CaO + 3C = CaC_2 + CO$$

Invented by T.L. Willson in 1892 and first practiced commercially at Niagara Falls in 1896.

Chem. Eng. (N.Y.), 1950, **57**(6), 129.

Wilman A metallurgical process for removing manganese from steel scrap. Developed at the Electricity Council Research Centre, Capenhurst, United Kingdom, and first commercialized in 1988.

Winkler Also called Fritz Winkler. A process for gasifying coal, using oxygen (or air) and steam in a fluidized bed at atmospheric pressure. Introduced by F. Winkler of IG Farbenindustrie, Germany, in 1922. It was developed in the 1920s and used mainly in Germany; the first plant was built at Leuna in 1926; by 1979, 36 units had been built. The largest was 33 m high, 6 m in diameter. The Flesch–Winkler process is a modification that permits the use of relatively unreactive coals, which produce ash having a low melting point. *See also* HTW.

German Patent 437,970.
Dainton, A.D., in *Coal and Modern Coal Processing*, Pitt, G.J. and Millward, G.R., Eds., Academic Press, London, U.K., 1979, 138.
Cornils, B., in *Chemicals from Coal: New Processes*, Payne, K.R., Ed., John Wiley & Sons, Chichester, U.K., 1987, 13.

Winkler–Koch An early mixed-phase petroleum *cracking process.

WINWOX [**WIN**frith **W**et **OX**idation] A process for oxidizing hazardous organic wastes by wet oxidation (*WOX) with hydrogen peroxide and a catalyst containing a transition metal such as iron or copper. Developed in 1987 by the Winfrith Technology Centre of the U.K. Atomic Energy Authority, originally for destroying ion-exchange resins containing radioactive isotopes, but later proposed for hazardous organic wastes generally. A pilot plant was built in 1989.

Eur. Chem. News, 1989, **53**(1389), 24.
Wilks, J.P. and Holt, N.S., *Waste Manage.*, 1990, **10**, 197.

WIP [**W**aste **I**mmobilization **P**lant (or **P**rocess)] A process for immobilizing nuclear waste by incorporation in a borosilicate glass for long-term disposal. Developed in the 1970s in India for use at the waste immobilization plant at Tarapur.

Lutze, W., in *Radioactive Waste Forms for the Future*, Lutze, W. and Ewing, R.C., Eds., North-Holland, Amsterdam, the Netherlands, 1988, 11.

Wisconsin A thermal process for fixing atmospheric nitrogen. Air is heated to over 2000°C by contact with a bed of magnesia pebbles and then cooled rapidly by contact with a bed of cold pebbles. The resulting air, containing 1%–2% of nitric oxide, is passed through beds of silica gel to dry it, to permit the nitric oxide to be oxidized to dinitrogen tetroxide, and to concentrate the dinitrogen tetroxide before desorbing it and dissolving it in water. Developed by F. Daniels at the University of Wisconsin in 1953 and piloted in Kansas, but subsequently abandoned.

Gilbert, N. and Daniels, F., *Ind. Eng. Chem.*, 1948, **40**, 1719.
Ermenc, E.D., *Chem. Eng. Prog.*, 1956, **42**(4), 149.
Chem. Eng. Prog., 1956, **42**(11), 488.
Chilton, T.H., *Strong Water: Nitric Acid, Its Sources, Methods of Manufacture, and Uses*, MIT Press, Cambridge, MA, 1968, 116.

Wisner *See* carbonization.

Witten A process for making dimethyl terephthalate. Also called Hercules-Witten. *See* Katzschmann.

WLP [Wasserstoff-Lichtbogen-Pyrolyse, German, meaning *hydrogen arc pyrolysis*] A process for converting gasoline into a mixture of acetylene and ethylene by injecting a jet of it into a hydrogen plasma. Piloted by Knapsack-Griesheim in Germany in the 1960s.

Miller, S.A., *Acetylene: Its Properties, Manufacture and Uses*, Vol. 1, Ernest Benn, London, U.K., 1965, 407.

Wohlwill An electrolytic process for refining gold. The crude gold, which may be made by fusing the anode slimes from the *Balbach process, is used as the anode, the cathode is of pure gold, and the electrolyte is a solution of gold chloride in hydrochloric acid. Gold deposits on the cathode. Silver deposits as a sediment of silver chloride. The process is relatively slow, so the interest lost on the inventory of metal in process is significant. Developed by E. Wohlwill at the Norddeutsche Affinerie in Hamburg in 1874, it became the principle method of gold refining in the world. It was largely superseded by the *Miller chlorine process at the end of the nineteenth century.

Dennis, W.H., *A Hundred Years of Metallurgy*, Gerald Duckworth, London, U.K., 1963, 281.
Yannopoulos, J.C., *The Extractive Metallurgy of Gold*, Van Nostrand Reinhold, New York, 1991, 243.

Woodall–Duckham Also called the Babcock W-D process. A process for recovering hydrochloric acid and metal oxides from spent metal chloride solutions, such as those obtained from metal pickling and ilmenite beneficiation. The liquor is first concentrated by evaporation and then atomized in a heated spray tower. Water evaporates from the droplets in the upper part of the tower, and chlorides are converted to oxides in the hotter, lower part. Developed by Woodall–Duckham in the 1960s; by 1992, over 150 installations were in use worldwide. Now offered by Babcock Woodall–Duckham, United Kingdom.

Woolwich [Named after the British Government laboratory at Woolwich Arsenal, where it was invented] A process for making the explosive RDX by nitrating hexamethylene tetramine.

WORCRA [**WOR**ner **C**onzinc **R**iotinto of **A**ustralia] A family of continuous smelting and refining processes developed by Conzinc Riotinto of Australia in the 1960s. Invented by H.K. Worner. The copper smelting process was piloted in Port Kembla, New South Wales, in 1968, but later abandoned.

U.S. Patent 3,326,671.
Worner, H.K., in *Advances in Extractive Metallurgy*, Institution of Mining & Metallurgy, London, U.K., 1968, 245.
Worner, H.K., *Eng. Min. J.*, 1971, **172**(8), 64.
Worner, H.K., Reynolds, J.O., Andrews, B.S., and Collier, A.W.G., in *Advances in Extractive Metallurgy and Refining*, Institution of Mining & Metallurgy, London, U.K., 1972, 18.

Workman *See* Dual-Spectrum.

WOX [**W**et **OX**idation] A process for destroying organic materials by catalyzed oxidation with hydrogen peroxide. Developed by ASEA Atom, Sweden. *See also* WINWOX.

WPO A *CWAO process for destroying organic wastes in water.

Pintar, A., *Catal. Today*, 2003, **77**(4), 451.

WSA [**W**et gas **S**ulphuric **A**cid] A process for recovering sulfur from flue gases and other gaseous effluents in the form of concentrated sulfuric acid. It can be used in conjunction with the *SCR process if oxides of nitrogen are present too. The sulfur dioxide is catalytically oxidized to sulfur trioxide, and any ammonia, carbon monoxide, and carbonaceous combustibles are also oxidized. The sulfur trioxide is then hydrolyzed to sulfuric acid under conditions that produce commercial-quality 95% acid. Developed by Haldor Topsoe. In 2005, it was used in 49 plants worldwide. *See also* SNOX.

Eur. Chem. News, 2005, **83**(2160), 17.

WSA-2 A variation of the WSA process, developed in 1989 but abandoned in 1994 in favor of the original WSA process.

WSA-SNOx A combined flue-gas treatment process that converts the sulfur dioxide to sulfuric acid and the nitrogen oxides to nitrogen. Developed by Snamprogetti and Haldor Topsoe, based on the *WSA process. A large trial was conducted in Ohio by ABB and Snamprogetti from 1992 to 1993.

Wulff A two-stage process for making acetylene by the pyrolysis of saturated aliphatic hydrocarbons. The feed gas is first pyrolyzed at approximately 1300°C and then passed into a refractory brick reactor at below 400°C. Developed by R.G. Wulff in California in 1927. Operated in the United States, Brazil, and Europe until the end of the 1960s. *See also* Ruhr Chemie.

U.S. Patents 880,308; 917,627; 1,843,965.
Bixler, G.H. and Coberly, C.W., *Ind. Eng. Chem.*, 1953, **45**, 2596.
Bogart, M.J.P. and Long, R.H., *Chem. Eng. Prog.*, 1962, **58**(7), 90.
Miller, S.A., *Acetylene: Its Properties, Manufacture and Uses*, Vol. 1, Ernest Benn, London, U.K., 1965, 384.
Tedeschi, R.J., *Acetylene-based Chemicals from Coal and Other Natural Resources*, Marcel Dekker, New York, 1982, 25.
Weissermel, K. and Arpe, H.-J., *Industrial Organic Chemistry*, 4th ed., translated by C.R. Hindley and S. Hawkins, Wiley-VCH, Weinheim, Germany, 2003, 96.

Wünsche An electrolytic process for liberating bromine from a bromide solution. It uses carbon electrodes and a porous clay separator. Developed in Germany in 1902. *See also* Kossuth.

German Patent 140,274.
Yaron, F., in *Bromine and Its Compounds*, Jolles, Z.E., Ed., Ernest Benn, London, U.K., 1966, 16.

WWT *See* Chevron WWT.

Wyandotte A process for making a mixture of ethylene and propylene glycols from propane for use as antifreeze. The propane is cracked to a mixture of ethylene and propylene; these are not separated but are converted to the corresponding glycols by *chlorohydrination. Developed by the Wyandotte Chemicals Corporation.

X

XCAT A process for making polypropylene, using a metallocene catalyst. Developed by Univation Technologies and licensed to Equistar Chemicals.

> *Eur. Chem. News*, 2004, **81**(2128), 29.

Xentia A process for extracting flavorings and fragrances using the fluorinated solvent HFC 134a. Developed by Ineos Fluor, Runcorn, United Kingdom, and now commercialized. The name is used also for a range of flavorings and fragrances made by the process.

> *Eur. Chem. News*, 2002, **76**(2000), 26.

XIS [**X**ylene **IS**omerization] A process for isomerizing *p*-xylene to the equilibrium mixture of C_8 aromatic hydrocarbons. Developed by Maruzen Oil in the United States.

> Weissermel, K. and Arpe, H.-J., *Industrial Organic Chemistry*, 4th ed , translated by C.R. Hindley and S. Hawkins, Wiley-VCH, Weinheim, Germany, 2003, 334.

Xonon Cool Combustion A catalytic technology that combusts fuel flamelessly. Incorporated inside a gas turbine engine, it reduces the production of oxides of nitrogen to <3 ppm by volume. Developed by Catalytica Energy Systems, CA. First demonstrated in 2002 in cooperation with Kawasaki Gas Turbines— Americas in Sonoma. The development was abandoned in 2006 because of unfavorable gas turbine market conditions.

Xylenes-plus A catalytic process for isomerizing toluene to a mixture of benzene and xylenes. A silica–alumina catalyst is used in a moving bed. It is unlike the related *Tatoray process, in that no extra hydrogen is required. Developed by Sinclair Research in 1964 and then licensed by Atlantic Richfield.

> U.S. Patents 3,116,340; 3,350,469; 3,437,709.
> Verdol, J.A., *Oil Gas J.*, 1969, **67**(23), 63.

Xylofining [**Xylo**l re**fining**] A process for isomerizing a petrochemical feedstock containing ethylbenzene and xylenes. The xylenes are mostly converted to the equilibrium mixture of xylenes; the ethylbenzene is dealkylated to benzene and ethylene. This is a catalytic, vapor-phase process, operated at approximately 360°C. The catalyst (Encilite-1) is a ZSM-5-type zeolite in which some of the aluminum has been replaced by iron. The catalyst was developed in India in 1981, jointly by the National Chemical Laboratory and Associated Cement Companies. The process was piloted by Indian Petrochemicals Corporation in 1985 and commercialized by that company at Baroda in 1991.

> Indian Patent 155,892.

XyMax A selective process for making *p*-xylene and *o*-xylene by isomerizing C_8 aromatics-rich hydrocarbon streams, using a proprietary zeolite catalyst. Developed by ExxonMobil and first used in 2000 in its refinery on Jurong Island, Singapore. Five more units have subsequently been licensed. Retrofit applications are licensed by ExxonMobil Chemical, grassroots applications by Axens.

Hydrocarbon Process. Int. Ed., 2001, **80**(7), 32; 2002, **81**(7), 75.
Eur. Chem. News, 2001, **74**(1947), 27.

XyMax-2 An improved version of *XyMax that uses a different zeolite catalyst.

Chem. Eng. (N.Y.), 2003, **110**(13), 16.

Z

Z3A [**Z**ero environmental damage, **3**rd **A**lternative] A process for pyrolyzing waste tires using microwave energy. Developed and piloted in the 1990s by Advanced Molecular Agitating Technology (AMAT), Crewe, United Kingdom. Not yet commercialized.

> U.S. Patent 6,693,265.
> Ludlow-Palafox, C. and Chase, H.A., in *Feedstock Recycling and Pyrolysis of Waste Plastics*, Scheirs, J. and Kaminsky, W., Eds., John Wiley & Sons, Chichester, U.K., 2006, 586.

Zadgaonkar A process for converting waste plastics to hydrocarbon fuels. Chlorine from PVC is first removed by heating the molten material with coal and additives. The molten product is then passed over a surface at 350°C in the presence of coal and other additives. The hydrocarbon products volatilize and are then condensed and fractionated. Operated on a commercial scale in Nagpur, India, since 2005.

> Zadgaonkar, A., in *Feedstock Recycling and Pyrolysis of Waste Plastics*, Scheirs, J. and Kaminsky, W., Eds., John Wiley & Sons, Chichester, U.K., 2006, 724.

Zadra A process for extracting gold from its ores. After *cyanidation and adsorption on activated carbon, the gold is reextracted into a hot alkaline cyanide solution and stripped from it by electrolysis using a steel wool cathode.

> Yannopoulos, J.C., *The Extractive Metallurgy of Gold*, Van Nostrand Reinhold, New York, 1991, 201.

ZECA [**Z**ero **E**mission **C**oal **A**lliance] A process for making electricity from coal. The coal is reacted with steam and hydrogen to produce methane, which is reformed to produce pure hydrogen, which is used in a solid oxide fuel cell to produce electricity. Developed from 1999 by Los Alamos National Laboratory and an alliance of industrial research laboratories, now organized as the ZECA Corporation.

Zeisberg An energy-efficient process for concentrating nitric acid in its manufacture. Developed by F.C. Zeisberg at DuPont.

> Chilton, T.H., *Strong Water: Nitric Acid, Its Sources, Methods of Manufacture, and Uses*, MIT Press, Cambridge, MA, 1968, 125.

Zenith A process for refining vegetable oils by passing droplets of them down a column of dilute aqueous sodium hydroxide.

> Braae, B., *J. Am. Oil Chem. Soc.*, 1976, **53**, 353.

Zeoforming A process for converting light paraffinic feedstocks to high-octane gasoline components. The catalyst is zeolite ZSM-5 and the process does not require hydrogen. Developed in the CIS, engineered by KTI, and first installed by Lurgi in Gorlice, Poland, in 1997.

> *Ch12em. Eng. (N.Y.)*, 1998, **105**(3), 23.
> *CATTECH*, 2002, **6**(3), 96.

z-forming A process for making aromatic hydrocarbons from aliphatic hydrocarbons. Developed jointly by Chiyoda and Mitsubishi Oil and operated in a demonstration plant in Kawasaki until it was closed in 1992.

The catalyst was an acidity-adjusted zeolite that was selective for BTX and resistant to coking. The process was used by Nippon Oil but abandoned by them in 2008 because the product yield was too low.

Eur. Chem. News, CHEMSCOPE, 1994, **61**, 7.
Giannetto, G., Morque, R., and Galiasso, R., *Catal. Rev. Sci. Eng.*, 1994, **36**, 271.
Jpn. Chem. Week, 2009, **50**(2498), 10.

Ziegler (1) A process for polymerizing ethylene under moderate temperatures and pressures, catalyzed by a mixture of titanium tetrachloride and a trialkyl aluminum such as triethyl aluminum. Invented in 1953 by K. Ziegler at the Max Planck Institut für Kohlenforschung, Mülheim/Ruhr, Germany. The first U.S. plant was opened in 1957. Now operated worldwide on a very large scale. *See also* Ziegler–Natta.

Belgian Patent 533,362.
Ziegler, K., Holzkamp, E., Breil, H., and Martin, H., *Angew. Chem.*, 1955, **67**, 426.
Natta, G., *Angew. Chem.*, 1956, **68**, 393.
Ziegler, K., *Angew. Chem.*, 1959, **71**, 623; 1960, **72**, 829.
Raff, R.A.V., in *Ethylene and Its Industrial Derivatives*, Miller, S.A., Ed., Ernest Benn, London, U.K., 1969, 335.
Wittcoff, H.A., Reuben, B.G., and Plotkin, J.S., *Industrial Organic Chemicals*, 2nd ed., Wiley-Interscience, Hoboken, NJ, 2004, 489–493.

Ziegler (2) A process proposed for making tetraethyl lead by electrolyzing the molten complex of ethyl potassium with triethyl aluminum, $KAl(C_2H_5)_4$, using a lead electrode. Invented in 1963 by K. Ziegler and H. Lehmkühl but not commercialized.

U.S. Patent 3,372,097.

Ziegler–Natta Also called Z–N. A general name for the family of olefin polymerization processes invented by K. Ziegler and G. Natta in the 1950s. Ziegler and Natta were jointly awarded the Nobel Prize for Chemistry in 1963 for their discoveries. *See* Natta, Ziegler (1).

Boor, J., Jr., *Ziegler-Natta Catalysts and Polymerizations*, Academic Press, New York, 1979.
James, L.K., Ed., *Nobel Laureates in Chemistry 1901–1992*, American Chemical Society and Chemical Heritage Foundation, Washington, DC, 1993, 442, 449.
Fink, G., Mülhaupt, R., and Brintzinger, H.H., Eds., *Ziegler Catalysts: Recent Scientific Innovations and Technical Improvements*, Springer Verlag, Berlin, Germany, 1995.

Zimmermann *See* Zimpro.

Zimpro [**Zim**mermann **pro**cess] Also called the Zimmermann process and wet air oxidation. A thermal process for oxidizing organic wastes in aqueous solution and for conditioning sewage sludge. Raw sewage sludge is pressurized with air and heated with steam to 150°C–250°C in a pressure vessel; the product is sterile and easy to filter. Invented by J.F. Zimmermann in the United States in 1954, first operated in Chicago in 1957, and now offered by USFilter. As of 1991, more than 200 units had been installed worldwide. *See also* SCWO.

Teletzke, G.M., *Chem. Eng. Prog.*, 1964, **60**(1), 33.
Pradt, L.A., *Chem. Eng. Prog.*, 1972, **68**(12), 72.
Metcalf and Eddy, Inc., *Wastewater Engineering: Treatment, Disposal, Re-use*, 2nd ed., McGraw-Hill, New York, 1979, 636.
Hydrocarbon Process. Int. Ed., 1996, **75**(8), 109.
Luck, F., *Catal. Today*, 1999, **53**(1), 82.

Zincex [**Zinc ex**traction] A process for extracting zinc from pyrite cinder leachate, using organic solvents. The chloride leachate is first extracted with a secondary amine and then with di-(2-ethylhexyl) phosphoric acid to remove iron. Developed by Tecnicas Reunidas, first commercialized in 1976, and now used in Spain and Portugal.

Nogueira, E.D., Regife, J.M., and Arocha, A.M., *Eng. Min. J.*, 1979, **180**(10), 92.
Nogueira, E.D., Regife, J.M., and Blythe, P.M., *Chem. Ind. (London)*, 1980, (2), 63.

Zinclor A development of the *Zincex process that uses di-pentylpentylphosphonate (DPPP) as the extractant. Developed by Tecnicas Reunidas.

Cox, M., in *Developments in Solvent Extraction*, Alegret, S., Ed., Ellis Horwood, Chichester, England, 1988, 181.

Zincote *See* metal surface treatment.

Zirpro A process for flame-proofing textiles by treating them with aqueous solutions of zirconium complexes. Wool is treated with aqueous potassium hexafluorozirconate and citric acid. Developed by the International Wool Secretariat, Yorkshire, now based in Melbourne, Australia.

Benisek, L., *J. Textile Inst.*, 1974, **65**, 102.
Ingham, P.E. and Benisek, L., *J. Textile Inst.*, 1977, **68**, 176.

Z–N *See* Ziegler–Natta.

Z-Sorb A process for removing hydrogen sulfide and other sulfur compounds from gas streams by absorption in a proprietary granular absorbent containing zinc oxide. The process can be operated at temperatures between 315°C and 555°C. Developed by Phillips Petroleum Co.

Kohl, H.L. and Nielsen, R.B., *Gas Purification*, 5th ed., Gulf Publishing, TX, 1997, 1329.
Chem. Eng. (N.Y.), 1998, **105**(2), 25.

Bibliography

Aftalion, F., *A History of the International Chemical Industry*, translated by O.T. Benfey, University of Pennsylvania Press, Philadelphia, PA, 1991.

Buchel, K.H., Moretto, H.-H., and Werner, D., *Industrial Inorganic Chemistry*, 2nd ed., Wiley-VCH, Weinheim, Germany, 2000.

Cornils, B., Herrmann, W.A., Wong, C.-H., and Zanthoff, H.-W., *Catalysis from A to Z: A Concise Encyclopedia*, 4th ed., 4 vols., Wiley-VCH, Weinheim, Germany, 2013.

East, M.B. and Ager, D.J., *Desk Reference for Organic Chemists*, Krieger Publishing, Malabar, FL, 1995, Chap. 4.

Haber, L.F., *The Chemical Industry during the Nineteenth Century*, Oxford University Press, New York, 1958.

Haber, L.F., *Chemical Industry, 1900–1930*, Oxford University Press, Oxford, U.K., 1971.

Hardie, D.W.F. and Pratt, J. Davidson, *A History of the Modern British Chemical Industry*, Pergamon Press, Oxford, U.K., 1966.

Hassner, A. and Namboothiri, I., *Organic Syntheses Based on Name Reactions: A Practical Guide to 750 Transformations*, 3rd ed., Elsevier, Amsterdam, the Netherlands, 2012.

Kirk-Othmer's Encyclopedia of Chemical Technology, 4th ed., John Wiley & Sons, New York, 1991–1996.

Kirk-Othmer's Encyclopedia of Chemical Technology, 5th ed., John Wiley & Sons, New York, 2004–2007.

Kirk-Othmer's Encyclopedia of Chemical Technology, 6th ed., on-line version, Wiley Interscience, Bognor Regis, U.K.

Laue, T. and Plagens, A., *Named Organic Reactions*, 2nd ed., John Wiley & Sons, New York, 2005. Published originally in 1998 as *Namen und Schlagwort-Reaktionen der Organische Chemie*.

Li, J.J., *Name Reactions*, 2nd ed., Springer Verlag, Berlin, Germany, 2003.

McKetta, J.J. and Cunningham, W.A., Eds., *Encyclopedia of Chemical Processing and Design*, Marcel Dekker, New York, 69 vols., 1976–2001.

Morris, P.J., Named reactions—Forgotten chemists, *Chem. Ind. (London)*, 1999, (24), 968.

Mundy, B.P., Ellerd, M.G., and Favaloro, Jr., F.G., *Name Reactions and Reagents in Organic Synthesis*, 2nd ed., Wiley-Interscience, Hoboken, NJ, 2005.

Neumüller, O.-A., Ed., *Römpps Chemie-Lexicon*, 9th ed., Georg Thieme Verlag, Stuttgart, Germany. Published on line since 2002. *Encyclopedia Römmp CD 2006*, Version 3.33 (2013), Georg Thieme Verlag, Stuttgart.

Russell, C.A., Ed., *Recent Developments in the History of Chemistry*, Royal Society of Chemistry, London, U.K., 1985.

Russell, C.A., Ed., *Chemistry, Society and Environment: A New History of the British Chemical Industry*, Royal Society of Chemistry, Cambridge, U.K., 2000.

Tarbell, D.S. and Tarbell, T., *Essays on the History of Organic Chemistry in the USA, 1875–1955*, Folio Publishers, Nashville, TN, 1986.

Tottle, C.R., *An Encyclopedia of Metallurgy and Materials*, The Metals Society and MacDonald & Evans, London, U.K., 1984.

Ullmann's Encyclopedia of Industrial Chemistry, 40 vols., 6th ed., on-line, Wiley Interscience, Bognor Regis, U.K.

Witcoff, H.A., Reuben, B.G., and Potkin, J.S., *Industrial Organic Chemicals*, 3rd ed., Wiley-Blackwell, New York, 2013.

Appendix: Key to Products

acetaldehyde Grünstein, Hoechst-Wacker, Wacker (1), Walker.

acetic acid Acetica, AO Plus, BUTOX, Cativa, Celanese LPO, DF, Glacido, Llangwell, Monsanto (3), Rhône-Poulenc/Melle Bezons, Shawinigan, Suida.

acetic anhydride Eastman-Halcon, Hoechst-Shawinigan.

acetone ABE, Cumene, Cumox, Hock, Roka, Squibb, Weizmann.

acetylene Avco, CCOP, Hoechst HTP, Hoechst-WLP, Hüls, Ruhrchemie, Sachsse, SBA, Schoch, WLP, Wulff.

acrylic acid Reppe.

acrylic esters Reppe.

acrylonitrile ammoxidation, Andrussov, Knapsack, Kurtz, OSW, PETROX, SNAM (2), SOHIO.

adiponitrile CANDID, EHD, hydrocyanation, Monsanto, UCB-MCI.

aggregate Neutralysis.

alcohols, aliphatic Alfol, Bashkirov, Dobanol, Epal, Octamix.

aldehydes Consortium, OXO, RCH/RP.

alkanes *See* Hydrocarbons, Aliphatic.

alkenes *See* Olefins.

alkyl anilines Thoma.

alkylate Alkar, AlkylClean, FBA, Detal, Detergent Alkylate, Ethalk, HAT, HETACAT, Mobil/Badger, SP-Isoether, Stratco, Thoma.

alkyl cyanides ammoxidation.

Alnico Magnicol.

alumina AEROSIL, Alumet, Anortal, Bayer, Blanc, Deville-Pechiney, Grzymek, Hall (4), M2M.

aluminum alloys Cowles.

aluminum casting INOTEC.

aluminum chloride Toth.

aluminum chlorohydrate ACH (2).

aluminum extraction Aloton, Alumet, Anortal, Blanc, Bretsznajder, Büchner, Calsinter, Hichlor, H-Plus, Kalunite, Nuvalon, Pechiney H+, Pedersen, Peniakoff, Reynolds Metal, Séailles-Dyckerhoff, Toth, VAW Lurgi, VAW-Sulfite, Walthal.

aluminum metal AIAG Neuhausen, ALCOA, ALUREC, Compagnie AFC, ARP, Deville (1), Grätzel, Hall-Héroult, Hoopes, Netto, Pechiney (1), Thermical.

aluminum nitride Serpek, SHS.

aluminum oxide *See* Alumina.

aluminum trichloride Hichlor, H-Plus, Toth, Willhoft.

amines, aliphatic Ashland.

amines, aromatic Béchamp.

amino acids Acylase.

o-**aminothiophenols** Herz.

ammonia AMV, Braun, BYAS, Casale, Claude (1), Claude-Casale, Dual Pressure, Fauser, Haber, KAAP, KAAP plus, LCA, LEAD, Megammonia, Mond Gas, Mont Cenis, NEC, PARC, Purifier, Serpak, U2A.

ammonium alum Aloton.

ammonium chloride Engechlor.

ammonium nitrate NSM, Stengel, Uhde-Hibernia, Victor.

ammonium phosphates Cros, Gardinier, Minifos, Norsk-Hydro, Swift.

ammonium sulfate Cominco, Erdölchemie, GEESI, Merseburg, OSAG.

ammonium thiosulfate ATS.

t-**amyl methyl ether** NExTAME.

antimony extraction BRGM.

argon HARP.

arsenic extraction Arseno.

ascorbic acid Bertrand, Reichstein-Grüssner.

Bakelite Baekeland.

benzene Aromax (1), Benzorbon, DETOL, Dynaphen, HDA, Houdry-Litol, Hydeal, hydrodealkylation, Hytoray, Kellogg-Hydrotreating, Lignol, Litol, MHC, MHD, MSTDP, MTDP, Newton Chambers, PX-Plus, Pyrotol, Tatoray, T2BX, TDP, THD, THDA, Xylenes-plus.

benzole BASF/Scholven, Ibuk, Thiofex.

benzonitrile Ammoxidation.

beryllium extraction Copaux, Copaux-Kawecki, Fuse-quench, Kjellgren, Kjellgren-Sawyer, Perosa, Schwenzfeier-Pomelée.

beryllium metal Kjellgren.

biodiesel bioliq, Bio-Synfining, BioTherm, BIOX (2), Butamax, Ecofining, Esterfip-H.

bisphenol-A CT-BISA.

brass Cementation (2).

bromine Dow bromine, Kossuth, Kubierschky, Wünsche.

BTX Alpha Plant, ART (2), ATA.

BUNA Buna.

butadiene Aldol, BB-flex, CAA, Catadiene, DIFEX, GPB, KLP, Lebedev, Ostromislenski, O-X-D, Oxo-D, Phillips (3), Reppe.

butane Krupp-Koppers (2).

1,4-butanediol Geminox, Linde/Yukong.

butanetriol trinitrate Biazzi.

n-**butanol** Weizmann.

t-**butanol** Oxirane.

1-butene Alphabutol, Idemitsu, IFP-SABIC, Isopol, SHP, Sorbutene.

2-butene Arco, BUTACRACKING, ISOMPLUS, Isopol, Trolefine.

butenes Krupp-Koppers (2), MTO.

butyraldehyde RCH/RP.

γ-butyrolactone FAST, Metabolix.

cadmium recovery Tetronics.

calcium ammonium nitrate CAN.

calcium carbide Willson.

calcium carbonate Kemgas.

calcium cyanamide Frank-Caro, Polzeniusz-Krauss.

calcium hypochlorite Mathieson (2), Pennsalt/Pennwalt, Perchloron, PPG, Thann.

calcium nitrate Cerny.

caprolactam ALTAM, HPO, HPO-plus, HSO, PNC.

carbides, cemented Coldstream, ROC.

carbon black Ayers, Acetylene Black, Channel Black, Enasco, Furnace Black, Gas Black, Jones, Lampblack, Thermal Black, Thermatomic.

carbon dioxide Backus, CORRS, GREENOX, KM CDR, Reich (1).

carbon disulfide Folkins.

carbon monoxide Calcor, COPISA, COPSA, Cosorb.

carboxylic acids Amoco, Armour (2), Bernardini, SOT.

cellulose acetate Acetate.

cellulose fiber Acetocell, Alceru, Bemberg, Carbacell, Chardonnet, Courtaulds Tencel, Cross-Bevan-Beadle, Cuprammonium, Lyocell, Newcell, Viscose, SINI.

cement Grzymek.

ceramics, nonoxide SHS.

ceramics, oxide Pechini, Sol-Gel, Stöber, Supramics.

charcoal, activated Urbain.

chlorine Airco, Castner-Kellner, Chlor Alkali, Deacon, De Nora, diaphragm cell, Downs, Glanor, Griesheim (1), Grosvenor-Miller, Hargreaves-Bird, Hasenclever, Hoechst-Uhde (1), Kel-Chlor, LeSeur, membrane cell, Salt, Schroeder, Shell Deacon, Weldon, Weldon-Pechiney, Westvaco (2).

chlorine dioxide Erco, Fröhler, Holst, Kesting, Munich, Mathieson (1), Persson, R-2, SVP.

chlorobenzene Auger.

chromium metal Thermite, van Arkel and de Boer.

coal tar Ab der-Halden, Heinrich Koppers.

cobalt extraction Caron.

cobalt metal Thermit.

coke CFDI, Desulco.

copper extraction ACTIVOX, AMAR, Anatread, Arbiter, BioCOP, CLEAR, Cuprex, Cymet (1), Cymet (2), Electroslurry, Falconbridge, Henderson, HydroCopper, LM, Longmaid-Henderson, Mannhès, MECER, NENATECH. Noranda, Orford, Outokumpu, Parker, RLE, Sherritt-Cominco, Sirosmelt, Treadwell, Welsh, WORCRA.

copper metal AMAR, Actimag.

cotton mercerization, Zirpro.

cresols Cresex.

cumene 3-DDM, CD-Cumene, Mobil/Badger Cumene, Q-max.

cyclohexane HB Unibon, Hydrar, Hytoray.

cyclohexane oxime Ammoximation, Nixan.

cyclohexylbenzene ARODIS.

cyclopropane Hass (2).

p-**cymene** Cymex.

cymenes Attisholz.

detergents Chemithon, Detal, Detergent Alkylate.

deuterium G-S.

1,2-dichloroethane CER, HTC (1), LTC, oxychlorination.

1,2-dichloropropane Hass-McBee.

1,4-dicyanobutane hydrocyanation.

dienes FEAST.

diesel fuel Akzo-Fina CFI, ARODIS, Biox (2), Blowdec, Carbo-V, CED, Dieselmax, Esterfip-H, GMD, HyCycle, HyTail, Polymer-Engineering, REDAR, Reentech, Smuda, ThermoFuel (1).

di-*iso*propyl ether Oxypro (1).

di*iso*propyl naphthalene Kureha.

dimethyl ether Mega TME, TIGAS.

dimethyl terephthalate Katzschmann, Hercules (2), Imhausen, Witten.

dinitrogen pentoxide Dipen.

Dual Magnum Metolachlor.

EDTA Geigy.

emeralds Espig.

epichlorohydrin Epicerol.

ethanol AFEX, Agrol, Arkenol, Biostil, BPX, BUTOX, EA, Emert, ENSOL, Gulf, Keyes, Madison-Scholler, Multicont, OxyNol, PROESA, Sunliquid, TCX, Van Ruymbeke (2).

ethyl acetate Avada, EA, Tischenko.

ethylbenzene Albene, Alkar, Ebex, EBmax, Ethylbenzene, Mobil/Badger.

ethylene ALCET, ARS, Benson, CCOP, Dianor, Ethoxene, GTE, Hoechst HTP, MOI, SBA-Kellogg, Score, Triolefin, WLP.

ethylene chlorohydrin Chlorohydrination.

ethylene glycol EHD, GO, Meteor, Wyandotte.

ethylene oxide Halcon (2), Lefort, Meteor, Scientific Design.

ethylene propylene copolymer Catalloy, Flexomer.

fatty acids *See* Carboxylic Acids.

flax Korte.

fluorine Fluorodec.

fluoro-organic compounds *See* Organofluorine Compounds.

formaldehyde Adkins-Peterson, Formox, Gutehoffnungshütte, Hibernia, Walker.

formic acid Biofine.

fructose Sarex.

furfural Attisholz, Biofine.

gallium extraction Beja, De la Breteque.

glass CRISFER.

glucose Sarex (1), Scholler.

glue Scheidemandel.

glycerol Garrigue, Neuberg, Shell Glycerol, Twitchell, Van Ruymbeke.

gold extraction APOL, Arseno, AuPLUS, Betts, BIOPRO, Calmet, Cashman, CGA, CIL, CIP (1), Cyanide, Haber Gold, K-Process, GOLDOX, Magchar, Merrill-Crowe, Moebius, PAL, Plattner, Rose (2), Salsigne, Zadra.

gold metal Miller, parting, Thum, Wohlwill.

graphite Acheson (2), Castner (1).

grout Joosten, Siroc.

gypsum Cerphos, Donau Chemie, Guillini, Knauf.

hexachlorocyclopentadiene Lidov.

hexane Hexall.

hexenes Dimersol G.

hydrazine Bayer Ketazine, Hoffman, Ketazine, Raschig (1).

hydrocarbons, aliphatic, C_4 AH Unibon, Alco, Anglo-Jersey, BUTENEX.

hydrocarbons, aliphatic, C_5–C_8 Isomate.

hydrocarbons, aliphatic, *iso* Hysomer, Penex.

hydrocarbons, aliphatic, linear ENSORB, IsoSiv, Molex, MS2, N-ISELF, Nurex, Parex (2), TSF.

hydrocarbons, alkyl aromatic Detal, Detergent Alkylate.

hydrocarbons, aromatic Alpha (1), ARIS, Aroforming, Aromax (1), Carom, Cyclar, Hydroforming, Hall (3), Koch, MTA, Platforming, Rittman, z-Forming.

hydrochloric acid Aman, EARS, Hargreaves, Hargreaves-Robinson, Mannheim (1), Woodall-Duckham.

hydrogen APR, Hoechst-Uhde (1), Hydrogen Polybed PSA, Hydrothermolysis (2), Hypro, HYSEC, HyTex, Lane, IS, LO-FIN, MRH, Proximol, PyStR, Rincker-Wolter, RKN, SER, SMART H_2, Sumitomo-BF, Westinghouse.

hydrogen cyanide Ammoxidation, Andrussov, BMA, Degussa, Fluohmic.

hydrogen peroxide AO, Barium, Huron-Dow, Krutzsch, Loewenstein-Riedel, Pietzsch and Adolph, Weissenstein.

hydroquinone Hock.

α-hydroxy*iso*butyric acid Escambia (1), Lonza (1).

hydroxylamine Olin Raschig, Raschig (3).

ilmenite beneficiate Becher, Benilite, ERMS, Musro/Murso, Reptile, SREP, SRET, TSR (2), Wah Chang, Wendell Dunn.

insulin Gas.

iron and steel Accar, Ajax, AOD, Armco, ASEA-SKF, Aston, Bloomery, Brassert, Catalan, Chenot, Corex, Cort, DR, DRC, EOF, FINMET, Fior, Ghaem, Hall (1), Hoerde, Hojanas, Imatra, ITmk3, Jindal, Kaldo, KHD-Contop, Larkin, NSC, Osmond, Pedersen, Puddling. *See also* DR, steelmaking.

iron oxide pigment Laux, Penniman (2).

***iso*amylene** Trans4m S.

***iso*butane** Butamer, Butomerate, I-Forming.

***iso*butanol** Butamax.

*iso*butene Adib, Arco, BLISS, BUTACRACKING, C_4 Butesom, Cold Acid, FBD-4, ISOFIN, ISOMPLUS, Isopure, OlefinSiv, Olex, SKIP.Trans4m S, *iso*octane. Nexoctane.

*iso*pentenes C_5 Pentesom.

isoprene IDAS.

*iso*propyl alcohol Ellis.

di-*iso*propyl ether Oxypro.

Lanxide Lanxide.

l-DOPA Monsanto (2).

lead carbonate, basic Bischof, Carter, Dutch, Thénard, Thompson-Stewart.

lead extraction Boliden (2), Carinthian, Flintshire, Huntington-Heberlein, Imperial Smelting, Isasmelt, Kaldo, Kivcet, QSL, Waelz.

lead metal Betterton (1), Betterton (2), Betterton-Kroll, Betts, Britannia, Davey, Harris, liquation, Parkes, Pattinson, Tetronics.

lead monoxide Barton.

lead tetraethyl Ziegler (2).

leather Liritan, TAL.

levoglucosenone Furacell.

levulinic acid Biofine.

lignin LignoBoost, SLRP.

lignosulfonates Howard, Magnefite.

lithium isotopes COLEX, ELEX, OREX.

l-lysine Toray (1).

magnesium extraction Chesney, Dow Seawater, MAGRAM, Pattinson (2).

magnesium metal Elektron, Gardner, Hansgirg, Magnetherm, MagSonic, Norsk-Hydro, Pidgeon, Radenthein.

magnetite Pirox.

maleic anhydride ALMA, Petro-Tex.

malononitrile Lonza (2).

manganese extraction Dean, MHO.

manganese metal Chemetals, Pidgeon, Thermite.

mannitol Creighton.

menthol Symrise.

mercury fulminate Chandelon.

methacrylates Escambia.

methacrylic acid Lonza.

methane ANTHANE/ANODEK, Binax, Biogas, Catalytica, HCM, Hydrane, Kryosol, Laran.

methanol Attisholz, CAMERE, Carnol, Catalytica, CTM, DMO, HYDROCARB, Hynol, ICI Low-Pressure Methanol, ILPM, LCM, LPMEOH, MAS, Mega-Methanol, Pier-Mittasch, POX, Remet, Walker.

methylamines Leonard (1).

methyl ethyl ketone Hoechst-Wacker, Maruzen (3).

methyl methacrylate ACH (1), Alpha (2), Aveneer, MIGAS.

methyl *t*-butyl ether catalytic distillation, CDETHEROL, Ethermax, Etherol, Isotex, NExETHERS.

mica Bardet, Samica.

molds, foundry Isocure, Pep Set, SAPIC.

monoethylene glycol Omega (2).

mustard gas Levinstein, Runcol.

naphthalene Brodie, Sulzer-MWB, UNIDAK, Unionfining.

naphthoquinone Kawasaki Kasei.

nickel catalyst Raney, SURECAT.

nickel extraction BioNIC, Caron, Falconbridge, Hybinette, INCO, Mond Nickel, Nicaro, Orford, Outokumpu, PAL II, Sherritt-Gordon.

niobium carbide Menstruum.

niobium extraction Fray-Farthing-Chen.

niobium metal Balke, Kroll (1).

nitrate esters Biazzi, Gyttorp.

nitric acid Arc, Bamag-Meguin, Birkeland-Eyde, CNA, CONIA, DSN, DWN, Grande Paroisse, Häusser, HOKO, Hycon (2), Ostwald, SABAR, Schönherr, SOLNOX, Wisconsin.

nitro-compounds, organic Bofors, Hass (1), Hercules.

nitrogen Bergbau-Forschung, DWN, Generon, KURASEP, Linde, MOLPSA-Nitrogen, NitroGEN, Serpek.

nitroglycerine Biazzi, Nitro Nobel, Schmidt.

nitroguanidine Marqueyrol and Loriette, Welland.

nuclear fuel IFR, Pyro A, Pyro B.

nuclear waste AVM, ESTER, FINGAL, FIPS, HARVEST, PAMELA, PHOTHO, PIVER, Synroc, VERA, WINWOX, WIP.

octenes Dimersol X, Octol.

oils, lubricating Bensmann, Chlorex (1), Mohawk.

oils, vegetable Behr, Zenith.

olefins Alphene, Alpha-Sablin, Alfa-Select, CATOFIN, Catpoly, DCC, DeFine, Hoechst Coker, Idemitsu, Indmax, Linear-1, MAXOFIN, MTO, OCP, Oleflex, Olex, Pacol, Petrofin, Polynaphta Essence, PYROCAT, SHOP, SOR, Superflex, TRC, Vniios.

oleum Squire and Messel.

organofluorine compounds Halex, La-Mar, Simons.

oxalic acid Rhône-Poulenc.

oxide catalysts GREENCAT.

oxygen Bosch, Brin, DWO, Linde, Mallet, Moltox, OxyGEN, Oxy-Rich, OXYWELL.

ozone MEMBREL, POZONE.

palmitic acid Emersol.

paper and pulp Acedox, Acetosolv, AhlStage, Alcell, ASAM, BFR, Burkheiser, Celdecor, DegOX, ECF, EnZone, Green liquor, GreenOx, Kraft, Kramfors, Lignox, Lyocell, Magnefite, Macrox, Middox, Milox, MOXY, NSSC, Organocell, Organosolv, Oxypro (2), Prenox, Sapoxal, SAPPI-Air Liquide-Kamyr, SCA-Billerand, Sirola, Solvay (2), Solv-X, SRP, Stora, Sulfate, Sulfite, TCF.

paraffin wax Krupp-Kohlechemie, Sharple.

pentanes Pentafining.

peptides Merrifield.

perchloric acid Pernert.

phenol AlphOx, Cumene, Cumox, Dennis–Bull, Dow-Phenol, Dynaphen, Guyot, Halcon (1), HDA, Hercules-BP, Hock, Lignol, Phenosolvan, Phenoraffin, Raschig (2), Tyrer.

phenols Noguchi.

phosphate fertilizer Davidson, Den, LETS, Oberphos, Odda.

phosphoric acid Adex, Central-Prayon, CFB, Dorr, Dorr-Oliver, Gulf-Swenson, Haifa, HDH, Jacobs-Dorr, KPA, Maunsell, Nordac, Ozark-Mahoning, Phorex, Prayon, Progil, SIAPE, Wet Process.

phosphorus Furnace.

phthalic acid Kawasaki Kasei.

phthalic anhydride Gibbs, Heyden-Wacker, LAR, Sapper, von Heyden.

plastic foams MuCell.

plastic waste Klobbie.

platinum metals Deville and Debray, Footing, Leidie, Matthey, Moebius, parting, slow-cooled matte, Sobolevsky, Solvex.

plutonium Butex, Purex, Recuplex, Redox, Tramex, Truex.

polybutene Mobil-Witco-Shell.

polyether polyols CAOS, Impact.

polyethylene ACP, Addipol, ATOL, Borstar, Catalloy, CP, CX, Dowlex, Energx, Evolue, Exxpol, High Productivity, Hi-Zex, Hostalen, Innovene, Lupotech G, Naphtachimie, Phillips (1), Prodigy, Sclair, Sclairtech, Selexsorb, Spherilene, Stamylex, Standard Oil, UNIPOL, Ziegler (1).

polyethylene terephthalate AIST, IntegRex, MTR, UnPET.

poly*iso*butene Cosden.

polymer gasoline Selectopol.

polymethyl benzenes Koch.

polypropylene Addipol, Borstar, E-flex, GPP, Horizone, Hypol, LIPP-SHAC, Meta-4, Metocene, MPC, MTP, Natta, Novolen, Sherpol, Spheripol, Spherizone, Rexene, UNIPOL, XCAT.

polystyrene NORSOLOR, SDS, Styro-Plus.

polyvinyl chloride MSP3, Vinnolit.

poly-*p*-xylene Paralene.

potassium carbonate Engel, Engel-Precht, Precht.

potassium cyanide Beilby, Erlenmeyer, Rodgers.

potassium hydroxide Griesheim (1).

potassium metal Griesheim (2).

potassium permanganate Carus.

potassium sulfate Alumet.

propanediol Tokuyama.

propylene E-Flex, Maxene, META-4, MTO, MTP, OCP (2), OCP (3), OCT, Omega, OTP, Propylur, Superflex.

propylene glycol GPTG, Wyandotte.

propylene oxide Cetus, Daicel, Escambia (2), HPO Plus, HPPO, Montoro, Oxirane, POSM, Propylox, SMPO.

protein Pekilo, Probion, Provesteen, Pruteen, Rothamsted, Scholler-Tornesch.

pulp *See* Paper and Pulp.

PVC Vinyloop.

quartz Heraeus.

rayon *See* Cellulose Fiber.

RDX Bachmann, KA, Woolwich.

resorcinol Hock.

rubber, foam Dunlop, Talalay.

rubber, natural Kaysam, Peachy, Vulcanization.

rubber, synthetic Aldol, Alfin, Buna, GRS.

rutile, synthetic *See* Ilmenite Beneficiate.

silica AEROSIL, Cab-o-Sil, HAL, SiVARA, Soot, Stöber.

silica gel Lentz.

silicon carbide Acheson (1).

silicon metal Siemens.

silicon nitride Satin.

silk, artificial *See* Cellulose Fiber.

silver extraction Boss, Britannia, Cazo, Clandot, Kiss, Patera, Patio, Russell, Washoe.

silver metal Balbach, Betts, cupellation, Davey, liquation, Luce-Rozan, Miller, Moebius, Parkes, parting, Pattinson (1), Rozan, Thum, Wohlwill.

soap Armour (1), Bradshaw, Clayton, Crosfield, De Laval Centripure, Kettle, Mazzoni, Mon Savon, saponification, Sharples, Twitchell.

sodium carbonate ammonia soda, Fresnel, Hou, Leblanc, Schloesing-Rolland, Solvay (1).

sodium chloride Alberger, Grainer, Recrystallizer, Salex.

sodium cyanide Bucher, Castner (2), neutralization, Raschen, Röhm, Schlempe.

sodium dithionite Amalgam, Formate.

sodium hydroxide Castner-Kellner, causticization, Chlor-Alkali, De Nora, diaphragm cell, Ferrite, Glanor, Hargreaves-Bird, Kiflu, lime-soda, Löwig, membrane cell.

sodium metal Castner (3), Castner (4), Deville (2), Downs, Tekkosha, Shanks.

sodium nitrate Guggenheim.

sodium perborate Acid, Duplex.

sodium salicylate Wacker (2).

sodium sulfate Climax, Hargreaves-Robinson, Mannheim, Salt Cake.

sorbitol Creighton, Dynatol.

stannic oxide Tin Sol.

stearic acid Emersol.

steel *See* Iron and Steel.

styrene ExSyM, Fina/Badger, GT-Styrene Recovery, Mark and Wulff, Montoro, Oxirane, SMART SM, SNOW, Styro-Plus.

styrene copolymers POSTech.

sucrose Bergius (2), Bergius-Rheinau, Boivan-Loiseau, Deguide, Harloff, Madison, NRS, Quentin, Scheibler, Scholler-Tornesch, Steffen, Sucro-Blanc, Talafloc, Talodura.

sulfur Boliden (1), Catasulf, CLINSULF, Claus, DOXOSULFREEN, ENsulf, EUROCLAUS, HySpec, Hysulf, Mond, RESOX, Selectox, SPOR, Sulpel, Trail.

sulfur dioxide BAYQIK.

sulfur extraction Chemico, Frasch, Orkla.

sulfuric acid Bayer-Bertrams, BOSAC, Cat-Ox, Chamber, Contact, IPA, Knietsch, Mannheim (2), Müller-Kühne, NoTICE, Pauling-Plinke, OSW/Krupp, PERCOS, SAR, SARP, Schröder-Grillo, Satco, Schaffner, Tenteleff, Winkler.

superphosphate *See* Phosphate Fertilizer.

syngas AGC-21, ATR (1), ATR (2), CAR, CXR, Ebara Ube, Electropox, ExSyte, Fauser-Montecatini, FTC, GasCat, GHR, Hycar (1), HYCO, Hydrocol, HydroMax, HyTex, ICAR, ICI Low-Pressure Methanol, Koppers Hasche, KRES, K-T, MRG, Octamix, ODR, Onia-Gegi, POX, PRENFLO, reforming, SGP, SBA-HT, Selox, SGP, SMDS, THERMOSELECT, HGP.

synroc Synroc.

tantalum Kroll (1).

tantalum carbide Menstruum.

tar acids Phenosolvan.

terephthalic acid Amoco, Henkel, Maruzen (1), Mid-Century, Raecke, Teijin, Toray (2).

tetraethyl lead Ziegler (2).

textiles Proban.

thiophene Socony-Vacuum.

tin extraction Ashcroft-Elmore, Berzelius, Caveat, Goldschmidt, liquation.

titanium carbide SHS.

titanium diboride SHS.

titanium dioxide AEROSIL, AHP, Blumenfeld, Chloride, CTL, Fletcher, ICON, Leonard (2), Monk-Irwin, Oxychloride, Polar, Reptile, Sulfate.

titanium extraction *See* Ilmenite Beneficiate.

titanium metal Armstrong, Chinuka, Fray-Farthing-Chen, Hunter, Kroll (1), Metalysis, TiRO.

titanium tetrachloride Chloride, Wendell Dunn.

toluene Aromax (1).

transuranic elements DIAMEX, Tramex, Truex. *See also* Plutonium.

triethylene glycol nitrate Biazzi.

trimethylol ethane trinitrate Biazzi.

trinitrotoluene Meissner.

tungsten extraction Fan steel, Shoppler.

tungsten metal Cooledge, van Arkel and de Boer.

ultramarine Guimet.

uranium dioxide ADU, AUC, Comurhex, IDR (2), Sol-Gel.

uranium extraction AMEX, Bufflex, Cogema, Dapex, DEPA-TOPO, Eluex, PhosEnergy, Purlex, Thorex, Tramex.

uranium isotopes Chemex, SILEX.

uranium metal Ames (1).

uranium tetrafluoride Excer.

urea ACES 21, Bosch-Meiser, HR, IDR (1), Mitsui-Toatsu, Pechiney (2), SNAM (1), Urea 2000plus, UTI.

vanadium metal McKechnie-Seybolt, van Arkel and de Boer.

vanillin Howard, Rhodia, Riedel.

vinyl acetate Bayer-Hoechst, Hoechst-Uhde (2), Leap, Loop, Vacido, VAntage, Wacker-Hoechst.

vinyl chloride BPR, DOC, Transcat.

wax, paraffin *See* Paraffin Wax.

white lead *See* Lead Carbonate, Basic.

wood Belmadur, Bethell, Empty-Cell, Full-Cell, Heiskenkjold, Iotech, Rueping.

wood pulp *See* Paper and Pulp.

wool Chlorine/Hercosett, Hercosett, Kroy, Zirpro.

***m*-xylene** JGCC, MGCC, MS Sorbex.

***p*-xylene** Aromax (2), BioForming, Chevron (2), Eluxyl, GT-Cryst PX, GT-STDP, GT-TolAlk, Krupp-Koppers (1), Maruzen (2), MHTI, MSTDP, MTPX, MVPI, Octafining, Parex (1), Phillips (2), PROABD MSC, Pxmax, PX-Plus, STDP, TAC-9.

xylenes, mixed Isarom, Isolene II, Isomar, MLPI, MTDP, Octafining, STDP, Tatoray, T2BX, TDP, TransPlus, XIS, Xylenes-plus, Xylofining.

xylitol Creighton.

zinc extraction BioZINC, Champion, English, Ezinex, Goethite, Haematite, HydroZinc, Jarosite, Imperial Smelting, Lacell, liquation, New Jersey, QSL, St. Joseph, Sherritt-Gordon, Tetronics, Waelz, Warner, Zincex, Zinclor.

zinc oxide American, Belgian, Direct, French, Fricker, Indirect, Wetherill.

zircon HAL.

zirconium diboride SHS.

zirconium extraction De Wet.

zirconium metal Kroll (1).